ELEMENTS OF POWER SYSTEM

For

B.Tech. (Electrical Engineering and Electrical and Electronics Engineering) Students

By

J.B. Gupta

B.Sc. (Elec. Engg.); C. Eng. (I);
F.I.E. (India); M.I.S.T.E.

S.K. KATARIA & SONS®

Publisher of Engineering Books

4885/109, Prakash Mahal, Dr. Subhash Bhargav Lane,
Opposite Delhi Medical Association, Daryaganj, New Delhi–110002 (INDIA)
Phone: +91-11-43551243; *Mobile:* +91-9871775858, 7678425209
e-mail: katariabook@gmail.com, katariabooks@yahoo.com
Website: www.skkatariaandsons.com

(e-books, subscriptions and bulk buying also available on our website)

Published by:

S.K. Kataria & Sons®
4885/109, Prakash Mahal, Dr. Subhash Bhargav Lane,
Opposite Delhi Medical Association, Daryaganj, New Delhi-110002 (INDIA)
Phone: +91-11-43551243; *Mobile:* +91-9871775858, 7678425209
e-mail: katariabook@gmail.com, katariabooks@yahoo.com
Website: www.skkatariaandsons.com

Edited By : **Rajeev Manglik**
B.E. (Elec.) Pb. Univ.
Rohit Manglik
B.E. (Elec.) Univ. of Roorkee

First Edition: July, 2010; Reprint: 2026

Laser Typesetting by :
AISA ENTERPRISES

Printed at :
New A.S. Offset Printers

PREFACE

The primary aim of presenting this book is to meet the needs of Third Year (FifthSemester) students of B.Tech. (Electrical Engineering and Electrical and Electronics Engineering) studying in Engineering Colleges affiliated to U.P. Technical University, Lucknow. The book has been prepared keeping in view the latest revised Syllabus and University Question Papers of previous years. The subject matter has been presented in an easy comprehensive manner, logical sequence using simple language and to provide a detailed exposition to fundamentals. A large number of workedout examples, theoretical questions, short-answer type questions with answers and numerical problems included in the text are from latest question papers of U.P. Technical University.

The book consists of 14 chapters divided into 5 units according to the latest syllabus (W.E.F. Session 2010-11) enforced in Engineering Colleges affiliated to U.P.T.U.

The solution to illustrative examples covering almost all topics and subtopics of different chapters makes the entire presentation easy to follow. Selected exercises and numerical problems with answers given at the end of each chapter help the students in evaluating their understanding of the subject. Short answer type questions with answers help the students in preparing for viva-voce/practical examinations.

Every care has been taken to eliminate misprints, omissions and errors but it is too much to expect that no inaccuracy/error/misprint has crept in and the author would be grateful to the readers for bringing to notice any of such errors/misprints/omissions they may come across while going through the book.

The author shall feel satisfied if the book meets the needs of the students for whom it is meant. Suggestions and criticism for the further improvement of the book shall be welcomed.

The author will like to appreciate the sincere efforts and contribution made by his sons Rajeev Manglik and Rohit Manglik in presentation of the book in this form.

The author will like to thank all the scientists, engineers, authors and publishers whose works and text have been the source of enlightenment, inspiration and guidance in presenting this book.

Lastly, but not the least, the author is thankful to his publisher, Sanjeev Kumar Kataria for bringing out the book in such a short time and for pricing it moderately despite heavy cost of paper and printing.

July 14, 2010 **J.B. Gupta**

To My Loving Grand Daughter

RUSHIL

OTHER USEFUL BOOKS

J.B. Gupta *A Course in Electrical Technology*
J.B. Gupta *Basic Electrical Engineering [A Course in Electrical Technology Vol. 1]*
J.B. Gupta *Electrical Machines [A Course in Electrical Technology Vol. 2]*
J.B. Gupta *Electronic Devices and Circuits [A Course in Electrical Technology Vol. 3]*
J.B. Gupta *Theory and Performance of Electrical Machines [DC Machines, AC Machines and Polyphase Circuits]*
J.B. Gupta *Electrical Machines-I [DC Machines]*
J.B. Gupta *Electrical Machines-II [AC Machines]*
J.B. Gupta *An Integrated Course in Electronics Engineering*
J.B. Gupta *Fundamentals of Electrical and Electronics Engineering*
J.B. Gupta *A Course in Electrical Power [Generation & Economic Considerations; Transmission & Distribution; Switchgear & Protection; Utilization of Electrical Power & Electric Traction]*
J.B. Gupta *Generation and Economic Considerations [Electrical Power Systems Vol. 1]*
J.B. Gupta *Transmission and Distribution [Electrical Power Systems Vol. 2]*
J.B. Gupta *Switchgear and Protection [Electrical Power Systems Vol. 3]*
J.B. Gupta *Electrical Power Systems [Vol. 1 of A Course in Electrical Power]*
J.B. Gupta *Utilization of Electrical Power and Electric Traction [Vol. 2 of A Course in Electrical Power]*
J.B. Gupta *A Course in Electrical and Electronic Measurements and Instrumentation*
J.B. Gupta *Question Bank in Electrical Engineering*
J.B. Gupta *An Integrated Course in Electrical Engineering [Objective Type Questions With Answers & Notes]*
J.B. Gupta *Electrical Installation Estimating & Costing*
J.B. Gupta *Electrical and Electronics Engineering [Second Semester R.T.U.]*
J.B. Gupta *Electrical Engineering [G.B.T.U.; M.M.T.U. and U.T.U.]*
J.B. Gupta *Electrical Science [G.G.S.I.P. Univ.]*
J.B. Gupta *Basic Electrical and Electronics Engineering [Pb. Technical University]*
J.B. Gupta *A Textbook of Electrical and Electronics Engineering [RGTU, Bhopal]*
J.B. Gupta *Electrical Technology [M.D. Univ. and Kurukshetra Univ.]*
J.B. Gupta *Electrical Science [AMITY Univ.]*
J.B. Gupta *Electronics Engineering [G.B.; M.M. and Uttarakhand Technical Universities]*
J.B. Gupta *Electrical Measurements and Measuring Instruments [M.D.U., P.T.U.; G.B.; M.M. and Uttarakhand Technical Universities]*
J.B. Gupta *Electronic Measurements and Instrumentation [3rd Semester Pb. Technical Univ.]*
J.B. Gupta *Electronic Instrumentation and Measurements [4th Semester G.B. and M.M. Technical Universities]*
J.B. Gupta *Electronic Devices and Circuits [3rd Semester Pb. Technical Univ.]*
J.B. Gupta *Electrical Machines*
J.B. Gupta *Electrical and Electronics Engineering Materials [4th Semester G.B. and M.M. Technical Universities]*
J.B. Gupta *Electromechanical Energy Conversion-I [4th Semester G.B. and M.M. Technical Universities]*
J.B. Gupta *Analog Electronics [4th Semester Pb. Technical Univ.]*
J.B. Gupta *Pulse and Digital Switching Circuits [5th Semester Pb. Technical Univ.]*
J.B. Gupta *Linear Integrated Circuits [5th Semester Pb. Technical Univ.]*
J.B. Gupta *Electromechanical Energy Conversion-II [5th Semester G.B. and M.M. Technical Universities]*
J.B. Gupta *Electronics Measurements and Instrumentation [5th Semester M.D. Univ.]*
J.B. Gupta *Electromechanical Energy Conversion [3rd Semester M.D.U.]*
J.B. Gupta *Analog Electronics [4th Semester M.D. Univ.]*
J.B. Gupta *Analog Electronic Circuits [5th Semester M.D. Univ.]*
J.B. Gupta *Basic Electronics [M.D. Univ.]*
J.B. Gupta *Fundamentals of Electronic Devices [3rd Semester G.B. and M.M. Technical Universities]*
J.B. Gupta *A Text Book of Analog Electronics [4th Semester R.T.U.]*
J.B. Gupta *Elements of Electrical Engineering [Gujarat Technical Univ.]*
J.B. Gupta *Electronic Devices and Circuits [3rd Semester M.D. Univ.]*
J.B. Gupta *Electrical Machines–I [3rd Semester M.D. Univ.]*
J.B. Gupta *Analog Electronics [3rd Semester G.G.S.I.P. Univ. Delhi]*
J.B. Gupta *Elements of Power System [5th Semester G.B. and M.M. Technical Universities]*
J.B. Gupta *Elements of Electronic Devices and Circuits [3rd Semester R.T.U.]*
J.B. Gupta *Power System Analysis [6th Semester G.B. and M.M. Technical Universities]*
J.B. Gupta *Switchgear and Protection [7th Semester G.B. and M.M. Technical Universities]*
J.B. Gupta *Electrical Machines-II [5th Semester M.D. University]*

CONTENTS

UNIT–I : POWER SYSTEM COMPONENTS, SUPPLY SYSTEM AND TRANSMISSION LINES 1–76

1. POWER SYSTEM COMPONENTS .... 3—35

1.1. Power System Structure 3
1.2. Electrical Supply System 4
1.3. Typical AC Electrical Power Supply System 5
1.4. Representation of Power Systems 7
1.4.1. Single Line Diagram 7
1.5. Percentage Resistance and Reactance and Base kVA and kV 8
1.6. Per-Unit (PU) Method of Representing Quantities 9
1.6.1. Advantages and Drawbacks of Per-Unit (PU) System 9
1.6.2. Selection of Bases 10
1.6.3. Determination of Base Impedance 10
1.6.4. Change of Base 11
1.6.5. Per-Unit Impedance of a Two Winding Transformer 12
1.7. Power System Components 13
1.8. Synchronous Generators 15
1.8.1. Hydroelectric Generators 15
1.8.2. Turbo Alternators 16
1.8.3. Diesel-Engine Driven Synchronous Generators 17
1.9. Power Transformers 17
1.10. Switchgear 18
1.10.1. Circuit Breakers 19
1.10.2. Isolators 20
1.10.3. Earthing Switch 21
1.11. Bus-Bars 22
1.12. Lightning Arresters 24
1.13. Bus-Bar Arrangements 25
1.14. Transmission Lines 31
■ **Exercises** 32
■ **Short Answer Type Questions With Answers** 33
■ **Problems** 35

2. SUPPLY SYSTEM .... 36—59

2.1. Introduction 36
2.1.1. Comparison Between DC and AC Systems of Transmission and Distribution 36
2.1.2. Comparison Between Overhead and Underground Systems of Transmission and Distribution 38
2.2. Influence of Working Voltage on Size of Feeders and Distributors in DC System 39
2.3. Various Systems of Transmission of Electrical Power 41
2.4. Comparison of Cost of Conductors-Overhead Systems 42
2.5. Comparison of Cost of Conductors-Underground Systems 47
2.6. Results of Comparison of Transmission Systems 51
2.7. Choice of Working Voltage for Transmission 54
■ **Exercises** 57
■ **Short Answer Type Questions With Answers** 58
■ **Problems** 59

3. TRANSMISSION LINES .... 60—76

3.1. Introduction 60
3.2. Conductor Configurations 61
3.3. Types of Conductors 63
3.4. Resistance of Line 66
3.5. Skin Effect 67
3.6. Proximity Effect 68
3.7. Economical Size of Line Conductor—Kelvin's Law 68
3.7.1. Practical Limitations to Application of Kelvin's Law 69
■ **Exercises** 74
■ **Short Answer Type Questions With Answers** 75
■ **Problems** 76

UNIT–II : OVERHEAD TRANSMISSION LINES 77—182

4. INDUCTANCE AND CAPACITANCE OF TRANSMISSION LINES .... 79—119

4.1. Line Inductance 79
4.1.1. Flux Linkages of a Conductor 79
4.2. Inductance of Single Phase Overhead Lines 81
4.2.1. Inductance of a Two-Wire Single Phase Line ...81
4.2.2. Flux Linkages of One Conductor in a Group of Conductors 82
4.2.3. Inductance of Composite Conductor Lines—Self and Mutual GMDs 83
4.3. Inductance of Three-Phase Overhead Lines 87
4.4. Inductance of Three Phase Lines With More Than One Circuit 91
4.4.1. Inductance of a 3-Phase Double Circuit Line With Symmetrical Spacing 92
4.4.2. Inductance of a 3-Phase Double Circuit With Unsymmetrical Spacing But Transposed 92
4.5. Bundled Conductors 96
4.6. Capacitance of a Transmission Line 99
4.6.1. Electric Field and Potential Difference 99
4.7. Capacitance of a Single Phase Overhead Line 101
4.8. Capacitance of 3-Phase Overhead Lines 102
4.8.1. Unsymmetrically Spaced Line 102
4.8.2. Unsymmetrical Line With Transposed Conductors 104
4.8.3. Equilaterally Spaced Line 105
4.9. Capacitance of Double Circuit Three Phase Overhead Lines 106
4.9.1. Symmetrically Spaced Line 106
4.9.2. Flat Vertically Spaced Line 107
4.10. Capacitance of Bundled Conductor Line 110
4.11. Effect of Earth on Capacitance of Transmission Line 112
4.11.1. Effect of Earth on the Capacitance of a Single Phase Transmission Line 113
4.11.2. Effect of Earth on the Capacitance of a Three Phase Line 113
4.12. Transposition of Three Phase Lines 115
- **Exercises** 116
- **Short Answer Type Questions With Answers** 117
- **Problems** 118

5. REPRESENTATION AND PERFORMANCE OF SHORT AND MEDIUM TRANSMISSION LINES .... 120—161

5.1. Introduction 120
5.2. Classification of Transmission Lines 121
5.3. Regulation and Efficiency of a Transmission Line 122
5.4. Short Transmission Lines 123
5.4.1. Effect of Load Power Factor on Transmission Efficiency and Voltage Regulation of Line 125
5.4.2. Mixed Sending-and Receiving-end Conditions 126
5.4.3. Maximum Power With a Given Regulation ... 127
5.5. Medium Transmission Lines 132
5.5.1. End Condenser Method 133
5.5.2. Nominal T-Method (or Middle Condenser Method) 134
5.5.3. Nominal-π Method (or Split-Condenser Method) 135
5.6. Ferranti Effect 144
5.7. Charging Current and Line Losses in Open-Circuited Line 145
5.8. Effect of Capacitance on Performance of Loaded Line 145
5.9. Generalised Circuit Constants 146
5.9.1. Proof For Relation AD – BC = 1 149
5.9.2. Two Networks in Tandem 149
5.9.3. Two Networks in Parallel 150
- **Exercises** 157
- **Short Answer Type Questions With Answers** 158
- **Problems** 159

6. REPRESENTATION AND PERFORMANCE OF LONG TRANSMISSION LINES .... 162—182

6.1. Introduction 162
6.2. Analysis of Long Transmission Lines (Rigorous Method of Solution of Long Transmission Lines) 163
6.3. Evaluation of Constants ABCD 165
6.4. Surge Impedance 166
6.5. Surge Impedance Loading 167
6.6. Transmission Line With Series Impedance At The Receiving End 171
6.7. Transmission Line With Series Impedance At The Sending End 171

6.8. Transmission Line With Series Impedances At Both Ends 172
6.9. Transmission Line With Transformers At Both Ends 172
6.10. Equivalent T-Network of a Long Transmission Line 174
6.11. Equivalent π-Network of a Long Transmission Line 174
6.12. Constant Voltage Transmission 175
6.13. Synchronous Phase Modifiers or Compensators 175
6.14. Tuned Power Lines 179
■ **Exercises** 181
■ **Short Answer Type Questions With Answer** 182
■ **Problems** 182

UNIT–III : CORONA AND INTERFERENCE ELECTROSTATIC AND ELECTROMAGNETIC INTERFERENCE WITH COMMUNICATION LINES AND OVERHEAD LINE INSULATORS 183—233

7. CORONA ..... 185—200

7.1. Introduction 185
7.2. Factors Affecting Corona 186
7.3. Triple-Harmonic Currents Due to Corona 187
7.4. Advantages and Disadvantages of Corona 187
7.5. Dielectric Strength of Air 187
7.6. Disruptive Critical Voltage 188
7.7. Visual Critical Voltage 191
7.8. Corona Power Loss 192
7.9. Factors Affecting Corona Loss 197
7.10. Methods of Reducing Corona Effects 198
7.11. Voltage limitations of Lines 198
■ **Exercises** 198
■ **Short Answer Type Questions With Answers** 199
■ **Problems** 200

8. ELECTROSTATIC AND ELECTROMAGNETIC INTERFERENCE WITH COMMUNICATION LINES ..... 201—207

8.1. Introduction 201
8.2. Electromagnetic Effect 201
8.2.1. Single Phase Single Circuit Line and Telephone Line 201
8.2.2. Three Phase Single Circuit Line and Telephone Line 202
8.3. Effects of Earth Fault Currents 204
8.4. Electrostatic Effect 204
8.5. Transposition of Lines 206
■ **Exercises** 206
■ **Problem** 207

9. OVERHEAD LINE INSULATORS .. 208—233

9.1. Introduction 208
9.2. Insulator Materials 209
9.3. Factors Involved in Insulator Design 210
9.4. Types of Insulators 210
9.5. Failure of Insulators 216
9.6. Distribution of Potential Over a String of Suspension Insulators 216
9.7. Methods of Improving String Efficiency 225
9.8. Arcing Horn and Grading Ring 231
■ **Exercises** 231
■ **Short Answer Type Questions With Answers** 232
■ **Problems** 233

UNIT–IV : MECHANICAL DESIGN OF TRANSMISSION LINE AND INSULATED CABLES 235—296

10. MECHANICAL DESIGN OF TRANSMISSION LINES ..... 237—261

10.1. Introduction 237
10.2. Conductor Configurations, Spacings and Clearances 238
10.2.1. Conductor Configurations 238
10.2.2. Conductor Spacings 239
10.2.3. Conductor Clearances 239
10.3. Span Lengths 239
10.4. Sag and Tension 240
10.4.1. Sag 240
10.4.2. Calculation of Sag and Tension 241
10.5. Stringing Chart 254
10.6. Sag Template 256
10.7. Equivalent Span 257

10.8. Vibrations and Dampers 257
■ **Exercises** 258
■ **Short Answer Type Questions With Answers** 259
■ **Problems** 260

11. INSULATED CABLES .... 262—296

11.1. Introduction 262
11.2. General Construction of Cable 262
11.3. Requirements of Cable 263
11.4. Cable Conductors 264
11.5. Insulating Materials For Cables 264
11.6. Classification of Cables 266
11.7. Insulation Resistance of a Single Core Cable 273
11.8. Capacitance of a Single Core Cable 274
11.9. Dielectric Stress in a Single Core Cable 275
11.10. The Most Economical Diameter of Conductor 277
11.11. Grading of Cables 278
11.12. Capacitance of Three-Core Belted Type Cables 285
11.13. Heating of Cables 288
11.14. Operating Problems Associated With Underground Cables 292
■ **Exercises** 293
■ **Short Answer Type Questions With Answers** 294
■ **Problems** 295

UNIT–V : NEUTRAL GROUNDING AND EHVAC AND HVDC TRANSMISSION 297—353

12. NEUTRAL GROUNDING .... 299—319

12.1. Power System Earthing 299
12.2. Definitions 300
12.3. Neutral Earthing 301
12.4. Isolated Neutral or Ungrounded Systems 301
12.5. Grounded Neutral Systems 304
12.6. Effectively Grounded System 305
12.7. Solid or Effective Grounding 305
12.8. Resistance Grounding 306
12.9. Reactance Grounding 308
12.10. Arc Suppression Coil Grounding (or Resonant Grounding) 309
12.11. Voltage Transformer Grounding 312
12.12. Grounding Transformer 312
12.13. Harmonic Suppressors 314
12.14. Neutral Grounding Practice 315
■ **Exercises** 316
■ **Short Answer Type Questions With Answers** 317
■ **Problems** 319

13. EXTRA HIGH VOLTAGE (EHV) AC TRANSMISSION .... 320—334

13.1. Introduction 320
13.2. Necessity of EHV Transmission 321
13.3. Problems Associated With EHV Transmission 322
13.4. Limitations and Design Aspects of EHV-AC Power Transmission Systems 324
13.5. Flexible AC Transmission System (FACTS) .. 331
■ **Exercises** 333
■ **Short Answer Type Questions With Answers** 333

14. HIGH VOLTAGE DIRECT CURRENT (HVDC) TRANSMISSION .... 335—353

14.1. Introduction 335
14.2. Principle of HVDC System Operation 336
14.2.1. Power Control in HVDC System 337
14.3. Advantages of HVDC Transmission 338
14.4. Limitations of HVDC Transmission 340
14.5. Comparison Between EHV-AC and HVDC Transmission 341
14.6. Types of DC Links 342
14.7. Ground Return 343
14.8. Earth Electrode and Station Earth 344
14.9. HVDC Transmission Systems 344
14.10. Converter Station Equipment 346
14.10.1. Thyristor Valves 346
14.10.2. Converter Transformers 348
14.10.3. DC Reactor 348
14.10.4. Harmonic Filtering Equipment 348
14.10.5. Control Equipment 349
14.10.6. Reactive Power Compensation Equipment 349
14.11. Transient Overvoltages and Insulation Coordination 350
14.12. Corona in HVDC Lines 350
14.13. Protection Aspect 351
14.14. HVDC Circuit Breaking 351
■ **Exercises** 352
■ **Short Answer Type Questions With Answers** 353

UNIVERSITY EXAMINATION PAPERS .. 355—384

INDEX .... 385—390

UNIT-I

POWER SYSTEM COMPONENTS, SUPPLY SYSTEM AND TRANSMISSION LINES

Contents in This Unit

POWER SYSTEM COMPONENTS : Single Line Diagram of Power System, Brief Description of Power System Elements : Synchronous Machine, Transformer, Transmission Line, Bus Bar, Circuit Breaker and Isolator.

SUPPLY SYSTEM : Different Kinds of Supply System and Their Comparison, Choice of Transmission Voltage.

TRANSMISSION LINES : Configurations, Types of Conductors, Resistance of Line, Skin Effect, Kelvin's Law, Proximity Effect.

CHAPTER 1. Power System Components.

CHAPTER 2. Supply Systems.

CHAPTER 3. Transmission Lines.

1

CHAPTER

POWER SYSTEM COMPONENTS

INSIDE THIS CHAPTER

1.1. Power System Structure **1.2.** Electrical Supply System **1.3.** Typical AC Electrical Power Supply System **1.4.** Representation of Power Systems **1.5.** Percentage Resistance and Reactance and Base kVA and kV **1.6.** Per-Unit (PU) Method of Representing Quantities **1.7.** Power System Components **1.8.** Synchronous Generators **1.9.** Power Transformers **1.10.** Switchgear **1.11.** Bus-Bars **1.12.** Lightning Arresters **1.13.** Bus-Bar Arrangements **1.14.** Transmission Lines.

1.1. POWER SYSTEM STRUCTURE

Power is the basic need for the economic development of any country. Availability of electricity has been the most powerful vehicle of introducing economic development and social change throughout the world. The process of modernisation, increase in productivity in industry and agriculture and the improvement in the standard of living of the people basically depend upon the adequate supply of electrical energy. Appropriately the programmes relating to the generation, transmission and distribution of electrical energy have been the highest priority in the national planning process.

Generating stations, transmission lines and distribution systems are the main components of an electrical power system. Generating stations and distribution systems are connected through transmission lines, which also connect one power system to another. A distribution system connects all the loads in a particular area to the transmission line.

The transmission system of an area (or state) is called the *grid*. The different grids are interconnected through tie lines to form a regional grid and the different regional grids are further interconnected to form a national grid. Each grid operates independently. However, power can be transmitted from one grid to another, via the tie lines under conditions of sudden loss of generation or increase in load.

The distribution system in India is quite extensive. The voltages used in distribution systems are 11, 6.6 and 3.3 kV at ht level (or for primary distribution) and 415 volts at lt level (or for secondary distribution).

The country has been divided into five regions (Northern, Western, Eastern, Southern and Northern-Eastern), each with a regional electricity board, so as to promote integrated operation of the constituent power systems. Each State has a State Electricity Board, responsible for generation, transmission and distribution of electrical power in their respective states. Central Government has also control over many generating plants, transmission lines and substations through central organisations like NTPC (National Thermal Power Corporation), NHPC (National Hydroelectric Power Corporation), PGCIL (Power Grid Corporation of India Ltd.) etc. Regional load despatch centres have been established for coordinating the activities of state load despatch centres so as to ensure optimum utilization of generation and transmission facilities in the regions. Setting-up of a National Load Despatch centre is under active consideration.

Electrical energy is normally generated at the power stations far away from the urban areas where consumers are located. The problem, therefore, is to transport (or more usually called transmit) the large blocks of power over long distances economically. In this volume of the book we will deal with the type of equipment, techniques and problems involved in transmitting large blocks of power over long distances from the generating stations to the load centres.

1.2. ELECTRICAL SUPPLY SYSTEM

Electrical energy is generated at the power stations (hydroelectric, thermal or nuclear) which are usually situated far away from the load centres. Hence an extensive network of conductors between the power stations and the consumers is required. This network of conductors may be divided into two main components, called the *transmission system and the distribution system*. The generation, transmission and distribution system of electrical power is called the *electrical power supply system*. The pictorial diagram of a typical electrical power supply system is shown in Fig. 1.1.

The *transmission system is* to deliver bulk power from power stations to the load centres and large industrial consumers beyond the economical service range of the regular primary distribution lines where as *distribution system* is to deliver power from power stations or substations to the various consumers.

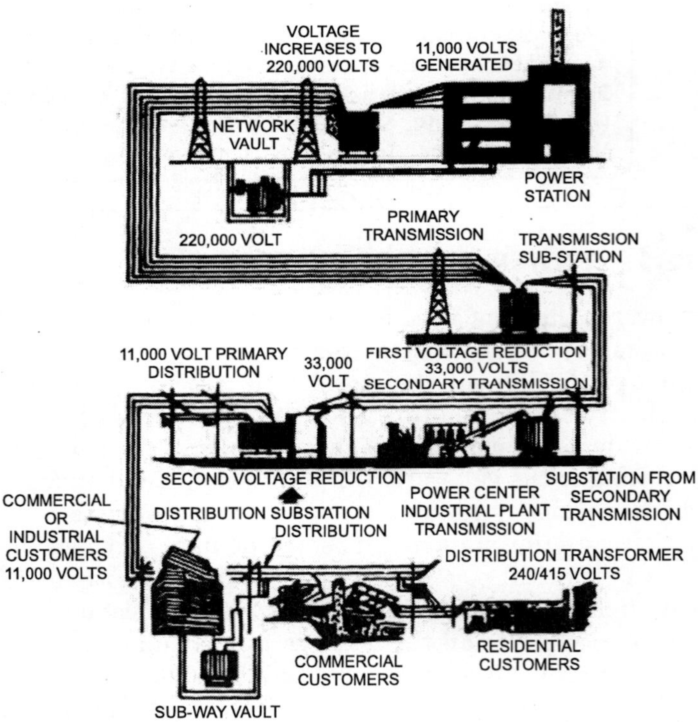

Fig. 1.1. ***Pictorial Diagram of a Typical Electrical Power Supply System***

Although electrical power can be transmitted and distributed by either ac or dc but in practice 3-phase 3-wire ac system is universally adopted for transmission of large blocks of power and 3-phase 4-wire ac system is usually adopted for distribution of electrical power.

The transmission system may be further divided into primary and secondary (or sub) transmission. Similarly the distribution system may be divided into primary and secondary distribution.

Generation voltages are 3·3, 6.6, 11 or 33 kV, most usual value adopted in practice is 11 kV. The primary transmission voltages are 110, 132, 220, 400 or 765 kV depending upon the distance, the amount of power to be transmitted and the system stability. Secondary transmission voltage is normally of the order of 33 or 66 kV. The voltages for primary distribution are 11, 6.6 or 3.3 kV depending upon the requirements of the bulk consumers and for secondary distribution usable voltage is 415/240V*.

The distribution system may further be divided into feeders, distributors and service mains.

(*i*) *Feeders.* Feeders are the conductors which connect the stations (in some cases generating stations) to the areas to be fed by those stations. Generally from feeders no tapping is taken to the consumers, therefore, current loading of a feeder remains the same along its length. It is designed mainly from the point of view of its current carrying capacity.

(*ii*) *Distributors.* Distributors are the conductors from which numerous tappings for the supply to the consumers are taken. The current loading of a distributor varies along its length. Distributors are designed from the point of view of the voltage drop in them.

(*iii*) *Service Mains.* Service mains are the conductors, which connect the consumer's terminals to the distributor. Elements of a distribution system are shown in Fig. 1.2.

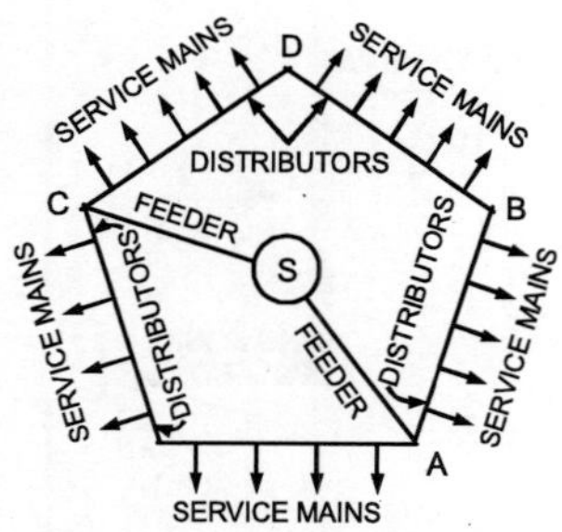

Fig. 1.2. ***Elements of a Distribution System***

A distributor is subject to the legal requirement that the voltage at the consumer's terminals should be maintained within ± 6% of the declared (or rated) voltage. However, there is no such legal restriction on a transmission line and the voltage can vary as much as 10% or even 15% due to variations in loads.

Modern ac electrical power supply system usually consists of the following elements:

(*i*) generating stations (*ii*) step-up transformer substations (*iii*) primary transmission lines (*iv*) primary transmission substations (*v*) secondary (or sub) transmission lines (*vi*) secondary transmission substations (*vii*) primary (high voltage) distribution lines (or networks) (*viii*) service transformer banks (*ix*) secondary (or low voltage) distribution lines (or networks). Essentially elements (*ii*), (*iii*), (*iv*), (*v*) and (*vi*) fall in the transmission system while the elements (*vii*), (*viii*) and (*ix*) fall in the distribution system.

All systems of power transmission and distribution may or may not include all elements enumerated above; for example, some systems may have no primary transmission, some may not have secondary transmission and the others may not have transmission at all, being very small and so on.

1.3. TYPICAL AC ELECTRICAL POWER SUPPLY SYSTEM

Block diagram of a typical ac electrical power supply system is shown is Fig. 1.3. Electrical power is generated by 3-phase alternators operating in parallel at the generating stations. The generation voltage is usually 11 kV. It may be 3.3, 6.6 or even 33 kV in certain cases.

* 415 V for 3-phase loads and 240 V for single phase loads.

For economy reasons this voltage is stepped up from 11 kV to 220 kV* at the generating stations by means of 3-phase transformers. The primary (or high voltage) transmission lines transmit electrical power from the generating stations to primary transmission substations. At these primary substations the voltage is stepped down to 33 kV (or 66 kV) by means of stepdown transformers. From the primary substations electrical power is transmitted through secondary (or sub) transmission lines at 33 kV to various secondary transmission substations. At the secondary substations the voltage is further stepped down to 11 kV (or 6.6 or 3.3 kV, as per needs) by means of 3-phase transformers and the power is supplied to distribution substations through primary (or high voltage) distribution lines at 11 kV. The

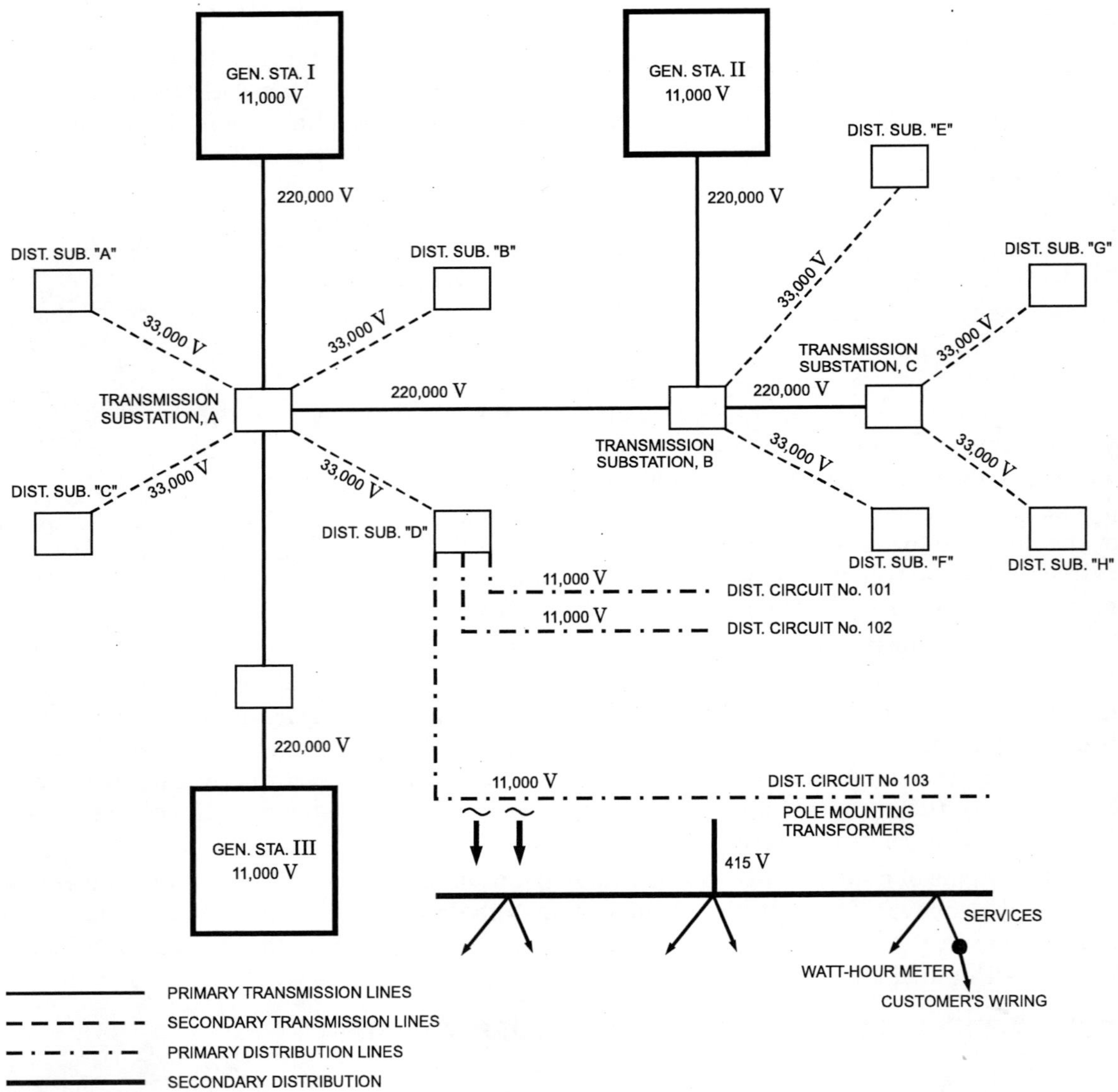

Fig. 1.3. ***Block Diagram of a Typical AC Electrical Power Supply System***

* This voltage may be 110, 132, 400 or 765 kV depending upon the length of transmission line and the amount of electrical power to be transmitted.

distribution substations are located at suitable places in the area in which power is to be supplied. At these substations the voltage is stepped down to 415 V. The secondary (or low voltage) distribution lines, called the *distributors,* are laid along the roads and service connections to the consumers are tapped off from the distributors.

Most of the domestic, commercial and small scale industrial consumers are supplied power at low voltage (at 415 volts for 3-phase loads and 240 volts for single phase loads). However, large commercial organisations, public buildings and industries having load in excess of 100 kW prefer to have bulk power supply at 11 kV (or even at 33 kV in case of large industries). This is accomplished by taking 11 kV line directly from secondary transmission substations right up to the premises of such consumers and connecting it to an 11 kV/415 V stepdown transformer.

1.4. REPRESENTATION OF POWER SYSTEMS

A complete diagram of power system representing all the three phases becomes too complicated and cumbersome for a system of practical size, so much so that it may no longer convey the information it is intended to convey. It is much, more practical to represent a power system by means of simple symbols for each component resulting in what is called a single line diagram.*

1.4.1. Single Line Diagram

The single line diagram of a power system network shows the *main connections* and *arrangements* of the system components along with their data (such as output rating, voltage, resistance and reactance etc.). In case of transmission lines sometimes the conductor size and spacings are given. It is not necessary to show all the components of the system on a single line diagram, e.g., circuit breakers need not be shown in a load flow study but are must for a protection study. In a *single line diagram*, the system components are usually drawn in the form of their symbols. Generators and transformer connections-star, delta and neutral earthing are indicated by symbols drawn by the side of the representation of these elements. Circuit breakers are represented by rectangular blocks. Figure 1.4 represents the single line diagram of a typical power system. The ratings of generator, motor and transformers are given below the diagram.

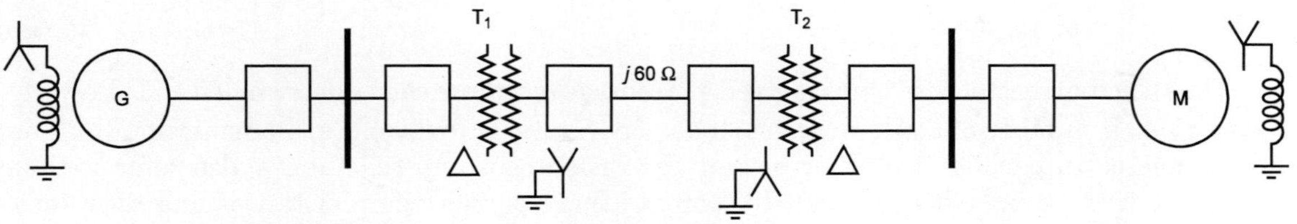

Generator	:	40 MVA, 11 kV, X" = 20 %
Motor	:	30 MVA, 11 kV, X" = 30 %
Transformer T_1	:	40 MVA, 11/220 kV, X = 15 %
Transformer T_2	:	40 MVA, 220/11 kV, X = 15 %

Fig. 1.4. ***Single Line Representation of a Typical Power System***

* Any 3-phase network consisting of generators, transformers, transmission lines etc., can be solved as a single phase network composed of one of the three phases and a neutral return, if it is balanced under normal operation. Many times, the components of the system are shown in a single line diagram omitting the neutral also.

1.5. PERCENTAGE RESISTANCE AND REACTANCE AND BASE kVA AND kV

Resistance and reactance (and consequently impedance) may be expressed in percentage or ohmic terms. In the case of rotating machines and transformers, manufacturer values are always in percentage terms while tables for cables and overhead lines are always in ohmic terms. Calculations can be made using either but not a mixture. For short-circuit calculations, generally the percentage values are employed and therefore their understanding is essential.

Percentage Resistance. It is the voltage drop across the given resistance expressed as percentage of normal voltage when carrying full-load current pertaining to normal rating, *i.e.*,

$$\% R = \frac{IR}{V} \times 100 \qquad ...(1.1)$$

where R is the resistance in ohms, I is the full-load current and V is the rated voltage.

Percentage Reactance. Percentage reactance can be defined in the same way as percentage resistance. It is the voltage drop across the given reactance expressed as percentage of normal voltage when carrying full-load current pertaining to normal rating, *i.e.*,

$$\% X = \frac{IX}{V} \times 100 \qquad ...(1.2)$$

where X is the reactance in ohms, I is the full-load current and V is the rated voltage.
From Eq. (1.2) we have

$$X = \frac{(\% X) \times V}{I \times 100} \text{ ohms}$$

Multiplying and dividing the right hand expression by V, we have

$$X = \frac{(\% X) \times V \times V}{V \times I \times 100} = \frac{(\% X) V^2}{(\text{Output in VA}) \times 100} \text{ ohms} \qquad ...(1.3)$$

when the voltage and the output are expressed in kV and kVA respectively, then

$$X = \frac{(\% X) \times \left(\frac{V}{1,000}\right) \times \left(\frac{V}{1,000}\right) \times 1,000}{100 \times \left(\frac{V}{1,000}\right) \times I} = \frac{(\% X)\,(kV)^2 \times 10}{kVA} \text{ ohms} \qquad ...(1.4)$$

$$\text{or} \quad \% X = \frac{(X)\,(kVA)}{10 (kV)^2} \qquad ...(1.5)$$

Thus if actual reactance in ohms is given, percentage reactance can be determined and vice-versa.

Base kVA. If number of equipments such as generators, transformers, transmission lines etc are connected in parallel and their percentage resistances and reactances also refer to their respective kVA ratings, it is difficult to compare these percentage resistances and reactances and their combined effect until and unless they are all referred to a common kVA. This common kVA which is taken as an arbitrary one is known as the *base* kVA of the system.

A base kVA may be chosen in the following manner :

(*i*) Equal to the kVA rating of the largest unit connected in the network.

(*ii*) Equal to the sum of the kVA ratings of all the units connected in the network.

(*iii*) Any arbitrary value.

It must, however, be clearly understood that the value of the base kVA, has no bearing whatsoever on the results; since in the ultimate formula for the calculation of short-circuit current base MVA is to be taken into consideration.

The conversion of percentage reactance at rated kVA to the percentage reactance at base kVA can be made by using the following expression:

$$\text{Percentage reactance at base kVA} = \frac{\text{Base kVA}}{\text{Rated kVA}} \times \text{percentage reactance at rated kVA} \quad ...(1.6)$$

Thus if a transformer is rated for 10,000 kVA and has percentage reactance of 7.5 %, then the percentage reactance at the base kVA of 25,000 shall be

$$= \frac{25,000}{10,000} \times 7.5\ \% = 18.75\ \%$$

Base kV. In some cases it is convenient to work in ohmic values of the various reactances rather than in percentage values. The method would become simple if all the reactances relate to the same voltage but if step-up or step-down transformers or other equipments operating at different voltages are also included, all the ohmic values will have to be reduced to a common base voltage.

Reactances can be converted from one operating voltage to the other by the following relation

$$X_2 = \left(\frac{E_2}{E_1}\right)^2 \times X_1 \quad ...(1.7)$$

where X_1 is the reactance at voltage E_1 and X_2 is the reactance at voltage E_2.

It must, however, be remembered clearly that all the values discussed above*i.e.*, resistances, reactances and voltages refer to the phase values and not to the line values.

1.6. PER-UNIT (PU) METHOD OF REPRESENTING QUANTITIES

The quantities involved in power system are kVA, voltage, current and impedance of the equivalent circuits of the various system components. The equivalent circuits are at different voltages and are connected together in the system by means of transformers and interconnections. Each apparatus is rated in kVA and its impedance in actual ohms or in percentage value* referred to its rated kVA and rated voltage, as already explained in Art. 1.5. In power system analysis, it is usual to express voltage, current, kVA and impedance in per unit of base or reference values of these quantities. Such a method simplifies the calculations.

$$\text{The per-unit value of any quantity is defined as } \frac{\text{the actual value of the quantity in any unit}}{\text{the base or reference value in the same unit}}$$

1.6.1. Advantages and Drawbacks of Per-Unit (PU) System

Advantages. 1. Calculations are simplified.

2. The characteristics of machines (generators, transformers, motors etc.) when described in per-unit system are specified by almost the same number, regardless of the rating of the machines. In other words, the characteristics (or parameters) tend to fall in relatively narrow range, making erroneous values conspicuous. Thus per-unit system provides a method of comparison.
3. For circuits connected by transformers, per-unit system is particularly suitable. By choosing suitable base kV's for the circuits the per unit reactance remains the same, referred to either sides of the transformer. Therefore, the various circuits can be connected in the reactance diagram.
4. This method is useful to eliminate ideal transformers as circuit components since the typical power system contains hundreds, if not thousands of transformers, and this is a non-trivial savings.

---

* Per cent value = Per-unit value × 100

Per cent value is not convenient for use as the factor of 100 has to be carried in computations.

Drawbacks. 1. Some equations that hold in the unscaled cases are modified when scaled into per-unit. Factors such as $\sqrt{3}$ and 3 are removed or added by this method.

2. Equivalent circuits of the components are modified, making them somewhat more abstract. Sometimes phase shifts that are clearly present in the unscaled circuit vanish in the per-unit circuit.

1.6.2. Selection of Bases

For a common representation, base kVA and base voltage are to be chosen. Then the base current and base impedance can be expressed as follows :

$$\text{Base current, } I_B = \frac{\text{Base kVA}\,(kVA_B)}{\text{Base kV}\,(kV_B)} \text{ amperes} \qquad ...(1.8)$$

$$\text{Base impedance, } Z_B = \frac{\text{Base voltage in volts, } V_B}{\text{Base current in amperes, } I_B} \text{ ohms}$$

$$= \frac{V_B \times V_B}{I_B \times V_B} = \frac{{V_B}^2}{V_B I_B} = \frac{\left(\frac{V_B}{1,000}\right) \times \left(\frac{V_B}{1,000}\right) \times 1,000}{\frac{V_B I_B}{1,000}} = \frac{(kV_B)^2 \times 1,000}{kVA_B}$$

$$= \frac{(kV_B)^2}{MVA_B} \text{ ohms} \qquad ...(1.9)$$

For a single phase, phase-to-neutral voltage, kVA per phase are taken as bases while in a three-phase system, three-phase line-to-line voltage and three-phase kVA are used as bases. This simplifies the calculations.

The base values in a system are so selected that the per-unit voltages and currents in the system are approximately unity. Sometimes the base kVA is chosen equal to the sum of the ratings of the various equipments in the system or equal to the kVA rating of the largest unit connected in the system.

If in the network there is no transformer present, the same base voltage is used throughout, but if the transformers are present, the rule is to change the base voltage in proportion to the transformation ratio of the transformer when transformer is reached. Thus, all impedances in the network before the transformer is reached, including the transformer primary leakage impedance, are expressed in per unit, to the voltage base chosen for the primary side ; all impedances beyond the transformer, including the transformer secondary leakage impedance, are expressed in per-unit to a new base voltage which is the primary side base voltage multiplied by the transformer transformation ratio. This is very important.

1.6.3. Determination of Base Impedance

Voltage, current, volt-amperes and impedances are so related that selection of base values for any two of them determines the base values of the remaining two. For example for a **single phase circuit** we have the following relations

$$S_B = V_B I_B \qquad ...(1.10)$$

$$Z_B = \frac{V_B}{I_B} \qquad ...(1.11)$$

where S_B, V_B, I_B and Z_B represent the base power, base voltage, base current and base impedance expressed in volt-amperes, volts, amperes and ohms respectively. Thus from above two equations, if any two of the four quantities are specified, the remaining two may be determined without any problem. As already mentioned in Art 1.6.2, base kVA and base kV are usually specified. Let kVA_B and kV_B be the base kVA and base kV, then

$$\text{Per-unit kV} = \frac{\text{Actual kV}}{\text{Base kV}} = \frac{kV_{actual}}{kV_B} \qquad ...(1.12)$$

$$\text{Base current, } I_B = \frac{\text{Base kVA}}{\text{Base kV}} = \frac{kVA_B}{kV_B} \text{ amperes} \qquad ...(1.13)$$

$$\text{Per-unit current, } I_{pu} = \frac{\text{Actual current}}{\text{Base current}} = \frac{\text{Actual current}}{kVA_B} \times kV_B \qquad ...(1.14)$$

$$\text{Base impedance, } Z_B = \frac{\text{Base kV} \times 1{,}000}{\text{Base current}} = kV_B \times \frac{kV_B}{kVA_B} \times 1{,}000$$

$$= \frac{(kV_B)^2 \times 1{,}000}{kVA_B} \text{ ohms ...already workedout Eq. (1.9)}$$

$$\text{Base power, } S_B = \text{Base kVA} \qquad ...(1.15)$$

$$\text{Per-unit impedance, } Z_{pu} = \frac{\text{Actual impedance}}{\text{Base impedance}} = \text{Actual impedance} \times \frac{kVA_B}{(kV_B)^2 \times 1{,}000} \qquad ...(1.16)$$

In a **3-phase system** rather than obtaining the per-unit values using per phase quantities, the per-unit values can be obtained directly by using three-phase base quantities (total kVA, line-to-line voltage and line currents). Let kVA_B be the base kVA (total output of three phases) and kV_B be the base line-to-line voltage in kV.

Assuming star-connection (equivalent star can always be found)

$$\text{Base current, } I_B = \frac{kVA_B}{\sqrt{3}\, kV_B} \text{ amperes}$$

$$\text{Base impedance, } Z_B = \frac{kV_B \times 1{,}000}{\sqrt{3} \times I_B} = \frac{kV_B}{\sqrt{3}} \times \frac{\sqrt{3}\, kV_B}{kVA_B} \times 1{,}000 = \frac{(kV_B)^2 \times 1{,}000}{kVA_B} \text{ ohms} \qquad ...(1.17)$$

$$\text{Per-unit impedance, } Z_{pu} = \frac{\text{Actual impedance}}{\text{Base impedance}} = \frac{\text{Actual impedance} \times kVA_B}{(kV_B)^2 \times 1{,}000} \qquad ...(1.18)$$

It should be noted that the same expression is obtained for a single phase and three-phase systems. In power system problems the data are given in terms of 3-phase kVA, line-to-line kV. The machine kVA and machine phase-to-phase (or line-to-line) kV rating. Further the direct-axis synchronous reactance is also known as positive sequence reactance of the machine.

1.6.4. Change of Base

Normally the per-unit impedance of various components corresponding to its own rating voltage and kVA are given and since we choose one common base kVA and base kV for the whole system, therefore, it becomes imperative to determine the per-unit impedance of the various equipments corresponding to the common base kV and base kVA. If the individual quantities are $Z_{pu\ old}$, kVA_{old}, kV_{old} and the common base quantities are $Z_{pu\ new}$, kVA_{new} and kV_{new}, then making use of Eq. (1.16), according to which per-unit impedance is directly proportional to the base kVA and inversely proportional to the square of the base kV, we have

$$\text{Per-unit impedance to new base, } Z_{pu\ new} = Z_{pu\ old} \times \frac{kVA_{new}}{kVA_{old}} \times \frac{(kV_{old})^2}{(kV_{new})^2} \qquad ...(1.19)$$

This is very important relation used in power system analysis.

1.6.5. Per-Unit Impedance of a Two Winding Transformer

The approximate equivalent circuit of a two winding transformer with all impedances referred to primary (low-voltage) side is illustrated in Fig. 1.5.

According to Eq. (1.16)

$$Z_{01\ pu} = \frac{Z_{01} \times kVA_B}{(kV_{B1})^2 \times 1{,}000} \qquad ...(1.20)$$

where Z_{01} is the total impedance of the transformer referred to primary side in ohms, kVA_B is the rated kVA of the transformer and kV_{B1} is the rated voltage of the transformer on the primary side in kV.

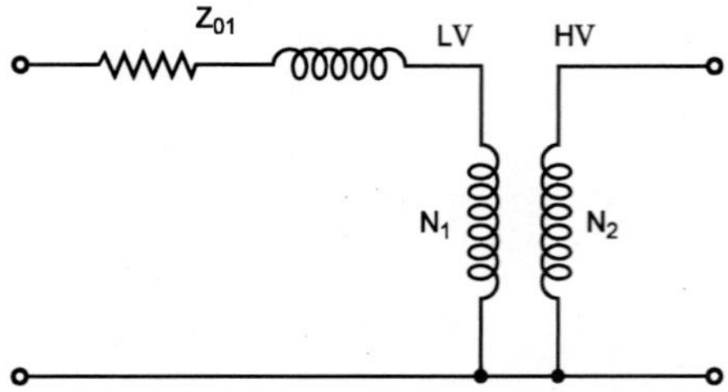

Fig. 1.5. ***Simplified Equivalent Circuit of a Single Phase Two Winding Transformer***

Now let us consider the transformer with all its impedances referred to secondary side. The total impedance of the transformer referred to secondary (high voltage side).

$$Z_{02} = Z_{01} \times \left(\frac{N_2}{N_1}\right)^2 = Z_{01} \frac{(kV_{B2})^2}{(kV_{B1})^2} \qquad ...(1.21)$$

Thus the impedance of the transformer in per-unit viewed from the secondary (high voltage) side,

$$Z_{02\ pu} = \frac{Z_{02} \times kVA_B}{(kV_{B2})^2 \times 1{,}000} \qquad ...(1.22)$$

where Z_{02} is the total impedance of the transformer referred to secondary side in ohms and kV_{B2} is the rated voltage of the transformer on the secondary side in kV.

Substituting the value of Z_{02} from Eq. (1.21) in Eq. (1.22) we have

$$Z_{02\ pu} = Z_{01} \times \frac{(kV_{B2})^2}{(kV_{B1})^2} \times \frac{kVA_B}{(kV_{B2})^2 \times 1{,}000}$$

$$= \frac{Z_{01} \times kVA_B}{(kV_{B1})^2 \times 1{,}000} \qquad ...(1.23)$$

Comparing Eqs. (1.20) and (1.23) we conclude that the per-unit impedance is the same regardless of the side from which it is viewed. Thus the simple equivalent circuit for the two-winding transformer in which $Z_{01\ pu} = Z_{02\ pu} = Z_{pu}$ is depicted in Fig. 1.6.

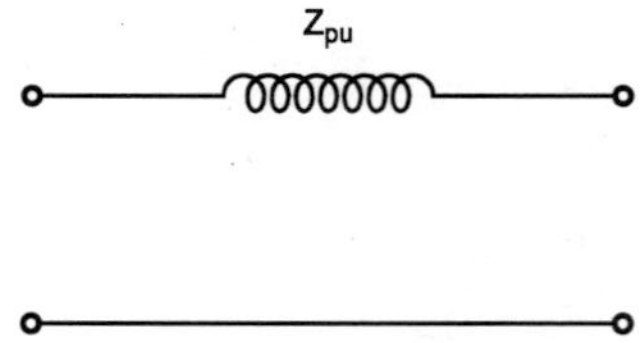

Fig. 1.6. ***Simplified Equivalent Circuit of a Two Winding Transformer With Impedance Expressed in Per-Unit***

Example 1.1. Base voltage = 1,100 volts. Base kVA = 10^6. What is base impedance ?

[*Madurai Kamaraj Univ. Power System Analysis Nov. 1996*]

Solution :

$$\text{Base kVA} = 10^6$$

$$\text{Base kV} = \frac{1{,}100}{1{,}000} = 1.1 \text{ kV}$$

$$\text{Base current, } I_B = \frac{kVA_B}{kV_B} = \frac{10^6}{1.1}$$

$$\text{Base impedance, } Z_B = \frac{kV_B \times 1{,}000}{I_B} = \frac{1.1 \times 1{,}000 \times 1.1}{10^6} = 0.00121\ \Omega \textbf{ Ans.}$$

Base impedance can also be determined directly by substituting the values of base kVA and base kV in Eq. (1.9) and we get

$$Z_B = \frac{(kV_B)^2 \times 1{,}000}{kVA_B} = \frac{(1.1)^2 \times 1{,}000}{10^6} = 0.00121\ \Omega$$ **Ans**, Same as above.

Example 1.2. If the resistance in ohms is 5 Ω, find the per unit value. Given base kVA = 10 and base kV = 11.
[Madras Univ. Power System Analysis October 1996]

Solution : Resistance, $R = 5\ \Omega$

$$\text{Base resistance, } R_B = \frac{(kV_B)^2 \times 1{,}000}{kVA_B} = \frac{(11)^2 \times 1{,}000}{10} = 12{,}100\ \Omega$$

$$\text{Per-unit resistance, } R_{pu} = \frac{R}{R_B} = \frac{5}{12{,}100} = 0.000413 \text{ pu}$$ **Ans.**

Example 1.3. A single phase transformer is rated as 2.5 kVA, 11/0.4 kV. If the leakage reactance is 0.96 Ω when referred to low-voltage side, then determine its leakage reactance in per unit.

Solution : Base voltage, $kV_B = 0.4$ kV

Base kVA, $kVA_B = 2.5$ kVA

$$\text{Low-voltage side base impedance, } Z_{BLV} = \frac{(kV_B)^2 \times 1{,}000}{kVA_B} = \frac{(0.4)^2 \times 1{,}000}{2.5} = 64\ \Omega$$

$$\text{Leakage reactance in per-unit, } Z_{pu} = \frac{\text{Actual reactance}}{\text{Base impedance}} = \frac{0.96}{64} = 0.015 \text{ pu}$$ **Ans.**

Example 1.4. Determine the per-unit impedance of a transmission line having an impedance of (30 + *j* 110) Ω on 100 MVA and 132 kV base voltage.

Solution : Impedance of line, $Z = (30 + j\,110)\ \Omega$

Base kVA, $kVA_B = 100$ MVA or 100,000 kVA

Base kV, $kV_B = 132$ kV

$$\text{Per-unit impedance, } Z_{pu} = \frac{Z \times kVA_B}{(kV_B)^2 \times 1{,}000} = \frac{(30 + j110) \times 100{,}000}{(132)^2 \times 1{,}000} = (0.172 + j\,0.631) \text{ pu}$$ **Ans.**

Example 1.5. A 30 MVA, 11 kV generator has a reactance of 0.2 pu referred to its ratings as bases. Determine the per-unit reactance when referred to base kVA of 50,000 kVA and base kV of 33 kV.

Solution : Old base kVA = 30,000 kVA

Old base kV = 11 kV

Old per-unit impedance, $Z_{pu\ old} = 0.2$ pu

New base kVA = 50,000 kVA

New base kV = 33 kV

$$\text{New per-unit impedance, } Z_{pu\ new} = Z_{pu\ old} \times \frac{\text{New base kVA}}{\text{Old base kVA}} \times \left(\frac{\text{Old base kV}}{\text{New base kV}}\right)^2$$

$$= 0.2 \times \frac{50{,}000}{30{,}000} \times \left(\frac{11}{33}\right)^2 = 0.037 \text{ pu}$$ **Ans.**

1.7. POWER SYSTEM COMPONENTS

Simplified single line diagram for a typical power system is shown in Fig. 1.7. Major components of a power system are : synchronous generators, synchronising equipment, circuit breakers, isolators, earthing switches, bus-bars, transformers, transmission lines, current transformers,

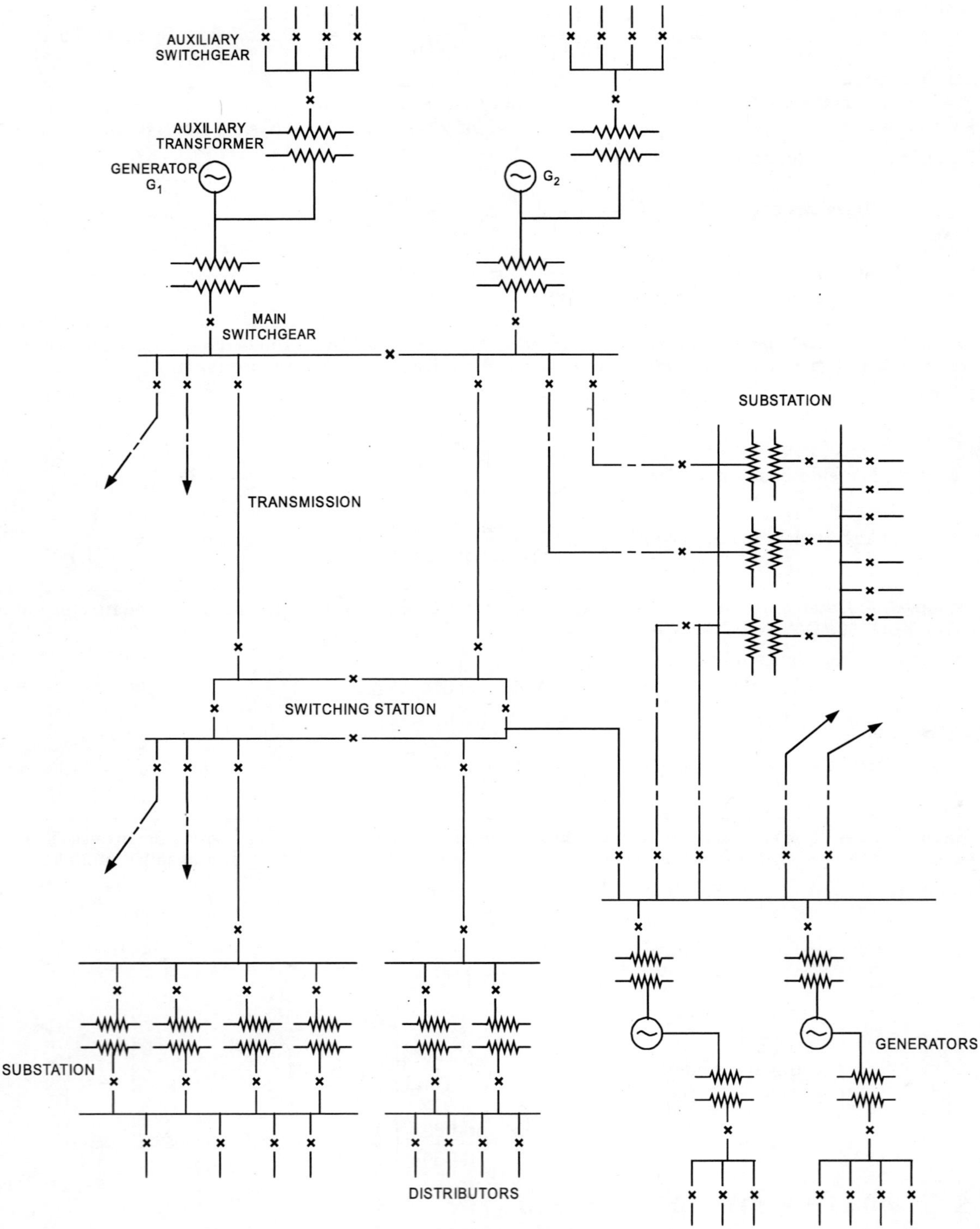

Fig. 1.7. ***Simplified Single Line Diagram For a Typical Power System***

potential transformers, relay and protection equipment, lightning arresters, station transformer, motors for driving auxiliaries in power station. Some of the components will be discussed here.

1.8. SYNCHRONOUS GENERATORS

The synchronous generators used in generating stations are revolving field type owing to its inherent advantages.

The synchronous generators, based on the type of prime movers to which they are mechanically coupled, may be classified as (*i*) hydro-generators (*ii*) turbo-generators, (*iii*) diesel engine driven generators.

1.8.1. Hydroelectric Generators

The generators employed in hydroelectric power plants are three phase alternating current synchronous generators, called the alternators. Alternator consists essentially of two parts namely armature and field magnet system.

The armature of an alternator is an iron ring, formed of laminations of special magnetic iron or steel alloy (silicon steel) having slots on its inner periphery to accommodate armature conductors and is known as stator. The whole structure is held in a frame which may be of cast iron or welded steel plates. For minimising eddy current losses due to rotation of field structure in between the stator, the stator core is laminated. The laminations (usually of thickness 0.5 mm or less) are stamped out in complete rings (for small machines or in segments (for larger machines) and insulated from each other with paper or varnish. The stampings also have openings which make axial and radial ventilating ducts to provide efficient cooling. The open slots are more commonly used because the coils can be form wound and insulated prior to being placed in the slots giving least expenditure and more satisfactory winding method. Such slots also facilitates in removal and replacement of defective coils.

The field structure is the largest and heaviest component of alternator (in large machines, it may be 15 m in diameter and 1,000 ton in weight), and is called the rotor. The rotor houses the dc excitation winding and the exciting current is supplied to the rotor through two slip- rings and brushes. The field windings are connected in series to form the excitation winding which is supplied with dc at 110/220/300 V. Earlier, the excitation was usually supplied by small dc generators (shunt or compound wound type) driven through pilot exciters from the turbo-generator shaft. The recent generators make use of static excitation systems.

The alternators employed in hydroelectric power plants have the following features.

The machines may have horizontal or vertical configuration. The alternators employed in conjunction with impulse turbines are usually of horizontal configuration while those employed with Francis and Kaplan turbines are of vertical configuration. Low speeds (about 50—500 rpm in case of vertical machines and 100—1,000 rpm for the horizontal machines). Protection against runaway speeds are to be provided. The machines are of large diameter and small lengths. The machines are usually of salient pole type and the number of poles, they consist of , varies from 6—120. These machines should be capable of supplying heavy line charging currents since hydroelectric power plants are usually located at considerable distance from load end. A higher value of short-circuit ratio (around unity) is therefore required. The machines are usually air cooled. Such machines are built in sizes varying from 0.5 MW to over 1,000 MW.

1.8.2. Turbo Alternators

In central power stations, the steam turbine and alternator are directly coupled to avoid transmission losses. Turbo-alternators are high speed machines (3,000 or 1500 rpm)* for 50 Hz systems. These machines have horizontal configurations and smooth cylindrical (or non-salient pole) type field structure wound usually for 2 or 4 poles. To reduce the peripheral speed (maximum peripheral speed should not exceed 175 m/s) the diameter of the rotor is kept small and axial length is increased. The ratio of diameter to axial length ranges from 1/3 to 1/2.

Due to high peripheral speed, the rotating part of the turbo-alternator is subjected to high mechanical stresses. As a result the rotor of large turbo-alternator is normally built from solid steel forging. Chromium-nickel-steel or special chrome-nickel-molybdenum steel is used for rotors of turbo-alternators. The forging has radial slots in which the field copper, usually in the form of strips, is placed. The coils are held in place by steel or bronze wedges and the coil ends are fastened by metal rings. Normally two-thirds of the rotor is slotted for the field winding and one-third is left without slots so as to form the pole faces. Rectangular slots with tapered teeth are milled out in the rotor so that rectangular conductors can be used for the field windings. Each slot is provided with a ventilation hole at the bottom. To reduce harmful tooth ripples, either stator slots or poles are skewed.

The non-salient field structure used in turbo-alternators has the following special features:

(*i*) They are of smaller diameter (maximum 1m in 2-pole machine) and of very long axial length.

(*ii*) Robust construction and noiseless operation.

(*iii*) Less windage (air-resistance) loss.

(*iv*) Better in dynamic balancing.

(*v*) High operating speed (3,000 or 1,500 rpm).

(*vi*) Nearly sinusoidal flux distribution around the periphery, and therefore, gives a better emf waveform than obtainable with salient pole field structure.

(*vii*) There is no need of providing damper windings (except in special cases to assist in synchronising) because the solid field poles themselves act as efficient dampers.

500 MW units generally use hollow stator conductors. The short-circuit ratio is 0.4 to 0.6.

Turbo-alternators are usually rated at 11 kV with 3-phase star-connection of stator windings. Standard frequency of generation is 50 Hz. Generally turbo-alternators are rated at 0.8 power factor lagging.

Main exciters are dc compound generators of 125 or 250 V rating capable of supplying required excitation for main generators on full load and with an overload of about 20% at rated power factor. Minimum of 10% overload capacity under normal operating conditions is provided for in the case of turbo-generators. Static exciters are also now in much use.

The small machines (of rating up to 50 MW) are air cooled. However, hydrogen cooling is invariably used for both stator and rotor of medium and large size turbo-generators. There are several advantages of using hydrogen as coolant in place of air in closed circuit ventilation system such as improved efficiency, increased output rating, increased life, elimination of fire hazard, less noise, smaller size of coolers etc.

The machines are built in size from 10 MW to over 1,500 MW. 500 MW units have been installed at Singrauli, Anpara, Chandarpur, Ramagundem, Trombay, Farakka and Korba.

* 3,000 rpm for central power stations and 1,500 rpm for industrial power stations.

1.8.3. Diesel-Engine Driven Synchronous Generators

Diesel-engine driven synchronous generators are of small ratings from 100 kVA or less to 1,000 kVA. The speed is generally between 375 to 1,500 rpm, *i.e.* the number of poles is 16 to 4. Generally 4 to 6 pole alternators are common for diesel driven generators. The diesel engines are manufactured as horizontal type, and therefore, both the diesel engine and the synchronous generator are mounted horizontally and are connected by a horizontal shaft. Such generators employ salient pole type construction. The torque of the diesel engine is not uniform and this makes the synchronous generator sensitive to torque variations.

For more details refer to Electromechanical Energy Conversion-II by J.B. Gupta.

1.9. POWER TRANSFORMERS

Power transformers are used for stepping-up the voltage for transmission at generating stations and for stepping-down voltage for further distribution at main stepdown transformer substations. Usually naturally cooled, oil immersed, known as ON type, two winding, three-phase transformers, are used up to the rating of 10 MVA. The transformers of rating higher than 10 MVA, are usually air blast cooled. For very high rating, the forced oil, water cooling and air blast cooling may be used. For regulating the voltage the transformers used are provided with on load tap changer.

Fig. 1.8. ***400/220 kV, 240 MVA Power Transformer***

They are put in operation during load hours and disconnected during light load hours *i.e.* they are usually operated at approximately full load. This is possible because they are arranged in banks and can be thrown in parallel with other units or disconnected at will. So power transformers are designed to have maximum efficiency at or near full load (*i.e.* with iron loss to full-load copper loss ratio of 1 : 1). Power transformers are designed to have considerable leakage reactance than is permissible in distribution transformers because in power transformers inherent voltage regulation is not as much important as current limiting effect of the higher leakage reactance. Power transformers usually make use of flux density of 1.5 to 1.77, have percentage impedance ranging from 6-18% and regulation 6-10%.

The transformer specifications cover the following:

1. KVA rating; 2 Rated voltage; 3. Number of phases (single or three phase); 4. Rated frequency; 5. Connections (Δ or λ in case of 3-phase transformer); 6. Tappings if any; 7 Type of core (core or shell); 8 Type (power or distribution); 9. Ambient temperature (generally average 40°C); 10. Type of cooling (*a*) cooling medium-air, oil or water (*b*) circulation type-natural or forced (*c*) simple or mixed cooling; 11. Temperature rise above ambient in °C depending upon the class of winding insulation; 12 Voltage regulation [(*a*) Per cent or pu at full load at 75°C unity pf or 0.8 pf lag (*b*) Impedance-per cent or pu (*c*) Reactance-per cent or pu]; 13 No-load current in amperes or per cent of rated current at rated voltage and rated frequency; 14. Efficiency-in per cent or pu at full load, 1/2 load, 3/4 load at unity power factor and 0.8 pf.

Power transformers are covered under IS 2026-1962.

The transformers are generally installed upon lengths of rails fixed on concrete slabs having foundation 1 to 1(1/2) metre deep.

For more details refer to Electromechanical Energy Conversion-I by J.B. Gupta.

1.10. SWITCHGEAR

Everyone is familiar with low voltage switches and rewirable fuses. A switch is used for opening and closing of an electric circuit while a fuse is used for over–current protection. Every electric circuit needs a switching device and protective device. Switching and protective devices have been developed in different forms. *Switchgear* is a general term covering a wide range of equipment concerned with switching and protection.

In a power system switchgear serves two basic purposes :

(*i*) Switching during normal operating conditions for the purpose of operation and maintenance.

(*ii*) Switching during abnormal conditions such as short–circuits and interrupting the fault currents.

The first of above could be served by relatively simple switches because it is relatively simple as it involves normal currents which are easy to interrupt. The second function is, however, complex. With the advancement of electrical power system the lines and other equipment operate at very high voltage and carry large currents. Whenever a short–circuit occurs, a heavy current flows through the equipment causing considerable damage to the equipment and interruption of service. In order to avoid such a damage every part of the power system is provided with a protective relaying system and an associated switching device. The function of protective relaying system is to cause the prompt removal from service of any element of a power system when it suffers a short–circuit, or when it starts to operate in any abnormal manner that might cause damage or otherwise interfere with the effective operation of the rest of the system. The relaying equipment is aided in this task by *circuit breakers* that are capable of disconnecting the faulty element when they are called upon to do so by the relaying equipment. In addition to circuit breakers and protective relays, the associated equipment for controlling, regulating and measuring can also be considered as switchgear devices. Thus the apparatus including its associated auxiliaries employed for controlling, regulating or switching on or off the electrical circuits in the electrical power system is known as*switchgear.* It includes switches, fuses, circuit breakers, isolators, relays, current and potential transformers, indicating instruments, lightning arresters and control panels.

A circuit breaker is a device that interrupts the abnormal or fault currents and in addition performs the function of a switch. Circuit breakers and fuses both possess a time element of operation, whereby they operate practically instantaneously on short–circuits but with definite time lag on overloads. Circuit breakers can be reset in less time and with less trouble that are required to replace blown fuses and spare parts are seldom required. For low voltage circuits fuses may be used to isolate the faulty circuits; but for higher voltages, say from 3.3 kV upwards, isolation is achieved by circuit breaker. Circuit breakers may also be preferred where continuity of service is an important consideration or where frequent fuse replacement may be expected. However, for low capacities a fuse with circuit breaking arrangement is quite useful and economical.

As the circuit breakers have to be accessible for inspection and maintenance, the Indian Standard regulations require the provision for separation of the breaker from the live parts by means of isolating switches. *Isolating switches* are employed for separating the disconnected element from the live portion of the system. They should be employed only after the corresponding circuit breakers have been opened. Isolating switches are employed for isolating a piece of equipment from adjacent live elements for purposes of safety.

1.10.1. Circuit Breakers

Circuit breakers are mechanical devices designed to close or open contact members, thus closing or opening of an electrical circuit under normal or abnormal conditions.

Automatic circuit breakers, which are usually employed for the protection of electrical circuits, are equipped with a trip coil connected to a relay or other means, designed to open the breaker automatically under abnormal conditions, such as over–current. The automatic circuit breakers perform the following duties :

(*i*) It carries the full–load current continuously without overheating or damage,

(*ii*) It opens and closes the circuit on no load,

(*iii*) It makes and breaks the normal operating current and

(*iv*) It makes and breaks the short–circuit currents of magnitude up to which it designed for.

The circuit breaker performs first three duties satisfactorily but in performing fourth duty *i.e.,* when it is to make or break short–circuit currents, it is subjected to mechanical and thermal stresses. The circuit breakers are rated in terms of maximum voltage, number of poles, frequency, maximum continuous current carrying capacity, maximum interrupting capacity and maximum momentary and 4–s current carrying capacity.

The interrupting or rupturing capacity of a circuit breaker is the maximum value of current which can be interrupted by it without any damage. The circuit breakers are also rated in MVA which is the product of interrupting current, rated voltage and 10^{-6}.

Operating Principle : A circuit breaker is a switching and current interrupting device. It consists, essentially, of fixed and moving contacts, which are touching each other and carry the current under normal conditions *i.e.,* when circuit breaker is closed. When the circuit breaker is closed, the current carrying contacts, called the electrodes, engage each other under the pressure of a spring.

During the normal operating condition the circuit breaker can be opened or closed by a station operator for the purpose of switching and maintenance. To open the circuit breaker, only a small pressure is required to be applied on a trigger. Whenever a fault occurs on any part of the power system, the trip coils of the breaker get energized and the moving contacts are pulled apart by some mechanism, thus opening the circuit. The separation of current carrying contacts produces an arc. The current is thus able to continue until the discharge ceases. The production of arc not only delays the current interruption process but it also generates enermous heat which may cause damage to the system or to the breaker itself. Therefore, the main problem in a circuit breaker is to extinguish the arc within the shortest possible time so that heat generated by it may not reach a dangerous value.

The basic construction of a circuit breaker requires the separation of contacts in an insulating fluid which serves two functions :

1. Extinguishes the arc drawn between the contacts when the circuit breaker opens.
2. Provides insulation between the contacts and from each contact to earth.

The insulating fluids commonly used for this purpose are as follows :

1. Air at atmospheric pressure.
2. Compressed air.
3. Oil producing hydrogen for arc extinction.
4. Ultra high vacuum.
5. Sulphur hexa–fluoride (SF_6).

The fluids used in circuit breakers should have the properties of high dielectric strength, non–inflammability, high thermal stability, arc extinguishing ability, chemical stability, and commercial availability at moderate cost.

Of the simple gases air is the cheapest and most widely used for circuit breaking. Hydrogen has better arc extinguishing property but it has lower dielectric strength as compared to air. Also if hydrogen is contaminated with air, it forms an explosive mixture. Nitrogen has similar properties as air. CO_2 has almost the same dielectric strength as air but is a better arc extinguishing medium at moderate currents. Oxygen is a good extinguishing medium but is chemically active. SF_6 has outstanding arc–quenching properties and good dielectric strength. Of all these gases SF_6 and air are used in commercial gas blast–circuit breakers.

Air-circuit breakers are often used instead of oil up to 15 kV in these units and oil reclousures are also sometimes used to cut cost in small rural substations.

1.10.2. Isolators

Since isolators (or isolating switches) are employed only for isolating circuit when the current has already been interrupted, they are simple pieces of equipment. They ensure that the current is not switched into the circuit until everything is in order.

Isolators or disconnect switches operate under no load condition. They are not equipped with arc-quenching devices. They do not have any specified current breaking capacity or current making capacity. The isolators in some cases are used for breaking charging current of transmission line.

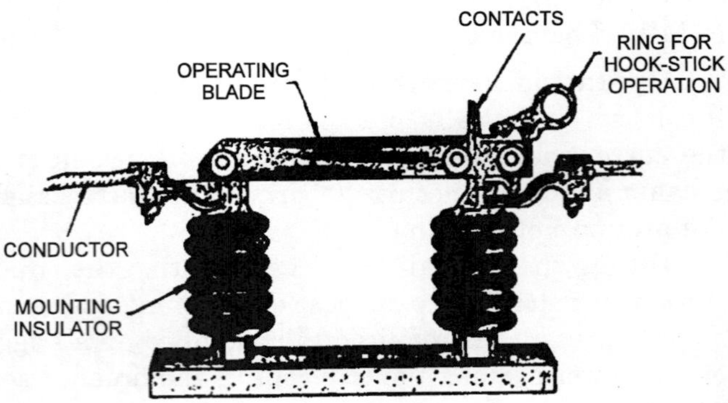

Fig. 1.9. Isolators.

Isolators are employed in addition to circuit breakers, and are provided on each side of every circuit breaker to provide isolation. While opening a circuit, the circuit breaker is opened first, then isolator. If an isolator is opened carelessly, when carrying a heavy current, the resulting arc could easily cause a flash-over to ground. This may shatter the supporting insulators and may even cause a fatal accident to the operator, particularly in *hv* circuits. While closing a circuit, the isolator is closed first, then circuit breaker. Isolators are necessary on the supply side of the circuit breakers in order to ensure isolation (disconnection) of the circuit breaker from the live parts for the purpose of maintenance. Automatic switching of isolators is preferred.

Isolators employed in power systems are usually 3-pole isolators, each having three identical poles. Each pole consists of two or three insulator posts mounted on a fabricated support. The fixed and moving conducting parts are of copper or aluminium rods. During the opening operation the conducting rods swing apart and isolation is obtained. The simultaneous operation of three poles is obtained by mechanical interlocking of the three poles.

To prevent the mal-operation, the isolator is provided with the following interlockings :

(*i*) Interlocking between three poles for simultaneous operation.

(*ii*) **Interlocking with circuit breakers.** *Isolator cannot be opened unless the circuit breaker is opened and circuit breaker cannot be closed unless the isolator is closed.*

(*iii*) **Load Interrupter Switches.** In addition to isolators and circuit breakers, there is one more device, called the **load-interrupting switch**, which combines the functions of the isolator and a switch. They are designed only for breaking and making the load currents.

These switches are designed and used to close and open high voltage circuits under normal working conditions (at normal load). The arc extinguishing device of the load interrupter is made in the form of a split, moulded plastic chute fitted with organic glass inserts. This chute surrounds the moving knife of the arc extinguishing system. The stationary arcing contact is located in the lower part of the chute.

Load interrupter switches are intended only for handling low-energy arcs resulting from the interruption of load current and altogether unsuitable for extinction of high-energy fault current arcs.

When the switch is opened, the working contacts between which the arc is drawn separate. Acted upon by the high temperature of arc, the walls of the organic material insert generated gases (mainly hydrogen), which create a longitudinal blast serving to extinguish the arc. Lever-arm manually operating mechanisms are employed for closing and opening the load interrupter switches.

It is a wide practice to install load-interrupting switches in low-capacity installations like industrial-shop, urban and rural, and like distribution substations where it is possible to provide short-circuit current protection with high-voltage fuses and where the only duty of load-interrupting switch is to make and break the load current.

The installation of a load-interrupting switch, including the high-voltage fuses, is not only cheaper, but also usually requires less space than the installation of a high-voltage circuit breaker.

1.10.3. Earthing Switch

Earthing switch is connected between the line conductor and earth. Normally it is open and it is closed to discharge the voltage trapped on the isolated or disconnected line. When the line is disconnected from the supply end, there is some voltage on the line to which the capacitance between the line and earth is charged. This voltage is significant in *hv* systems. Before commencement of maintenance work it is necessary that these voltages are discharged to earth by closing the earthing switch. Normally, the earthing switches are mounted on the frame of the isolator.

1.11. BUS-BARS

Bus-bar (or bus in short) term is used for a main bar or conductor carrying an electric current to which many connections may be made.

Bus-bars are merely convenient means of connecting switches and other equipment into various arrangements. The usual arrangement of connections in most of the substations permits working on almost any piece of equipment without interruption to incoming or outgoing feeders.

In some arrangements two buses are provided to which the incoming or outgoing feeders and the principal equipment may be connected. One bus is usually called the "*main*" bus and the other "auxiliary" or "transfer" bus. The main bus may have a more elaborate system of measuring instruments, relays etc. associated with it. The switches used for connecting feeders or equipment to one bus or the other are called "*selector*" or "transfer" switches.

Bus-bars may be of copper, aluminium or steel. *Copper* has a comparatively low resistivity and also the advantage of relatively high mechanical strength; this makes it economical to use copper bus-bars in installations of very large capacity where the currents are particularly heavy. During 1960's the need for substituting the copper with aluminium became very urgent, particularly in countries like India where copper is imported. Now aluminium is being increasingly used for various switchgear installations due to its numerous advantages over copper such as higher conductivity on weight basis, lower cost for equal current carrying capacity, excellent corrosion resistance and ease of formability. For proper reliable electrical connections aluminium buses are coated with silver. The **aluminium** used for bus-bars should have high conductivity, good mechanical properties, high softening temperature etc.**Steel bus-bars** have a comparatively high specific resistance (about 7 times greater than that of copper). Furthermore, the losses due to hysteresis and eddy currents when carrying ac are also considerable. The primary advantage is that steel bus-bars cost very little. The influence of the above factors is seen in the wide application of steel bus-bars in low-capacity installations where the load currents do not exceed 200-300 A.

The bus-bars used in substations are usually bare rectangular x-section bars (but they can be of other shapes such as round tubes, round solid bars, or square tubes) as they are more economical in comparison with round solid bus-bars. This is explained by the fact that *rectangular-section* bus-bars of the same x-sectional area have a higher rate of heat dissipation due to their greater cooling surface. Furthermore, the ac (or effective) resistance of a round bus-bar is greater than that of a rectangular-section bus-bars because of the skin effect. Because of these two facts rectangular-section bus-bars are able to carry larger load currents than round solid bus-bars (for the same cross-sectional area and the same temperature rise).

For the same cross-sectional area and the same temperature rise, copper bus-bars will have the greatest and steel bus-bars the least-permissible current-carrying capacity because of the difference in their resistance. As the size of the bus-bars increases, their heat dissipation capacity falls off and the permissible current density must therefore be reduced.

If the load current to be carried exceeds the permissible current for a single-strip bus-bar of the greatest available size, each phase of the bus-bars shall have to be assembled of several strips arranged in a stack and clamped on post insulators. The air-gaps left between the strips in a bus-bar stack are usually made equal to the thickness of the strips, this is necessary for adequate cooling.

As the number of strips in a bus bar stack is increased, the current with which it is permissible to load the bus-bar cannot be raised in direct proportion to the number of strips;

the increase in permissible current must be appreciably less because the conditions for cooling become more unfavourable. Moreover, when dealing with ac it is necessary to take another factor into consideration, *i.e.*, the **proximity effect**. The result is that the bus-bar metal in a stacked bus-bar is much less efficiently utilised than in a single-strip bus-bar.

It is for the above reason that bus-bars for ac are designed with not more than two, and, rarely, with three strips. For dc, the bus-bars may consist of a large number of strips because there is no proximity effect and current distribution is uniform.

In large-capacity ac installations with extremely heavy working currents it is more effective to install box-shaped types of bus-bars of aluminium or copper. In this case the bus-bar takes the form of a hollow square conductor in which the metal is much more efficiently utilized than it would be in a stacked-strip bus-bar. The box shape also makes for better cooling than is obtained with stacked-strip bus-bars.

When installations operate at 33 kV and high voltages, the bus-bars have to be designed with due consideration for corona effects.

All rigid types of bus-bars mounted on support insulators are coated with enamel paints of the following colours :

Three-phase systems—Red, yellow and blue to indicate different phases.

DC systems—Positive bus-bars-claret-coloured; negative bus-bars—blue.

Coating bus-bars with paint improves their rates of cooling to some extent and therefore permits them to carry a larger load current. By coating steel bus-bars with paint we protect them from corrosion. The use of different colours is important because it helps the operating personnel to distinguish between the different phases of the installation at a glance.

Flexible bus-bars (bare stranded conductors) are not coated with paint. To identify the phases of the bus-bars, disks painted with the respective phase colour are hung from the bus-bars.

The most common sizes of bus-bars are 25 × 6 (150 mm^2); 50 × 6 (300 mm^2), 75 × 6 (450 mm^2); 100 × 6 (600 mm^2); 125 × 6 (750 mm^2); 50 × 10 (500 mm^2); 75 × 10 (750 mm^2); 100 × 10 (1,000 mm^2); 125 × 10 (1,250 mm^2); 150 × 10 (1,500 mm^2); 200 × 10 (2,000 mm^2); 75 × 12 (900 mm^2; 100 × 12 (1,200 mm^2); 125 × 12 (1,500 mm^2); 150 × 12 (1,800 mm^2); 200 × 12 (2,400 mm^2); 250 × 12 (3,000 mm^2). The bus-bars are of 5 or 6 metres in length.

The early substations were generally with *flexible bus design*. A flexible bus consists of flexible ACSR or all-aluminium alloy stranded conductors supported by strain insulators from each end. The flexible bus is held at higher level above the different substation equipment. The connections between the flexible bus and the terminals of substation equipment are made by flexible conductors held in vertical or inclined plane.

1. *Rigid bus-bars* are easy to maintain. They are at lower height. Connections to substation equipment are easy. Aluminium tubes are preferred for rigid bus-bars.

A substation usually has a combination of rigid bus-bars and flexible bus-bars. ACSR conductors are preferred for flexible bus-bars.

The bus-bars are designed to carry certain normal current continuously. The x-section of conductors is designed on the basis of rated normal current and permissible temperature rise. The value of x-section so obtained is verified for temperature rise under short-time short-circuit current. The bus-bar conductors are supported on post insulators or strain insulators.

1.12. LIGHTNING ARRESTERS

The lightning arrester is a surge diverter and is used for the protection of power system against the high voltage surges. It is connected between the line and earth and so diverts the incoming high voltage wave to the earth.

Lightning arresters act as safety valves designed to discharge electric surges resulting from lightning strokes, switching or other disturbances, which would otherwise flash-over insulators or puncture insulation, resulting in a line outage and possible failure of equipment. They are designed to absorb enough transient energy to prevent dangerous reflections and to cut off the flow of power-frequency follow (or dynamic) current at the first current zero after the discharge of the transient. They include one or more sets of gaps to establish the breakdown voltage, aid in interrupting the power follow current, and prevent any flow of current under normal conditions (except that gap shunting resistors, when used to assure equal distribution of voltage across the gaps, permits a very small leakage current). Either resistance (valve) elements to limit the power follow current to values the gaps can interrupt, or an additional arc extinguishing chamber to interrupt the power follow current are connected in series with gaps. Arresters have a short time lag of breakdown compared with the insulation of apparatus, the breakdown voltage being nearly independent of the steepness of the wavefront.

Lightning protection by means of lightning arresters and gaps and overhead ground wires is a means of reducing outages and preventing damage to station equipment from lightning disturbances. The amount and kind of protection vary in different applications, depending upon the exposure of the lines, the frequency and the severity of lightning storms, the cost of the protection as an insurance value against damage to equipment and the value of reduced line outages.

Transmission line is protected from direct strokes by running a conductor, known as *ground wire,* over the towers or poles and earthed at regular intervals preferably at every pole/tower.

Substations, interconnectors and **power houses** are protected from direct strokes by *earthing screen* that consists of a network of copper conductors, earthed at least on two points, overall the electrical equipment in the substation.

The ground wire or earthing screen does not provide protection against the high voltage waves reaching the terminal equipment, so some protective devices are necessary to provide protection to power stations, substations and transmission lines against the voltage wave reaching there. The most common device used for the protection of the power system against the high voltage surge is *surge diverter*, which is connected between line and earth and so diverts the incoming high voltage wave to earth. Such a diverter is also called the *lightning arrester*.

Different types of gaps and lightning arresters (rod gap, horn gap, electrolytic arrester, oxide film arrester, thyrite arrester, expulsion type lightning arrester, valve type lightning arrester etc. will be discussed in details in another book titled "Switchgear and Protection".

Rod gap is a very inferior type of surge diverter and is usually employed as a second line of defence in view of its low cost. **Horn gap arrester** was one of the earliest type of surge diverters to be developed, and is still used to a certain extent on low voltage lines on account of its great simplicity. Like rod gap arrester it is also employed as an auxiliary protection. **Electrolytic arrester** operates on the fact that a thin film of aluminium hydroxide deposited on the aluminium plates immersed in electrolyte acts as a high resistance to low voltage but a low resistance to voltage above a critical value. Such arresters are very delicate, require daily supervision, and the film is required to be reformed whenever destroyed, therefore these arresters have become obsolete nowadays. **Oxide film arrester** operates on the fact that certain chemicals (such as lead peroxide) have the property to change rapidly from a good conductor to almost perfect insulator when slightly heated. The great advantage of such an arrester is that it does not require daily charging, and it may be installed at points on transmission systems where daily attendance is difficult or expensive to provide. **Thyrite arrester** is the most common and is mostly used for the protection against high dangerous voltages. It operates on the fact that thyrite, a dense organic compound of ceramic nature, has high resistance decreasing rapidly from high value to low value for currents of low value to those of high value. The **expulsion type arrester** consists of a tube made of fibre (which is very effective gas evolving material), an isolating spark gap (or external series gap) and an interrupting spark gap inside the fire tube. Expulsion type arresters can be considered as economical means of surge protection for small rural transformers, where valve type arresters may prove expensive and the application of airgaps yield inadequate protection. These arresters can also be used on special transmission towers of extra height on river crossings where the possibilities of lightning strokes are relatively high. Such arresters can be considered very favourably for use on systems operating at voltages up to 33 kV. **Valve type lightning arrester** is very cheap, effective and robust and is, therefore, extensively used nowadays for high voltage systems. This consists of a number of discs of a porous material stacked one above the other and separated by their mica rings. The mica rings provide insulation during normal operation.

1.13. BUS-BAR ARRANGEMENTS

There are several types of bus-bar arrangements. The choice of a particular arrangement depends upon different factors such as system voltage, position of substation in the system, reliability of supply, flexibility and cost. The other technical considerations governing the selection of any one particular arrangement are:

1. Simplicity which is a key note of a dependable system.
2. Maintenance possibility without interruption of supply or danger to the operating personnel.
3. Provision of extension with the load growth.
4. The installation should be as economical as possible keeping in view the needs and continuity of supply.
5. Availability of alternative arrangements in the event of an outage of any of the apparatus.
6. Importance of load and local conditions. Freedom from total shutdown and its period desired.

For a small and medium sized station where shutdown can be permitted at times, simple, single bus-bar arrangement can be favoured. But in case of major plants an elaborate bus-bar arrangement is essential in order that the duplicate supply is always available in case of fault, which might otherwise cause interruption of supply to a large area. Under these circumstances the heavy initial cost is justified. Economy is most important. The extra-high voltage equipment such as circuit breakers, isolators are usually costly. Hence unnecessary equipment should not be provided. Additional equipment to take care of remote contingencies should not be provided as it, in addition to increasing of installation cost will prove a source of weakness.

Different bus-bar arrangements will be discussed here. All the diagrams refer to 3-phase arrangement but are shown in single phase for simplicity.

1. *Single Bus-bar Arrangement.* This is the simplest arrangement consisting of a single set of bus-bars for the full length of the switchboard and to this set of bus-bars are connected all the generators, transformers and feeders, as illustrated by single line diagram in Fig. 1.10. Each generator and feeder is controlled by a circuit breaker. The isolators permit isolation of generators, feeders and circuit breakers from the bus-bars for maintenance. The chief advantages of such a bus-bar arrangement are low initial cost, less maintenance and simple operation.

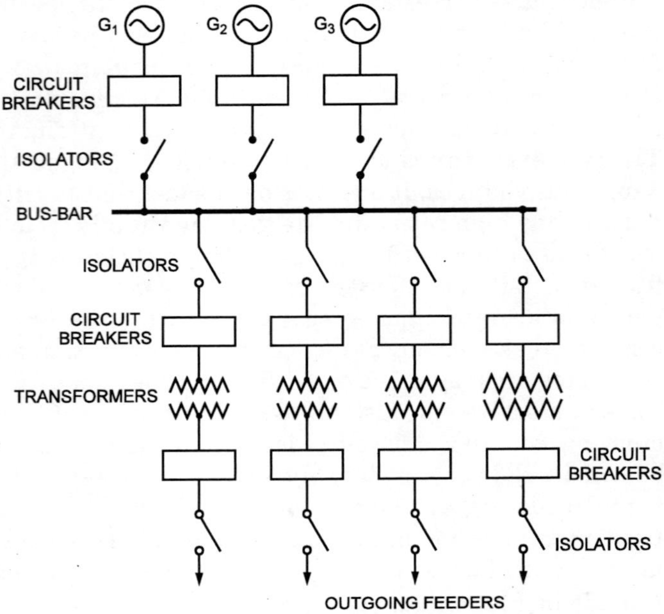

Fig. 1.10. ***Single Bus-Bar Arrangement***

The glaring drawback of this system is that in case of fault on the bus-bars, whole of the supply is affected and all the healthy feeders are disconnected. Moreover, when maintenance is to be carried out on any of the feeder sections or on a part of bus-bar the whole supply is to be disrupted. Thus such an arrangement provides least flexibility and immunity from total shutdown.

Such bus-bar arrangement is employed for switchboards, small and medium sized substations, small power stations and dc stations.

2. *Single Bus-Bar Arrangement With Bus Sectionalization.* The bus-bar may be sectionalized by a circuit breaker and isolating switches so that a fault on one part does not cause a complete shutdown. In large generating stations, where several units are installed, it is a common practice to sectionalize the bus as illustrated in Fig. 1.11. Normally the number of sections of a bus-bar are 2 to 3 in a substation, but actually it is limited by the short-circuit current to be handled. In a sectionalized bus-bar arrangement only one additional circuit breaker is required

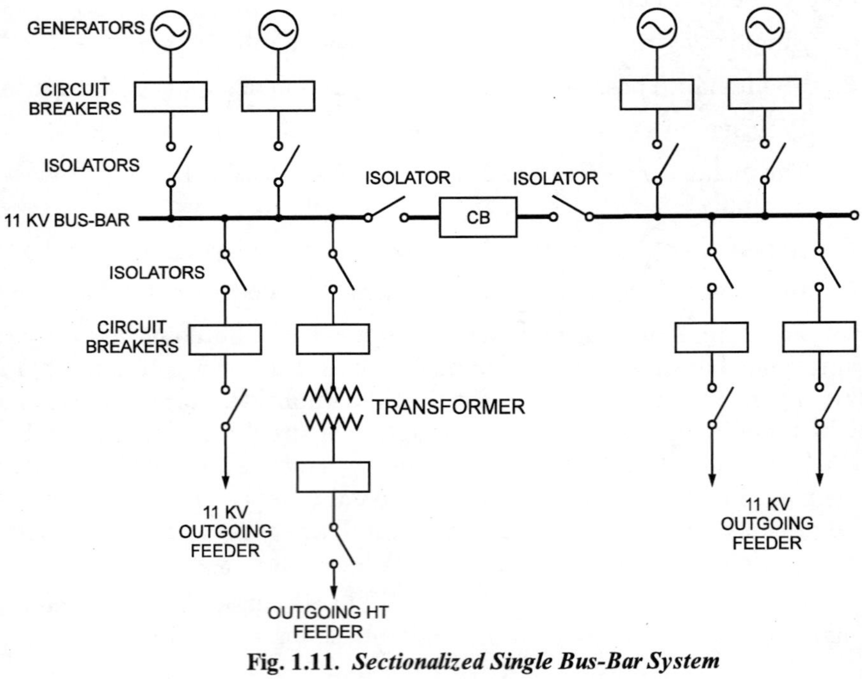

Fig. 1.11. ***Sectionalized Single Bus-Bar System***

which does not cost much in comparison to the total cost of the bus-bar system. Such an arrangement provides three main advantages over simple single bus-bar arrangement. Firstly, in the event of occurrence of fault on any section of the bus-bar, the faulty section can be isolated without affecting the supply of other section or sections. Secondly, one section can be completely shutdown for maintenance and repairs without affecting the supply of the other section (s). Thirdly, by adding a current limiting reactor between the sections the fault level (MVA) can be reduced thereby circuit breakers of lower capacity can be used.

At times air-break isolators were used in place of circuit breakers as bus-sectionalizer due to economy, but it must be remembered that any isolation affected by them must be affected under off-load conditions otherwise it may cause spark. It will be preferable to provide circuit breaker as a sectionalizing switch so that uncoupling of bus-bar may be carried out safely during load transfer. A double isolation is however necessary when the circuit breaker is employed as sectionalizing switch so that the maintenance work can be carried out on circuit breaker while the bus-bars are alive.

3. *Main and Transfer Bus Arrangement.* This arrangement has been quite frequently adopted where the loads and continuity of supply justify additional costs. This arrangement provides additional flexibility, continuity of supply and allows periodic maintenance without total shutdown. Such an arrangement is suitable for highly interconnected power network in which flexibility is very important.

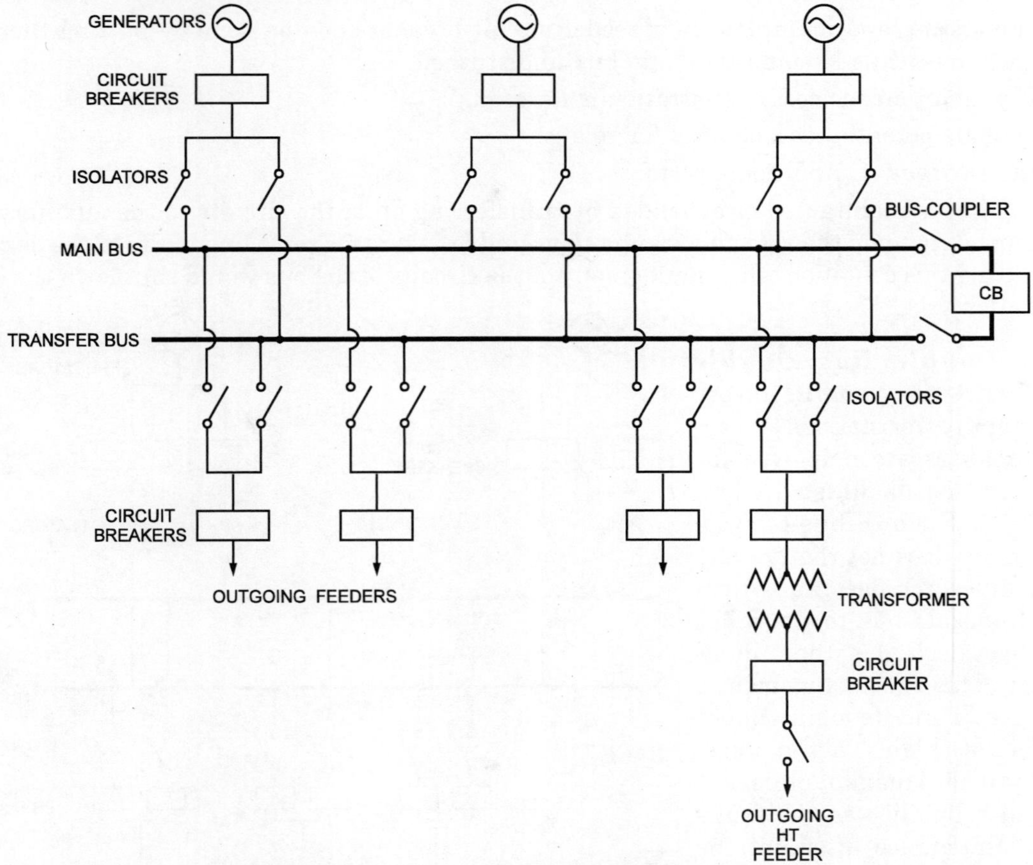

Fig. 1.12. ***Main and Transfer Bus Arrangement***

Figure 1.12 illustrates the main and transfer bus arrangement in a generating station. Such an arrangement consists of two bus-bars, known as *main bus-bar* and *transfer bus-bar* used as an auxiliary bus-bar. Each generator and feeder may be connected to either bus-bar with the help of bus coupler which consists of a circuit breaker and isolating switches. In this arrangement a bus coupler is usually used so that change-over from one bus-bar to the other can be carried out under load conditions. While transferring the load to the reserve bus, the following steps may be taken.

1. Close the bus-coupler (circuit breaker) so as to make the two buses at the same potential.
2. Close isolators on the reserve bus.
3. Open isolators on the main bus.

The load is now transferred to the reserve or auxiliary bus and main bus is disconnected.

The advantages and disadvantages of the arrangement are gives below.

Advantages. 1. It ensures continuity of supply in case of bus fault. In the event of occurrence of fault on one of the bus, the entire load can be transferred to the other bus.

2. Repair and maintenance can be carried out on the main bus without interrupting the supply as the entire load can be transferred to the auxiliary bus.
3. Each load can be supplied from either bus.
4. The infeed and load circuit may be divided into two separate groups if required from operational considerations.
5. The testing and maintenance of feeder circuit breakers can be done by putting them on spare bus, thus keeping the main bus undisturbed.
6. The maintenance cost of substation is lowered.
7. The bus potential can be used for relays.

Disadvantages. 1. Additional costs.

2. The bus is maintained or expanded by transferring all of the circuits to the auxiliary bus depending upon the remote backup relays and breakers for removing faults of the circuits. During this condition a line fault on any of the circuits of the bus would shutdown the entire station.

4. ***Double Bus Double Breaker Arrangement.*** In very important power stations two circuit breakers are employed for each circuit, as illustrated in Fig. 1.13. Such a bus-bar arrangement does not require any bus-coupler and permits switch-over from one bus to the other whenever desired, without interruption. This bus arrangement is very costly and its maintenance cost is also high. This arrangement provides maximum flexibility and reliability as the faults and maintenance interrupt the supply to the minimum. A circuit breaker can be opened for

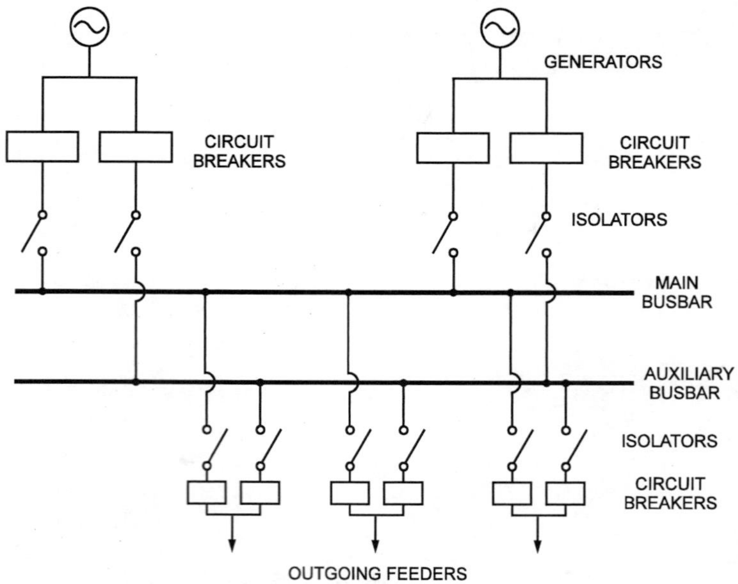

Fig. 1.13. ***Double Bus Double Breaker Arrangement***

repairs and usual checks and the load can be shifted on the other circuit breaker easily. But because of its higher cost, this arrangement is seldom used at the substations.

For 400 kV switchyards two main buses plus one transfer bus scheme is preferred. The transfer bus is used for transferring power from main bus 1 to main bus 2 and vice-versa.

5. *Sectionalized Double Bus Arrangement.* In this arrangement duplicate busbars are used with the main bus-bar in sections connected through a bus-coupler, as illustrated in Fig. 1.14. In this arrangement, any section of bus-bar can be isolated for maintenance, while any section may be synchronised with any other through the auxiliary bus-bar. Sectionalization of auxiliary bus-bar is not required and would increase the cost if done.

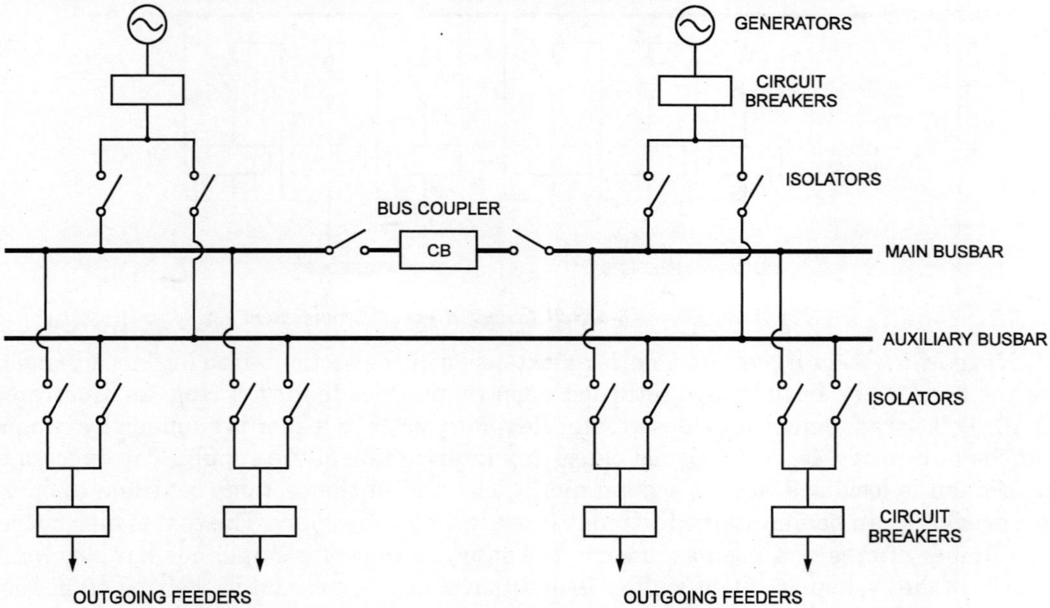

Fig. 1.14. ***Sectionalized Double Bus Arrangement***

6. *One-and a Half Breaker Arrangement.* This is an improvement over double bus double breaker arrangement and it affects saving in the number of circuit breakers. This arrangement needs three circuit breakers for two circuits. The number of circuit breakers per circuit comes out to be $1\frac{1}{2}$ hence the name. This arrangement is preferred in important large stations where power handled per circuit is large. This arrangement is shown in Fig. 1.15. This arrangement provides high security against loss of supply as a fault in a bus or in a breaker will not interrupt the supply. Possibility of addition of circuits to the system is another advantage. The bus potential can be used as supply to relays, however, at the time of bus fault such potential to the relay should be thrown off.

The main drawback of this arrangement is complications in relaying system because at the time of fault two breakers are to be opened. The other drawback is that for maintenance of circuit breakers if the load shedding is not used, two breakers are to be opened in which case the other circuit in the line up will be operating with one breaker from one bus only. At the time of fault in that bus supply to the other circuit is also interrupted. The maintenance cost is higher.

The above arrangement has been used in important 400 kV and 750 kV substations.

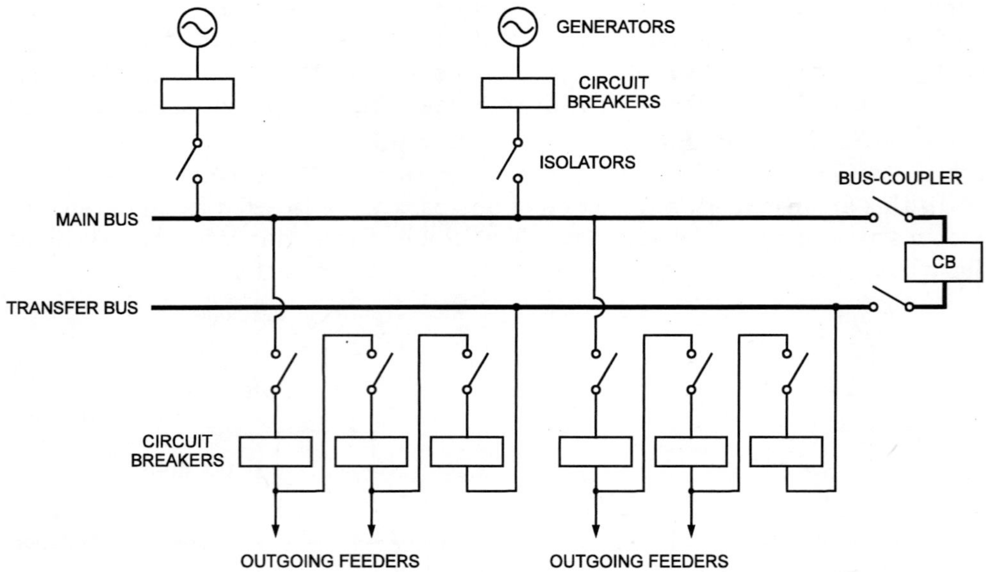

Fig. 1.15. ***One-and-a-Half Circuit Breaker Arrangement***

7. *Ring Main Arrangement.* This is an extension of the sectionalized busbar arrangement where the ends of the busbars are returned upon themselves to form a ring, as illustrated in Fig. 1.16. This arrangement provides greater flexibility as each feeder is supplied by two paths, so that the failure of a section does not cause any interruption of the supply. The effect of fault in one section is localised to that section alone. The rest of the sections continue to operate. Circuit breakers can be maintained without interrupting the supply. The cost is also not much as the number of breakers used are nearly the same as that of a single bus-bar system. The drawbacks of the system are (*i*) difficulties in addition of any new circuit in the ring (*ii*) possibility of overloading of the circuits on opening of any section of the breaker and (*iii*) necessity of supplying potential to relays separately to each of the circuit.

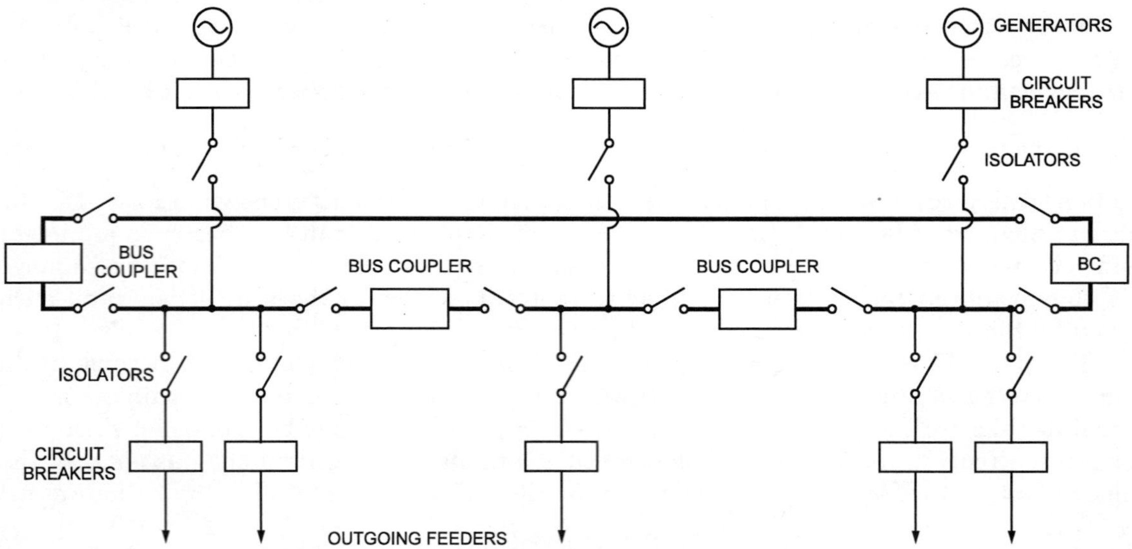

Fig. 1.16. ***Ring Main Arrangement***

8. ***Mesh Arrangement.*** This is another arrangement making economical use of circuit breakers in a subtation. In this bus-bar arrangement, the circuit breakers are installed in the mesh formed by the buses, as illustrated in Fig. 1.17. The circuits are tapped from the node points of the mesh. In Fig. 1.17 eight circuits are controlled by four circuit breakers. When fault occurs on any section, two circuit breakers have to open, resulting in opening of the mesh. This arrangement provides security against bus-bar faults but lacks switching facility. It needs fewer circuit breakers than that required by one-and a half breaker arrangement. It is preferred for substations having large number of circuits.

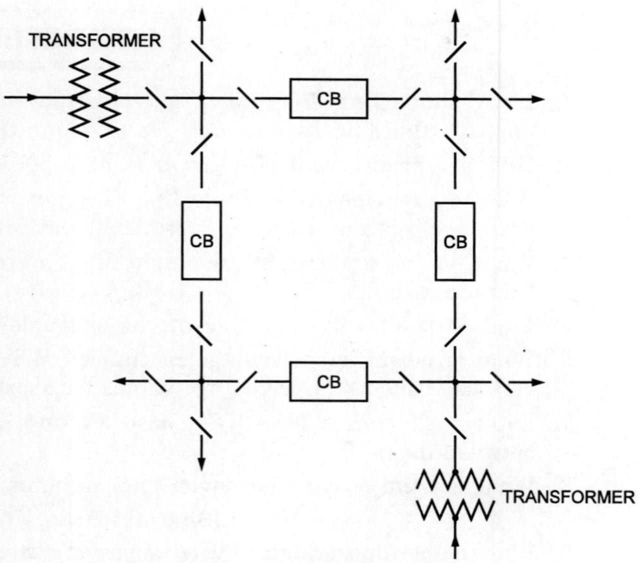

Fig. 1.17. ***Mesh Arrangement***

1.14. TRANSMISSION LINES

Transmission lines will be discussed later on (Refer to Art. 3.1).

Example 1.6. Draw single line diagram of a three-bus system having generator G_1 connected to bus-1 through transformer T_1, generator G_2 connected to bus-2 through transformer T_2, four synchronous motors M_1 to M_4 connected to bus-3 through transformer T_3, transmission lines TL_1, TL_2 and TL_3 connected between bus 1-2, 2-3 and 1-3 respectively.

[*U.P. Technical Univ. Elements of Power System 2008-09*]

Solution : Single line diagram of the given system is shown in Fig. 1.18.

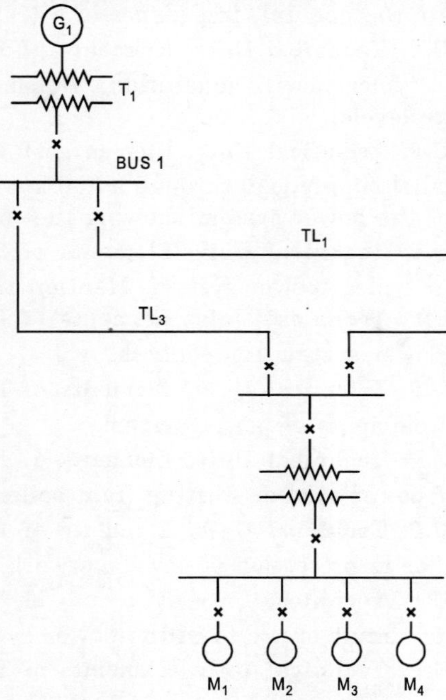

Fig. 1.18

EXERCISES

1. Draw single line diagram of a typical power system. Locate various sections of it and mention there on typical voltages of generation, transmission and distribution.
2. Give a typical layout of a power system between generation and use of electrical power.
3. What do you mean by single line diagram of power system ? What is the usefulness of drawing line-line diagram ? **[U.P. Technical Univ. Elements of Power System 2007-08]**
4. What do you understand by single line diagram of a power system ? What is the need of this diagram ? **[U.P. Technical Univ. Elements of Power System 2009-10]**
5. Explain the merits and limitations of single line diagrams in power systems.
6. What is meant by percentage reactance ? Why is it preferable to express the reactances of various elements in percentage values for short-circuit calculations ?
7. Explain the terms base kVA, base kV and percentage reactance. State the relationship between them.
8. What are the merits and demerits of per-unit systems ?
[Bharathidasan Univ. Power System Analysis November 1996]
9. Why the per-unit system is used in power system analysis ? Which of the electrical quantities are chosen as base values ?
10. Explain the advantages of per-unit system and derive the fundamental relations between the actual values and per-unit values on chosen base values for a three phase system.
11. Give the advantages and drawbacks of per-unit system. Prove that the per-unit impedance of a transformer is the same regardless of the side from which it is viewed.
12. Derive formula to convert the per unit values from given old base to new base values.
[U.P. Technical Univ. Elements of Power System 2002-03]
13. Show that the per unit equivalent impedance of a 2-winding transformer is the same whether it is referred to high voltage side or low voltage side.
[U.P. Technical Univ. Elements of Power System 2003-04]
14. Derive an expression to calculate the per unit impedance of a transformer.
[U.P. Technical Univ. Elements of Power System 2005-06]
15. Draw a single line diagram of a typical power generation, transmission and distribution system including various voltage levels.
[U.P. Technical Univ. Elements of Power System 2003-04]
16. Two alternators running in parallel supply load through a 250 km long transmission line. Draw the single line diagram of the power system showing the important components.
[U.P. Technical Univ. Elements of Power System 2002-03]
17. Draw the one line diagram of a typical power system. Mention the various components and their voltage levels. **[U.P. Technical Univ. Elements of Power System 2005-06]**
18. Describe in brief the different power system components.
[U.P. Technical Univ. Elements of Power System 2006-07]
19. Describe the various important elements of power system.
[U.P. Technical Univ. Elements of Power System 2004-05]
20. Draw the schematic diagram of power system starting from boiler to the grid structure.
[U.P. Technical Univ. Elements of Power System 2005-06]
21. Identify the role of circuit breaker in protection of power network.
[U.P. Technical Univ. Elements of Power System 2005-06]
22. Describe the location and role of circuit breakers in the power system.
[U.P. Technical Univ. Elements of Power System 2004-05]

23. What is the difference between isolator and circuit breaker ?

[U.P. Technical Univ. Elements of Power System 2008-09]

24. Differentiate between isolator and circuit breaker.

[U.P. Technical Univ. Elements of Power System 2007-08]

25. Justify the need of using an isolator in addition to a circuit breaker ? How are they operated while breaking and while making a circuit ?

[U.P. Technical Univ. Elements of Power System 2009-10]

26. Explain the concept of bus-bar in power network alongwith suitable diagram.

[U.P. Technical Univ. Elements of Power System 2005-06]

27. What is the role of bus-bar in power system ? Discuss any two types of bus-bar arrangements.

SHORT ANSWER TYPE QUESTIONS WITH ANSWERS

Q. 1. What is an electric power supply system ?

Ans. The system of generation, transmission and distribution of electrical power is called the electrical power supply system.

Q. 2. What are voltage levels for different components of power system ?

Ans. The power system network has various voltage levels for generation, transmission and distribution, which are mentioned below :

Generation : 3.3, 6.6, 11 or 33 kV (3-phase) most usual value adopted in practice is 11 kV.

Primary Transmission : 110, 132, 220, 400 or 765/800 kV (3-phase) depending upon the distance, the amount of power to be transmitted and the system stability.

Secondary Transmission : 33 or 66 kV (3-phase)

Primary Distribution : 3.3, 6.6 or 11 kV (3-phase) depending upon the requirements of the bulk consumers.

Secondary Distribution : 415/240 V (415 V for three phase loads and 240 V for single phase loads).

Q. 3. On what factors does the primary distribution voltage depend ?

Ans. The voltage employed for primary distribution system depends upon the amount of power to be supplied and distances of the substations required to be supplied.

Q. 4. What is/are the primary distribution voltage/voltages adopted in India ?

Ans. The usual value of primary distribution voltage in India is 11 kV but in case of high load density area it may be higher (33 or 66 kV) even.

Q. 5. What is the declared consumer voltage as per IE rules ?

Ans. The declared consumer voltage, according to IER is 415/240 V (*i.e.* 415 V between phases and 240 V between phase and neutral) with a permissible voltage variation of $\pm 6\%$.

Q. 6. What is a feeder ?

Ans. Feeder is a line which supplies a point of distribution network without being tapped at any intermediate point. In simple terms, the feeder may be defined as the line carrying current from the secondary substation to the distribution substations or as a primary distribution line.

Q. 7. What is a distributor ?

Ans. Distributor is a line from which tappings are taken along its length for providing supply to the consumers.

Q. 8. Why is it necessary to keep voltage variation at the consumer's terminals as low as possible ?

Ans. All the appliances and motors are designed to be operated at a particular or declared value of voltage. Large voltage variation means either a too high or a too low voltage. Low voltage will cause not only in burning of motors/appliances/equipment and insufficient lighting but also loss of revenue to the supplier. On the other hand high voltage will cause the lamps to burn out permanently and also may cause failure of other appliances. So the primary requirement of a distribution system is that voltage variations at the consumer's terminals should be the lowest possible (not more than $\pm 6\%$ of the rated value).

Q. 9. Explain the difference between feeder, distributor and service main ?

[U.P. Technical Univ. 2005-06]

Ans. (*i*) *Feeders.* Feeders are the conductors which connect the stations (in some cases generating stations) to the areas to be fed by those stations. Generally from feeders no tapping is taken to the consumers, therefore, current loading of a feeder remains the same along its length. It is designed mainly from the point of view of its current carrying capacity.

(*ii*) *Distributors.* Distributors are the conductors from which numerous tappings for the supply to the consumers are taken. The current loading of a distributor varies along its length. Distributors are designed from the point of view of the voltage drop in them.

(*iii*) *Service Mains.* Service mains are the conductors, which connect the consumer's terminals to the distributor. Elements of a distribution system are shown in Fig. 1.2.

A distributor is subject to the legal requirement that the voltage at the consumer's terminals should be maintained within ± 6% of the declared (or rated) voltage. However, there is no such legal restriction on a transmission line and the voltage can vary as much as 10% or even 15% due to variations in loads.

Q. 10. Explain why the voltage drop is of primary concern in the design of the distributors ?

[U.P. Technical Univ. 2006-07]

Ans. The voltage regulation is very important in the design of distributor. It is because of the statutory regulation giving the voltage variations permissible at the consumer's terminals (± 6% of declared or rated value). The size of the distributor should be such that voltage at the consumer's terminals remains within the permissible limits.

According to IE rules the declared consumer voltage is 415/240 V (*i.e.* 415 V between phases and 240 V between phase and neutral). Thus a consumer near a distribution substation may have a voltage as high as 440/254 V during light load hours while a consumer at the farthest end from the distribution substation may have a voltage as low as 400/230 V.

Q. 11. What is the voltage that has been selected for HVDC transmission ?

Ans. ± 500 kV

Q. 12. What do you understand by switchgear ?

Ans. The apparatus employed for switching, controlling and protecting the electrical circuits and equipment is known as switchgear.

Q. 13. What is a circuit breaker ?

Ans. Circuit breaker is a mechanical device designed to close or open contact members, thus closing or opening an electrical circuit under normal or abnormal conditions.

Q. 14. How does a circuit breaker differ from a switch ?

Ans. A circuit breaker is a device that interrupts the abnormal or fault currents and in addition performs the function of a switch.

Q. 15. What for isolating switches are provided along with circuit breakers ?

Ans. Isolating switches are provided for separating the disconnected element from the live portion of the system for the purposes of safety.

Q. 16. In what way is an isolator different from an air-break switch ?

Ans. Isolators are used to connect or disconnect the line only under no-load condition while air-break switches are used for making and breaking of the circuit under loaded condition. Air-break switches are provided with arc control devices.

Q. 17. How does isolator differ from circuit breaker ?

Ans. A circuit breaker is a device that interrupts the abnormal or fault currents in addition to switching at no load or on full load.

Isolators or disconnect switches operate under no load condition. They are not equipped with arc-quenching devices. They do not have any specified current breaking capacity or current making capacity. The isolators in some cases are used for breaking charging current of transmission line.

Q. 18. Why are isolators necessarily provided on the supply side of the circuit breakers ?

Ans. Isolators are necessarily provided on the supply side of the circuit breaker in order to ensure isolation of the circuit breaker from the live parts for the purpose of maintenance.

Q. 19. Name the interlockings provided with isolators.

Ans. The interlocking provided with isolators are :

1. Interlocking between three poles for simultaneous operation.
2. Interlocking with circuit breakers.

Q. 20. Give the sequence of operation during opening and closing of a circuit.

Ans. Sequence of operation of circuit breaker, isolator and earthing switch during opening/closing of a circuit are given below.

While Opening: Open circuit breaker, open isolator and then close earthing switch if provided.

While Closing: Open earthing switch, close isolator and then close circuit breaker.

Q. 21. What is meant by the term "bus-bar" ?

Ans. Bus-bar (or bus in short) term is used for a main bar or conductor carrying an electric current to which many connections may be made. The term bus is derived from the word *omnibus,* meaning collector of things. Thus electrical bus-bar is the collector of electrical energy at one location. Bus-bars are merely convenient means of connecting switches and other equipment into various arrangements.

Q. 22. What is flexible bus ?

Ans. Flexible bus consists of flexible ACSR or all aluminium alloy stranded conductors supported by strain insulators from each end.

PROBLEMS

1. A single phase transformer is rated 100 kVA, 11/0.4 kV. Its leakage reactance is 0.15 Ω when referred to low-voltage side. Determine its leakage reactance in per cent and per unit. **[Ans** 9.375%; 0.09375 pu]
2. Convert an impedance of (4 + *j*24) into per-unit impedance with base voltage of 11 kV and base kVA of 500. **[Ans** (0.0165 + j 0.0992) pu]
3. The per-unit impedance of 100 MVA, 11/132 kV step-up transformer is 0.3 Ω on its own base. Determine (*i*) actual impedances of the transformer referred to hv as well as lv side and (*ii*) base impedance referred to both sides. **[Ans** (*i*) 52.272 Ω ; 0.363 Ω (*ii*) 174.24 Ω ; 1.21 Ω]
4. Two generators rated 10 MVA, 13.2 kV and 15 MVA, are connected to common bus-bars. Two synchronous motors are connected to these bus-bars. These two motors have 8 and 12 MVA capacity respectively. Both motors are rated 12.5 kV. Choosing 50 MVA and 13.8 kV as base quantities, compute reactances of all the machines referred to these base quantities. The generators have 15 % reactance each and motors have 20 % reactance each on the basis of the respective ratings of the machines. **[Ans.** 68.62%; 45.75%; 102.56%; 68.37%]
5. A synchronous generator G_1 is feeding power to bus 1 through transformer T_1. Bus 2 receives power through a tie-line L_1 connected between buses 1 and 2 and feeds a set of four motors M_1 to M_4 through transformer T_2. A local load $S_1 = P_1 + jQ_1$ is also fed by synchronous generator at bus 1. Draw single line diagram of the system.

[U.P. Technical Univ. Elements of Power System 2004-05]

❑❑❑

2

CHAPTER

SUPPLY SYSTEM

INSIDE THIS CHAPTER

2.1. Introduction **2.2.** Influence of Working Voltage on Size of Feeders and Distributors in DC System **2.3.** Various Systems of Transmission of Electrical Power **2.4.** Comparison of Cost of Conductors-Overhead Systems **2.5.** Comparison of Cost of Conductors-Underground Systems **2.6.** Results of Comparison of Transmission Systems **2.7.** Choice of Working Voltage for Transmission.

2.1. INTRODUCTION

The system transporting electrical energy from power stations to the consumers' premises is known as *electric supply system*. An electric supply system consists of three principal components viz. the power station, the transmission system and distribution system.

The electric supply system can be broadly classified into (*i*) dc or ac system. (*ii*) overhead or underground system.

Each system of transmission and distribution has its own advantages and disadvantages and it is, therefore, desirable to make comparison between (*i*) dc and ac systems and (*ii*) overhead and underground systems of transmission and distribution.

2.1.1. Comparison Between DC and AC Systems of Transmission and Distribution

Transmission of electric power by high voltage dc system is superior to that by high voltage ac system in many respects.

(*i*) It requires only two conductors for transmission and it is also possible to transmit the power through only one conductor by using earth as returning conductor, hence much copper is saved.

(*ii*) No inductance, capacitance, phase displacement and surge problem in dc transmission.

(*iii*) There is no skin effect in dc, cross-section of line conductor is, therefore, fully utilised.

(*iv*) The potential stress on the insulation in case of dc system is $1/\sqrt{2}$ times of that in ac system for the same working voltage. Hence for same working voltage less insulation is required.

(*v*) Charging current, which contributes to continuous loss even on no load, is eliminated.

(*vi*) A dc line has less corona loss and reduced interference with communication circuits.

(*vii*) Underground cables can be used because of less potential stress and negligible dielectric loss.

(*viii*) Since there is no inductance, the voltage drop in a dc transmission line due to inductive reactance does not exist. Hence for the same load and sending-end voltage, voltage regulation for the dc transmission line is better.

(*ix*) No stabilizer is required for transmission over long distances.

The only difficulty is of obtaining the necessary high voltage required for transmission as neither electrical power can be generated at high voltage because of commutating difficulties nor voltage can be stepped up.

In case of ac systems the main advantage is that the voltage can be stepped up at generating end by means of step-up transformers to the required value for transmission and then stepped down at distributing end by means of step-down transformers for distribution with ease and efficiency. The other advantages of ac systems are that electric power can be generated at high voltage easily and the maintenance of ac sub-stations is easier and cheaper. The disadvantages of ac systems are given below:

(*i*) In case of overhead lines spacing between the conductors is to be kept more in order to provide adequate insulation and to avoid corona loss.

(*ii*) More copper is required.

(*iii*) Greater the length of transmission line more is the capacitance of the line. There is a continuous loss on account of charging current even though the line is open.

(*iv*) The construction of transmission lines for ac systems is not so easy as for dc systems.

(*v*) The alternators are to be synchronised before putting them into parallel.

(*vi*) The variation in speeds of alternators are to be controlled within very low limits.

Distribution by ac system is undoubtedly superior to that by dc system as in ac system voltage control is easy by means of transformers.

The best method is to employ ac system for generation and distribution and dc system for transmission.

Transmission of electric power by dc system became possible because of introduction of mercury-arc rectifiers and thyratrons*, which can convert ac into dc and vice-versa efficiently and at reasonable cost.

Single line diagram of an hv dc transmission system is shown in Fig. 2.1. The electric power is generated as ac, its voltage is stepped up to high voltage by step-up transformer and converted into dc by some suitable rectifying device*. The transmission of electric power is carried at high dc voltage (500 kV). At receiving end, dc power is converted back into 3-phase ac using suitable converters and then stepped down to low voltage for distribution by suitable step-down transformers.

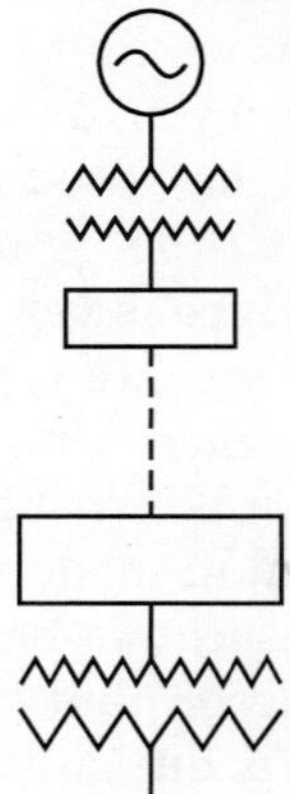

Fig. 2.1. ***Line Diagram of Typical HV DC Transmission System***

The favourable conditions resulting from the use of converters in the above system gives the following additional advantages:

1. Independence of frequencies at sending-end and receiving-end networks makes possible generation of power at a frequency different from that of load system.

* Since the early 1970s, all new HVDC schemes use thyristors instead of earlier used mercury-arc valves

2. With the use of suitable converters, power can be reversed, increased or decreased in the system at will and with rapidity *i.e.*, the control is easy and rapid.
3. Because of absence of charging and stability problems, there is no limit of distance over which dc power can be transmitted.
4. Keeping in view the circuit breaker capacities etc., the short-circuit currents are not transmitted through lines and hence a tremendous amounts of saving is affected by installing smaller circuit breakers.
5. The smaller amount of power can be transmitted much more economically.

2.1.2. Comparison Between Overhead and Underground Systems of Transmission and Distribution

Transmission and distribution of electric power can be carried out by overhead as well as underground systems. Comparison between the two is given below:

(*i*) *Public Safety:* Underground system is more safer than overhead system.

(*ii*) *Initial Cost:* Underground system is more expensive. For a particular amount of power to be transmitted at a given voltage the underground system costs almost double the cost of overhead system.

(*iii*) *Flexibility:* Overhead system is more flexible than underground system. In overhead system new conductors can be laid along the existing ones for load expansion. In case of underground system new conductors are to be laid in new channels.

(*iv*) *Working Voltage:* The underground system cannot be operated above 66 kV because of insulation difficulties but overhead system can be designed for operation up to 400 kV or higher even.

(*v*) *Maintenance Cost:* Maintenance cost of underground system is very low in comparison with that of overhead system.

(*vi*) *Frequency of Faults or Failures:* As the cables are laid underground, so these are not easily accessible. The insulation is also better, so there are very few chances of power failures or fault as compared to overhead system.

(*vii*) *Frequency of Accidents:* The chances of accidents in underground system are very low as compared to overhead system.

(*viii*) *Voltage Drop:* In underground system because of less spacing between the conductors inductance is very low as compared to overhead lines, therefore, voltage drop is low in underground system.

(*ix*) *Appearance:* Underground system of distribution or transmission is good looking because no wiring is visible. Due to its good looking, inspite of its higher cost it is adopted in modern cities like Chandigarh, Faridabad.

(*x*) *Fault Location and Repairs:* Though there are very rare chances of occurring fault in underground system, but if occurs it is very difficult to locate that fault and its repair is difficult and expensive.

(*xi*) *Charging Current:* On account of less spacing between the conductors the cables have much capacitance, so draw higher charging current.

(*xii*) *Jointing:* Jointing of underground cables is difficult so tapping for loads and service mains is not conveniently possible in underground system.

(*xiii*) *Damage Due to Lightning and Thunder Storm:* Underground system is free from interruption of service on account of thunder storm, lightning and objects falling across the wires.

(*xiv*) *Surge Effect:* In underground system surge effect is smoothened down as surge energy is absorbed by the sheath.

(*xv*) *Interference to Communication Circuits:* In underground system there is no interference to communication circuits.

Mostly the high voltage transmission is carried out by overhead system due to low cost. However, distribution in conjested areas and in modern cities, like Chandigarh, is carried out by underground cables.

The overhead line as a mean of transmitting electrical power over long distances is cheap and efficient (a 400 kV double-circuit line may transmit 2,000 MW over a distance of 160 km in either direction with an efficiency of 98%). It is reliable—there are only one or two breakdowns due to lightning, and one or two due to other reasons, such as frost and fog, per 160 km of a line a year. Most of the breakdowns which do occur are transient and none out of them can be rendered innocuous, as far as system is concerned, by using high-speed automatic-reclosing circuit-breakers. Repairs, where necessary, can easily and quickly be carried on site. The only real disadvantage of the overhead line is that it needs large electrical clearances and it is sometimes not possible to get a number of circuits into congested areas.

EHV power cables are used in congested or built-up areas, in the vicinity of air fields, line terminations and transformer banking connections where the use of overhead lines is not permitted because of space limitations, safety requirements, amenities, etc. The major factor against a greater use of insulated cables is the high price of the equipment.

2.2. INFLUENCE OF WORKING VOLTAGE ON SIZE OF FEEDERS AND DISTRIBUTORS IN DC SYSTEM

There are following advantages of adopting high voltage for transmission and distribution of electrical energy in dc system.

(*i*) *It reduces the size (area of* x-section *of the core carrying the current) of the feeders and distributors.*

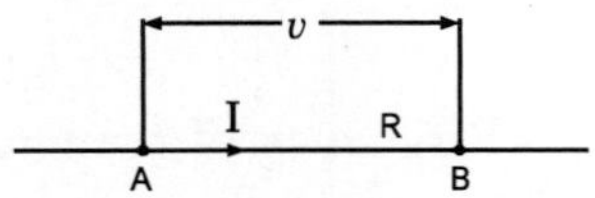

Fig. 2.2.

Consider first of all a feeder. As no consumer is taped off from a feeder so voltage drop in feeder will not be so important provided that it is not outside the range of compound generators in the generating stations. So the fundamental basis of designing of the size of a feeder is the current-carrying capacity.

Though the permissible current density for a given type of cable is not constant but decreases somewhat with the increase in size due to poor cooling facilities with higher sizes but for approximation we assume it as constant.

Now if supply voltage is increased n times, then for a given power to be delivered the current is reduced to $\frac{1}{n}$th so the size of the cable is also reduced approximately to $\frac{1}{n}$th.

Consider a distributor in which voltage drop is to be considered mainly while deciding its size. This is because the voltage drop in secondary distribution lines is to be limited to a specified value. Let distributor AB carry a current of I amperes and let permissible voltage drop be of v volts. Then the resistance of the distributor is given by $R = \frac{v}{I}$. Now if the supply voltage is increased n times, then for a given power to be delivered the current in the distributor becomes $\frac{I}{n}$ and permissible voltage drop increases for a given percentage of permissible voltage drop in the distributor and becomes nv volts. The resistance of the distributor with high voltage of distribution is given by $R' = \frac{nv}{1/n} = n^2R$.

Since for a given length, the area of x-section of a cable is inversely proportional to its resistance, so by increasing the supply voltage to n times, the area of cross-section can be reduced to $\frac{1}{n^2}$ times for supplying a given power with same per cent of voltage drop in feeder.

(ii) It increases the efficiency of transmission.

Let voltage at feeding point = V volts

Current supplied by the feeder = I amps

Resistance of each conductor of feeder = R ohms

Total voltage drop in feeder = 2 IR

Voltage at supply end = V + 2 I R

$$\text{Efficiency of transmission, } \eta_1 = \frac{VI}{(V + 2IR)I} = \frac{V}{(V + 2IR)} = \left(1 - \frac{2IR}{V + 2IR}\right) \qquad ...(2.1)$$

Now let the voltage at feeding point be raised to n V volts then for transmission of same power

$$\text{Current in feeder} = \frac{VI}{nV} = \frac{I}{n}$$

Assuming current density constant, the size of the cable of feeder is reduced to $\frac{1}{n}$th, hence resistance of each cable of feeder becomes n R

$$\text{Voltage drop in feeder} = \text{Current} \times \text{resistance} = \frac{I}{n} \times 2nR = 2IR \text{ (same as before)}$$

Voltage at supply end = n V + 2 I R

$$\text{Efficiency of transmission, } \eta_2 = \frac{\text{Output}}{\text{Input}} = \frac{nV \times \frac{I}{n}}{\frac{I}{n}(nV + 2IR)}$$

$$= \frac{V}{V + \frac{2IR}{n}} = 1 - \frac{2IR}{nV + 2IR} \qquad ...(2.2)$$

Comparing Eqs. (2.1) and (2.2) we conclude that transmission efficiency is sufficiently increased. For example if I = 50 A; R = 0.4 Ω; V = 100 V and n = 10 then

$$\eta_1 = 1 - \frac{2IR}{V + 2IR} = 1 - \frac{2 \times 50 \times 0.4}{100 + 2 \times 50 \times 0.4} = 0.714 \text{ or } 71.4\%$$

$$\eta_2 = 1 - \frac{2IR}{nV + 2IR} = 1 - \frac{2 \times 50 \times 0.4}{10 \times 100 + 2 \times 50 \times 0.4} = 0.9615 \text{ or } 96.15\%$$

Thus with the increase in transmission voltage, efficiency of transmission increases and cost of copper in cable decreases. For this reason it is advantageous to transmit power at high voltage.

Example 2.1. What is the percentage saving in feeder copper if the line voltage in a 2-wire dc system be raised from 220 volts to 400 volts for the same power transmitted over same distance and having the same power loss. [*U.P. Technical Univ. Elements of Power System 2006-07*]

Solution : Let the power delivered be P watts

$$\text{Load current, } I_1 = \frac{P}{220} \text{ A} \qquad \text{when line voltage is 220 V}$$

$$\text{Load current, } I_2 = \frac{P}{400}\text{ A} \qquad \text{when line voltage is 400 V}$$

Let the power be transmitted over length l and power loss be W watts.
When line voltage is 220 V

$$\text{Power losses in line, } W = I_1^2 \times 2R_1 = \left(\frac{P}{220}\right)^2 \times 2\rho\frac{l}{a_1}$$

where ρ is the resistivity of copper, and a_1 is the area of x-section of copper required

$$\text{or } a_1 = \frac{P^2\rho l}{24{,}200\text{ W}}$$

$$\text{Volume of copper required, } v_1 = 2a_1 l = \frac{P^2\rho l^2}{12{,}100\text{ W}} \qquad ...(i)$$

When line voltage is 400 V

$$\text{Power losses in line, } W = I_2^2 \times 2R_2 = \left(\frac{P}{400}\right)^2 \times 2\rho\frac{l}{a_2} = \frac{P^2\rho l}{80{,}000 a_2}$$

$$\text{or } a_2 = \frac{P^2\rho l}{80{,}000\text{ W}}$$

$$\text{Volume of copper required, } v_2 = 2a_2 l = \frac{P^2\rho l^2}{40{,}000\text{ W}} \qquad ...(ii)$$

$$\text{Percentage saving in copper} = \frac{v_1 - v_2}{v_1} \times 100 = \frac{\frac{P^2\rho l^2}{12{,}100\text{ W}} - \frac{P^2\rho l^2}{40{,}000\text{ W}}}{\frac{P^2\rho l^2}{12{,}100\text{ W}}} \times 100$$

$$= \left[1 - \frac{12{,}100}{40{,}000}\right] \times 100 = (1 - 0.3025) \times 100 = 69.75\% \textbf{ Ans.}$$

2.3. VARIOUS SYSTEMS OF TRANSMISSION OF ELECTRICAL POWER

Though in practice 3-phase 3-wire ac system is universally used for transmission and 3-phase 4-wire ac system is used for distribution of electric power but for special purposes other systems may also be used. The various systems of power transmission are :

(*a*) *DC Systems :*

- (*i*) DC two-wire system.
- (*ii*) DC two-wire system with mid point earthed.
- (*iii*) DC three wire-system.

(*b*) *Single Phase AC Systems:*

- (*i*) Single phase two-wire system.
- (*ii*) Single phase two-wire system with mid point earthed.
- (*iii*) Single phase three-wire system.

(*c*) *Two-Phase AC Systems:*

- (*i*) Two-phase four-wire system.
- (*ii*) Two-phase three-wire system.

(*d*) *Three Phase AC Systems:*

(*i*) Three-phase 3-wire system.

(*ii*) Three-phase 4-wire system.

From such a big list of possible systems of transmission it is difficult to decide the best system without making comparison. The basis for comparison between the various systems of power transmission is usually economy. Since in a transmission system, the cost of the conductor material accounts for a major part of the total cost, the best system for transmission of electrical power is that for which the volume of conductor material required is minimum. Thus the requirement of volume of conductor material forms the basis of comparison between various systems.

In making comparison of the volume of conductor material required for various transmission systems, the basis will be the *equal maximum stress on the dielectric*. This is because the voltage is only limited by the problem of insulating the conductors against disruptive discharge.

For comparing the amount of conductor material required for different systems two cases arise:

(*i*) *When overhead system is employed for transmission of power.* In overhead system the conductors are insulated from the x-arms and supporting towers and as the towers and x-arms are earthed so the maximum voltage between each conductor and earth forms the basis of comparison of volume of conductor material required.

(*ii*) *When underground cables are employed for transmission of power.* In underground cables the maximum disruptive stress is between the two conductors of the cable; so the maximum voltage between conductors forms the basis of comparison of volume of conductor material required.

2.4. COMPARISON OF COST OF CONDUCTORS-OVERHEAD SYSTEMS

The assumptions made for comparison are:

(*i*) In all cases power to be transmitted is same (say, P watts).

(*ii*) The distance over which the power is to be transmitted is same (say l).

(*iii*) The line losses are same (say, W watts).

(*iv*) The maximum voltage to earth is same (say, V_m volts).

(*a*) **DC Systems**

(*i*) *DC 2-Wire System With One Conductor Earthed* [Fig. 2.3 (*i*)].

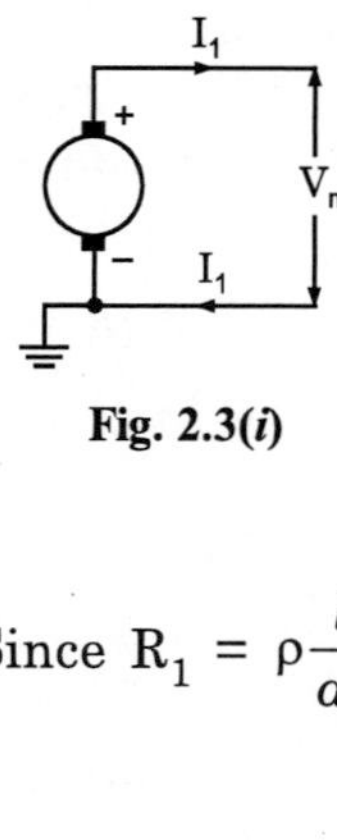

Fig. 2.3(*i*)

Maximum voltage between conductors $= V_m$ volts

Power to be transmitted $=$ P watts

$$\text{Load current, } I_1 = \frac{P}{V_m}$$

$$\text{Line losses, } W = 2\, I_1^2\, R_1 = 2\left(\frac{P}{V_m}\right)^2 \rho \frac{l}{a_1} \qquad \text{Since } R_1 = \rho \frac{l}{a_1}$$

$$\text{or Area of x-section of conductor, } a_1 = \frac{2\rho P^2 l}{W V_m^2}$$

$$\text{Volume of conductor material required} = 2\, a_1 l = \frac{4\rho P^2 l^2}{W V_m^2} \qquad ...(2.3)$$

This system is usually formed the basis for comparison with other systems. Thus the volume of conductor material required in this system shall be taken as the basic quantity *i.e.*

$$\frac{4\rho P^2 l^2}{W V_m^{\ 2}} = K \text{ (say)}$$

(*ii*) *DC Two-Wire System With Mid Point Earthed.* This system is shown in Fig. 2.3 (*ii*). The maximum voltage between any conductor and earth is V_m volts so

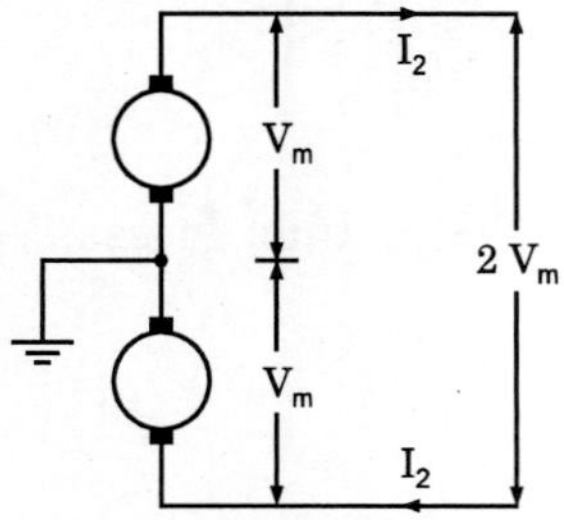

Fig. 2.3(*ii*)

Maximum voltage between conductors = 2 V_m volts

$$\text{Load current, } I_2 = \frac{P}{2V_m}$$

$$\text{Line losses, } W = 2\ I_2^2\ R_2 = 2\left(\frac{P}{2\ V_m}\right)^2 \rho \frac{l}{a_2} = \frac{P^2 \rho l}{2\ a_2\ V_m^2}$$

$$\text{or } a_2 = \frac{P^2 \rho l}{2\ W V_m^2}$$

$$\text{Volume of conductor material required} = 2\ a_2\ l = \frac{P^2 \rho l^2}{W\ V_m^2} = 0.25\ K \qquad ...(2.4)$$

Hence volume of conductor material required is one-fourth of that required in two-wire dc system with one conductor earthed.

(*iii*) *D C Three-Wire System.* In a 3-wire d c system, there are two outers and one middle wire, called the *neutral wire* which is grounded at the generator end, as illustrated in Fig. 2.3 (*iii*). For balanced load on the system, the middle wire carries no current.

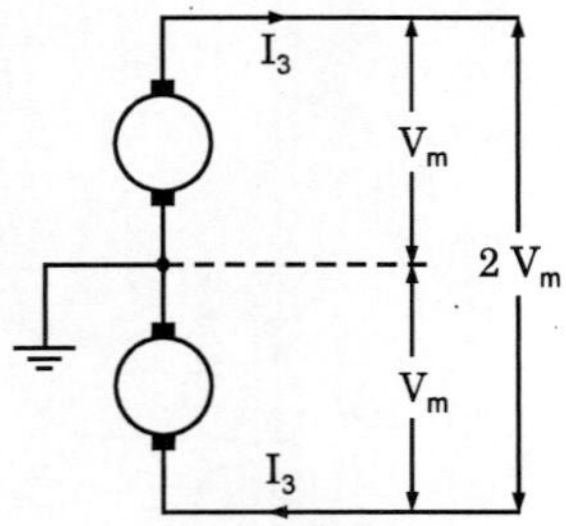

Fig. 2.3(*iii*)

Assuming load balanced

Maximum voltage between the outer and earth = V_m volts

$$\text{Load current, } I_3 = \frac{P/2}{V_m} = \frac{P}{2\ V_m}$$

$$\text{Line losses, } W = 2\ I_3^2 R_3$$

$$= 2\left(\frac{P}{2\ V_m}\right)^2 \frac{\rho l}{a_3} = \frac{P^2 \rho l}{2\ a_3 V_m^2} \quad \text{or} \quad a_3 = \frac{P^2 \rho l}{2\ W\ V_m^2}$$

Assuming area of x-section of neutral wire as half of that of any of the outers

$$\text{Volume of conductor material required} = 2.5\ a_3 l = \frac{2.5\ P^2 \rho\ l^2}{2\ W V_m^2} = 0.3125\ K \qquad ...(2.5)$$

Hence volume of conductor material required is 0.3125 times of that required in two-wire dc system with one conductor earthed.

(*b*) Single Phase AC Systems

(*i*) *AC Single Phase Two-Wire System With One Conductor Earthed* Fig. 2.3 (*iv*).

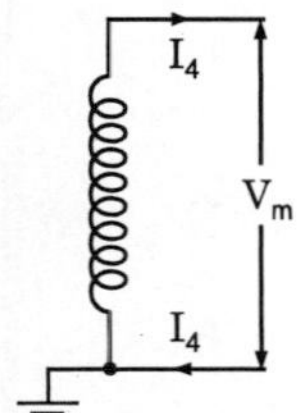

Fig. 2.3(*iv*)

Peak value of voltage between conductors = V_m volts

$$\text{RMS value of voltage between conductors} = \frac{V_m}{\sqrt{2}} \text{ volts}$$

$$\text{Load current, } I_4 = \frac{P}{\frac{V_m}{\sqrt{2}} \cos \phi} = \frac{\sqrt{2}\, P}{V_m \cos \phi}$$

where cos ϕ is the power factor of the load

$$\text{Line losses, } W = 2\, I_4^2 \;\; R_4 = 2 \left(\frac{\sqrt{2}\, P}{V_m \cos \phi} \right)^2 \frac{\rho l}{a_4} = \frac{4\, P^2 \rho\, l}{a_4\, V_m^2 \cos^2 \phi}$$

$$\text{or } \;\; a_4 = \frac{4\, P^2 \rho\, l}{W\, V_m^2 \cos^2 \phi}$$

$$\text{Volume of conductor material required} = 2\, a_4\, l = \frac{8\, P^2 \rho\, l^2}{W\, V_m^2 \cos^2 \phi} = \frac{2}{\cos^2 \phi} K \qquad ...(2.6)$$

Hence volume of conductor material required in this system is $\frac{2}{\cos^2 \phi}$ *times of that required in 2-wire dc system with one conductor earthed.*

(*ii*) *AC Single Phase Two-Wire System With Mid Point Earthed* [Fig. 2.3 (*v*)].

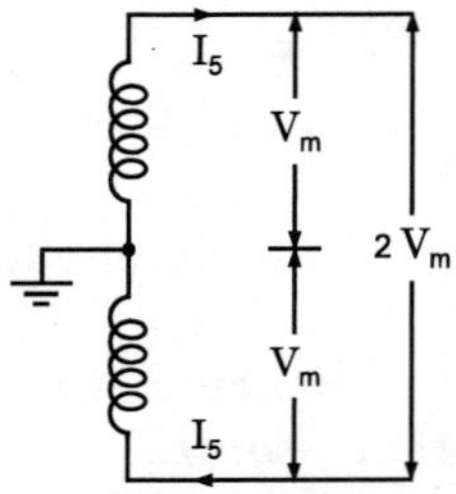

Fig. 2.3(*v*)

Peak value of voltage between conductors = 2 V_m volts

$$\text{RMS value of voltage between conductors} = \frac{2\, V_m}{\sqrt{2}} = \sqrt{2}\, V_m \text{ volts}$$

$$\text{Load current, } I_5 = \frac{P}{\sqrt{2}\, V_m \cos \phi}$$

$$\text{Line losses, } \;\; W = 2\, I_5^2\, R_5 = 2 \left(\frac{P}{\sqrt{2}\, V_m \cos \phi} \right)^2 \frac{\rho l}{a_5} = \frac{P^2 \rho l}{a_5\, V_m^2 \cos^2 \phi}$$

$$\text{or } \;\; a_5 = \frac{P^2 \rho l}{W\, V_m^2 \cos^2 \phi}$$

$$\text{Volume of conductor material required} = 2\, a_5\, l = \frac{2\, P^2 \rho l^2}{W\, V_m^2 \cos^2 \phi} = \frac{0.5\, K}{\cos^2 \phi} \qquad ...(2.7)$$

Hence volume of conductor material required in this system is $\frac{0.5}{\cos^2 \phi}$ *times of that required in 2-wire dc system with one conductor earthed.*

(*iii*) *AC Single Phase Three-Wire System* [Fig. 2.3 (*vi*)].

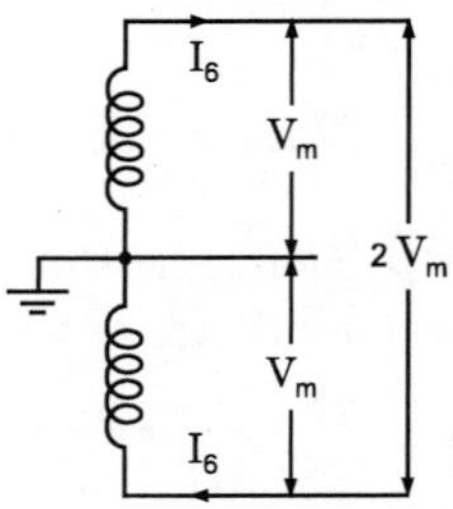

Fig. 2.3(*vi*)

Maximum voltage between outer conductors = 2 V_m volts

$$\text{RMS value of voltage between outer and earth} = \frac{V_m}{\sqrt{2}} \text{ volts}$$

$$\text{Load current, } I_6 = \frac{P/2}{\frac{V_m}{\sqrt{2}} \cos \phi}$$

$$= \frac{P}{\sqrt{2}\, V_m \cos \phi} \text{ load being assumed balanced}$$

$$\text{Line losses, } W = 2\, I_6^2\, R_6 = 2\left(\frac{P}{\sqrt{2}\, V_m \cos\phi}\right)^2 \frac{\rho l}{a_6} = \frac{P^2 \rho l}{V_m^2 \cos^2\phi\, a_6}$$

$$\text{or } a_6 = \frac{P^2 \rho l}{W V_m^2 \cos^2\phi}$$

Assuming x-section of neutral wire half of that of any of the outers
Volume of conductor material required

$$= 2.5\, a_6\, l = \frac{2.5\, P^2 \rho l^2}{W V_m^2 \cos^2\phi} = \frac{0.625\, K}{\cos^2\phi} \qquad \text{...(2.8)}$$

Hence volume of conductor material required in this system is $\frac{0.625}{\cos^2\phi}$ *times of that required in 2-wire dc system with one conductor earthed.*

(c) Two-Phase AC Systems

(i) AC Two-Phase Four-Wire System [Fig. 2.3 (vii)].

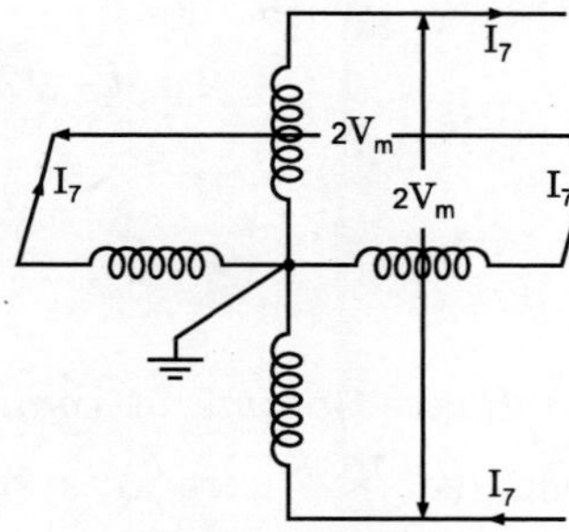

Fig. 2.3(*vii*)

$$\text{RMS value of voltage between the outers} = \frac{2V_m}{\sqrt{2}} = \sqrt{2}\, V_m$$

$$\text{Load supplied by each phase} = \frac{P}{2}$$

$$\text{Load current per phase, } I_7 = \frac{P}{2\sqrt{2}\, V_m \cos\phi}$$

$$\text{Line losses, } W = 4\, I_7^2\, R_7$$

$$= 4\left(\frac{P}{2\sqrt{2}\, V_m \cos\phi}\right)^2 \rho \frac{l}{a_7} = \frac{P^2 \rho l}{2\, V_m^2\, a_7 \cos^2\phi}$$

$$\text{or } a_7 = \frac{P^2 \rho l}{2\, V_m^2\, W \cos^2\phi}$$

$$\text{Volume of conductor material required} = 4\, a_7 l = \frac{2P^2 \rho l^2}{V_m^2 \cos^2\phi\, W} = \frac{0.5\, K}{\cos^2\phi} \qquad \text{...(2.9)}$$

Hence volume of conductor material required in this system is $\frac{0.5}{\cos^2\phi}$ *times of that required in two-wire dc system with one conductor earthed.*

(ii) AC Two-Phase Three-Wire System [Fig. 2.3 (viii)].

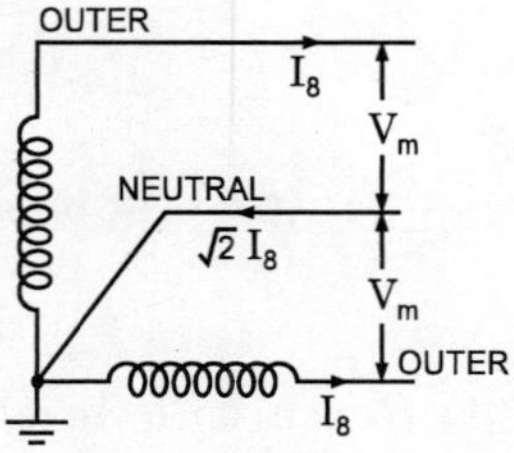

Fig. 2.3(*viii*)

Considering balanced load, RMS value of voltage between any outer and neutral

$$= \frac{V_m}{\sqrt{2}}$$

$$\text{Current in each of outer conductors, } I_8 = \frac{P/2}{\frac{V_m}{\sqrt{2}} \cos\phi} = \frac{P}{\sqrt{2}\, V_m \cos\phi}$$

Current in neutral wire $= \sqrt{\left(I_8^2 + I_8^2\right)} = \sqrt{2}\ I_8$

Assuming current density constant, area of x-section of neutral wire is $\sqrt{2}$ times of that of either of the outers.

So resistance of neutral wire $= \dfrac{R_8}{\sqrt{2}} = \dfrac{\rho l}{\sqrt{2}\, a_8}$ where $R_8 = \dfrac{\rho l}{a_8}$

Line losses, $W = 2\ I_8^2\ R_8 + \left(\sqrt{2}\ I_8\right)^2 \dfrac{R_8}{\sqrt{2}} = I_8^2 \dfrac{\rho l}{a_8}\left(2+\sqrt{2}\right) = \dfrac{P^2 \rho l}{2\ V_m^2 \cos^2\phi\, a_8}\left(2+\sqrt{2}\right)$

or $a_8 = \dfrac{P^2 \rho l}{2\ V_m^2 \cos^2\phi\ W}\left(2+\sqrt{2}\right)$

Volume of conductor material required

$$= 2\ a_8\ l + \sqrt{2}\ a_8\ l = a_8\ l\left(2+\sqrt{2}\right) = \frac{P^2 \rho l^2}{2\ V_m^2 \cos^2\phi\ W}\left(2+\sqrt{2}\right)^2$$

$$= \frac{P^2 \rho l^2}{V_m^2 \cos^2\phi\ W}\left(\sqrt{2}+1\right)^2 = \frac{1.457}{\cos^2\phi} K \quad ...(2.10)$$

Hence volume of conductor material required in this system is $\dfrac{1.457}{\cos^2\phi}$ *times of that required in 2-wire dc system with one conductor earthed.*

(d) 3-Phase AC Systems

(*i*) AC 3-Phase 3-Wire System *[Fig. 2.3 (ix)]*.

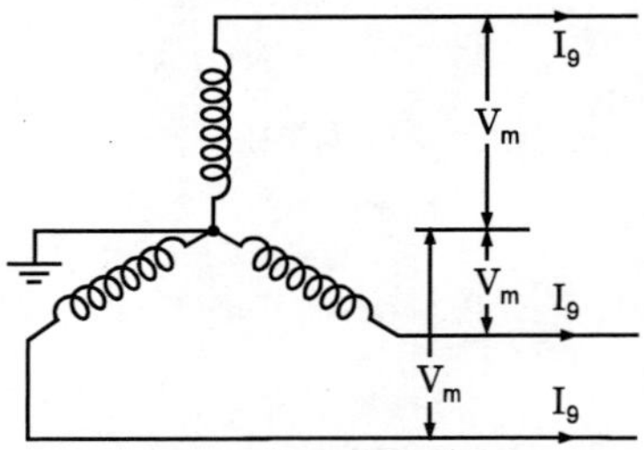

Fig. 2.3(*ix*)

RMS value of voltage per phase $= \dfrac{V_m}{\sqrt{2}}$

Load current per phase, $I_9 = \dfrac{P/3}{\dfrac{V_m}{\sqrt{2}} \cos\phi} = \dfrac{\sqrt{2}\ P}{3\ V_m \cos\phi}$

Line losses, $W = 3\ I_9^2\ R_9$

$$= 3\left(\frac{\sqrt{2}\ P}{3\ V_m \cos\phi}\right)^2 \frac{\rho l}{a_9} = \frac{2\ P^2 \rho l}{3 V_m^2 \cos^2\phi\ a_9}$$

or $a_9 = \dfrac{2\ P^2 \rho l}{3 V_m^2 \cos^2\phi\ W}$

Volume of conductor material required $= 3\ a_9\ l = \dfrac{2\ P^2 \rho l^2}{V_m^2 \cos^2\phi\ W} = \dfrac{0.5\ K}{\cos^2\phi}$ ...(2.11)

Hence volume of conductor material required in this system is $\dfrac{0.5}{\cos^2\phi}$ *times of that required in two-wire dc system with one conductor earthed.*

(ii) AC 3-Phase 4-Wire System. [Fig. 2.3 (x)].

Assuming balanced load, there will be no current in neutral wire and copper losses will be same as in 3-phase 3-wire system

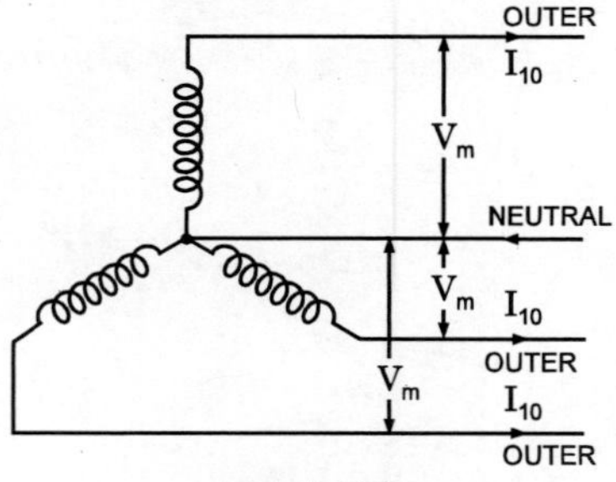

Fig. 2.3(*x*)

$$\text{i.e. } W = \frac{2P^2 \rho l}{3 V_m^2 \cos^2 \phi \, a_{10}}$$

$$\text{or } a_{10} = \frac{2 P^2 \rho l}{3 V_m^2 \cos^2 \phi \, W}$$

Taking x-section of neutral wire as half of either outer

$$\text{Volume of conductor material required} = 3.5 \, a_{10} \, l = \frac{7 \rho P^2 l^2}{3 \cos^2 \phi \, V_m^2 \, W} = \frac{0.583}{\cos^2 \phi} K \quad ...(2.12)$$

Hence volume of conductor material required in this case is $\frac{0.583}{\cos^2 \phi}$ *times of that required in case of two-wire dc system with one conductor earthed.*

2.5. COMPARISON OF COST OF CONDUCTORS-UNDERGROUND SYSTEMS

The assumptions made for comparison are:

(*i*) In all cases the power to be transmitted is same (say, P watts).
(*ii*) The distance over which the power is to be transmitted is same (say, *l*).
(*iii*) The line losses are same (say, W watts).
(*iv*) The maximum voltage between two conductors is same (say, V_m volts).

(*a*) DC Systems

(i) DC Two-Wire System [Fig. 2.4 (i)].

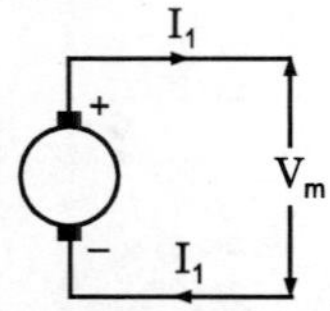

Fig. 2.4(*i*)

Voltage between two conductors $= V_m$ volts

$$\text{Load current, } I_1 = \frac{P}{V_m}$$

$$\text{Line losses, } W = 2 I_1^2 R_1 = \frac{2 P^2 \rho l}{V_m^2 \, a_1}$$

$$\text{or } a_1 = \frac{2 P^2 \rho l}{W V_m^2}$$

$$\text{Volume of conductor material required} = 2 \, a_1 \, l = \frac{4 P^2 \rho l^2}{W V_m^2} = K \text{ (say)} \quad ...(2.13)$$

(ii) DC Two-Wire System With Mid Point Earthed [Fig. 2.4 (ii)].

This system is the same as a 2-wire dc system, so volume of conductor material required for this system is the same as that in a 2-wire dc system.

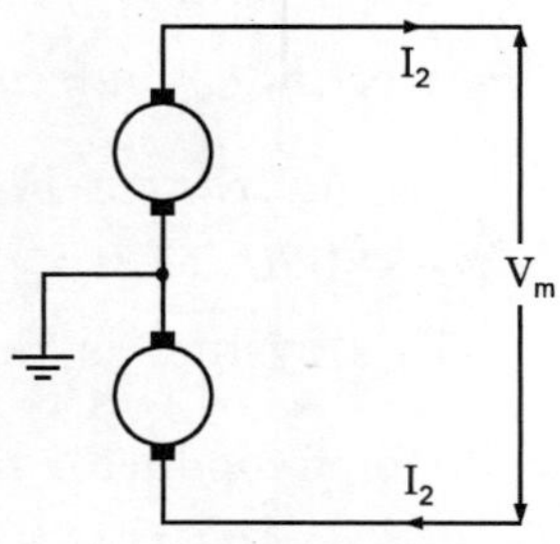

Fig. 2.4(*ii*)

(iii) DC Three-Wire System [Fig. 2.4 (iii)].

Assuming balanced load, the current through the neutral or mid wire is zero.

Maximum voltage between outers $= V_m$ volts

$$\text{Load current, } I_3 = \frac{P}{V_m}$$

$$\text{Line losses, } W = 2\,I_3^2\,R_3 = \frac{2P^2\rho l}{V_m^2\,a_3}$$

$$\text{or } a_3 = \frac{2P^2\rho l}{V_m^2\,W}$$

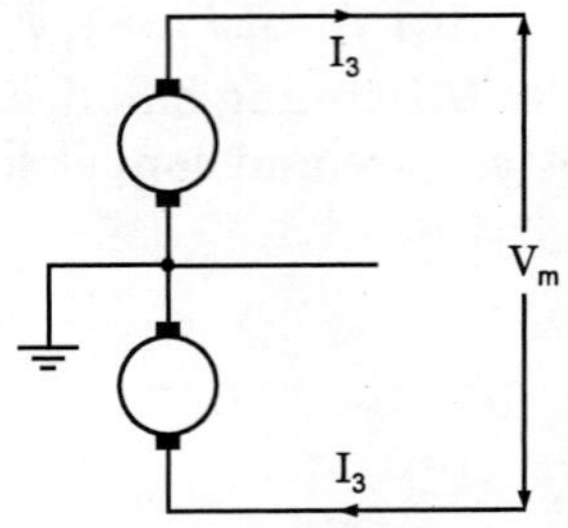

Fig. 2.4(*iii*)

Assuming x-section of neutral as half of that of either outer
Volume of conductor material required

$$= 2.5\,a_3\,l = \frac{5\,P^2\rho l^2}{V_m^2\,W} = 1.25\text{ K} \qquad ...(2.14)$$

Hence volume of conductor material required in this system is 1.25 times of that required in a two-wire dc system.

(*b*) Single Phase AC Systems

(*i*) *AC Single Phase Two-Wire System [Fig 2.4 (iv)].*

Maximum value of voltage between outers $= V_m$ volts

$$\text{RMS value of voltage between outers} = \frac{V_m}{\sqrt{2}} \text{ volts}$$

$$\text{Load current, } I_4 = \frac{P}{\frac{V_m}{\sqrt{2}}\cos\phi} = \frac{\sqrt{2}\,P}{V_m\cos\phi}$$

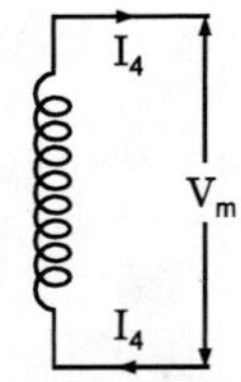

Fig. 2.4(*iv*)

where $\cos\phi$ is the power factor of the load

$$\text{Line losses, } W = 2I_4^2\,R_4 = 2I_4^2\frac{\rho l}{a_4} = \frac{4\,P^2\rho l}{V_m^2\cos^2\phi\,a_4}$$

$$\text{or } a_4 = \frac{4\,P^2\rho l}{V_m^2\cos^2\phi\,W}$$

$$\text{Volume of conductor material required} = 2\,a_4\,l = \frac{8\,P^2\rho l^2}{V_m^2\cos^2\phi\,W} = \frac{2K}{\cos^2\phi} \qquad ...(2.15)$$

Hence volume of conductor material required in this system is $\frac{2}{\cos^2\phi}$ *times of that required for a two-wire dc system.*

(*ii*) *AC Single Phase Two-Wire System With Mid Point Earthed [Fig. 2.4 (v)].*

This system is the same as a 2-wire single phase ac system, so volume of conductor material required in this case is also $\frac{2}{\cos^2\phi}$ times of that required in a two-wire dc system.

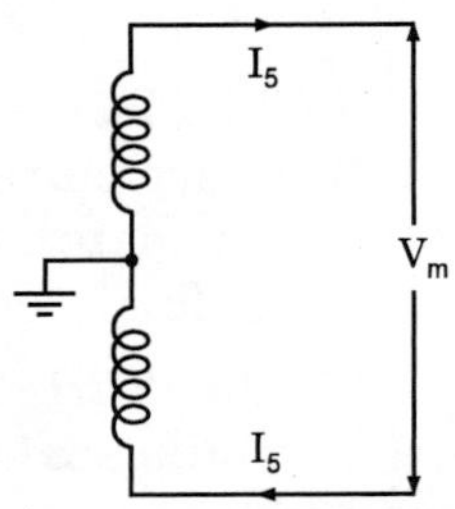

Fig. 2.4(*v*)

(iii) AC Single Phase Three-Wire System [Fig. 2.4 (vi)]. Assuming balanced load, the system reduces to a single phase, 2-wire ac system except that a neutral conductor of half the x-section is provided in addition.

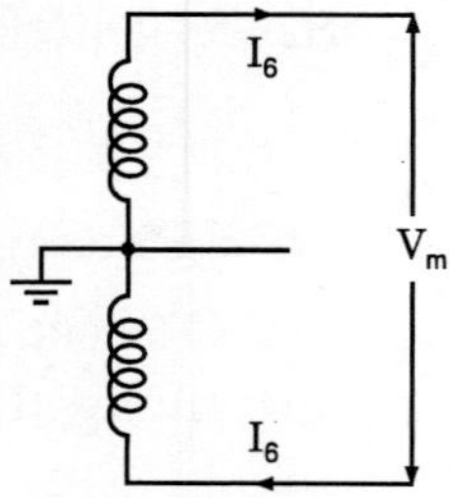

Fig. 2.4(*vi*)

So volume of conductor material required

$$= 2.5\ a_4 l = \frac{10\,\rho\,P^2\,l^2}{V_m^2\cos^2\phi\,W} = \frac{2.5}{\cos^2\phi}K \qquad ...(2.16)$$

Hence volume of conductor material required in this case is $\frac{2.5}{\cos^2\phi}$ *times of that required in a two-wire dc system.*

(c) Two-Phase AC Systems

(i) AC Two Phase 4-Wire System [Fig. 2.4 (vii)].

In this system each phase shares the half of the total load. This system is equivalent to two-wire ac system. In this case cross-section area of each conductor is taken half of that of single phase two-wire ac system but four wires are required in place of two wires, so the same volume of conductor material is required *i.e.* $\frac{2}{\cos^2\phi}$ times of that required in case of two-wire dc system.

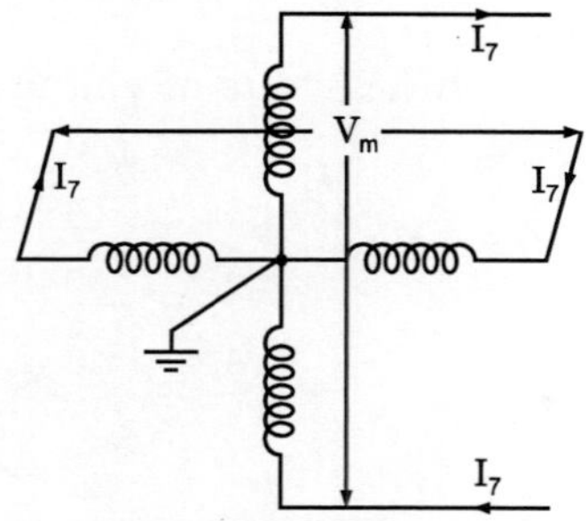

Fig. 2.4(*vii*)

(ii) AC Two-Phase Three-Wire System [Fig. 2.4 (viii)].

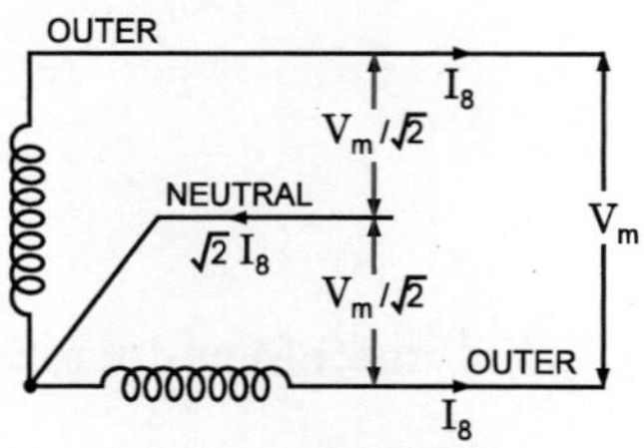

Fig. 2.4(*viii*)

$$\text{RMS value of voltage between outers} = \frac{V_m}{\sqrt{2}}$$

$$\text{RMS value of voltage per phase} = \frac{V_m}{\sqrt{2}\,\sqrt{2}} = \frac{V_m}{2}$$

$$\text{Load current per phase, } I_8 = \frac{P}{2\times\frac{V_m}{2}\cos\phi} = \frac{P}{V_m\cos\phi}$$

$$\text{The current in middle wire} = \sqrt{2}\,I_8$$

The x-section of middle wire is taken $\sqrt{2}$ times of either outer so that current density may remain the same.

$$\text{Line losses, } W = 2I_8^2\frac{\rho l}{a_8} + \left(\sqrt{2}\,I_8\right)^2\frac{\rho l}{\sqrt{2}\,a_8} = \frac{P^2\rho l}{V_m^2\cos^2\phi\,a_8}\left(2+\sqrt{2}\right)$$

$$\text{or } a_8 = \frac{P^2\rho l}{V_m^2\cos^2\phi\,W}\left(2+\sqrt{2}\right)$$

Volume of conductor material required $= 2\,a_8\,l + \sqrt{2}\,a_8\,l$

$$= a_8\,l\left(2+\sqrt{2}\right) = \frac{P^2\rho l^2}{V_m^2\cos^2\phi\,W}\left(2+\sqrt{2}\right)^2 = \frac{2.914\,K}{\cos^2\phi} \qquad ...(2.17)$$

Hence conductor material required in this system is $\frac{2.914}{\cos^2 \phi}$ *times of that required in a 2-wire dc system.*

(d) 3-Phase AC Systems

(i) AC Three-Phase Three-Wire System [Fig. 2.4 (ix)].

Maximum value of voltage between the conductors $= V_m$

$$\text{RMS value of voltage between the conductors} = \frac{V_m}{\sqrt{2}}$$

$$\text{RMS value of voltage per phase} = \frac{V_m}{\sqrt{2}\sqrt{3}} = \frac{V_m}{\sqrt{6}}$$

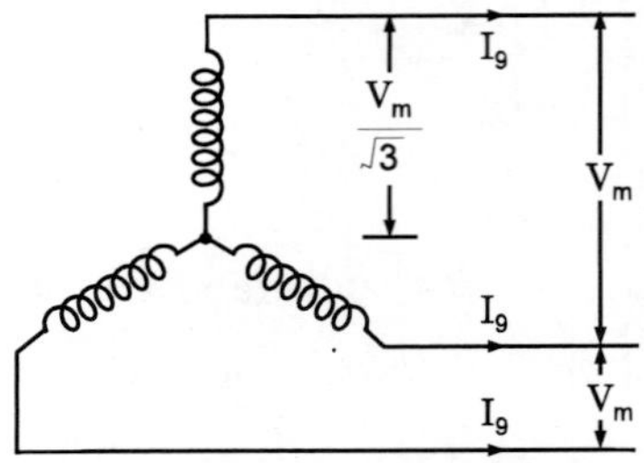

Fig. 2.4(*ix*)

$$\text{Load current, } I_9 = \frac{P}{\frac{3V_m}{\sqrt{6}} \cos \phi} = \frac{\sqrt{6}\,P}{3V_m \cos \phi}$$

$$\text{Line losses, } W = 3I_9^2 \frac{\rho l}{a_9} = \frac{3 \times 6 \times P^2 \rho l}{9V_m^2 \cos^2 \phi \, a_9}$$

$$\text{or} \quad a_9 = \frac{2P^2 \rho l}{V_m^2 \cos^2 \phi W}$$

$$\text{Volume of conductor material required} = 3a_9 l = \frac{6P^2 \rho l^2}{V_m^2 \cos^2 \phi W} = \frac{1.5}{\cos^2 \phi} K \qquad ...(2.18)$$

Hence volume of conductor material required in this case is $\frac{1.5}{\cos^2 \phi}$ *times of that required in case of a two-wire dc system.*

(ii) AC Three-Phase Four-Wire System [Fig. 2.4 (x)].

Assuming balanced load, this system is reduced to a 3-ϕ, ac system except that an additional wire, called the *neutral wire,* is provided of the half the cross-section of that either outer

So volume of conductor material required

$$= 3.5\, a_9\, l$$

$$= \frac{7P^2 \rho l^2}{V_m^2 \cos^2 \phi W} = \frac{1.75\, K}{\cos^2 \phi} \qquad ...(2.19)$$

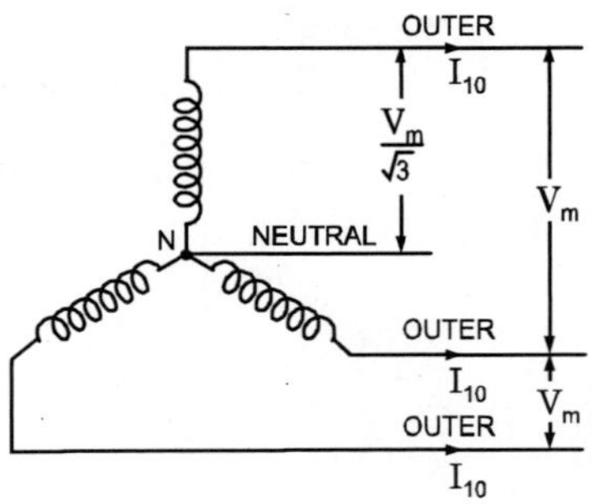

Fig. 2.4(*x*)

Hence volume of conductor material required in this system is $\frac{1.75}{\cos^2 \phi}$ *times of that required in case of a 2-wire dc system.*

2.6. RESULTS OF COMPARISON OF TRANSMISSION SYSTEMS

The results obtained above are summarised as below :

Systems	Volume of conductor material required on the basis of	
	Same maximum voltage between conductor and earth	Same maximum voltage between any two conductors
(*a*) *DC systems*		
(*i*) Two-wire	1	1
(*ii*) Two-wire with mid point earthed	0.25	1
(*iii*) Three-wire	0.3125	1.25
(*b*) *AC Single Phase Systems*		
(*i*) Two-wire	$\frac{2}{\cos^2 \phi}$	$\frac{2}{\cos^2 \phi}$
(*ii*) Two-wire with mid point earthed	$\frac{0.5}{\cos^2 \phi}$	$\frac{2}{\cos^2 \phi}$
(*iii*) Three-wire	$\frac{0.625}{\cos^2 \phi}$	$\frac{2.5}{\cos^2 \phi}$
(*c*) *AC Two-phase Systems*		
(*i*) Two-phase four-wire	$\frac{0.5}{\cos^2 \phi}$	$\frac{2}{\cos^2 \phi}$
(*ii*) Two-phase three-wire	$\frac{1.457}{\cos^2 \phi}$	$\frac{2.914}{\cos^2 \phi}$
(*d*) *AC Three-phase Systems*		
(*i*) Three-phase three-wire	$\frac{0.5}{\cos^2 \phi}$	$\frac{1.5}{\cos^2 \phi}$
(*ii*) Three-phase four-wire	$\frac{0.583}{\cos^2 \phi}$	$\frac{1.75}{\cos^2 \phi}$

From the above table it is obvious that for transmission dc system is an ideal one from economic point of view, particularly when it is remembered that the power factor of an ac system is usually considerably less than unity.

Two-phase three-wire system is obviously quite unsuitable for long distance transmission and needs no further consideration.

Among the remaining ac systems for overhead systems there is no difference in single phase, 2-wire and three-phase 3-wire systems and for underground systems there is a decided saving in conductor material with the three phase system. Considering other factors such as efficiency of operation and convenience three-phase three-wire system is usually adopted.

Example 2.2. A 3-phase 4-wire system is used for lighting. Compare the amount of conductor material required with that needed for a 2-wire direct current system with the same lamp voltage. Assume the same losses and balanced load. The neutral is one-half the x-section of one of the respective outers.

Solution : Let the lighting load to be supplied be P watts at voltage V volts (between line and neutral in case of 3-phase, 4-wire system) and line losses be W watts in both of the cases:

(*i*) *Two-wire dc system*

$$\text{Load current, } I_1 = \frac{P}{V} \text{ amperes}$$

$$\text{Line losses, } W = 2I_1^2 R_1 = 2\frac{P^2}{V^2}\rho\frac{l}{a_1} = \frac{2P^2\rho l}{V^2 a_1}$$

where ρ is the resistivity of conductor material, l is the length of line, a_1 is the x-section of conductor in case of 2-wire dc system.

$$\text{or Area of x-section of conductor, } a_1 = \frac{2P^2\rho l}{V^2 W}$$

$$\text{Volume of conductor material required} = 2a_1 l = \frac{4P^2\rho l^2}{V^2 W} \quad ...(i)$$

(*ii*) *Three-Phase Four-Wire System*

$$\text{Power supplied by each phase} = \frac{P}{3} \quad (\because \text{ load is balanced})$$

R M S voltage between line and neutral = V volts

Power factor of lighting load be assumed unity

$$\text{Load current, } I_2 = \frac{P}{3V} \text{ amperes}$$

$$\text{Line losses } W = 3I_2^2 R_2 = 3\left(\frac{P}{3V}\right)^2 \rho\frac{l}{a_2} = \frac{P^2\rho l}{3V^2 a_2}$$

where a_2 is the x-section of conductor in case of 3-phase 4-wire system

$$\text{or } a_2 = \frac{P^2\rho l}{3V^2 W}$$

Since the neutral is one-half of the x-section of the outer,

$$\text{Volume of conductor material required} = 3.5\, a_2 l = \frac{3.5P^2\rho l^2}{3V^2 W} = \frac{3.5}{12} \times \frac{4P^2\rho l^2}{V^2 W}$$

$$= 0.292 \times \frac{4P^2\rho l^2}{V^2 W}$$

i.e. In case of 3-phase 4-wire system conductor material required will be 0.292 times of that required in a 2-wire dc system. **Ans.**

Example 2.3. An existing single phase ac system comprising of two overhead conductors is to be converted into a 3-phase, 3-wire system by providing an additional conductor of same size. Calculate the percentage of additional load that can be transmitted by the three phase system if the operating line voltage and percentage line losses remain the same in both the systems.

Solution : Let the operating line voltage and power factor in both the systems be V volts and cos ϕ respectively. If I_1 is single phase current, I_2 is the three phase current and R is the resistance of each conductor, then

in single phase ac system

$$\text{Power transmitted, } P_1 = VI_1 \cos\phi \text{ watts}$$

$$\text{Line losses, } W_1 = 2I_1^2 R \text{ watts}$$

$$\text{Percentage line losses} = \frac{W_1}{P_1} \times 100 = \frac{2I_1^2 R}{VI_1\cos\phi} \times 100 \quad ...(i)$$

and in three phase ac system

$$\text{Power transmitted, } P_2 = \sqrt{3}\, VI_2 \cos\phi$$

$$\text{Lines losses, } W_2 = 3\, I_2^2\, R$$

$$\text{Percentage line losses} = \frac{W_2}{P_2} \times 100 = \frac{3\, I_2^2\, R}{\sqrt{3}\, V\, I_2 \cos\phi} \times 100 \quad ...(ii)$$

For the same percentage line losses in both the cases, we have

$$\frac{2\, I_1^2\, R}{V\, I_1 \cos\phi} \times 100 = \frac{3\, I_2^2\, R}{\sqrt{3}\, V\, I_2 \cos\phi} \times 100$$

$$\text{or } I_2 = \frac{2}{\sqrt{3}} I_1$$

So power transmitted in case of 3-phase, 3-wire system,

$$P_2 = \sqrt{3}\ V \times \frac{2}{\sqrt{3}}\ I_1 \cos\phi = 2\ V\, I_1 \cos\phi = 2P_1$$

$$\text{Additional load transmitted} = P_2 - P_1 = 2\, P_1 - P_1 = P_1$$

$$\text{Percentage of additional load} = \frac{P_2 - P_1}{P_1} \times 100 = \frac{P_1}{P_1} \times 100 = 100\%$$

Thus 100 per cent additional load can be transmitted by converting single phase line to three-phase line. **Ans.**

Example 2.4. An existing dc three-wire system is to be converted into a 3-phase 4-wire system by adding a fourth wire equal in x-section to each outer of the dc system. Assuming the same supply and load voltages to neutral and balanced conditions find the extra power at unity power factor that can be supplied by the ac system.

Solution : Let the supply and load voltages to neutral be V_S and V_L volts respectively and resistance of each conductor be R ohms.

$$\text{Line current in 3-wire dc system, } I_1 = \frac{V_S - V_L}{R}$$

$$\text{Power supplied, } P_1 = 2\, V_L I_1 = \frac{2\, V_L}{R}\ (V_S - V_L) \quad ...(i)$$

$$\text{Line current in 3-phase 4-wire ac system, } I_2 = \frac{V_S - V_L}{R}$$

$$\text{Power supplied, } P_2 = 3\, V_L I_2 \cos\phi = \frac{3\, V_L}{R}\ (V_S - V_L) \times 1 \quad ...(ii)$$

Comparing expressions (*i*) and (*ii*) we get

$$P_2 = 1.5\ P_1$$

i.e. 50 per cent extra power can be supplied at unity pf by 3-phase, 4-wire ac system **Ans.**

Example 2.5. A 3-wire dc system is converted to a 4-wire 3-phase ac system by the addition of another wire equal in section to one of the outers. For the same effective voltage between outers and neutral at the consumer's terminals and the same percentage loss, find the percentage additional load that can be supplied. Assume balanced load and in the ac system a power factor of 0.9.

Solution : Let the voltage between outers and neutral be V volts in both of the cases and currents in 3-wire dc system and 3-phase 4-wire system be I_1 and I_2 respectively.

3-Wire dc system

$$\text{Power supplied, } P_1 = 2\ I_1 V$$

$$\text{Power loss} = 2\, I_1^2\ R$$

$$\text{Percentage power loss} = \frac{2\, I_1^2\ R}{2\, I_1\ V} \times 100 = \frac{100\, I_1\ R}{V} \quad ...(i)$$

3-phase 4-wire ac system

$$\text{Power supplied, } P_2 = 3\, I_2\, V \cos\phi = 3\, I_2 V \times 0.9 = 2.7\, I_2 V$$

$$\text{Power loss} = 3\, I_2^2\, R$$

$$\text{Percentage power loss} = \frac{3\, I_2^2\, R}{2.7\, I_2\, V} \times 100 = \frac{100\, I_2\, R}{0.9\, V} \quad ...(ii)$$

Since load is balanced, therefore neutral wire will carry no current

Since percentage loss is same, therefore comparing Eqs. (*i*) and (*ii*) we get

$$\frac{100\, I_1\, R}{V} = \frac{100\, I_2\, R}{0.9\, V}$$

$$\text{or } I_2 = 0.9\, I_1$$

Hence power supplied in 3-phase 4-wire system,

$$P_2 = 2.7 \times 0.9\, I_1 V = 2.43\, I_1 V$$

$$\text{Percentage increase in load supplied} = \frac{P_2 - P_1}{P_1} \times 100 = \frac{2.43\, I_1 V - 2\, I_1 V}{2\, I_1 V} \times 100 = 21.5\% \textbf{ Ans.}$$

2.7. CHOICE OF WORKING VOLTAGE FOR TRANSMISSION

Before discussing the method of determination of the most economical working voltage for transmission we will discuss here the advantages and limitations of high voltage transmission.

The advantages of adopting high voltage for transmission are given below:

(*i*) With the increase in transmission voltage the size of the conductor (area of x-section of the core carrying the current) is reduced. This further reduces the cost of the supporting structure materials.

(*ii*) With the increase in transmission voltage, line current is reduced, which results in reduction of line losses.

(*iii*) With the increase in transmission voltage reduction in line losses results in higher efficiency.

(*iv*) Due to low current at high transmission voltage, the voltage drop in the lines is low. This leads to better voltage regulation.

The above facts can be shown very clearly as follows:

Let the power delivered by *n*-phase system be P watts per phase at a phase voltage of V volts and power factor $\cos\phi$.

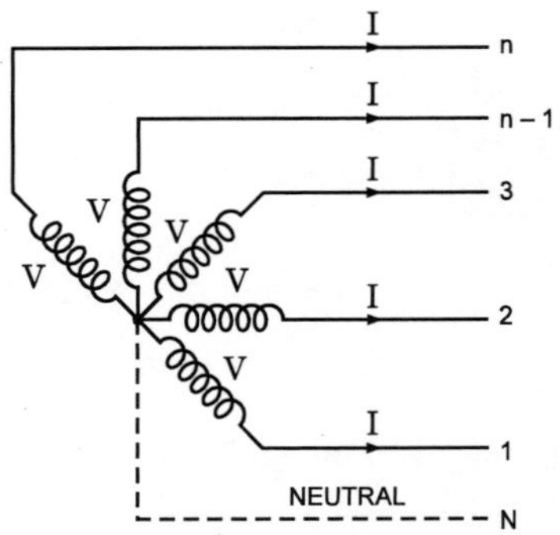

Fig. 2.5. ***Polyphase Transmission System***

$$\text{Load current per phase, } I = \frac{P}{V\cos\phi}$$

$$\text{Resistance of conductor, } R = \rho\frac{l}{a}$$

where l is the length of line, a is the x-section of conductor and ρ is the resistivity of conductor material.

$$\text{Power loss in the line (per phase), } W = I^2 R = \frac{P^2}{V^2\cos^2\phi}\,\frac{\rho l}{a} \quad ...(2.20)$$

Also cross-section of conductor, $a = \dfrac{I}{\delta} = \dfrac{P}{V\cos\phi\,\delta}$ where δ is the current density

So substituting $a = \dfrac{P}{V \cos \phi \, \delta}$ in Eq. (2.20) we get

$$W = \frac{P^2}{V^2 \cos^2 \phi} \times \rho l \times \frac{V \cos \phi \, \delta}{P} = \frac{P \rho l \delta}{V \cos \phi} \quad \text{...(2.21)}$$

$$\text{Transmission efficiency, } \eta = \frac{\text{Output}}{\text{Input}} = \frac{\text{Output}}{\text{Output + line losses}}$$

$$\text{or } \eta = \frac{P}{P + \dfrac{P l \delta \rho}{V \cos \phi}} = 1 - \frac{\rho \delta l}{V \cos \phi} \quad \text{...(2.22)}$$

$$\text{Resistance drop per line} = I R = \frac{P \rho l \times V \cos \phi \, \delta}{V \cos \phi \times P} = \rho \, l \, \delta \quad \text{...(2.23)}$$

$$\text{Volume of conductor material required per phase} = a \, l = \frac{P l}{\delta V \cos \phi} \quad \text{...(2.24)}$$

Substituting $\delta = \dfrac{W V \cos \phi}{P \rho l}$ from Eq. (2.21) in Eq. (2.24) we get

Volume of conductor material required per phase

$$= \frac{P l}{V \cos \phi} \times \frac{\rho P l}{W V \cos \phi} = \frac{P^2 l^2 \rho}{W V^2 \cos^2 \phi} \quad \text{...(2.25)}$$

Hence from Eqs. (2.21), (2.22), (2.23), and (2.24) it is obvious that for constant values of P, l, δ, $\cos \phi$ and ρ

(*a*) With the increase in transmission voltage line losses are reduced since line losses are inversely proportional to supply voltage, V.

(*b*) The transmission efficiency increases with the increase in transmission voltage, V.

(*c*) Resistance drop per line is constant. Hence percentage resistance drop decreases with the increase in transmission voltage V.

(*d*) The volume of conductor material required is inversely proportional to the supply voltage.

(*e*) From Eq. (2.25) it is obvious that for the same line losses and for transmission of given amount of power over a given distance through the conductors of given material and at a given power factor *i.e.*, for constant values of W, P, l, ρ and $\cos \phi$ conductor material required is inversely proportional to the square of supply voltage.

The above formula, besides showing the great saving of conductor material by employing a high transmission voltage, indicates also the importance of working as near unity power factor as possible. The effect of operating at, say, 0.707 power factor instead of unity, is to increase the amount of conductor material used in the ratio of 1: $\dfrac{1}{(0.707)^2}$ or an increase of 100%. A very considerable economy can be affected in the conductor material if the transmission line is operated at high power factor.

Where there are several advantages of high voltage for transmission, there are some limitations also, which are given below:

(*a*) With the increase in voltage of transmission, the insulation required between the conductors and the earthed tower increases. This increases the cost of line supports.

(*b*) With the increase in voltage of transmission, more clearance is required between conductors and ground. Hence higher towers are required.

(*c*) With the increase in voltage of transmission, more distance is required between the conductors. Therefore, longer cross-arms are required.

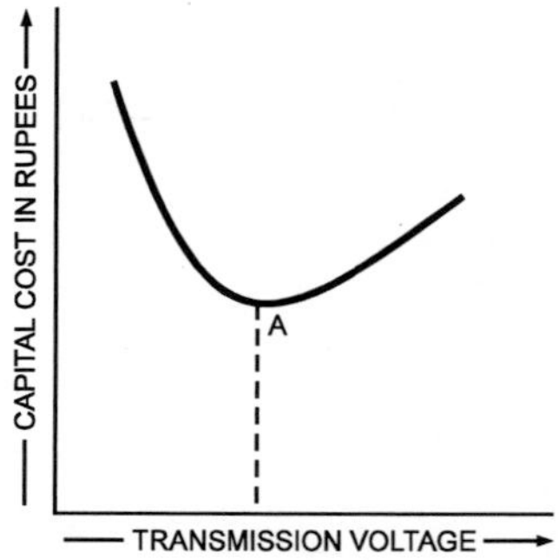

Fig. 2.6.

For every transmission line there is a superior limit fixed for the voltage to be employed, beyond which nothing is gained in the matter of economy. This limit is reached when the cost of conductors, cost of insulators, supports, transformers, switchgear, lightning arrestors and also the cost of erection is minimum. The method of finding the optimum voltage is to choose a certain standard voltage and calculate the cost of (*i*) transformers to be used at the generating and receiving stations, (*ii*) switchgear consisting of isolators, bus-bars, circuit breakers, relays etc, (*iii*) lightning arresters, (*iv*) insulators, (*v*) supports including the cross-arms and other fittings, (*vi*) conductors and cost of erection for the given voltage of generation, given power to be transmitted and given length of the transmission system. The total sum of all the costs gives the total capital cost of transmission. Thus for various standard voltages the total capital cost is worked out and a curve of capital cost against the transmission voltage is plotted. The curve plotted is of the form shown in Fig. 2.6. From the curve it is observed that the optimum voltage from economy point of view will be voltage corresponding to point A on the curve. In fact higher voltage would be choosen as with large working voltage, power loss and voltage drop are reduced, load can be increased if required and control becomes easier.

In practice it is not possible to determine the economic voltage by the method explained above. To avoid the complications and labour *Cable Research Hand Book* gives an empirical formula for determining the optimum voltage for lines more than 30 km long. The formula is

Approximate economical voltage for transmission in kV

$$= 5.5\sqrt{\frac{\text{km}}{1.6} + \frac{\text{load in kVA}}{150}} \qquad ...(2.26)$$

The worthnoting point is that power to be transmitted and distance of transmission have been taken into account in the above formula, because both of these factors have considerable influence in arriving at economic voltage for transmission, as explained below.

With the increase in distance of transmission, the cost of terminal apparatus is reduced resulting in higher economic transmission voltage. Similarly if the power to be transmitted becomes large, the cost per kW of the terminal station equipment is reduced.

As a rough guide the voltage for transmission is chosen as 625 volts per km though in practice the voltage per km varies from about 400 to 900 volts for longer to shorter distances. The choice is usually limited in practice by the requirement of standardisation and for satisfactory regulation without excessive equipment cost.

The voltages normally adopted for transmission are given below.

Distance in km	15-30	30-60	60-100	100-200	200-300	300-400
Transmission voltage in kV	11	33	66	132	220	400

As already mentioned, the most common transmission voltages employed in India are 33 kV, 66 kV, 132 kV, 220 kV and 400 kV.

Example 2.6. Estimate the weight of copper required to supply a load of 100 MW at UPF by a 3-phase, 380 kV system over a distance of 100 km. The neutral point is earthed. The resistance of the conductor is 0.045 Ω/cm²/km. The weight of copper is 0.01 kg/cm³. The efficiency of transmission can be assumed to be 90 per cent.

Solution : Line current, $I_L = \dfrac{\text{P in MW} \times 10^6}{\sqrt{3} \times V_L \text{ in kV} \times 10^3 \times \cos\phi} = \dfrac{100 \times 10^6}{\sqrt{3} \times 380 \times 10^3 \times 1.0} = 152 \text{ A}$

Power losses in line, $W = P(1 - \eta) = 100\,(1\text{–}0.9) = 10 \text{ MW}$

Line loss per conductor $= I_L^2 R = \dfrac{W}{3} = \dfrac{10 \times 10^6}{3}$ watts

Resistance per conductor, $R = \dfrac{10 \times 10^6}{3 \times (152)^2} = 144.4\ \Omega$

or Resistance per conductor per km $= \dfrac{144.4}{100} = 1.444\ \Omega$

Conductor area, $a = \dfrac{0.045}{1.444} = 0.03 \text{ cm}^2$

Volume of copper required per km run $= 0.03 \times 100 \times 1{,}000 = 3{,}000 \text{ cm}^3$

Weight of copper required per km run $= 0.01 \times 3{,}000 = 30$ kg

Weight of copper required for 3 conductors, each of 100 km length

$= 3 \times 100 \times 30 = 9{,}000$ kg **Ans.**

EXERCISES

1. Compare the advantages of using ac transmission system over dc system. **[U.P. Technical Univ. Elements of Power System 2002-03]**
2. Give a comparison of an overhead line with underground cable as a medium of power transmission. **[U.P. Technical Univ. Elements of Power System 2005-06]**
3. Discuss the advantages of using overhead lines compared to underground cables. **[U.P. Technical Univ. Elements of Power System 2006-07]**
4. Discuss the effect of transmission voltage on line performance. Derive any formula to support your answer. **[U.P. Technical Univ. Elements of Power System 2005-06]**
5. Distinguish between a feeder, distributor and service main in distribution scheme. Show that the x-section of a feeder and distributor would be reduced to 1/k and $1/k^2$ of their respective values with an increase in working voltage by k times. **[U.P.S.C. 1972]**
6. Discuss the effect of increase in voltage on the transmission line efficiency. **[U.P. Technical Univ. Elements of Power System 2006-07]**
7. Derive formula to calculate the ratio of copper volume used in single phase two-wire system and a two-wire dc system. **[U.P. Technical Univ. Elements of Power System 2002-03]**
8. Compare the copper efficiencies of single phase and three-phase three wire supply systems, assuming equal power loss in transmitting same power over a given distance. **[U.P. Technical Univ. Elements of Power Systems 2005-06]**

9. Compare the conductor requirement for 1-ϕ ac system with 3-ϕ ac system. Assume same maximum potential difference between conductor and earth.

[U.P. Technical Univ. Elements of Power System 2008-09]

10. Find the ratio of volume of copper required to transmit a given power over a given distance by overhead system using (*i*) dc two wire system (*ii*) three phase three wire ac system. **[U.P. Technical Univ. Elements of Power System 2006-07]**

11. Compare the conductor cost of a 3-phase, 4-wire system of distribution with single phase, 2-wire system. **[U.P. Technical Univ. Elements of Power System 2003-04]**

12. Show that at higher voltages and higher power factor, the efficiency of transmission line will be increased. What are the limiting factors of high voltage transmission lines ?

[U.P. Technical Univ. Elements of Power System 2007-08]

13. Compare the different supply systems with their choice of transmission voltage.

[U.P. Technical Univ. Elements of Power System 2006-07]

14. Explain on which factors, the choice of transmission voltage depends ?

[U.P. Technical Univ. Elements of Power System 2007-08]

15. What criterion is adopted for selection of transmission line operating voltage in case of long transmission lines ? **[U.P. Technical Univ. Elements of Power System 2005-06]**

16. Discuss about the factors affecting choice of transmission voltage level.

[U.P. Technical Univ. Elements of Power System 2009-10]

SHORT ANSWER TYPE QUESTIONS WITH ANSWERS

Q. 1. Why does the maximum voltage between each conductor and earth form the basis of comparison of volume of conductor material required in overhead system ?

Ans. In overhead system the conductors are insulated from the x-arms and supporting towers and as the towers and x-arms are earthed so the maximum voltage between each conductor and earth forms the basis of comparison of volume of conductor material required in overhead system.

Q. 2. Why does the maximum voltage between conductors form the basis of comparison of volume of conductor material required in underground system ?

Ans. In underground cables the maximum disruptive stress is between the two conductors of the cable; so the maximum voltage between conductors forms the basis of comparison of volume of conductor material required in underground system.

Q. 3. What is the percentage saving in feeder conductor if the line voltage in a 2-wire dc system be raised from 200 V to 500 V for the same power to be transmitted ?

Ans. 84%

Q. 4. What is the ratio of volume of conductor material required for transmission of a given amount of power over a given distance by overhead system using

(*i*) dc 2-wire system and (*ii*) 3-ϕ, 3-wire ac system ?

Ans. $0.5/\cos^2\phi$

Q. 5. State the empirical formula for determining the system voltage of transmission line.

Ans. Approximate economical voltage for transmission in kV $= 5.5\sqrt{\dfrac{\text{km}}{1.6} + \dfrac{\text{Load in kVA}}{150}}$

Q. 6. What is the ratio of volume of conductor material required for transmission of a given amount of power over a given distance by underground system using (*i*) Two-phase 4-wire system, (*ii*) Three-phase three-wire system.

Ans. Three-fourth.

PROBLEMS

1. What is the percentage saving in feeder copper if the line voltage in a 2-wire dc system be raised from 220 V to 500 V for the same power transmitted over the same distance and having the same power loss ? **[Ans.** 80.64%]
2. A single phase load of 10,000 kW is transmitted by a pair of overhead conductors. If a third conductor of the same x-section be added and a 3-phase supply be thus substituted for the original single phase, calculate the 3-phase load which can be now transmitted if voltage between the wires and percentage loss in the line remain unchanged. **[Ans.** 20,000 kW]
3. A dc 2-wire overhead distribution system is to be converted into a 3-ϕ, 3-wire ac system by installing a third conductor of the same section as the other two. Calculate the additional percentage load which can be distributed by the ac system, assuming that the voltage between the conductors and the percentage power loss remain the same. Assume a balanced 3-ϕ load at (*i*) unity pf and (*ii*) 0.8 lagging pf. **[Ans.** (*i*) 100% (*ii*) 28%]
4. A 3-wire dc transmission system is converted into a 4-wire, 3-phase balanced ac transmission system by addition of another conductor, equal in x-section to other conductors. Assume that the middle (grounded) conductor in dc system carries negligible current and the voltage between the outer conductors is 2 V. Assume that the rms voltage to ground for ac system is V. For the same percentage power loss find the additional power that can be transmitted by the ac system at 0.9 pf. **[Ans.** 21.5%]
5. A 3-wire dc supply system is converted to a 3-phase, 4-wire ac system by the addition of another conductor identical to the outers. For the same voltage at consumer's terminals and the same percentage copper loss, find the percentage additional load that can be supplied. Assume balanced loads at unity power factor. **[Ans.** 50%]

❑❑❑

3

CHAPTER

TRANSMISSION LINES

INSIDE THIS CHAPTER

3.1. Introduction **3.2.** Conductor Configurations **3.3.** Types of Conductors **3.4.** Resistance of Line **3.5.** Skin Effect **3.6.** Proximity Effect **3.7.** Economical Size of Line Conductor—Kelvin's Law.

3.1. INTRODUCTION

A transmission line consists of a set of conductors carrying electrical energy in bulk and transmitting it from power stations to primary substations. The conductors are parallel to each other and are carried on supports which provide insulation between the different conductors and between each conductor and earth. Figs. 3.1(*a*), (*b*) illustrate the transmission line in which line conductors are strung on insulators fixed to the cross-arms. Single circuit tarnsmission line [Fig. 3.1(*a*)] carries 3 line conductors pertaining to three phases R, Y and B and one ground wire running over the top of the towers. Double circuit transmission line [Fig. 3.1(*b*)] carries 6 line conductors. These 6 line conductors constitute two separate circuit, each consists of three line conductors pertaining to phases R, Y and B. In case of double circuit transmission line two wires, known as ground wires, run over the top of the towers.

Transmission lines are basically electrical circuits having distributed constants—resistance, inductance, capacitance and shunt conductance. The shunt conductance is mostly due to leakage over the insulators and is so small that it can be neglected. These line constants, also called the *line parameters,* are uniformly distributed along the entire length of the line and are usually expressed as resistance, inductance and capacitance per unit length. Concentration of all such parameters for the complete length of line at a single point is not possible. The performance of a transmission line depends to a considerable extent upon these parameters–series resistance causes a power loss in the conductor and so affects the transmission efficiency, series inductance mainly governs the power transmission capacity of the line, and shunt capacitance causes a charging current to flow in the line.

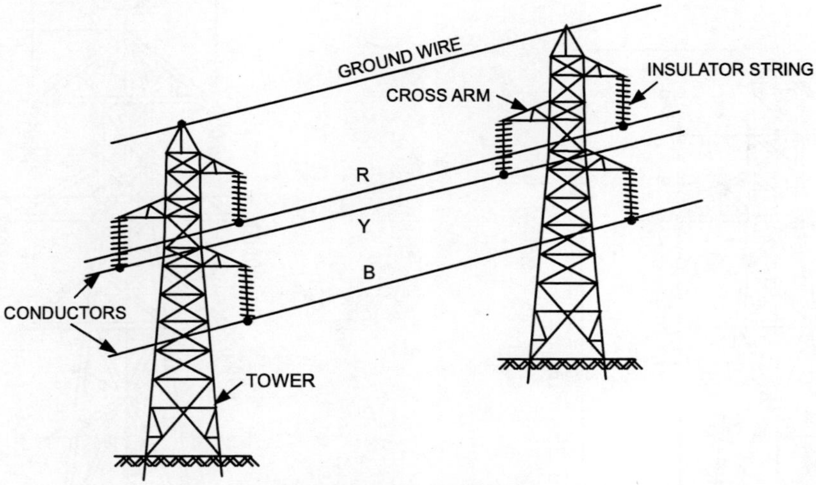

(a) Single Circuit Transmission Line

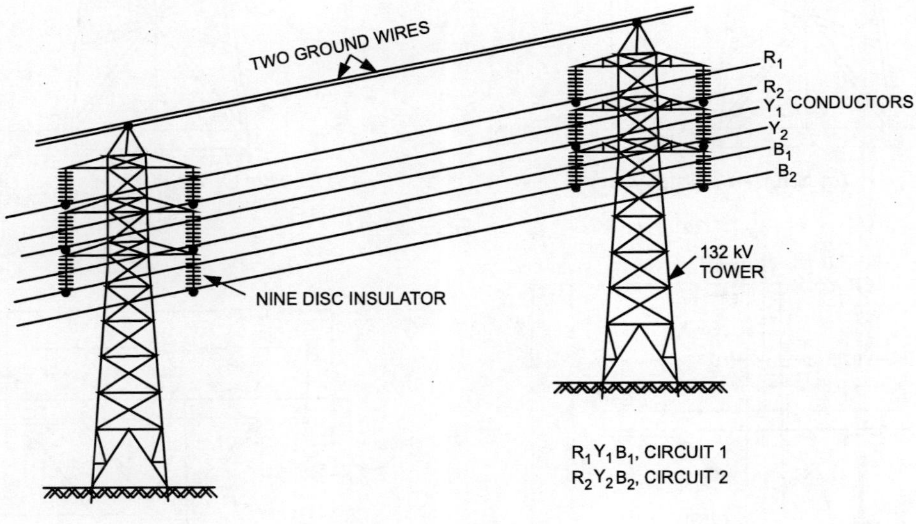

(b) Double Circuit Transmission Line

Fig. 3.1.

3.2. CONDUCTOR CONFIGURATIONS

Several conductor configurations are possible, but three configurations are the most common *i.e.* horizontal configuration (or horizontal disposition of conductors), vertical configuration and triangular configuration.

There is no special advantage in using the symmetrical *delta* or *triangular configuration* [Fig. 3.3(*a*)] and in most cases flat horizontal or vertical configurations are employed from mechanical considerations, particularly when suspension insulators are used. In *horizontal configuration,* all the conductors are mounted over one cross-arm, as shown in Fig. 3.3(*b*).

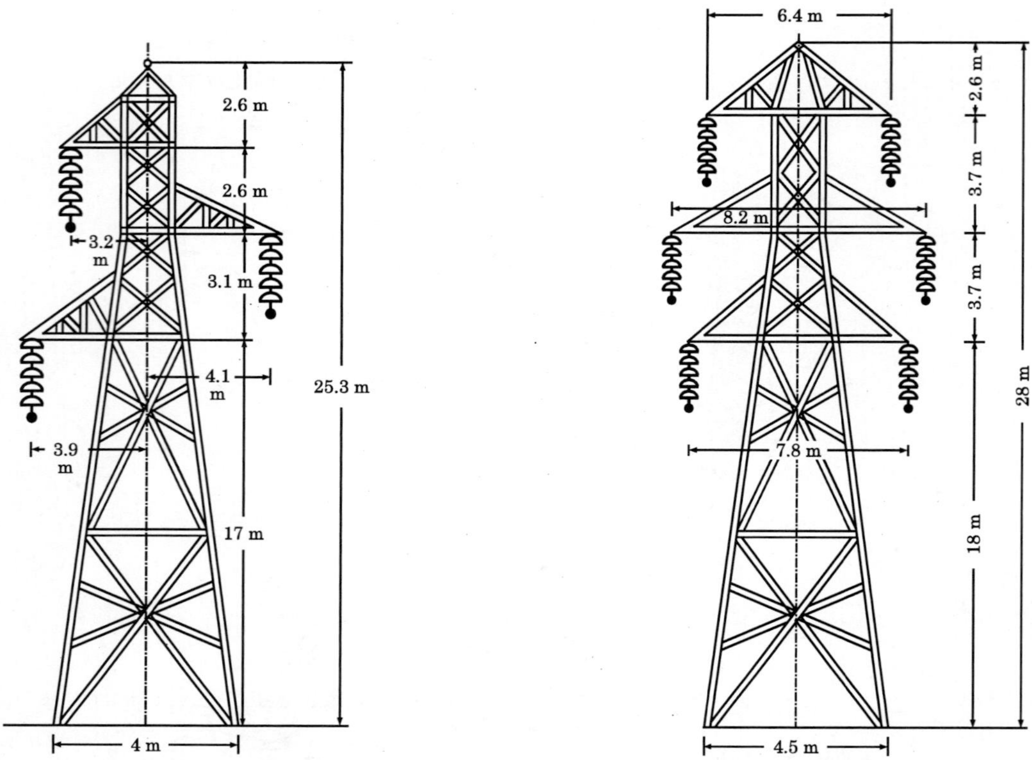

(a) Single Circuit Tower (b) Double Circuit Tower

Fig. 3.2

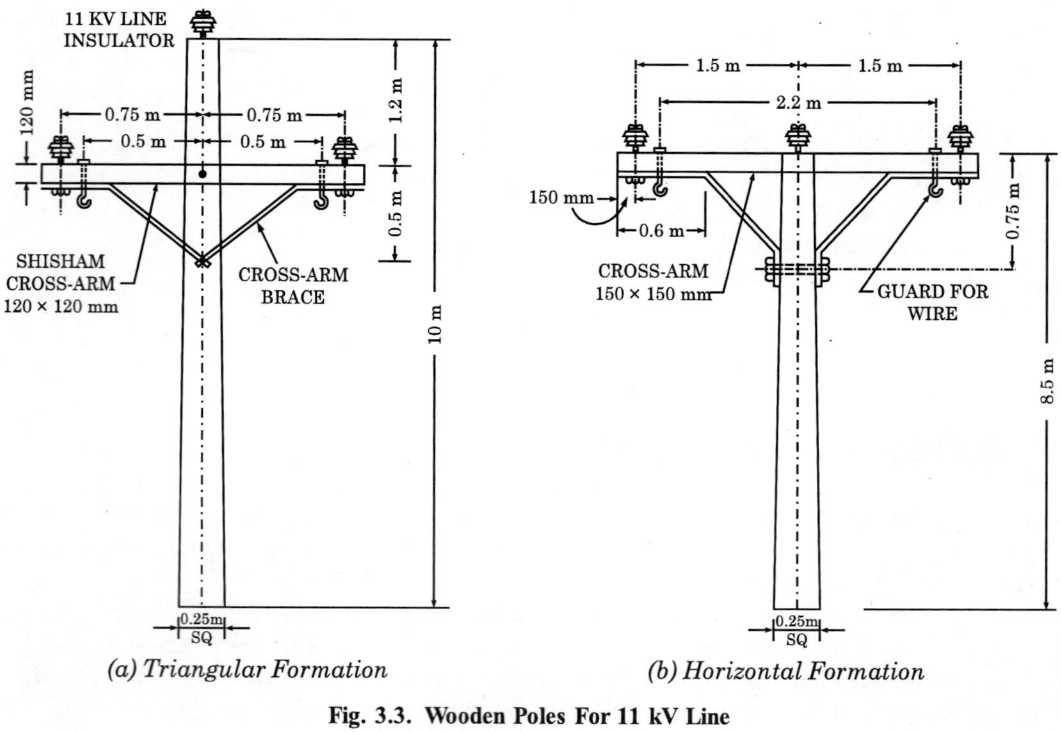

(a) Triangular Formation (b) Horizontal Formation

Fig. 3.3. Wooden Poles For 11 kV Line

Though such an arrangement of conductors needs supports of smaller height but needs a wider right of way. In certain congested areas where it is not possible to have horizontal arrangement of conductors, the conductors are placed in vertical formation (along the length of pole one below the other). The drawbacks of vertical formations are taller towers and more lightning hazards. There are places where both horizontal and vertical formations are applied.

In unsymmetrical arrangement of conductors, the conductors are usually transposed at regular intervals in order to balance the electrical characteristics of various phases, and prevent inductive interference with neighbouring communication circuits.

Experience shows that a vertical configuration is the most economical for double circuit lines and horizontal or L-type configuration for single circuit lines.

3.3. TYPES OF CONDUCTORS

The conductor is one of the important items as in a transmission and distribution system of electric power, the cost of the conductor material accounts for a major part of the total cost. So proper choice of conductor material and size of the conductor is of utmost importance. The conductor material used for transmission and distribution of electric power must have the following characteristics:

(*i*) High electric conductivity *i.e.* low specific resistance.

(*ii*) High tensile strength in order to withstand the mechanical stresses.

(*iii*) Low specific gravity in order to give low weight per unit volume.

(*iv*) Low cost in order to be used over long distances usually involved in transmission lines.

(*v*) Easy availability.

(*vi*) Should not be brittle.

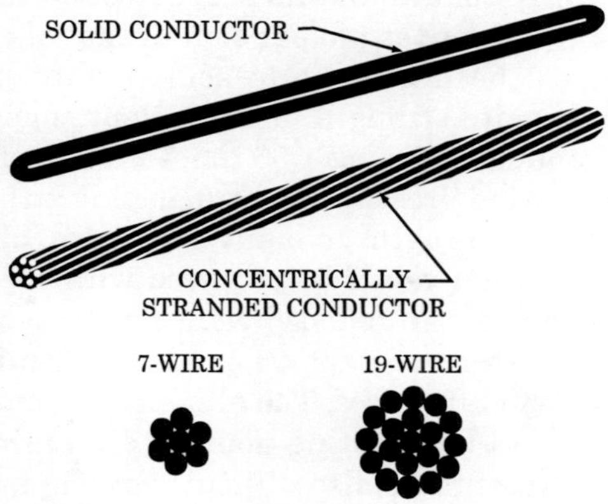

Fig. 3.4. ***Types of Conductors***

No single conductor material meets all of the above requirements. Hence a compromise will have to be made between the cost and the desired electrical and mechanical properties in the selection of a conductor material for a given situation.

The most commonly used conductor materials for overhead lines are copper, aluminium, steel-cored aluminium, galvanized steel and cadmium copper.

All conductors used for overhead lines are preferably stranded in order to increase the flexibility. Solid wires, except of smaller sizes, are difficult to handle and when employed for long spans tend to crystallise at the points of support because of swinging in winds.

Stranded conductors usually have a central wire around which there are successive layers of 6, 18, 36 wires. For n layers, the total number of individual wire is $3n\ (n + 1) + 1$. If the diameter of each strand is d then diameter of the stranded conductor will be $(2n + 1)\ d$. In the process of manufacture adjacent layers are spiralled in opposite directions so that the layers are bound together. The method of construction is called as '*concrete lay*'.

With conductors of large cross-section, however, another method known as '*rope lay*' is sometimes employed in order to give more flexibility.

1. Stranded Hard Drawn Copper. Hard drawn copper conductor is the best conductor owing to its high electrical conductivity and great tensile strength for all types of transmission. Though hard drawing (cold rolling and drawing) reduces the conductivity slightly but increases the tensile strength considerably. It does not corrode in normal atmosphere and is not subjected to electrolytic troubles. It has higher current density so lesser x-sectional area of conductor is required and so lesser area is subjected to wind loads. It is quite homogeneous, durable and of high scrap value. The other advantages of hard drawn copper are long life and ease of jointing.

However, medium hard-drawn copper conductor is suitable for distribution lines and soft-drawn copper wires are suitable for secondary distribution circuits, and for service connections to buildings.

Though copper is ideally suitable for the transmission and distribution, but owing to the difficulty of importing it due to lack of foreign exchange in our country, the trend nowadays is to use aluminium in place of copper. The use of copper is being restricted for the manufacturing of the machines only.

2. Aluminium. Aluminium is cheaper in cost and lighter in weight but is poor in conductivity and tensile strength as compared to copper. Its conductivity is 60% to that of copper and density 0.303 time that of copper. An aluminium conductor has a diameter about 1.26 times that of copper conductor of equal resistance but due to its low density only half weight of aluminium is required to that of copper. The tensile strength is much lower than that of copper (45 per cent to that of copper), but the larger sectional area of metal neutralizes the difference to some extent, and an aluminium conductor has about 75% of the ultimate strength of equivalent copper conductor. For the same conductivity aluminium conductor having 1.66 times the x-section of copper is required thus causes a greater surface for wind pressure and supporting structures are required to be designed for a greater transverse strength. In many cases also, high towers must be employed with aluminium conductors than would be required with copper conductors having the same length of span. As a result of the reduced working stress permissible and also owing to the fact that the linear coefficient of expansion of aluminium is 1.4 times that of copper, the sag is greater in aluminium wires. The aluminium conductor being liable to swing, requires larger cross-arms. Due to low melting point of aluminium it cannot withstand short-circuits etc. Jointing of aluminium is also difficult as compared to that of copper.

These days there is a great trend towards the aluminium as conductor material, because of its greater availability and cheapness in comparison to copper. All aluminium stranded conductor (AAC) weighs only half as the equivalent copper conductor and cost per unit length at prevailing market rates, is considerably less. All aluminium stranded conductors are mainly used for low voltage distribution overhead lines having short spans of up to 65 m.

3. Steel-Cored Aluminium. Steel-cored aluminium conductor, abbreviated as **ACSR** (*aluminium conductor steel reinforced*), consists of a core of galvanised steel strand surrounded by a number of aluminium strands. The steel conductors used are galvanised in order to prevent rusting and electrolytic corrosion, the x-section of two metals are in the ratio 1:6 but in case of high strength conductors their ratio may be 1:4. The steel core takes a greater percentage of mechanical stresses while the aluminium carries the bulk of current. ACSR conductor being of high tensile strength and lighter in weight produces small sag and therefore longer spans can be used. The ACSR conductor has a largest diameter than any other type of conductor of same resistance, so corona losses are reduced, but stronger supports are required for a given span. The ACSR conductor gets deteriorated in service to the atmospheric corrosion due to bimetallic action of zinc and the aluminium, electrochemical in nature. The skin effect is very predominating in ACSR conductors hence the resistance of the composite conductor is taken equal to that of aluminium covering alone. The reactance of ACSR conductor is taken to be equal to that of a non-magnetic conductor having diameter equal to that outside of the ACSR conductor.

Another advantage with ACSR conductor is that, because of reduced corona losses, critical voltage limit of the conductor can be raised by 30 to 50 % as compared to copper conductors.

Because of use of larger span, the number of line supports may be reduced by about 25%. Thus the overall cost of supports, foundations, insulators and erection is considerably reduced. In addition, there is a substantial saving in maintenance cost. Further, since most faults occur at the line supports, so the frequency of occurrence of faults is automatically reduced.

Composite type of conductor, being costlier than the all-aluminium type, is chiefly used in those cases where an all aluminium conductor would require a much higher and more costly support than copper. This occurs principally where spans are 100 metres and over are involved.

4. Galvanised Steel. Galvanised steel conductors have been used to advantage for extremely long spans, or for short line sections exposed to normally high stresses due to climatic conditions. These conductors are found most suitable for lines supplying rural areas and operating at voltages of about 11 kV, where cheapness is the main consideration. Iron or steel wire use is most advantageous for transmission of small power over a short distance, where the size of copper conductor desirable from economical considerations comes out to be smaller than 8 SWG, which cannot be used because of poor mechanical strength. This conductor is not suitable for EHT lines for the purpose of transmitting large amounts of power over a long distance due to its following properties:

(*i*) Poor conductivity, 13% that of copper.

(*ii*) High internal reactance.

(*iii*) It is subjected to eddy current and hysteresis.

Nowadays use of galvanised steel wire is limited to telecommunication lines, stay wires, earth wires and guard wires. Galvanised steel wires in small sizes are employed as binding wires. Stranded galvanised steel wires are used as guy wires and earth/ground wires. In *lt*, 11 kV and 33 kV lines, the earth wire is provided below the line conductors for the safety of personnals, cattles and material moving under the line as a precaution in case a line conductor snaps. In case of extra-high voltage lines of 66, 132 or 220 kV, the ground wire is provided above the line conductors for protection against lightning.

5. Cadmium Copper. The conductor being used in certain cases is copper alloyed with cadmium. Addition of 1 or 2% of cadmium in copper increases the tensile strength by about 40% and reduces the conductivity only by 17% below that of pure copper. However, cadmium copper is costlier than pure copper. Use of cadmium copper will be economical for a line with long spans and small cross-section i.e. where the cost of conductor material is comparatively small in comparison to that of supports etc. Cadmium-copper conductors are also employed for telephone and telegraph lines where currents involved are quite small. However, owing to scarcity of copper, cadmium-copper conductors on communication lines are being replaced by ACSR conductors.

6. Copper-Clad Steel. A composite wire, known as *copper-clad* or *copper-weld steel wire,* is obtained by welding a copper coating on a steel wire core. Line conductors made of copper-clad steel are preferably stranded, and have a considerably large tensile strength than the equivalent all-copper conductors. The proportion of copper and steel is so chosen that the conductivity of composite wire is 30 to 40% of that of copper conductor of equal diameter. Such material appears to be very suitable for river-crossings or other places where an extremely long span is involved.

7. Phosphor Bronze. When harmful gases such as ammonia are present in atmosphere and the spans are extremely long, phosphor bronze is most suitable material for an overhead line conductor. In this conductor some strands of phosphor bronze are added to the cadmium copper.

Example 3.1. Electric power of 30 MW at a pf of 0.8 lagging is to be transmitted by a 132 kV, 3-phase, 3-wire transmission line over a distance of 120 km. The efficiency of transmission is 90 per cent. Calculate:

(*i*) Weight of material required in case of copper having resistivity of 1.78 × 10^{-8} Ω-m and specific gravity of 8.9.

(*ii*) Weight of material required in case of aluminium having resistivity of 2.6 × 10^{-8} Ω-m and specific gravity of 2.

Solution : Power to be transmitted, $P = 30 \text{ MW} = 30 \times 10^6$ watts

Line voltage, $V_L = 132 \text{ kV} = 132 \times 10^3$ volts

Length of 3-phase, 3-wire line, $l = 120 \text{ km} = 120 \times 10^3$ metres.

$$\text{Line current, } I_L = \frac{P}{\sqrt{3}\, V_L \cos\phi}$$

$$\text{Power loss, } W = 3\, I_L^2 R = 3 \times \left(\frac{P}{\sqrt{3}\, V_L \cos\phi}\right)^2 \times \rho \frac{l}{a} = \frac{P^2 \rho l}{V_L^2 \cos^2\phi\, a}$$

$$\text{or } a = \frac{P^2 \rho l}{V_L^2 \cos^2\phi\, W} = \frac{P^2 \rho l}{V_L^2 \cos^2\phi\, (1-\eta) P}$$

∵ Power loss, $W = (1 - \eta)P$

$$\text{Volume of conductor material required} = 3\, a\, l = \frac{3 \rho l^2 P}{(1-\eta)\, V_L^2 \cos^2\phi} \text{ m}^3$$

$$\text{Weight of conductor material required} = \frac{3 \rho l^2 P}{(1-\eta) V_L^2 \cos^2\phi} \times D$$

$$\text{Weight of copper required} = \frac{3 \rho l^2 P}{(1-\eta) V_L^2 \cos^2\phi} \times 8{,}900$$

$$\because \text{ specific gravity of copper} = \frac{8.9 \times 10^6}{1{,}000} = 8{,}900 \text{ kg}/m^3$$

$$= \frac{3 \times 1.78 \times 10^{-8} \times \left(120 \times 10^3\right)^2 \times 30 \times 10^6}{(1-0.9) \times \left(132 \times 10^3\right)^2 \times (0.8)^2} \times 8{,}900 = 1{,}84{,}000 \text{ kg } \textbf{Ans.}$$

$$\text{Weight of aluminium required} = \frac{3 \times 2.6 \times 10^{-8} \times (120 \times 10^3)^2 \times 30 \times 10^6}{(1-0.9) \times (132 \times 10^3)^2 \times (0.8)^2} \times 2{,}000 = 60{,}400 \text{ kg } \textbf{Ans.}$$

3.4. RESISTANCE OF LINE

Every electric conductor offers opposition to the flow of current and this opposition is called the *resistance*. The resistance of transmission line conductors is the most important cause of power loss (I^2R) in a transmission line. The ohmic resistance R of a conductor of length l and uniform x-section a is given by the expression

$$R = \rho \frac{l}{a} \quad \text{...(3.1)}$$

where the resistivity or specific resistance of the conductor (ρ) depends not only on the conductor material but also on its temperature.

If ρ_1 and ρ_2 are the values of resistivity corresponding to two different temperatures t_1 and t_2 then

$$\rho_2 = \rho_1 [1 + \alpha (t_2 - t_1)] \quad \text{...(3.2)}$$

where α is the temperature coefficient of resistance of the material. The value of temperature coefficient of resistance is also not constant but depends upon the initial temperature.

The temperature coefficient of resistance at any temperature t_1 is given by the expression

$$\alpha_1 = \frac{\alpha_0}{1+\alpha_0 t_1} \quad \text{...(3.3)}$$

where α_0 is the temperature coefficient of resistance at 0 °C.

In a single phase line or 2-wire d c line, the loop resistance is taken as double of resistance of either conductor but in 3-phase system the resistance per phase is the resistance of either conductor.

3.5. SKIN EFFECT

The distribution of current over the cross-section of the conductor is uniform for dc only. An alternating current flowing through a conductor does not distribute uniformly but tends to concentrate near the surface of the conductor. In fact, in ac system no current flows through the core and the entire current is concentrated at the surface regions. The result is that the effective area of the conductor is reduced causing an increase in its ac resistance. The effective ac resistance is usually referred as the *effective resistance* of the conductor. The phenomenon is called the *skin effect* as it causes concentration of current at the skin of the conductor.

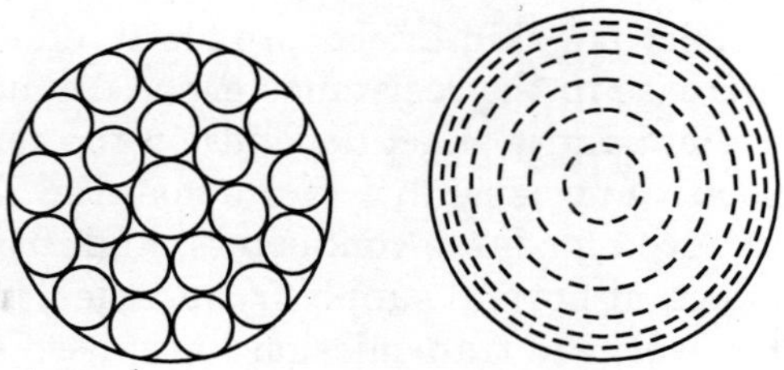

(a) (b)

Fig. 3.5

The skin effect can be easily explained by considering a solid conductor to be composed of a large number of annular filaments, each carrying a fraction of total current. The flux linkages due to the filaments lying at the surface link the whole of the conductor while the flux set up due to the inner filaments does not link with the surface or outer filaments. Thus the filaments near the centre are of larger inductance than that near the outer surface. The high reactance of the inner filaments causes the current to distribute in such a way that the current density is less in the interior of the conductor than at the surface. The distribution of current over the section will, therefore, be non-uniform, as shown in Fig. 3.5(*b*).

The skin effect depends upon (*i*) type of material (*ii*) frequency (*iii*) diameter of conductor, and (*iv*) shape of conductor.

At low frequencies the effect is very small, in fact it is only of importance with high frequencies or with solid conductors of larger cross-section. For commercial frequency of 50 Hz or less the increase in effective resistance is inappreciable for solid copper conductors up to 1 cm in diameter; about 2.5 percent for 2 cm diameter and 8 per cent for 2.5 cm diameter. In an aluminium wire the effect is the same as in a copper wire of equal conductivity. Thus since the resistivity of copper is 0.6 times that of aluminium, the increased resistance due to skin effect on an aluminium wire of a square mm in cross-section will be of the same percentage as on a 0.6 a square mm in copper wire.

The skin effect is much smaller with stranded conductors than with solid conductors. It increases with the increase of cross-section, permeability and supply frequency.

In practice stranded conductors are invariably used for transmission and distribution lines and hollow conductors for solid bus-bar. This is done in order to overcome the adverse effects of skin effect.

3.6. PROXIMITY EFFECT

The inductance and, therefore, the current distribution in a conductor is also affected by the presence of other conductors in its vicinity. This effect is known as the *proximity effect.*

While considering the skin effect it was assumed that there is no other current carrying conductor nearby but when a current carrying conductor is nearby, its flux will link with the conductor under consideration and its effect to the nearer half of the conductor will be more than with the farther half.

If conductors carry currents in the opposite directions the magnetic fields set up will tend to cause an increase in the current density in the adjacent portions of the conductors while if the currents are in the same direction, the current density is increased in remote parts of the conductor.

So like skin effect, proximity effect affects the current distribution and results in an increase in the resistance of the conductor and decrease of self reactance. Like skin effect, the proximity effect depends on the conductor size, frequency of the supply, resistivity and relative permeability of the material. This phenomenon (*i.e.*, proximity effect) is more pronounced for large conductors, high frequencies and close proximity. The magnitude of the effect, at normal supply frequencies, in the case of the wide spacing of conductors required for overhead transmission lines, is so small that it can be ignored. However, the proximity effect is pronounced in case of cables where the spacing between the conductors is small.

In the case of stranded conductors, each wire traverses alternately weaker and stronger portions of the magnetic field caused by the external current carrying conductor. Thus the average value of the field along the path of any wire remains the same, and assuming that the currents in the conductor follow to a large extent the paths of the individual wires, the proximity effect is substantially eliminated.

3.7. ECONOMICAL SIZE OF LINE CONDUCTOR—KELVIN'S LAW

In case of a transmission system carrying bulk power over a long distance, the question of voltage regulation is unimportant; in some very long lines a regulation of 40 per cent is considered satisfactory. The only consideration is the economy. Hence the cross-section of the conductor of a transmission line (or feeder) is decided on the basis of its current carrying capacity and, where practicable, of minimum cost, both fixed and operating.

The fixed cost is on account of annual interest and depreciation on capital cost of conductors, supports and insulators and the cost of their erection in case of overhead systems. For underground systems the fixed cost is on account of annual interest and depreciation on cost of conductors, insulation and the cost of laying the cables. For a particular voltage cost of insulation is practically constant and does not change with the x-section of the conductor, but the cost of conductors varies directly as the x-section of the conductor irrespective of the system of transmission. In case of overhead system the cost of supports and their crection partly varies as the x-section of the conductors and partly constant. Thus total annual fixed cost may be represented as Rs ($P_1 + P_2 a$) where P_1 and P_2 are constants and a is the area of x-section of conductor.

The annual running cost is on account of energy lost in the conductor due to its ohmic resistance *i.e* $I^2 R$ losses; losses in insulating material and in metallic sheaths (for insulated cables). The losses in insulating material and in metallic sheaths are very small and in comparison to ohmic losses may be neglected for the low voltage, but at higher voltages these are considerable. The ohmic resistance R is inversely proportional to area of x-section of conductor, and therefore for a given curve of demand for current through out the year *i.e.* for a given annual load curve, the energy lost in the conductor will be proportional to the resistance and therefore, inversely proportional to area of x-section. Hence annual running cost on account of energy lost in transmission line may be represented as $\frac{P_3}{a}$ where P_3 is a constant.

$$\therefore \text{ Total annual cost, } C = P_1 + P_2 a + \frac{P_3}{a} \quad ...(3.4)$$

This total annual cost will be minimum if the differentiation of it with respect to a is zero

$$i.e. \text{ if } \frac{dC}{da} = 0$$

$$\text{or } \frac{d}{da}\left(P_1 + P_2 a + \frac{P_3}{a}\right) = 0$$

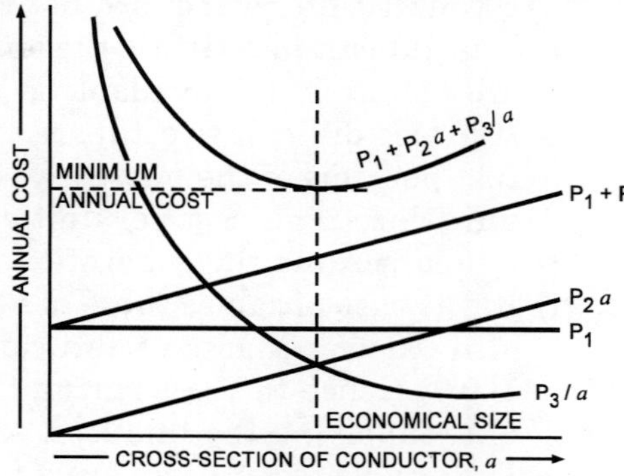

Fig. 3.6. ***Illustration of Kelvin's Law***

$$\text{or } P_2 - \frac{P_3}{a^2} = 0 \text{ or } P_2 a = \frac{P_3}{a} \quad ...(3.5)$$

i.e. if variable part of annual cost on account of interest and depreciation on the capital outlay is equal to the annual cost of electrical energy wasted in the conductors, the total annual cost will be minimum and the corresponding size of conductor will be most economical. This statement is modified version of *Kelvin's law**. Though theoretically it holds good but in actual working condition number of difficulties are faced in its application.

3.7.1. Practical Limitations To Application of Kelvin's Law

Though theoretically Kelvin's law is true but in actual working, an economical cross-section of the conductor determined by Kelvin's law may not suit because of the following factors.

(*i*) The assumption that the total annual cost on account of interest and depreciation on the capital outlay is of the form $P_1 + P_2 a$ is strictly speaking not true. For example, neither the cost of cable dielectric and sheath in underground cables vary according to this simple law, nor the cost of laying.

* According to Kelvin's law the most economical x-section of a conductor is that which makes the annual cost of electrical energy wasted in the conductor equal to the annual cost of interest and depreciation on the capital cost of the conductor. This law was based on the assumption that the cost of the supporting structures e.g. towers, insulators and the cost of erection-cum-commisioning *i.e.* stringing of the conductors etc. are fixed and independent of the area of x-section of the conductor. But this is not true in practice. So the Kelvin's law was modified in order to take the above fact into account by assuming that the initial capital cost on the complete line installation can be split-up into two parts, one independent of the area of x-section of the conductor and the other dependent upon it, directly. Thus total annual fixed cost can be represented as Rs ($P_1 + P_2 a$).

(*ii*) The size of the conductor determined may be of such a small x-section that it may cause too much voltage drop in the line.

(*iii*) The size of the conductor determined may be too weak or too strong from mechanical consideration. In case the size determined is too weak, it is advisable to go to the higher size irrespective to economy.

(*iv*) In case of cables the sizes of conductors determined by Kelvin's law usually gives higher current density thereby giving excessive heating. The only remedy is to decide the size of conductors in case of cables on the basis of current carrying capacity.

(*v*) The diameter of the conductor may be so small as to cause high corona loss.

(*vi*) Interest and depreciation on the capital outlay depends upon commercial conditions and estimated life of conductors and also the probable scrap value at the end of estimated life, which are not known with a fair degree of accuracy.

(*vii*) It is difficult to estimate the energy loss in the line without actual load curves, which are definitely not available at the times of estimation.

(*viii*) It is also difficult to estimate the cost per unit of the energy wasted in the line. The cost per unit of the energy wasted is not the same as that of cost of generation per unit since their cost per unit depends upon load factors which are different for the line losses and the generation.

(*ix*) In the case of cables there are sheath losses and with high voltages dielectric losses also. Dielectric loss occurs continuously therefore load factor of dielectric loss is 100%. Hence the cost per unit of energy lost as dielectric loss is less than cost per unit supplying the line loss.

It is impossible to derive an exact formula for the conductor x-section which will take into account all the above factors.

In case of overhead lines the question of temperature rise is of less importance and there are no losses other than I^2R loss so cross-sections derived from Kelvin's law have more chances of being acceptable. In the case of underground systems it is advisable to examine the available cross-sections above and below the calculated section and to ensure that temperature rise will not be excessive.

Example 3.2. The cost of a two-core feeder cable, including installation is Rs (90 *a* + 20) per metre and the interest and depreciation charges are 10 per cent. The cable is 2 km in length and cost of energy is 4 p per unit. The maximum current in the feeder is 250 A and the demand is such that the copper loss is equivalent to what would be produced by the full current flowing for 6 months. If the resistance of a conductor of 1 sq cm and 1 km length is 0.173 Ω, calculate the most economical cross-sectional area.

Solution : Resistance of each conductor of area a cm^2 and 2 km in length,

$$R = \rho\frac{l}{a} = \frac{0.173 \times 2}{a} = \frac{0.346}{a}\ \Omega$$

$$\text{Power loss, } W = 2\,I^2R = 2 \times (250)^2 \times \frac{0.346}{a} = \frac{43,250}{a}\ \text{watts or } \frac{43.25}{a}\ \text{kW}$$

$$\text{Energy loss per annum} = \frac{43.25}{a} \times \frac{365 \times 24}{2} = \frac{189,435}{a}\ \text{kWh}$$

$$\text{Annual cost of energy loss, } \frac{P_3}{a} = \text{Rs } 0.04 \times \frac{189,435}{a} = \text{Rs } \frac{7,577.4}{a}$$

Capital cost (variable) of feeder cable = Rs 90 a × 2 × 1,000 = Rs 180,000 a

Annual charges on account of interest and depreciation on variable cost of feeder cable,

$$P_2 a = \text{Rs } 180,000\ a \times \frac{10}{100} = \text{Rs } 18,000\ a$$

For most economical x-section of conductor

$$18{,}000\ a = \frac{7{,}577.4}{a}$$

$$\text{or} \qquad a = \sqrt{\frac{7{,}577.4}{18{,}000}} = 0.649 \text{ cm}^2 \textbf{ Ans.}$$

Example 3.3. Determine the best current density in amperes/mm^2 for a three phase overhead line. The line is in use for 2,600 hours per year and if the conductor costs Rs. 3.0/kg. It has a specific resistance of 1.73 × 10^{-8} Ω-m and weighs 6,200 kg/m^3. Cost of energy is 10 paise/unit. Interest and depreciation is 12% of conductor costs. [*U.P. Technical Univ. Elements of Power System 2006-07*]

Solution : Let I be the line current and '*a*' be the x-sectional area of conductors in mm^2.

Weight of conductor of 1 km length = Volume of conductor × specific gravity of material

$$= \frac{a \times 1{,}000 \times 6{,}200}{1{,}000 \times 1{,}000} = 6.2\ a \text{ kg}$$

Cost of conductor of 1 km length = Rs 3 × 6.2 a = Rs 18.6 a

Annual interest and depreciation per conductor per km

$$= \text{Rs } 18.6\ a \times \frac{12}{100} = 2.232\ a$$

Energy lost per annum per conductor per km

$$= I^2 R \times \frac{\text{Time in hours}}{1{,}000}$$

$$= \frac{I^2 \times 1.73 \times 10^{-8} \times 1{,}000}{a \times 10^{-6}} \times \frac{2{,}600}{1{,}000} = 44.98\ \frac{I^2}{a}$$

$$\text{Annual cost due to energy losses} = \text{Rs. } \frac{10}{100} \times 44.98\ \frac{I^2}{a} = \text{Rs } 4.498\ \frac{I^2}{a}$$

For most economical x-section of conductor

$$2.232\ a = 4.498\ \frac{I^2}{a}$$

$$\text{or} \qquad \frac{I}{a} = \sqrt{\frac{2.232}{4.498}} = 0.705 \text{ A/mm}^2$$

∴ The best current density = 0.705 A/mm^2 **Ans.**

Example 3.4. A 3-core, 11 kV cable supplies a load of 1,500 kW at 0.8 pf lagging for 300 days in a year at an average of 8 hours per day. The capital cost per km of cable is Rs 8,000 + 20,000 *a* (*a* is the cross-sectional area of cable). The resistance per km of a cable of cross-sectional area 1 cm^2 is 0.173 Ω. If the energy loss cost is 2 paisa per unit, and the rate of interest and depreciation is 12%, calculate the most economical current density and diameter of the conductor.

Solution : Capital cost (variable) of cable = Rs 20,000 a per km length.

Annual charges on account of interest and depreciation on variable cost of cable,

$$P_2 a = \text{Rs } 20{,}000\ a \times \frac{12}{100} = \text{Rs } 2{,}400a \text{ per km length}$$

Resistance of each conductor of x-sectional area a cm^2,

$$R = \frac{0.173\Omega}{a} \text{ per km length}$$

$$\text{Current in each conductor, } I = \frac{\text{Power supplied in kW}}{\sqrt{3} \times \text{line voltage in kV} \times \text{pf}} = \frac{1{,}500}{\sqrt{3} \times 11 \times 0.8} = 98.412 \text{ A}$$

Total power loss in all the three conductors for length of 1 km,

$$W = 3\,I^2R = 3 \times (98.412)^2 \times \frac{0.173}{a} = \frac{5,026.47}{a} \text{ watts}$$

Energy loss per annum = W in kW × working hours/annum

$$= \frac{5,026.47}{a \times 1,000} \times 300 \times 8 = \frac{12,064}{a} \text{ kWh}$$

Annual cost of energy loss, $\frac{P_3}{a}$ = Rs $0.02 \times \frac{12,064}{a} = \frac{241.28}{a}$

For most economical cross-section of conductor,

$$2,400\ a = \frac{241.28}{a}$$

$$\text{or} \quad a = \sqrt{\frac{241.28}{2,400}} = 0.317 \text{ cm}$$

and diameter of conductor, $d = \sqrt{\frac{4a}{\pi}} = \sqrt{\frac{4 \times 0.317}{\pi}} = 0.635 \text{ cm} \quad \therefore a = \frac{\pi}{4} d^2$ **Ans.**

Current density, $\delta = \frac{I}{a} = \frac{98.412}{0.317} = 310.4 \text{ A/ cm}^2$ **Ans.**

Example 3.5. The cost of a 3-ϕ overhead line having x-section **area 'a' cm² is Rs (500 + 2,000a) per km. Calculate the most economical current density for the conductor if the rate of interest and depreciation is 12% per annum and the cost of energy wasted is 5 paise per kWh. The resistivity of conductor may be taken as 1.7 μΩ-cm. Take load factor of 12%.**

[U.P. Technical Univ. Elements of Power System 2008-09]

Solution : Capital cost (variable) of transmission line = Rs 2,000a per conductor per km length

Annual charges on account of interest and depreciation on variable cost of transmission line,

$$P_2 a = \text{Rs } 2,000a \times \frac{12}{100} = \text{Rs } 240a \text{ per conductor per km length.}$$

Resistance of each conductor of x-sectional area a cm²

$$R = \frac{\rho l}{a} = \frac{1.78 \times 10^{-8}}{a \times 10^{-4}} \times 1,000 = \frac{0.178}{a}\,\Omega \text{ per km length}$$

If I is the current in amperes in each conductor, then power loss in each conductor for a length of 1 km

$$W = I^2R = \frac{0.178}{a} I^2 \text{ watts}$$

Energy loss per annum per conductor for a length of 1 km,

$$= \text{W in kW} \times \text{Load factor of losses} \times 8,760 \text{ kWh}$$

$$= \frac{0.178}{a} I^2 \times \frac{0.12 \times 8,760}{1,000} = \frac{0.1871136 I^2}{a} \text{ kWh}$$

Annual cost of energy loss for a length of 1 km,

$$\frac{P_3}{a} = \text{Rs } \frac{5}{100} \times \frac{0.1871136 I^2}{a} = \text{Rs } \frac{0.00935568 I^2}{a}$$

For most economical cross-section of conductor,

$$P_2 a = \frac{P_3}{a}$$

$$\text{or} \quad 240a = \frac{0.00935568 I^2}{a}$$

$$\text{or} \quad \frac{I}{a} = \sqrt{\frac{240}{0.00935568}} = 160.2 \text{ A/cm}^2 \quad \textbf{Ans.}$$

Example 3.6. Describe the main limitations of Kelvin's law. The cost of a 3-phase overhead transmission line having cross-sectional area A cm^2 is Rs (500 + 2,000 A) per km. Calculate the most economical current density for the conductor if the rate of interest and depreciation is 12% per annum. The cost of energy wasted is Rs 0.05/kWh. The resistance of each conductor is $\frac{0.17}{A}$ ohm/km. Take load factor for losses = 12%. [*U.P. Technical Univ. Elements of Power System 2007-08*]

Solution : Annual charges on account of interest and depreciation on variable cost of transmission line is

$$P_2A = \text{Rs } 2{,}000A \times \frac{12}{100} = \text{Rs. } 240 \text{ A per conductor per km length}$$

$$\text{Resistance of each conductor, } R = \frac{0.17}{A} \; \Omega \text{ per km length}$$

Power loss in each conductor for a length of 1 km,

$$W = I^2R = \frac{0.17I^2}{A} \text{ watts}$$

where I is the current in amperes in each conductor.

Energy loss per annum per conductor for a length of 1 km

$$= \frac{0.17}{A} \frac{I^2 \times 0.12 \times 8{,}760}{1{,}000} = \frac{0.1787I^2}{A} \text{ kWh}$$

Annual cost of energy loss for a length of 1 km

$$\frac{P_3}{A} = \frac{\text{Rs } 0.05 \times 0.1787I^2}{A} = \text{Rs } \frac{0.08935I^2}{A}$$

For most economical cross-section of conductor

$$P_2A = \frac{P_3}{A}$$

$$\text{or } 240 \text{ A} = \frac{0.08935I^2}{A}$$

$$\text{Most economical current density} = \frac{I}{A} = \sqrt{\frac{240}{0.08935}} = 51.83 \text{ A/cm}^2 \textbf{ Ans.}$$

Example 3.7. Determine the most economical size of a 3-phase line which supplies the following loads at 10 kV.

(*i*) 1,000 kW at 0.8 pf (lag) for 10 hours,

(*ii*) 500 kW at 0.9 pf (lag) for 8 hours and

(*iii*) 100 kW at upf for 6 hours.

The above gives the daily load cycle. The cost per km of the completely erected line is Rs. (8,000 *a* + 1,500) where *a* is the area of x-section of each conductor in cm^2. The combined interest and depreciation is 10 per cent per annum of capital cost. Cost of energy losses is 80 paise per kWh. Resistivity of conductor material = 1.72 × 10^{-6} Ω-cm.

Solution : Let the length of the line be L km

$$\text{Capital cost of line} = \text{Rs. } (8{,}000\,a + 1{,}500) \times L$$

Annual cost on account of interest and depreciation on capital (variable) cost

$$= \text{Rs. } \frac{10}{100} \times 8{,}000aL = \text{Rs. } 800aL$$

The load currents

(*i*) at 1,000 kW, 0.8 pf (lag), $I = \dfrac{1,000 \times 1,000}{\sqrt{3} \times 10,000 \times 0.8} = \dfrac{125}{\sqrt{3}}$ A

(*ii*) at 500 kW, 0.9 pf (lag), $I = \dfrac{500 \times 1,000}{\sqrt{3} \times 10,000 \times 0.9} = \dfrac{500}{9\sqrt{3}}$ A

(*iii*) at 100 kW, upf, $I = \dfrac{100 \times 1,000}{\sqrt{3} \times 10,000 \times 1} = \dfrac{10}{\sqrt{3}}$ A

Resistance of line conductor, $R = \dfrac{\rho l}{a} = \dfrac{1.72 \times 10^{-6} \times 1,000 \times 100 \times L}{a}$ ohms $= \dfrac{0.172\, L}{a}$ ohms

Energy lost per annum $= \dfrac{3\, I^2\, Rt}{1,000}$ kWh

$$= \frac{3 \times 0.172\, L}{a} \times \frac{365}{1,000}\left[\left(\frac{125}{\sqrt{3}}\right)^2 \times 10 + \left(\frac{500}{9\sqrt{3}}\right)^2 \times 8 + \left(\frac{10}{\sqrt{3}}\right)^2 \times 6\right] = \frac{11,397\, L}{a} \text{ kWh}$$

Cost of energy wasted per annum = Rs $0.80 \times \dfrac{11,397L}{a}$ = Rs $\dfrac{9,117.6\, L}{a}$

For most economical cross-section of conductor

$$800aL = \frac{9,117.6L}{a}$$

$$\text{or } a = \sqrt{\frac{9,117.6}{800}} = 3.376 \text{ cm}^2 \textbf{ Ans.}$$

EXERCISES

1. Explain different types of conductors used for transmission lines. Give their properties. **[U.P. Technical Univ. Elements of Power System 2006-07]**
2. Which material is used for overhead line conductor ? Discuss the reasons for selecting that material. **[U.P. Technical Univ. Elements of Power System 2009-10]**
3. What are the advantages and disadvantages of using aluminium over copper in power transmission. **[U.P. Technical Univ. Elements of Power System 2002-03]**
4. What is necessity of having stranded conductor in transmission lines ? **[U.P. Technical Univ. Elements of Power System 2004-05]**
5. Describe ACSR conductors used in transmission lines. **[U.P. Technical Univ. Elements of Power System 2004-05]**
6. What advantage you get by using stranded conductors ? **[U.P. Technical Univ. Elements of Power System 2008-09]**
7. Explain what is meant by 'skin effect' in a conductor and state the factors on which it depends. **[U.P. Technical Univ. Elements of Power System 2003-04]**
8. Describe the influence of skin effect on the resistance of conductors. **[U.P. Technical Univ. Elements of Power System 2002-03]**
9. Show how 'skin effect' increases effective resistance of the conductor. **[U.P. Technical Univ. Elements of Power System 2004-05]**
10. State whether following sentence is true or false. Justify your answer. "DC resistance is more than ac resistance of a line". **[U.P. Technical Univ. Elements of Power System 2004-05]**
11. Explain with justification whether the following statement is TRUE or FALSE. "Skin effect increases effective resistance and reactance of conductor". **[U.P. Technical Univ. Elements of Power System 2008-09]**

12. Explain difference between skin effect and proximity effect.
[U.P. Technical Univ. Elements of Power System 2009-10]

13. Differentiate between skin effect and proximity effect.
[U.P. Technical Univ. Elements of Power System 2006-07, 2007-08]

14. Compare skin effect with proximity effect.
[U.P. Technical Univ. Elements of Power System 2004-05]

15. State and prove Kelvin's law for determining economic size of conductor for overhead lines and discuss its limitations in practice.
[U.P. Technical Univ. Elements of Power System 2003-04]

16. State Kelvin's law for obtaining the size of conductor for transmission. Discuss its limitations. **[U.P. Technical Univ. Elements of Power System 2006-07]**

17. What are the limitations of Kelvin's law ?
[U.P. Technical Univ. Elements of Power System 2004-05]

SHORT ANSWER TYPE QUESTIONS WITH ANSWERS

Q. 1. What are the main requirements of good conductor material used for overhead lines ?

Ans. Main requirements of good conductor material used for overhead lines are high electrical conductivity, high tensile strength, low specific gravity, low cost, easy availability, free from brittleness and the material should not react with atmospheric condition.

Q. 2. What for ACSR conductor stands ?

Ans. ACSR conductor stands for aluminium conductor steel reinforced conductor.

Q. 3. Why are conductors for transmission line stranded ?

Ans. All conductors used for overhead transmission lines are stranded so as to increase the flexibility. Solid wires, except of smaller sizes, are difficult to handle when used for long spans and tend to crystallise at the points of supports because of swinging in winds.

Q. 4. Why is galvanised steel wire not suitable for EHT lines for the purpose of transmitting large amounts of power over a long distance ?

Ans. The galvanised steel wire is not suitable for EHT lines for the purpose of transmitting large amounts of power over a long distance because of poor conductivity, high internal reactance and eddy current and hysteresis.

Q. 5. What is skin effect ?

Ans. The phenomenon of concentration of an electric current near the surface of the conductor is known as skin effect.

Q. 6. On what factors does the skin effect depend ?

Ans. The skin effect depends upon the type of material, frequency of current, diameter of conductor and shape of conductor. It increases with the increase of x-section, permeability and supply frequency.

Q. 7. What is proximity effect ?

Ans. The inductance and, therefore, the current distribution in a conductor is also affected by the presence of other conductors in its vicinity. This effect is known as the *proximity effect.*

Q. 8. What is Kelvin's law ?

Ans. According to Kelvin's law, the most economical x-section of a conductor is that which makes the annual cost of electrical energy wasted in the conductor equal to the annual cost of interest and depreciation on the capital cost of the conductor.

Q. 9. What is modified Kelvin's law ?

Ans. Modified Kelivn's law states that the most economical cross-sectional area of transmission line conductor makes the variable part of annual cost on account of interest and depreciation on the capital outlay equal to the annual cost of electrical energy wasted in the conductor.

PROBLEMS

1. Calculate the most economical x-section of the conductor of a km long transmission line if it delivers a maximum current of 300 A and annual copper losses are such as to be caused by a constant current of 100 A. The energy cost is 10 paise/kWh. The cost of line is Rs (50,000 a + 8,000) per km. Resistance per km per sq cm of copper conductor is 0.173 Ω. The annual rate of interest and depreciation is 10%, a is the area of x-section of conductor in cm^2. [**Ans.** 0.55 cm^2]
2. A 500 V, 2-core feeder 0.8 km long is required to supply a constant load of 100 kW. The cost of the cable including installation charges is Rs (6 a + 1.3) per metre where a is the cross-sectional area of each conductor in cm^2. Interest and depreciation total 10%. Determine the most economical size. Cost of energy is 2P per unit. Specific resistance of copper is 1.75×10^{-6} Ω per cm^2 cross-sectional area and 1 cm long. [**Ans.** 2.022 cm^2]
3. A 2-core, 11 kV cable is to supply 1 MW at 0.8 pf lag for 3,000 hours in a year. Capital cost of the cable is Rs (20 + 400 a) per metre where a is the x-sectional area of core in cm^2. Interest and depreciation total 10% and cost per unit of energy is 15 paise. If the length of the cable is 1 km, calculate the most economical x-section of the conductor. The specific resistance of copper is 1.75×10^{-6}Ω-cm. [**Ans.** 0.2255 cm^2]
4. Find the most economical x-section of a 3-core feeder cable 250 m long supplying a load of 80 kW, 0.8 pf (lagging) at 400 V for 4,000 hours per year. The cost of the cable including installation is Rs (150 a + 250) per metre where a is the x-section of each conductor in cm^2. Interest and depreciation total 10%. Cost of energy wasted is Rs 1 per unit. The resistance per km of conductor 1 cm^2 cross-section is 0.173 Ω. [**Ans.** 1.7 cm^2]
5. An industrial load is supplied by a 3-phase cable from a substation at a distance of 6 km. The voltage at the load is 11 kV. The daily load cycle for six days in a week for the entire year is as given below:

   (*i*) 700 kW at 0.8 pf for 7 hours.

   (*ii*) 400 kW at 0.9 pf for 3 hours.

   (*iii*) 88 kW at unity pf for 14 hours.

   Compute the most economical x-section of conductors for a cable whose cost is Rs. (5,000 A + 1,500) per km (including the cost of laying etc.). The tariff for the energy consumed at the load is Rs. 150 per annum per kVA of MD plus 5 paise per unit. Assume the rate of interest and depreciation as 15%. The resistance per km of conductor is (0.173/A) ohms.

   [**Ans.** 0.63 cm^2]
6. A 500-V, 2-core feeder cable 4 km long supplies a maximum current of 200 A and the demand is such that the copper loss per annum is such as would be produced by the full load current flowing for 6 months. The resistance of the conductor 1 km long and 1 sq cm. cross-sectional area is 0.17Ω. The cost of cable including installation is Rs. (120 A + 24) per metre where A is the area of x-section in cm^2 and interest and depreciation charges are 10%. The cost of energy is 4 paise per kWh. Determine the most economical x-section.

   [**Ans.** 44.6 mm^2]

❑❑❑

UNIT-II

OVERHEAD TRANSMISSION LINES

CONTENTS IN THIS UNIT

OVERHEAD TRANSMISSION LINES : Calculation of Inductance and Capacitance of Single Phase, Three Phase, Single Circuit and Double Circuit Transmission Lines.
Representation and Performance of Short, Medium and Long Transmission Lines, Ferranti Effect, Surge Impedance Loading.

CHAPTER 4. Inductance and Capacitance of Transmission Lines

CHAPTER 5. Representation and Performance of Short and Medium Transmission Lines

CHAPTER 6. Representation and Performance of Long Transmission Lines.

4

CHAPTER

INDUCTANCE AND CAPACITANCE OF TRANSMISSION LINES

INSIDE THIS CHAPTER

4.1. Line Inductance **4.2.** Inductance of Single Phase Overhead Lines **4.3.** Inductance of Three Phase Overhead Lines **4.4.** Inductance of Three Phase Lines With More Than One Circuit **4.5.** Bundled Conductors **4.6.** Capacitance of a Transmission Line **4.7.** Capacitance of a Single Phase Overhead Line **4.8.** Capacitance of 3-Phase Overhead Lines **4.9.** Capacitance of Double Circuit Three Phase Overhead Lines **4.10.** Capacitance of Bundled Conductor Line **4.11.** Effect of Earth on Capacitance of Transmission Line **4.12.** Transposition of Three Phase Lines.

4.1. LINE INDUCTANCE

Inductance is the property by virtue of which a circuit opposes changes in the value of a varying current flowing through it. While the resistance of a circuit opposes the flow of both steady as well as varying currents, the inductance causes opposition only to varying currents. Inductance does not cause any opposition to steady or direct current. In case of transmission and distribution lines, the current flowing is varying or alternating current, the effect of inductance, in addition to that of resistance, is therefore to be considered. The opposition to the flow of varying current owing to inductance is viewed as a voltage drop.

It is well known fact that a current carrying conductor is surrounded by concentric circles of magnetic lines. In case of ac system this field set up around the conductor is not constant but changing and links with the same conductor as well as with other conductors. Due to these flux linkages the line possesses inductance, defined as the flux linkages per unit current. Thus for determination of inductance of a circuit, determination of flux linkages is essential.

4.1.1. Flux Linkages of a Conductor

A long straight cylindrical conductor carrying a current is surrounded by a magnetic field. The magnetic lines of force will exist inside the conductor as well as outside the conductor. Both of these fluxes contribute to the inductance of the conductor.

Flux Linkages of a Conductor Due to Internal Flux. Consider a long straight cylindrical conductor of radius r metres and carrying a current of I amperes, as shown in Fig. 4.1.

In overhead lines it may be assumed without appreciable error that the current is uniformly distributed.

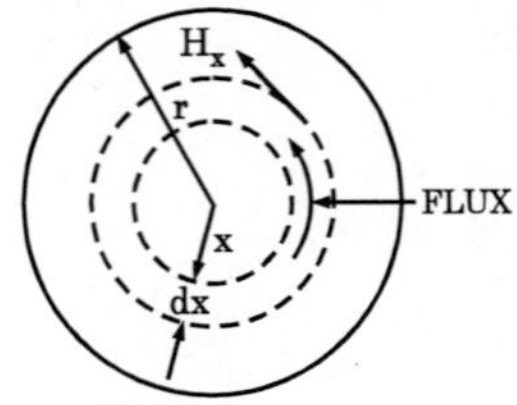

Fig. 4.1

The current inside a line of force of radius x,

$$I_x = \frac{I}{\pi r^2} \times \pi x^2 = \frac{I x^2}{r^2}$$

Field strength inside the conductor at a distance x from the centre,

$$H_x = \frac{\text{Current}}{2\pi \times \text{distance}} = \frac{I x^2}{r^2} \times \frac{1}{2\pi x} = \frac{I x}{2\pi r^2} \text{ AT/m}$$

Flux density, $B_x = \mu_0 \mu H_x$ Wb/m^2 or **T** where $\mu_0 = 4\pi \times 10^{-7}$ and μ is the relative permeability of the medium and for non-magnetic materials $\mu = 1$

$$\therefore B_x = \frac{\mu_0 I x}{2\pi r^2} \text{ Wb/m}^2$$

Now the flux through a cylindrical shell of radial thickness dx and axial length one metre,

$$d\phi = B_x \times 1 \times dx = \frac{\mu_0 x}{2\pi r^2} \cdot I \cdot dx \text{ webers}$$

But this flux links with only the current lying within the circle of radius x *i.e.* with I_x or $\frac{x^2}{r^2} I$

$$\therefore \text{Linkages of the shell, } d\psi = \frac{\mu_0 x}{2\pi r^2} \cdot dx \times \frac{x^2}{r^2} \times I = \frac{\mu_0 I x^3}{2\pi r^4} dx \text{ weber-turns}$$

Total flux linkages from centre of the conductor and up to the surface of the conductor,

$$\psi_{int} = \int_0^r \frac{\mu_0 I x^3}{2\pi r^4} dx \text{ Wb-turns/m} = \frac{\mu_0 I}{8\pi} \text{ Wb-turns/m} = \frac{4\pi \times 10^{-7} \times I}{8\pi} \text{ Wb-turns/m}$$

$$= \frac{1}{2} I \times 10^{-7} \text{ Wb-turns/m} \quad ...(4.1)$$

Flux Linkages of a Conductor Due To External Flux. Consider two points 1 and 2 distant d_1 and d_2 from the centre of the conductor. Since the flux paths are concentric circles around the conductor, whole of the flux between points 1 and 2 lies within the concentric cylindrical surfaces passing through these points 1 and 2.

The field strength at any distance x from the centre of the conductor ($x > r$),

$$H_x = \frac{I}{2\pi x} \text{ AT/m}$$

Flux density, $B_x = \mu_0 H_x = \dfrac{\mu_0 I}{2\pi x}$ Wb/m^2

So the flux through a cylindrical shell of radial thickness dx and axial length one metre

$$d\phi = \frac{\mu_0 I}{2\pi x} dx \text{ webers per metre}$$

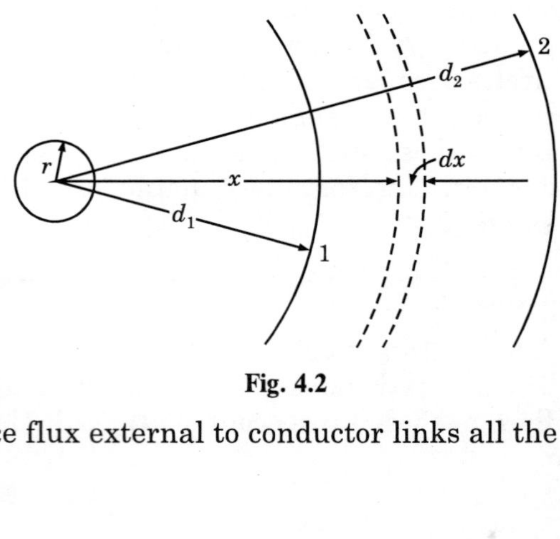

Fig. 4.2

Now flux linkages per metre is equal to $d\phi$ since flux external to conductor links all the the current in the conductor once and only once

$$\therefore \quad d\psi = \frac{\mu_0 I}{2\pi x} dx \text{ Wb-turns per metre}$$

Total flux linkages between points 1 and 2

$$\psi_{ext} = \int_{d_1}^{d_2} \frac{\mu_0 I}{2\pi x} dx = \frac{4\pi \times 10^{-7}}{2\pi} I [\log_e x]_{d_1}^{d_2}$$

$$= 2 \times 10^{-7} I \log_e \frac{d_2}{d_1} \text{ Wb-turns/m} \quad ...(4.2)$$

4.2. INDUCTANCE OF SINGLE PHASE OVERHEAD LINES

4.2.1. Inductance of a Two-Wire Single Phase Line

Consider a single phase line consisting of two parallel conductors A and B of radii r_1 and r_2 spaced d metres apart (d being very large as compared to r_1 and r_2). Conductors A and B carry the same current (*i.e.* $I_A = I_B$) in magnitude but opposite in directions, as one forms the return path for the other. The inductance of each conductor is due to internal flux linkages and external flux linkages and the following points are to be noted regarding external flux linkages :

1. A line of flux produced due to current in conductor A at a distance equal to or greater than $(d + r_2)$ from the centre of conductor A links with a zero net current, as the current flowing in the two conductors A and B are equal in magnitude but opposite in directions.

2. Flux lines at a distance $(d - r_2)$ link with a current I and those between $(d - r_2)$ and $(d + r_2)$ link with a current varying from I to zero.

As a simplifying assumption, it can be assumed that all the flux produced by current in conductor A links all the current up to the centre of conductor B and that the flux beyond the centre of conductor B does not link any current.

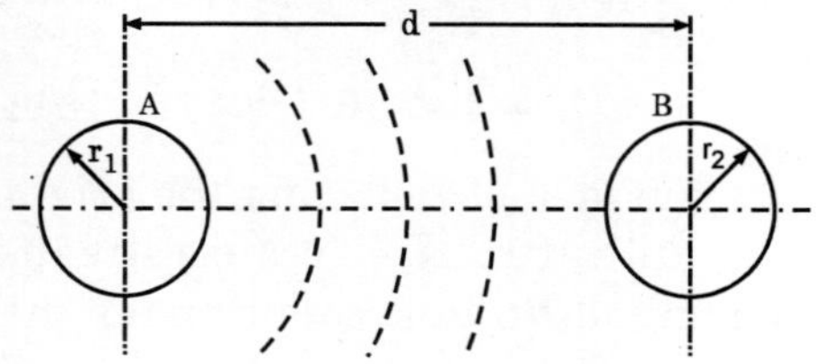

Fig. 4.3

The above assumption simplifies the calculations and results obtained are quite accurate specially when d is much greater than r_1 and r_2, as is usually the case in overhead lines.

Based on the above assumption flux linkages of conductor A due to external flux can be determined from Eq. (4.2) by substituting $d_2 = d$ and $d_1 = r_1$. Thus flux linkages of conductors A due to external flux only,

$$\psi_{A\,ext} = 2 \times 10^{-7} I \log_e \frac{d}{r_1} \text{ Wb-turns/metre} \quad ...(4.3)$$

Flux linkages of conductor A due to internal flux only,

$$\psi_{A\,int} = \frac{1}{2} I \times 10^{-7} \text{ Wb-turns/metre} \quad ...(4.4)$$

Total flux linkages of conductor A

$$\psi_A = \psi_{A\,ext} + \psi_{A\,int} = \left(2 \times 10^{-7} I \log_e \frac{d}{r_1} + \frac{1}{2} I \times 10^{-7}\right) = \left(0.5 + 2 \log_e \frac{d}{r_1}\right) I \times 10^{-7} \text{ Wb-turns/m}$$

Total inductance of conductor A,

$$L_A = \frac{\psi_A}{I} = \frac{\left(0.5 + 2\log_e \frac{d}{r_1}\right)}{I} \times I \times 10^{-7} \text{ H/m}$$

$$= 2 \times 10^{-7} \left(0.25 + \log_e \frac{d}{r_1}\right) \text{ H/m}$$

$$= 2 \times 10^{-7} \left(\log_e e^{1/4} + \log_e \frac{d}{r_1}\right) \text{H/m} = 2 \times 10^{-7} \log_e \frac{d}{r_1 e^{-1/4}} \text{ H/m}$$

The product ($r_1\, e^{-1/4}$) is known as *Geometric Mean Radius* (GMR) of the conductor and is equal to 0.7788 times the radius of the conductor. Let it be represented by r'_1 where

$$r'_1 = 0.7788\, r_1$$

$$\therefore \quad L_A = 2 \times 10^{-7} \log_e \frac{d}{r'_1} \text{ H/m} \qquad \text{...(4.5)}$$

Similarly inductance of conductor B, $L_B = 2 \times 10^{-7} \log_e \frac{d}{r'_2}$ H/m

Loop inductance of the line, $L = L_A + L_B = 2 \times 10^{-7} \left(\log_e \frac{d}{r'_1} + \log_e \frac{d}{r'_2}\right)$

If $r'_1 = r'_2 = r'$, the loop inductance of the line is given as

$$L = 4 \times 10^{-7} \log_e \frac{d}{r'} \text{ H/m} = 0.4 \log_e \frac{d}{r'} \text{ mH/km} \qquad \text{...(4.6)}$$

The idea of replacing the original conductor of radii r by a fictitious conductor of radii r' is quite attractive because streamlined equations for inductance can be developed without bogging down in accounting for the internal flux.

4.2.2. Flux Linkages of One Conductor in a Group of Conductors

Consider a group of parallel conductors 1, 2, 3, ... n carrying currents I_1, I_2, I_3 ... I_n respectively, as illustrated in Fig. 4.4. Let it be assumed that the sum of the currents in various conductors is zero *i.e.* $I_1 + I_2 + I_3 + ... I_n = 0$

Theoritically, the flux due to a conductor extends from the centre of the conductor right up to infinity but let us assume that the flux linkages extend up to a remote point P and the respective distances are as marked in Fig. 4.4.

The current in each conductor sets up a certain flux due its own current. The sum of all these fluxes is the total flux of the system and the total flux linkages of any one conductor is the sum of its linkages with all the individual fluxes set up by the conductors of the system.

Now let us determine the flux linkages of conductor 1 due to current I_1 carried by the conductor itself and flux linkages to conductor 1 due to currents carried by other conductors (2, 3, ... n)

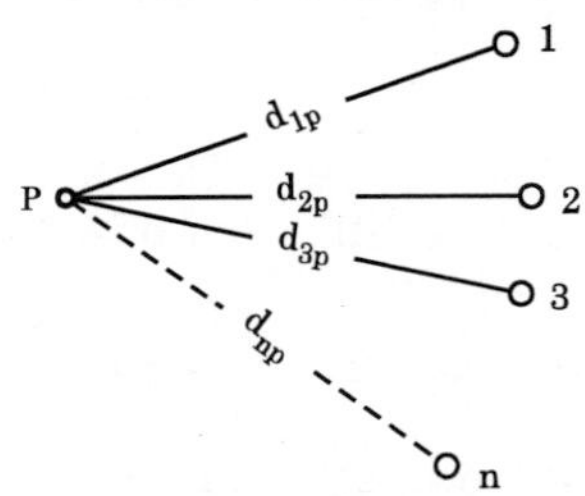

Fig. 4.4. ***Cross-sectional View of a Group of n Conductors and Distances of Conductors From a Remote Point P***

The flux linkages of conductor 1 due to its own current I_1 (internal and external), up to point P

$$\psi_{1p1} = 2 \times 10^{-7}\ I_1 \log_e \frac{d_{1p}}{r_1'} \text{ Wb-turns/metre}$$

The flux linkages of conductor 1 due to current in conductor 2

$$\psi_{1p2} = 2 \times 10^{-7}\ I_2 \log_e \frac{d_{2p}}{d_{12}}$$

Flux due to conductor 2 that lies between conductors 2 and 1 does not link conductor 1 and therefore the distances involved are d_{2p} and d_{12}.

Thus the expression for flux linkages of conductor 1 due to currents in all conductors can be written as

$$\psi_{1p} = 2 \times 10^{-7} \left[I_1 \log_e \frac{d_{1p}}{r_1'} + I_2 \log_e \frac{d_{2p}}{d_{12}} + I_3 \log_e \frac{d_{3p}}{d_{13}} + ... I_n \log_e \frac{d_{np}}{d_{1n}} \right] \text{ Wb-turns/m} \quad ...(4.7)$$

The above equation may be written as

$$\psi_{1p} = 2 \times 10^{-7} \left[I_1 \log_e \frac{1}{r_1'} + I_2 \log_e \frac{1}{d_{12}} + I_3 \log_e \frac{1}{d_{13}} + ... I_n \log_e \frac{1}{d_{1n}} \right]$$

$$+ 2 \times 10^{-7} [I_1 \log_e d_{1p} + I_2 \log_e d_{2p} + I_3 \log_e d_{3p} + ... I_n \log_e d_{np}] \text{ Wb-turns/m} \quad ...(4.8)$$

To account for the total flux linkages to conductor 1, the point P must approach infinity and in this condition

$$d_{1p} \simeq d_{2p} \simeq d_{3p} \simeq d_{4p} ... d_{np} \simeq d \text{ (say)}$$

Then $\lim_{d \to \infty} (I_1 + I_2 + I_3 + ... + I_n) \log_e d = 0 \qquad \because I_1 + I_2 + I_3 + ... I_n = 0$

This simplifies the Eq. (4.8) and the equation for the flux linkages to conductor 1 becomes

$$\psi_1 = 2 \times 10^{-7} \left[I_1 \log_e \frac{1}{r_1'} + I_2 \log_e \frac{1}{d_{12}} + I_3 \log_e \frac{1}{d_{13}} + ... I_n \log_e \frac{1}{d_{1n}} \right] \text{Wb-turns/m} \quad ...(4.9)$$

4.2.3. Inductance of Composite Conductor Lines—Self and Mutual GMDs

Consider a single phase line consisting of two parallel conductors A and B, conductor A consisting of x and conductor B of y strands, as illustrated in Fig. 4.5.

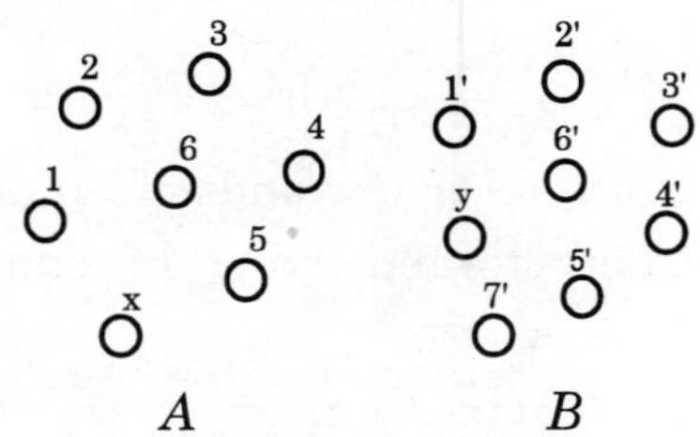

Fig. 4.5

Let the conductors A and B carry currents I and – I respectively (since conductors of a 2-wire line carry the same current but in opposite directions).

Assuming uniform current density in both the conductors the current carried by each strand of conductor A will be $\frac{I}{x}$ while that carried by each strand of conductor B will be $\frac{-I}{y}$

Using Eq. (4.9) the flux linkages of strand 1 in conductor A can be written as

$$\psi_1 = 2 \times 10^{-7}\frac{I}{x}\left[\log_e\frac{1}{r_1'} + \log_e\frac{1}{d_{12}} + \log_e\frac{1}{d_{13}} + ... + \log_e\frac{1}{d_{1x}}\right]$$

$$- 2 \times 10^{-7}\ \frac{I}{y}\left[\log_e\frac{1}{d11'} + \log_e\frac{1}{d12'} + \log_e\frac{1}{d13'} + ... + \log_e\frac{1}{d_{1y}}\right]$$

$$\text{or } \psi_1 = 2 \times 10^{-7}\ I \log_e\frac{\sqrt[y]{(d_{11'}\ d_{12'}\ d_{13'}\ ....\ d_{1y})}}{\sqrt[x]{r_1'\ d_{12}\ d_{13} .... d_{1x}}}\ \text{Wb-turns/m}$$

Inductance of strand 1 of conductor A

$$L_1 = \frac{\psi_1}{I/x} = 2 \times 10^{-7} x \log_e\frac{\sqrt[y]{(d_{11'}\ d_{12'}\ d_{13'}\ d_{14'}\ ....\ d_{1y})}}{\sqrt[x]{r'\ d_{12}\ d_{13}\ d_{14} .... d_{1x}}} \quad ...(4.10)$$

In above expression numerator of argument of $\log_e$ is written y th root of distances $d_{11'}$ $d_{12'}$, $d_{13'}$, $d_{14'}$ ..., d_1y multiplied together where distances $d_{11'}$, $d_{12'}$, $d_{13'}$, $d_{14'}$ ..., d_{1y} are the distances of strands 1′, 2′, 3′, 4' ...., y (all segments of conductor B) from segment 1 under consideration.

The denominator of argument of $\log_e$ is x th root of distances d_{12}, d_{13}, d_{14} ....d_{1x} and r' multiplied together where distances d_{12}, d_{13}, d_{14} .... d_1x are the distances of strands 2, 3, 4, ..., n (all strands of conductor A) from strand 1. r' can also be represented by distance d_{11} and the expression for inductance for conductor A becomes

$$L_1 = 2 \times 10^{-7} x \log_e\frac{\sqrt[y]{d_{11'}\ d_{12'}\ d_{13'}\ d_{14'}\ ....\ d_{1y}}}{\sqrt[x]{d_{11}\ d_{12}\ d_{13}\ d_{14}\ ....\ d_{1x}}}\ \text{H/m} \quad ...(4.11)$$

Similarly the expression for inductance for strand 2 can be written as

$$L_2 = 2 \times 10^{-7} x \log_e\frac{\sqrt[y]{d_{21'}\ d_{22'}\ d_{23'}\ d_{24'}\ ....\ d_{2y}}}{\sqrt[x]{d_{21}\ d_{22}\ d_{23}\ d_{24}\ ....\ d_{2x}}}\ \text{H/m} \quad ...(4.12)$$

Thus we see that the different strands of a conductor have different inductances.

∴ Average inductance of strands of conductor A,

$$L_{av} = \frac{L_1 + L_2 + L_3 + ...L_x}{x}$$

Since x such strands of conductor A are electrically in parallel,
Therefore inductance of conductor A,

$$L_A = \frac{L_{av}}{x} = \frac{L_1 + L_2 + L_3 + ...\ L_x}{x^2}$$

$$= 2\times 10^{-7}\log_e\left[\frac{\sqrt[xy]{(d_{11'}\ d_{12'}\ d_{13'}\ d_{14'}...d_{1y})(d_{21'}\ d_{22'}\ d_{23'}...d_{2y})...(d_{x1'}\ d_{x2'}\ d_{x3'}...d_{xy})}}{\sqrt[x^2]{(d_{11}\ d_{12}\ d_{13}\ d_{14}...d_{1x})(d_{21}\ d_{22}\ d_{23}...d_{2x})(d_{x1}\ d_{x2}\ d_{x3}...dxx)}}\right] \quad ...(4.13)$$

In the above expression the numerator of argument of $\log_e$ is called the GMD (often called the mutual GMD) between conductors A and B and the denominator of argument $\log_e$ is called GMR (often called self GMD). GMD (Geometric Mean Distance) and GMR (Geometric Mean Radius) are denoted by D_m and D_s respectively.

$$\therefore\ L_A = 2 \times 10^{-7} \log_e \frac{D_m}{D_{sA}} \text{ H/m}$$

$$\text{Similarly } L_B = 2 \times 10^{-7} \log_e \frac{D_m}{D_{sB}} \text{ H/m}$$

If conductors A and B are identical

$$\therefore\ D_{sA} = D_{sB} = D_s \text{ (say)}$$

$$\text{Loop inductance, } L = L_A + L_B = 2 \times 10^{-7} \log_e \frac{D_m}{D_s} + 2 \times 10^{-7} \log_e \frac{D_m}{D_s}$$

$$= 4 \times 10^{-7} \log_e \frac{D_m}{D_s} \text{ H/m} \quad \text{or } 0.4 \log_e \frac{D_m}{D_s} \text{ mH/km} \quad ...(4.14)$$

Example 4.1. Calculate the loop inductance per km of a single phase transmission line comprising of two parallel conductors one metre apart and 1.25 cm in dia. Also calculate the reactance of the transmission line. Frequency is 50 Hz.

Solution : Conductor of radius, $r = \frac{1.25}{2} = 0.625$ cm

Sapcing of conductors, $d = 1 \text{ m} = 100$ cm

Geometric mean radius (GMR) of conductor, $r' = 0.7788\, r = 0.7788 \times 0.625 = 0.48675$ cm

Loop inductance per km of a single phase line $L = 0.4 \log_e \frac{d}{r'}$ mH Refer to Eq. (4.6)

$$= 0.4 \log_e \frac{100}{0.48675} = 2.13 \text{ mH } \textbf{Ans.}$$

Reactance of transmission line, $X_L = 2\pi f L = 2\pi \times 50 \times 2.13 \times 10^{-3} = 0.669\ \Omega$/km **Ans.**

Example 4.2. Calculate inductance of a 500 m long, 1-ϕ, 2-wire transmission line. Each wire has identical area of cross-section of 10 cm^2 and conductors are separated by a distance of 5 m.

[*U.P. Technical Univ. Elements of Power System 2004-05*]

Solution : *Conductor radius,* $r = \sqrt{\frac{a}{\pi}} = \sqrt{\frac{10}{\pi}} = 1.784$ cm $\because a = \pi r^2$

Spacing of conductors, $d = 5$ m

Geometric mean radius (GMR) of conductor.

$$r' = 0.7788r = 0.7788 \times 1.784 = 1.39 \text{ cm}$$

Loop inductance of 500 m long single phase line,

$$L = 0.4 \log_e \frac{d}{r'} \times l = 0.4 \log_e \frac{500}{1.39} \times \frac{1}{2} = 1.177 \text{ mH } \textbf{Ans.}$$

Example 4.3. Calculate the loop inductance per km of a single phase line comprising of 2 parallel conductors 1 metre apart and 1 cm in diameter, when the material of conductor is (*i*) copper and (*ii*) steel of relative permeability 50. [*U.P. Technical Univ. Elements of Power System 2006-07*]

Solution : Conductor radius, $r = 0.5$ cm

Spacing of conductors, $d = 1$ metre $= 100$ cm

***(i)* Loop inductance per km of line with copper conductor**

$$= \left(1 + 4\log_e \frac{d}{r}\right) \times 10^{-7} \times 1{,}000 \text{ H} = \left(0.1 + 0.4\log_e \frac{100}{0.5}\right) \times 10^{-3} \text{ H} = 2.22 \text{ mH } \textbf{Ans.}$$

***(ii)* Loop inductance per km of line with steel conductor**

$$= \left(\mu + 4\log_e \frac{d}{r}\right) \times 10^{-7} \times 1{,}000 \text{ H} = \left(50 + 4\log_e \frac{100}{0.5}\right) \times 10^{-4} \text{ H} = 7.12 \text{ mH } \textbf{Ans.}$$

Example 4.4. Calculate the GMR of a 6/3 mm Al, 1/3 mm steel ACSR conductor.

Solution : Figure 4.6 shows an ACSR conductor with 6 outer conductors of aluminium each of radius r = 3 mm and one central conductor of steel also of radius r = 3 mm.

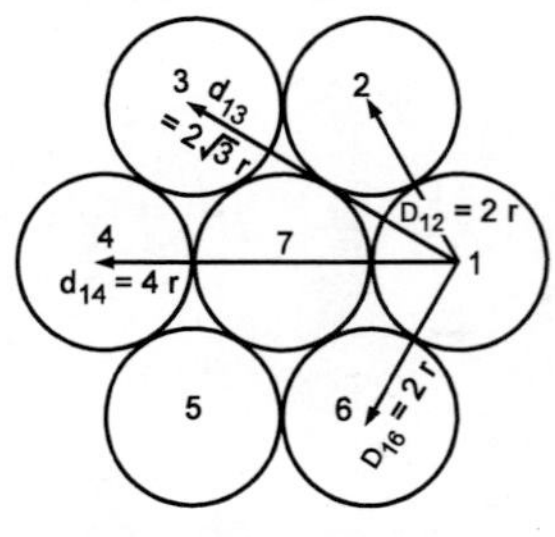

Fig. 4.6. ***Cross-section of a 7-Strand Conductor***

From Fig. 4.6

$d_{11} = r$; $d_{12} = d_{16} = d_{17} = 2\,r$; $d_{14} = 4\,r$ and

$$d_{13} = d_{15} = \sqrt{D_{14}^2 - D_{34}^2} = \sqrt{(4r)^2 - (2r)^2} = 2\sqrt{3}\ r$$

$$D_{s1} = D_{s2} = D_{s3} = D_{s4} = D_{s5} = D_{s6}$$

$$= \sqrt[7]{0.7788r \times 2r \times 2\sqrt{3}\,r \times 4r \times 2\sqrt{3}\,r \times 2r \times 2r} = \sqrt[7]{299}\ r$$

$$D_{s7} = \sqrt[7]{2r \times 0.7788r \times 2r \times 2r \times 2r \times 2r \times 2r} = \sqrt[7]{49.8432}\ r$$

Geometric mean radius,

$$D_s = \sqrt[7]{D_{s1} \cdot D_{s2} \cdot D_{s3} \cdot D_{s4} \cdot D_{s5} \cdot D_{s6} \cdot D_{s7}}$$

$$= r\ \sqrt[49]{299^6 \times 49.8432} = 2.176\ r \quad = 2.176 \times 3 \text{ mm} = 6.528 \text{ mm } \textbf{Ans.}$$

Note: The overall conductor radius is $3\,r$. So the ratio of GMR to overall conductor radius is $2.176\,r / 3\,r$ *i.e.* 0.7253. With the increase in number of strands, this ratio approaches 0.7788 which is that for a solid conductor. This ratio comes out to be 0.678 and 0.777 for 3 strands and 169 strands respectively.

Example 4.5. A split-phase, single phase transmission line is shown in Fig. 4.7. Conductors 1 and 2 in parallel form one path while conductors 1′ and 2′ in parallel form the return path. The current is equally shared by the two parallel conductors. Determine the total inductance per km of the line. The radius of each conductor is 1.2 cm.

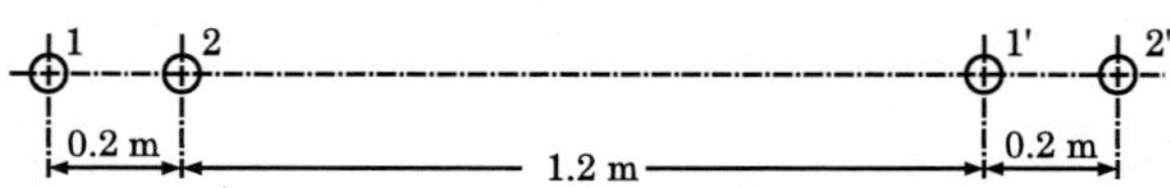

Fig. 4.7

Solution : Let the current flowing in each conductor be I amperes

Radius of each conductor, $r = 1.2$ cm

Geometric mean radius, $r' = 0.7788 \times 1.2 = 0.93456$ cm

Spacings of conductors, $d_{12} = 0.2$ m $= 20$ cm

$d_{11'} = (0.2 + 1.2)$ m $= 140$ cm; $d_{22'} = (0.2 + 1.2)$ m $= 140$ cm

$d_{12'} = (0.2 + 1.2 + 0.2)$ m $= 160$ cm; $d_{21'} = 1.2$ m $= 120$ cm

$$\text{Mutual GMD, } D_m = \sqrt[4]{d_{11'}\, d_{12'}\, d_{21'}\, d_{22'}} = \sqrt[4]{140 \times 160 \times 120 \times 140} = 139.28 \text{ cm}$$

$$\text{Self GMD, } D_s = \sqrt[4]{d_{11}\, d_{12}\, d_{21}\, d_{22}} = \sqrt[4]{0.93456 \times 20 \times 20 \times 0.93456} = 4.32 \text{ cm}$$

$$\text{Loop inductance of line, } L = 0.4 \log_e \frac{D_m}{D_s} \text{ mH/km} = 0.4 \log_e \frac{139.28}{4.32} = 1.389 \text{ mH/km } \textbf{Ans.}$$

Example 4.6. A single phase double circuit transmission line is shown in Fig. 4.8. Conductors 1 and 2 in parallel form one path while conductors 1′ and 2′ in parallel form the return path. The current is equally shared by the two parallel lines. Determine the total inductance per km of the line. The diameter of each conductor is 2.0 cm.

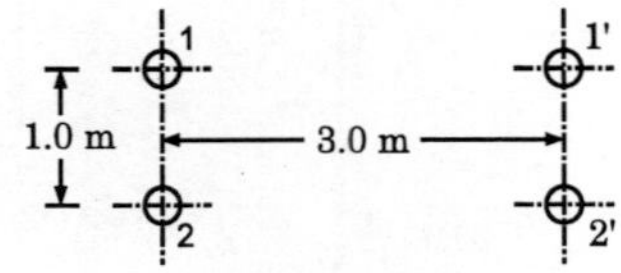

Fig. 4.8

Solution : Radius of each conductor, $r = \frac{2.0}{2} = 1.0$ cm

Geometric mean radius, $r' = 0.7788\ r = 0.7788$ cm

Spacing of conductors, $d_{11'} = 300$ cm

$$d_{12'} = \sqrt{(300)^2 + (100)^2} = 316.23 \text{ cm};$$

$$d_{21'} = d_{12'} = 316.23 \text{ cm};$$

$$d_{22'} = 300 \text{ cm}$$

$$d_{11} = d_{22} = r' = 0.7788 \text{ cm}; \ d_{12} = d_{21} = 100 \text{ cm}$$

$$\text{Mutual GMD, } D_m = \sqrt[4]{d_{11'}\, d_{12'}\, d_{21'}\, d_{22'}} \text{ cm} = \sqrt[4]{300 \times 316.23 \times 316.23 \times 300} = 308.0 \text{ cm}$$

$$\text{Self GMD, } D_s = \sqrt[4]{d_{11}\, d_{12}\, d_{21}\, d_{22}} \text{ cm} = \sqrt[4]{0.7788 \times 100 \times 100 \times 0.7788} = 8.825 \text{ cm}$$

$$\text{Loop inductance} = 0.4 \log_e \frac{D_m}{D_s} \text{ mH/km} = 0.4 \log_e \frac{308}{8.825} = 1.421 \text{ mH/km } \textbf{Ans.}$$

4.3. INDUCTANCE OF THREE-PHASE OVERHEAD LINES

With Unsymmetrical Spacing. Consider a 3-ϕ line with conductors A, B, and C; each of radius r metres. Let the spacing between them be d_1, d_2 and d_3 and the current flowing through them be I_A, I_B and I_C respectively.

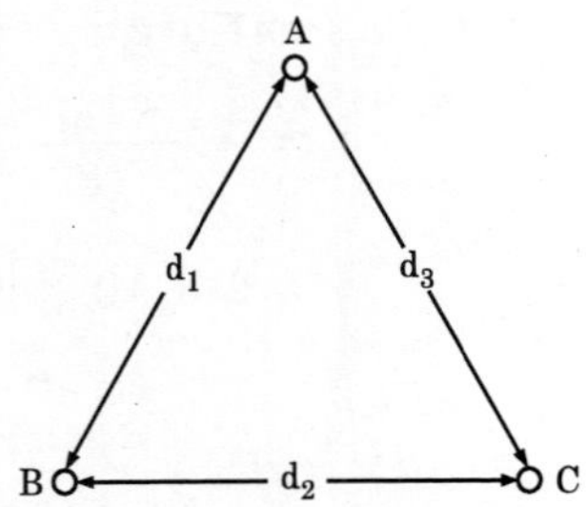

Fig. 4.9. ***3-ϕ Transmission Line With Unsymmetrical Spacing***

The flux linkages of conductor A due to its own current I_A and other conductor currents I_B and I_C

$$\psi_A = 2 \times 10^{-7}\left[I_A \log_e \frac{1}{r'} + I_B \log_e \frac{1}{d_1} + I_C \log_e \frac{1}{d_3}\right] \text{ Wb-turns/m}$$

Similarly

$$\psi_B = 2 \times 10^{-7}\left[I_B \log_e \frac{1}{r'} + I_A \log_e \frac{1}{d_1} + I_C \log_e \frac{1}{d_2}\right] \text{ Wb-turns/m}$$

and $$\psi_C = 2 \times 10^{-7}\left[I_C \log_e \frac{1}{r'} + I_A \log_e \frac{1}{d_3} + I_B \log_e \frac{1}{d_2}\right] \text{ Wb-turns/m}$$ ...Refer to Eq. (4.9)

If the system is balanced,

$$I_A = I_B = I_C = I \text{ (say) in magnitude}$$

Taking I_A as a reference phasor, the currents are represented, in symbolic form as

$$\mathbf{I_A} = I\ ;\ \mathbf{I_B} = I\,(-\,0.5 - j\,0.866) \text{ and } \mathbf{I_C} = I\,(-\,0.5 + j\,0.866)$$

Substituting these values of I_B and I_C in the expression for ψ_A we get

$$\psi_A = 2 \times 10^{-7}\left[I \log_e \frac{1}{r'} + I(-\,0.5 - j0.866)\log_e \frac{1}{d_1} + I(-\,0.5 + j0.866)\log_e \frac{1}{d_3}\right]$$

$$= 2 \times 10^{-7} I\left[\log_e \frac{1}{r'} + \log_e \sqrt{d_1 d_3} + j\sqrt{3}\log_e \sqrt{\frac{d_1}{d_3}}\right]$$

$$\text{and } L_A = \frac{\psi_A}{I_A} = 2 \times 10^{-7}\left[\log_e \frac{1}{r'} + \log_e \sqrt{d_1 d_3} + j\sqrt{3}\log_e \sqrt{\frac{d_1}{d_3}}\right] \text{ H/m} \quad ...(4.15)$$

$$\text{Similarly } L_B = 2 \times 10^{-7}\left[\log_e \frac{1}{r'} + \log_e \sqrt{d_1 d_2} + j\sqrt{3}\log_e \sqrt{\frac{d_2}{d_1}}\right] \text{ H/m} \quad ...(4.16)$$

$$\text{and } L_C = 2 \times 10^{-7}\left[\log_e \frac{1}{r'} + \log_e \sqrt{d_2 d_3} + j\sqrt{3}\log_e \sqrt{\frac{d_3}{d_2}}\right] \text{H/m} \quad ...(4.17)$$

Thus we see that when the conductors of a 3-phase transmission line are not equidistant from each other, *i.e. unsymmetrically spaced*, the flux linkages and inductances of various phases are different which cause unequal voltage drops in the three phases and transfer of power between phases (represented by imaginary terms of the expression for inductances) due to mutual inductances even if the currents in the conductors are balanced.

The unbalancing effect on account of irregular spacing of conductors is avoided by transposition of conductors, as shown in Fig. 4.10. In practice the conductors are so transposed that each of the three possible arrangements of conductors exist for one-third of the total length of the line.

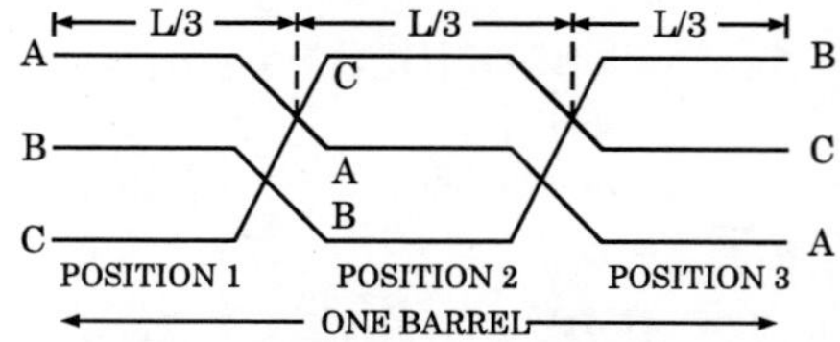

Fig. 4.10. ***Transposition of a 3-Phase Line***

(*i*) *The effect of transposition is that each conductor has the same average inductance, which is given as*

$$L = \frac{L_A + L_B + L_C}{3}$$

$$= 2 \times 10^{-7}\left[\log_e \frac{1}{r'} + \frac{1}{3}\left(\log_e \sqrt{d_1 d_3} + \log_e \sqrt{d_1 d_2} + \log_e \sqrt{d_2 d_3}\right)\right.$$

$$\left. + \frac{j\sqrt{3}}{3}\left(\log_e \sqrt{\frac{d_1}{d_3}} + \log_e \sqrt{\frac{d_2}{d_1}} + \log_e \sqrt{\frac{d_3}{d_2}}\right)\right] \text{ H/m}$$

$$= 2 \times 10^{-7}\left[\log_e \frac{1}{r'} + \log_e \sqrt[3]{d_1 d_2 d_3} + 0\right] = 2 \times 10^{-7} \log_e \frac{\sqrt[3]{d_1 d_2 d_3}}{r'} \text{ H/m} \quad ...(4.18)$$

(*ii*) *If the conductors are equispaced* (let the spacing be equal to *d*), as shown in Fig. 4.11, the inductance of each conductor will be same and can be obtained by substituting $d_1 = d_2 = d_3$ in above expression. So inductance of each conductor

$$L = 2 \times 10^{-7} \log_e \frac{\sqrt[3]{d \times d \times d}}{r'} \text{ H/m}$$

$$= 2 \times 10^{-7} \log_e \frac{d}{r'} \text{ H/m} \quad ...(4.19)$$

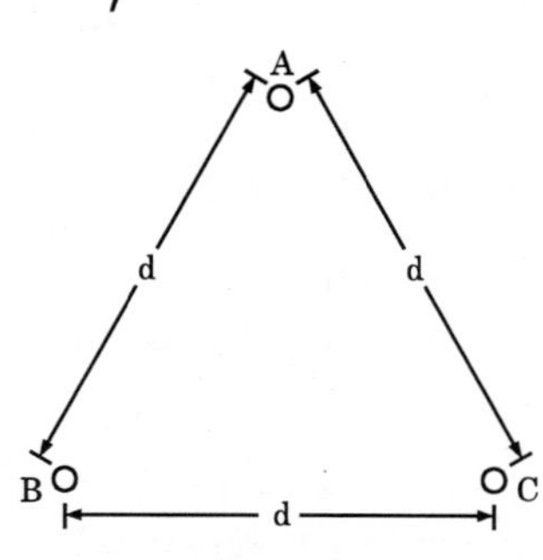

Fig. 4.11

For stranded conductor r' will be replaced by D_s (self GMD)

(*iii*) *When the conductors of three-phase transmission line are in the same plane, as shown in Fig. 4.12*

In this position $d_1 = d$, $d_2 = d$ and $d_3 = 2\,d$

Substituting $d_1 = d_2 = d$ and $d_3 = 2\,d$

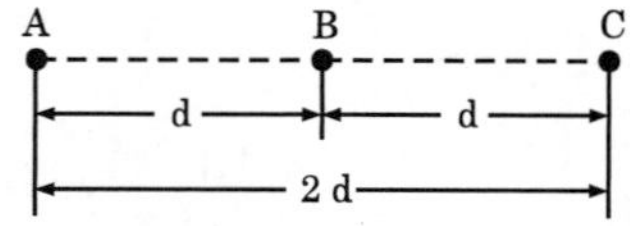

Fig. 4.12

in general Eqs. (4.15), (4.16), and (4.17) for L_A, L_B and L_C respectively we get

$$L_A = 2 \times 10^{-7} \left[\log_e \frac{1}{r'} + \log_e \sqrt{2d \times d} + j\sqrt{3} \log_e \sqrt{\frac{d}{2d}} \right]$$

$$= 2 \times 10^{-7} \left[\log_e \frac{d}{r'} + \frac{1}{2} \log_e 2 - j\, 0.866 \log_e 2 \right] \text{ H/m} \quad ...(4.20)$$

$$L_B = 2 \times 10^{-7} \left[\log_e \frac{1}{r'} + \log_e \sqrt{d \times d} + j\sqrt{3} \log_e \sqrt{\frac{d}{d}} \right] \text{ H/m} = 2 \times 10^{-7} \log_e \frac{d}{r'} \text{ H/m} \quad ...(4.21)$$

and
$$L_C = 2 \times 10^{-7} \left[\log_e \frac{1}{r'} + \log_e \sqrt{2d \times d} + j\sqrt{3} \log_e \sqrt{\frac{2d}{d}} \right] \text{ H/m}$$

$$= 2 \times 10^{-7} \left[\log_e \frac{d}{r'} + \frac{1}{2} \log_e 2 + j\, 0.866 \log_e 2 \right] \text{ H/m} \quad ...(4.22)$$

(*iv*) *When the conductors are at the corners of a right-angled triangled, as shown in Fig.* 4.13. In this position

$$d_1 = d_2 = d \text{ and } d_3 = \sqrt{2}\, d$$

Substituting $d_1 = d_2 = d$ and $d_3 = \sqrt{2}\, d$ in general Eq. (4.15), (4.16) and (4.17) for L_A, L_B and L_C respectively we get

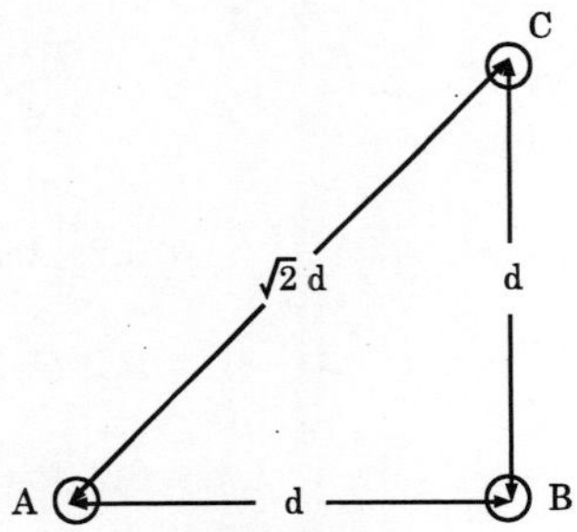

Fig. 4.13

$$L_A = 2 \times 10^{-7} \left[\log_e \frac{1}{r'} + \log_e \sqrt{\sqrt{2}\, d \times d} + j\sqrt{3} \log_e \sqrt{\frac{d}{\sqrt{2}d}} \right] \text{ H/m}$$

$$= 2 \times 10^{-7} \left[\log_e \frac{d}{r'} + \frac{1}{2} \log_e \sqrt{2} - j\, 0.866 \log_e \sqrt{2} \right] \text{ H/m} \quad ...(4.23)$$

$$L_B = 2 \times 10^{-7} \left[\log_e \frac{1}{r'} + \log_e \sqrt{d \times d} + \sqrt{3} \log_e \sqrt{\frac{d}{d}} \right] \text{ H/m} = 2 \times 10^{-7} \log_e \frac{d}{r'} \text{ H/m} \quad ...(4.24)$$

and
$$L_C = 2 \times 10^{-7} \left[\log_e \frac{1}{r'} + \log_e \sqrt{\sqrt{2}d \times d} + j\sqrt{3} \log \sqrt{\frac{\sqrt{2}d}{d}} \right] \text{ H/m}$$

$$= 2 \times 10^{-7} \left[\log_e \frac{d}{r'} + \frac{1}{2} \log_e \sqrt{2} + j\, 0.866 \log_e \sqrt{2} \right] \text{ H/m} \quad ...(4.25)$$

Example 4.7. Calculate the inductance per km of a 3-phase overhead transmission line using 1.24 cm diameter conductors when these are placed at the corners of an equilateral triangle of each side 2 metres. [*U.P. Technical Univ. Elements of Power System 2003-04*]

Solution : Conductor radius, $r = \dfrac{1.24}{2} = 0.62$ cm

Geometric mean radius $r' = 0.7788r = 0.7788 \times 0.62 = 0.4829$ cm

Spacing of conductors, $d = 2$ m $= 200$ cm

Inductance per phase per km $= 0.2 \log_e \dfrac{d}{r'}$ mH $= 0.2 \log_e \dfrac{200}{0.4829}$ mH $= 1.205$ mH **Ans.**

Example 4.8. Determine the inductance per km of a 3-ϕ transmission line using 20 mm diameter conductors when conductors are situated at the corners of a triangle with spacing of 4, 5 and 6 metres. Conductors are regularly transposed.

Solution : Conductor radius, $r = \dfrac{20}{2} = 10$ mm $= 1.0$ cm

Geometric mean radius, $r' = 0.7788 \times 1.0 = 0.7788$ cm

$$\text{Inductance per phase, L} = 0.2 \log_e \frac{(d_1 d_2 d_3)^{1/3}}{r'} \text{ mH/km}$$

$$= 0.2 \log_e \frac{(400 \times 500 \times 600)^{1/3}}{0.7788} \text{ mH} = 1.29 \text{ mH } \textbf{Ans.}$$

Example 4.9. The three conductors of a 3-phase transmission line are arranged in a horizontal plane and are 3 metres apart. The diameter of each conductor is 4 cm. Determine the inductance per km of each phase. Assume balanced load and R, Y, B phase sequence.

Solution : Radius of each conductor, $r = \frac{4}{2} = 2$ cm

GMR, $r' = 0.7788 \times 2 = 1.5576$ cm

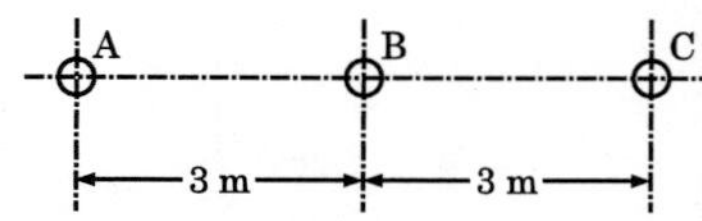

Fig. 4.14

From Eqs. (4.20), (4.21) and (4.22), respectively

$$L_A = 0.2 \left[\log_e \frac{d}{r'} + \frac{1}{2}\log_e 2 - j\,0.866 \log_e 2\right] \text{ mH/km}$$

$$= 0.2 \left[\log_e \frac{300}{1.5576} + \frac{1}{2}\log_e 2 - j\,0.866 \log_e 2\right] \text{ mH/km}$$

$$= 0.2\,[5.26 + 0.346 - j\,0.866 \times 0.6931] = (1.1212 - j\,0.12) \text{ mH } \textbf{Ans.}$$

$$L_B = 0.2 \log_e \frac{d}{r'} \text{ mH/km} = 0.2 \log_e \frac{300}{1.5576} = 1.052 \text{ mH } \textbf{Ans.}$$

$$L_C = 0.2 \left[\log_e \frac{d}{r'} + \frac{1}{2}\log_e 2 + j\,0.866 \log_e 2\right] \text{ mH}$$

$$= 0.2 \left[\log_e \frac{300}{1.5576} + \frac{1}{2}\log_e 2 + j\,0.866 \log_e 2\right] \text{mH} = (1.1212 + j\,0.12) \text{ mH } \textbf{Ans.}$$

Example 4.10. A three-phase overhead line is designed with an equilateral spacing of 3.5 m with a conductor diameter of 1.2 cm. If the line is constructed with horizontal spacing with suitably transposed conductors, find spacing between adjacent conductors which would give the same value of inductance as in the equilateral arrangement.

Solution : Conductor radius, $r = \frac{1.2}{2} = 0.6$ cm or 6 mm

Geometric mean radius, $r' = 0.7788 \times 6 = 4.6728$ mm

In equilateral arrangement of line conductors [Fig. 4.15 (a)]

Spacing of conductors, $d = 3.5$ m $= 3{,}500$ mm

$$\text{Inductance per phase, L} = 2 \times 10^{-7} \log_e \frac{d}{r'} \text{ H/m}$$

...Refer to Eq. (4.19)

$$= 2 \times 10^{-7} \log_e \frac{3{,}500}{4.6728}$$

$$= 1.324 \times 10^{-6} \text{ H/m}$$

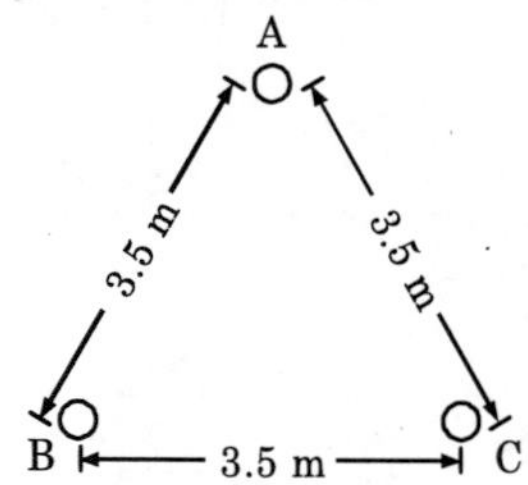

(a) Equilateral Arrangement of Line Conductors

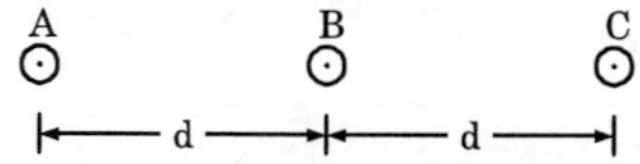

(b) Horizontally Spaced Line Conductors

Fig. 4.15

In the arrangement of conductors horizontally spaced [Fig. 4.15 (*b*)]

Let the spacing between adjacent conductors be *d* mm

The effect of transposition of line is that each conductor has the same average inductance which in given as

$$L_{av} = \frac{L_A + L_B + L_C}{3}$$

Substituting the values of L_A, L_B and L_C from Eqs. (4.20), (4.21) and (4.22) respectively

$$L_{av} = \frac{1}{3} \times 2 \times 10^{-7}\left[\log_e \frac{d}{r'} + \frac{1}{2}\log_e 2 - j0.866 \log_e 2 + \log_e \frac{d}{r'} + \log_e \frac{d}{r'} + \frac{1}{2}\log_e 2 + j0.866 \log_e 2\right] \text{H/m}$$

$$= 2 \times 10^{-7}\left[\log_e \frac{d}{r'} + \frac{1}{3}\log_e 2\right] \text{ H/m}$$

Subtituting $L_{av} = 1.324 \times 10^{-6}$ and $r' = 4.6728$ mm in above equation we have

$$1.324 \times 10^{-6} = 2 \times 10^{-7}\left[\log_e \frac{d}{4.6728} + \frac{1}{3}\log_e 2\right]$$

$$\text{or } d = 4.6728\, e^{6.389} = 2{,}781.5 \text{ mm or } 2.781 \text{ m} \textbf{ Ans.}$$

4.4. INDUCTANCE OF THREE PHASE LINES WITH MORE THAN ONE CIRCUIT

It is usual practice to run 3-phase transmission lines with more than one circuit in parallel on the same towers, because it gives greater reliability and a higher transmission capacity. If such circuits are so widely separated that the mutual inductance between them becomes negligible, the inductance of the equivalent single circuit would be half of each of the individual circuits considered alone. But in actual practice the separation is not very wide and the mutual inductance is not negligible. GMD method is used for determination of inductance per phase by considering the various conductors connected in parallel as strands of one composite conductor.

It is desirable to have a configuration that provides minimum inductance so as to have maximum transmission capacity. This is possible only with low GMD and high GMR. Therefore, the practice is to have the individual conductors of a phase widely separated to provide high GMR and the distance between the phases small to give low GMD. Thus in the case of a double circuit in vertical formation the arrangement of conductors would be as illustrated in Fig. 4.16 (*a*) and not as illustrated in Fig. 4.16 (*b*) because the arrangement of conductors given in Fig. 4.16 (*a*) results in low inductance in comparison to that given by the arrangement illustrated in Fig. 4.16 (*b*).

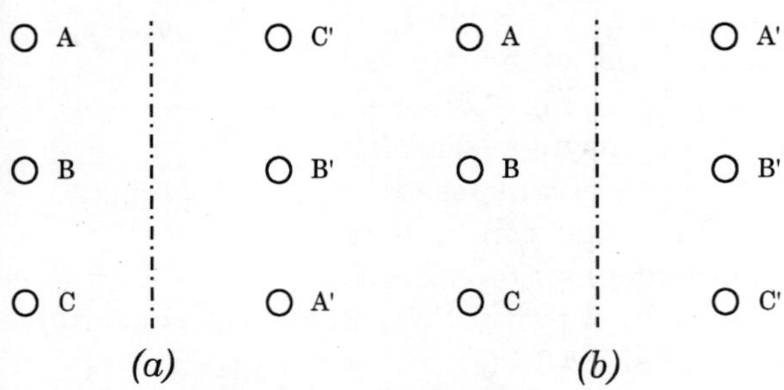

Fig. 4.16. ***Arrangements of Conductors in a Double Circuit 3-Phase Line***

4.4.1. Inductance of a 3-Phase Double Circuit Line With Symmetrical Spacing

Consider a 3-phase double circuit connected in parallel-conductors A, B, C forming one circuit and conductors A', B', C' forming the other one, as illustrated in Fig. 4.17.

Flux-linkages of phase A conductors

$$\psi_A = 2 \times 10^{-7} \left[I_A \left(\log_e \frac{1}{r'} + \log_e \frac{1}{2d} \right) + I_B \left(\log_e \frac{1}{d} + \log_e \frac{1}{\sqrt{3}\, d} \right) + I_C \left(\log_e \frac{1}{\sqrt{3}\, d} + \log_e \frac{1}{d} \right) \right]$$

$$= 2 \times 10^{-7} \left[I_A \log_e \frac{1}{2dr'} + (I_B + I_C) \log_e \frac{1}{\sqrt{3}\, d^2} \right]$$

$$= 2 \times 10^{-7} \left[I_A \log_e \frac{1}{2dr'} - I_A \log_e \frac{1}{\sqrt{3}\, d^2} \right] \qquad \because I_A + I_B + I_C = 0$$

$$= 2 \times 10^{-7} I_A \log_e \frac{\sqrt{3}\, d}{2r'} \text{ Wb-turns/m}$$

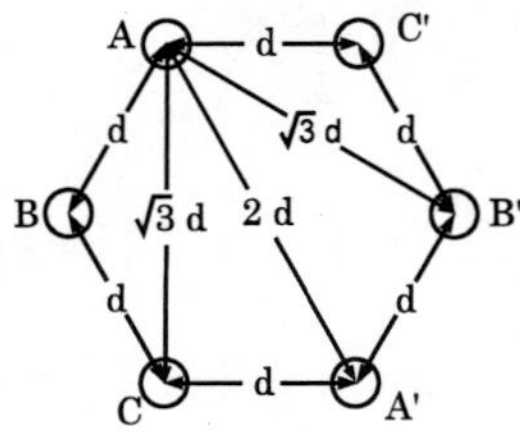

Fig. 4.17

$$\text{Inductance of conductor A, } L_A = \frac{\psi_A}{I_A} = 2 \times 10^{-7} \log_e \frac{\sqrt{3}\, d}{2r'} \text{ H/m} \qquad \text{...(4.26)}$$

Similarly inductance of remaining conductors can be worked out, which will be the same as L_A. This is due to the fact that the conductors of different phases are symmetrically placed.

Since conductors are electrically in parallel, inductance of each phase

$$= \frac{1}{2} L_A = 1 \times 10^{-7} \log_e \frac{\sqrt{3}\, d}{2r'} \text{ H/m} \qquad \text{...(4.27)}$$

4.4.2. Inductance of a 3-Phase Double Circuit With Unsymmetrical Spacing But Transposed

Now consider a 3-phase double circuit connected in parallel—conductors A, B and C forming one circuit and conductors A′, B′ and C′ forming the other one, as illustrated in Fig. 4.18. (conductors unsymmetrically spaced and transposed).

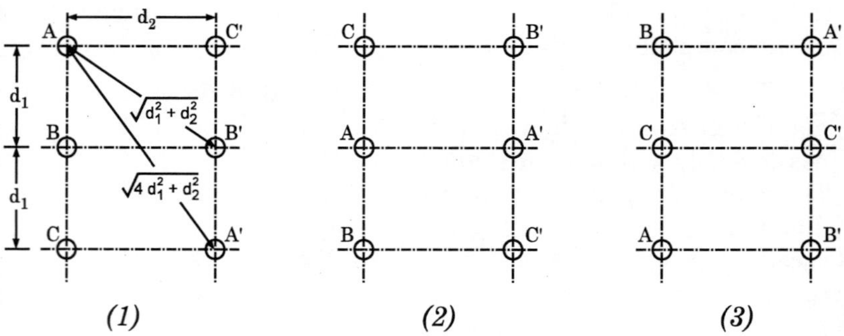

Fig. 4.18

Since the conductors are thoroughly transposed, the conductor situations in the transposition cycle would be, as illustrated in Figs. 4.18 (1), 4.18 (2) and 4.18 (3).

Flux linkages with conductor A in position (1)

$$\psi_{A1} = 2 \times 10^{-7} \left[I_A \left(\log_e \frac{1}{r'} + \log_e \frac{1}{\sqrt{4d_1^2 + d_2^2}} \right) + I_B \left(\log_e \frac{1}{d_1} + \log_e \frac{1}{\sqrt{d_1^2 + d_2^2}} \right) + I_C \left(\log_e \frac{1}{2d_1} + \log_e \frac{1}{d_2} \right) \right]$$

Similarly flux linkages with conductor A in position (2) and (3)

$$\psi_{A2} = 2 \times 10^{-7} \left[I_A \left(\log_e \frac{1}{r'} + \log_e \frac{1}{d_2} \right) + I_B \left(\log_e \frac{1}{d_1} + \log_e \frac{1}{\sqrt{d_1^2 + d_2^2}} \right) + I_C \left(\log_e \frac{1}{d_1} + \log_e \frac{1}{\sqrt{d_1^2 + d_2^2}} \right) \right]$$

and
$$\psi_{A3} = 2 \times 10^{-7} \left[I_A \left(\log_e \frac{1}{r'} + \log_e \frac{1}{\sqrt{4\, d_1^2 + d_2^2}} \right) + I_B \left(\log_e \frac{1}{2d_1} + \log_e \frac{1}{d_2} \right) + I_C \left(\log_e \frac{1}{d_1} + \log_e \frac{1}{\sqrt{d_1^2 + d_2^2}} \right) \right]$$

Average flux linkages with conductor A

$$\psi_A = \frac{\Psi_{A1} + \Psi_{A2} + \Psi_{A3}}{3}$$

$$= \frac{2 \times 10^{-7}}{3} \left[I_A \left(3 \log_e \frac{1}{r'} + \log_e \frac{1}{d_2} + \log_e \frac{1}{4\, d_1^2 + d_2^2} \right) + I_B \left(2 \log_e \frac{1}{d_1} + \log_e \frac{1}{2\, d_1} + \log_e \frac{1}{d_1^2 + d_2^2} + \log_e \frac{1}{d_2} \right) + I_C \left(\log_e \frac{1}{2\, d_1} + \log_e \frac{1}{d_1^2 + d_2^2} + \log_e \frac{1}{d_2} + 2 \log_e \frac{1}{d_1} \right) \right]$$

$$= \frac{2 \times 10^{-7}}{3} \left[I_A \left(3 \log_e \frac{1}{r'} + \log_e \frac{1}{d_2} + \log_e \frac{1}{4\, d_1^2 + d_2^2} \right) + (I_B + I_C) \left(\log_e \frac{1}{2\, d_1} + \log_e \frac{1}{d_1^2 + d_2^2} + \log_e \frac{1}{d_2} + 2 \log_e \frac{1}{d_1} \right) \right]$$

$$= \frac{2 \times 10^{-7}}{3} \left[I_A \left(3 \log_e \frac{1}{r'} + \log_e \frac{1}{d_2} + \log_e \frac{1}{4\, d_1^2 + d_2^2} \right) - I_A \left(\log_e \frac{1}{2\, d_1} + \log_e \frac{1}{d_1^2 + d_2^2} + \log_e \frac{1}{d_2} + 2 \log_e \frac{1}{d_1} \right) \right]$$

$$\because I_A + I_B + I_C = 0$$

or
$$\psi_A = \frac{2 \times 10^{-7}}{3} I_A \log_e \frac{2\, d_1^3 \left(d_1^2 + d_2^2 \right) d_2}{(r')^3\, d_2 \left(4\, d_1^2 + d_2^2 \right)}$$

or
$$\psi_A = 2 \times 10^{-7}\, I_A \log_e \frac{2^{1/3}\, d_1 \left(d_1^2 + d_2^2 \right)^{1/3}}{r' \left(4\, d_1^2 + d_2^2 \right)^{1/3}}$$ wb-turns per metre

and inductance, $L_A = \frac{\Psi_A}{I_A} = 2 \times 10^{-7} \log_e \frac{2^{1/3} d_1 \left(d_1^2 + d_2^2\right)^{1/3}}{r' \left(4 d_1^2 + d_2^2\right)^{1/3}}$ H/m

$$= 2 \times 10^{-7} \log_e 2^{1/3} \frac{d_1}{r'} \left(\frac{d_1^2 + d_2^2}{4 d_1^2 + d_2^2}\right)^{1/3} \text{ H/m} \quad ...(4.28)$$

Inductance of each phase, $L = \frac{1}{2} \times$ inductance per conductor $= \frac{1}{2} L_A$

$$= 2 \times 10^{-7} \log_e 2^{1/6} \left(\frac{d_1}{r'}\right)^{1/2} \left(\frac{d_1^2 + d_2^2}{4 d_1^2 + d_2^2}\right)^{1/6} \text{ H/m} \quad ...(4.29)$$

If the distance d_2 is too large as compared to d_1, $\frac{d_1^2 + d_2^2}{4 d_1^2 + d_2^2}$ would tend to be unity and inductance per phase,

$$L = 2 \times 10^{-7} \log_e 2^{1/6} \left(\frac{d_1}{r'}\right)^{1/2} \text{ H/m or } L = 0.2 \log_e 2^{1/6} \left(\frac{d_1}{r'}\right)^{1/2} \text{ mH/km} \quad ...(4.30)$$

Example 4.11. Determine the inductance per phase per km of a double circuit 3-phase line. The radius of each conductor is 20 mm and the conductors are placed on the circumference of an imaginary circle of radius 7 m forming a regular hexagonal figure.

Solution : Refer to Fig. 4.17.

$d = 7$ m or 7,000 mm

GMR, $r' = 0.7788\, r = 0.7788 \times 20 = 15.576$ mm

From Eq. (4.27), inductance per phase

$$L = 1 \times 10^{-7} \log_e \frac{\sqrt{3}}{2} \frac{d}{r'} \text{ H/m} = 0.1 \log_e \frac{\sqrt{3}}{2} \frac{d}{r'} \text{ mH/km} = 0.1 \log_e 0.866 \times \frac{7{,}000}{15.576} = 0.5964 \text{ mH} \quad \textbf{Ans.}$$

Example 4.12. A part of the transposition cycle of a 3-phase double circuit line is shown in Fig. 4.19. Radius of each conductor is 0.9 cm. The conductors are solid copper. Find the inductance per phase per km of line.

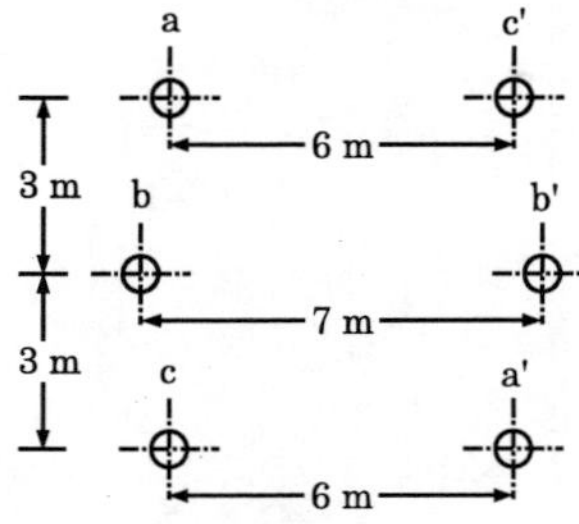

Fig. 4.19

Solution : GMR of each conductor

$$r' = 0.7788 \times 0.9 = 0.7 \text{ cm} = 0.7 \times 10^{-2}\ m$$

Distance between conductors a and a', $D_{aa}' = \sqrt{6^2 + 6^2} = 8.485\ m$

$$d_{bb}' = 7\ m \text{ and } d_{cc}' = d_{aa}' = 8.485\ m$$

GMR of conductors of phase a,

$$D_{sa} = \sqrt[4]{d_{aa}\, d_{aa'}\, d_{a'a'}\, d_{a'a}}$$

$$= \sqrt{0.7 \times 10^{-2} \times 8.485 \times 0.7 \times 10^{-2} \times 8.485} = 0.2437\ m$$

Similarly $D_{sb} = \sqrt{0.7 \times 10^{-2} \times 7} = 0.2214\ m$

and $D_{sc} = \sqrt{0.7 \times 10^{-2} \times 8.485} = 0.2437\ m$

Self GMD, $D_s = \sqrt[3]{D_{sa}\, D_{sb}\, D_{sc}} = \sqrt[3]{0.2437 \times 0.2214 \times 0.2437} = 0.236\ m$

Distance between conductors a and b, $d_{ab} = \sqrt{3^2 + 0.5^2}\ \ 3.041\ m$

Distance between conductors a and b', $d_{ab}' = \sqrt{3^2 + 6.5^2} = 7.159\ m$

$$D_{ab} = \sqrt[4]{d_{ab}\, d_{ab'}\, d_{a'b}\, d_{a'b'}} = \sqrt[4]{3.041 \times 7.159 \times 7.159 \times 3.041} = 4.666\ m$$

$$\text{Similarly } D_{bc} = \sqrt[4]{(3.041 \times 7.159)^2} = 4.666\ m \text{ and } D_{ca} = \sqrt[4]{(6 \times 6)^2} = 6 \text{ m}$$

$$D_{eq} \text{ or } D_m = \sqrt[3]{D_{ab}\, D_{bc}\, D_{ca}} = \sqrt[3]{4.666 \times 4.666 \times 6} = 5.074\ m$$

$$\text{Inductance per phase, } L = 0.2 \log_e \frac{D_m}{D_s} \text{ mH/km} = 0.2 \times \log_e \frac{5.074}{0.236} = 0.6136 \text{ mH/km } \textbf{Ans.}$$

Example 4.13. A 100 km double circuit transmission line with 7 strand copper conductors each strand having 5 mm diameter, has 6 conductors arranged in the form shown in Fig. 4.20. The line is transposed at regular intervals. Evaluate the GMD and GMR and calculate the overall inductance per phase of the line.

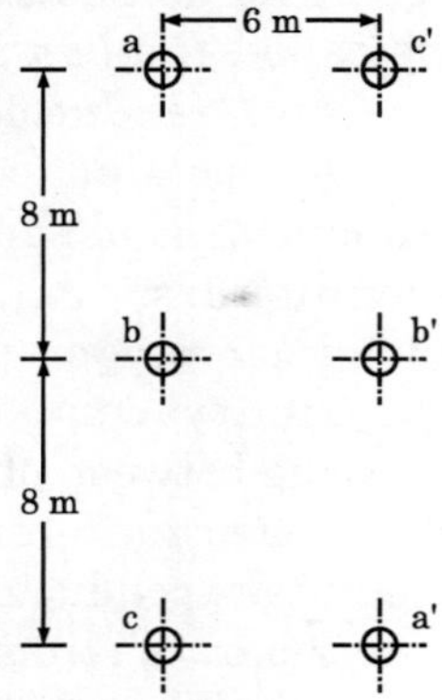

Fig. 4.20

Solution : Radius of each strand of conductor $= \frac{5}{2}$ mm $= 2.5\ mm$

GMR of conductor, $r' = 2.176 \times 2.5 \text{ mm} = 5.44 \times 10^{-3}\ m$

[Refer to Example 4.4]

Distance between conductors

$$d_{aa'} = \sqrt{6^2 + 16^2} = 17.089\ m;\ d_{bb'} = 6\ m$$

$$d_{cc'} = d_{aa'} = 17.089\ m;\ d_{ab} = 8\ m$$

$$d_{ab'} = \sqrt{6^2 + 8^2} = 10\ m$$

$$\text{GMR of phase a, } D_{sa} = \sqrt[4]{d_{aa}\, d_{aa'}\, d_{a'a'}\, d_{a'a}} = \sqrt{r' \times d_{aa'}}$$

$$= \sqrt{5.44 \times 10^{-3} \times 17.089} = 0.305\ m$$

$$\text{Similarly } D_{sb} = \sqrt{r' \times d_{bb'}} = \sqrt{5.44 \times 10^{-3} \times 6} = 0.1807\ m$$

$$\text{and } D_{sc} = \sqrt{r' \times d_{cc'}} = \sqrt{5.44 \times 10^{-3} \times 17.089} = 0.305\ m$$

$$\text{Self GMD or GMR, } D_{s'} = \sqrt[3]{D_{sa}\, D_{sb}\, D_{sc}} = \sqrt[3]{0.305 \times 0.1807 \times 0.305} = 0.2562\ m \textbf{ Ans.}$$

$$D_{ab} = \sqrt[4]{d_{ab}\, d_{ab'}\, d_{a'b}\, d_{a'b'}} = \sqrt{d_{ab}\, d_{ab'}} = \sqrt{8 \times 10} = 8.944\ m$$

$$\text{Similarly } D_{bc} = \sqrt{d_{bc}\, d_{bc'}} = \sqrt{8 \times 10} = 8.944\ m$$

$$\text{and } D_{ca} = \sqrt{d_{ca}\, d_{ca'}} = \sqrt{16 \times 6} = 9.8\ m$$

$$D_{eq} \text{ or } D_m = \sqrt[3]{D_{ab}\, D_{bc}\, D_{ca}} = \sqrt[3]{8.944 \times 8.944 \times 9.8} = 9.22\ m \textbf{ Ans.}$$

$$\text{Inductance per phase, } L = 0.2 \log_e \frac{D_m}{D_s} \text{ mH/km} = 0.2 \log_e \frac{9.22}{0.2562} = 0.717 \text{ mH/km}$$

$$\text{Inductance of 100 km line} = \frac{0.717 \times 100}{1{,}000} = 0.0717 \text{ H } \textbf{Ans.}$$

Inductance per phase per km can also be determined from Eq. (4.29) *i.e.*

$$L = 0.2 \log_e 2^{1/6} \left(\frac{d_1}{r'}\right)^{1/2} \left(\frac{d_1^2 + d_2^2}{4\, d_1^2 + d_2^2}\right)^{1/6} \text{ mH or } L = 0.1 \log_e 2^{1/3} \left(\frac{d_1}{r'}\right) \left(\frac{d_1^2 + d_2^2}{4\, d_1^2 + d_2^2}\right)^{1/3} \text{ mH}$$

where $d_1 = 8\ m$, $r' = 5.44 \times 10^{-3}\ m$; $d_1^2 + d_2^2 = 8^2 + 6^2 = 100\ m^2$ and $4\ d_1^2 + d_2^2 = 4 \times 8^2 + 6^2 = 292\ m^2$

$$\text{So } L = 0.1 \log_e 2^{1/3} \left(\frac{8}{5.44 \times 10^{-3}}\right)\left(\frac{100}{292}\right)^{1/3} = 0.1 \log_e 1.26 \times 1{,}470.6 \times 0.6996 = 0.717 \text{ mH}$$

Inductance of 100 km line $= 0.717 \times 10^{-3} \times 100 = 0.0717$ H **Ans.**

Same as before and as expected.

4.5. BUNDLED CONDUCTORS

The demand of electric power is increasing throughout the world and in many countries it is doubling every five to eight years. The power stations (hydro, thermal or nuclear) are usually located far away from the load centres. Thus transmission of large amounts of power over long distances, which can be accomplished most economically only by using extra high voltages (or simply EHV), is necessary. Voltages in excess of 230 kV fall in this category. An increase in transmission voltage results in reduction of electrical losses, increase in transmission efficiency, improvement of voltage regulation and reduction in conductor material requirement. At voltages above 300 kV, corona causes a significant power loss and interference with communication circuits, if a round single conductor per phase is used. Instead of going for a hollow conductor it is preferable to use more than one conductor per phase which is called the *bundling* of conductors. Lines of 400 kV and higher voltages invariably use *bundled conductors.*

A bundled conductor is a conductor made up of two or more conductors, called the *sub-conductors*, per phase in close proximity compared with the spacing between phases (Fig. 4.21). The basic difference between a composite conductor and a bundled conductor is that the sub-conductors of a bundled conductor are separated from each other by a constant distance varying from 0.2 m to 0.6 m depending upon designed voltage and surrounding conditions throughout the length of the line with the help of spacers whereas the wires of a composite conductor touch each other. The bundled conductors have filter material or air space inside so that the overall diameter is increased.

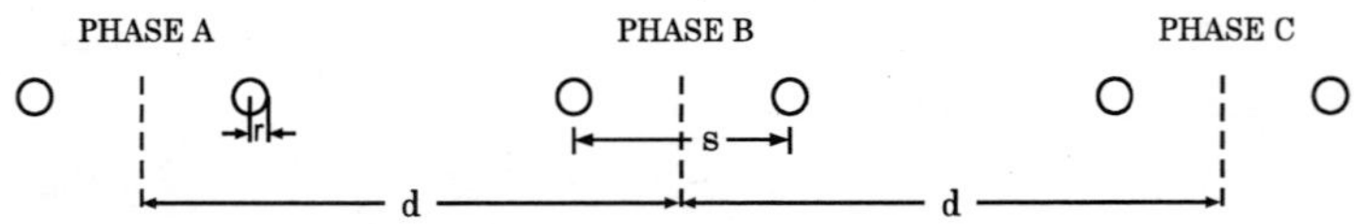

Fig. 4.21

The use of bundled conductors per phase reduces the voltage gradient in the vicinity of the line and thus, reduces the possibilities of the corona discharge.

Although the bundled conductors are used on EHV transmission lines primarily to reduce corona loss and radio interference, but they have several other advantages over single conductors such as given below :

1. The bundle conductor lines transmit bulk power with reduced losses, thereby giving increased transmission efficiency.
2. Since the bundle conductor lines have a higher capacitance to neutral in comparison with single conductor lines, they have higher charging current, which helps in improving the power factor.
3. Since by bundling, the self GMD or GMR is increased, the inductance per phase, in comparison with single conductor lines, is reduced. As a result reactance per phase is reduced.
4. Since surge impedance of a line is given by $Z_0 = \sqrt{\frac{L}{C}}$ and the bundle conductor lines have higher capacitance and lower inductance in comparison with single conductor lines, therefore, bundle conductor lines have comparatively lower surge impedance with a corresponding increase in the maximum power transfer capability. The following table represents relative power transfer with number of sub-conductors forming a bundle conductor.

Number of sub-conductors	1	2	4	8
Relative power transfer	1.0	1.3	1.6	1.7

It is to be noted that there is a little to be gained by using more than four sub-conductors per phase, two or three sub-conductors per phase are sufficient for most of the EHV lines.

The geometric mean radius of a bundled conductor for a two-conductor (duplex), a three-conductor (triplex) and four conductor (quadruplex) arrangements shown in Figs. 4.22 (*a*), 4.22 (*b*) and 4.22 (*c*) respectively can be determined from the following relations

$$D_s = \sqrt{r' s} \text{ for duplex arrangement} \quad ...(4.31)$$

$$D_s = \sqrt[3]{r' s^2} \text{ for triplex arrangement} \quad ...(4.32)$$

$$D_s = 1.09 \sqrt[4]{r' s^3} \text{ for quadruplex arrangement} \quad ...(4.33)$$

where r' is the geometric mean radius of each sub-conductor of bundle and s is the spacing between sub-conductors of a bundle.

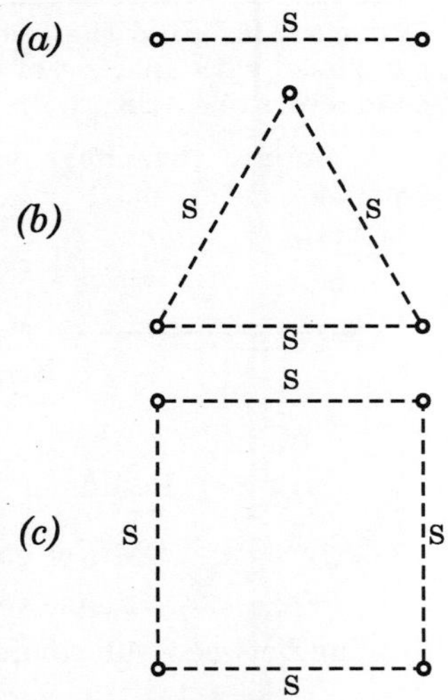

Fig. 4.22

GMD of a bundled conductor line can be determined by taking the root of the product of distances from each conductor of a bundle to every other conductor of the other bundles. However, if the distances from centre of one bundle to the centres of other bundles is taken as the distances d_{ab}, d_{bc}, d_{ca}, accuracy will not be much affected.

Example 4.14. Determine the inductance per km per phase of a single circuit 460 kV line using two bundle conductors per phase as shown in the figure. The diameter of each conductor is 5.0 cm.

[*U.P. Technical Univ. Elements of Power System 2007-08*]

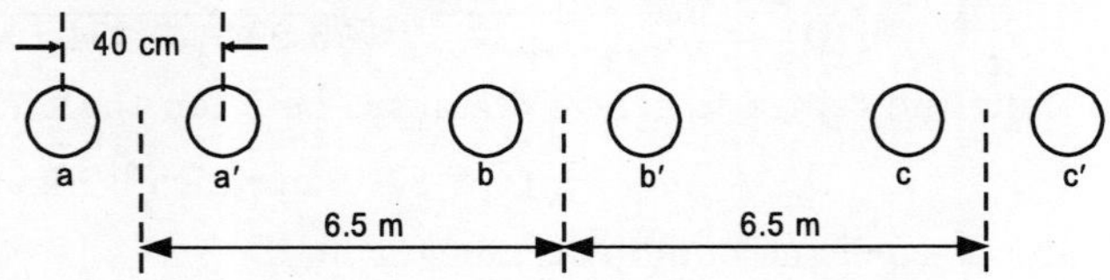

Fig. 4.23.

Solution : Radius of each sub-conductor, $r = \frac{5.0}{2} = 2.5$ cm

GMR of each sub-conductor, $r' = 0.7788r = 0.7788 \times 2.5 = 1.947 \text{ cm} = 1.947 \times 10^{-2}$ m

Phase-to-phase separation, $d = 6.5$ m

Spacing between sub-conductors of one phase, $s = 0.4$ m

$$\text{GMR, } D_s = \sqrt{r's} = \sqrt{1.947 \times 10^{-2} \times 0.4} = 0.088 \text{ m}$$

$$D_{ab} = D_{bc} = \sqrt[4]{d_{ab} d_{ab'} d_{a'b} d_{a'b'}} = \sqrt[4]{6.5 \times 6.9 \times 6.1 \times 6.5} = 6.494 \text{ m}$$

$$D_{ca} = \sqrt[4]{d_{ca} d_{ca'} d_{c'a} d_{c'a'}} = \sqrt[4]{13 \times 12.6 \times 13.4 \times 13} = 12.997 \text{ m}$$

$$D_{eq} \text{ or } D_m = \sqrt[3]{D_{ab} D_{bc} D_{ca}} = \sqrt[3]{6.494 \times 6.494 \times 12.997} = 8.184 \text{ m}$$

Taking the centre to centre distances between the bundles as D_{ab}, D_{bc} and D_{ca}

$$D_{eq} = \sqrt[3]{6.5 \times 6.5 \times 13} = 8.189 \text{ m}$$

which is almost the same, as the accurate value of D_{eq}.

Inductance of the bundled conductor line per phase,

$$L = 0.2 \log_e \frac{D_m}{D_s} \text{ mH/km} = 0.2 \log_e \frac{8.184}{0.088} = 0.907 \text{ mH/km} \quad \textbf{Ans.}$$

Example 4.15. Find the inductive reactance of a 3-phase bundled conductor line with 2 conductors per phase with spacing of 40 cm. Phase to phase separation is 7 m in horizontal configuration. All conductors are ACSR with dia of 3.5 cm.

Compare the above value with that of an 'equivalent' single-conductor line.

Solution :

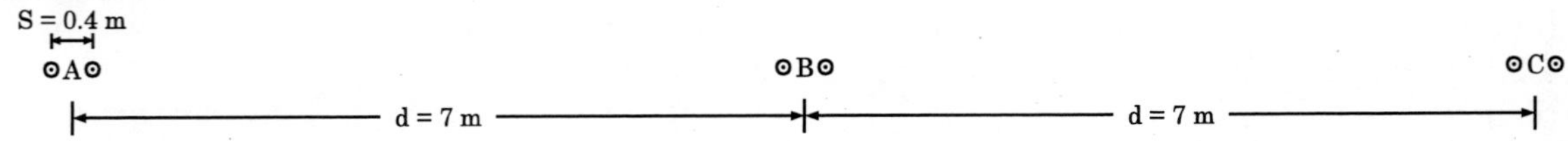

Fig. 4.24

Radius of each sub-conductor, $r = \frac{3.5}{2} = 1.75$ cm

GMR of each sub-conductor, $r' = 0.7788r = 0.7788 \times 1.75 \times 10^{-2} = 1.363 \times 10^{-2}$ m

Phase to phase separation, $d = 7$ m

Spacing between sub-conductors of one phase, $s = 40$ cm $= 0.4$ m

$$\text{GMR, } D_s = \sqrt{r' s} = \sqrt{1.363 \times 10^{-2} \times 0.4} = 0.0738 \text{ m}$$

$$D_{ab} = D_{bc} = \sqrt[4]{d_{ab}\, d_{ab'}\, d_{a'b}\, d_{a'b'}} = \sqrt[4]{7 \times 7.4 \times 6.6 \times 7} = 6.994 \text{ m}$$

$$D_{ca} = \sqrt[4]{d_{ca}\, d_{ca'}\, d_{c'a}\, d_{c'a'}} = \sqrt[4]{14 \times 13.6 \times 14.4 \times 14} = 13.997 \text{ m}$$

$$D_{eq} \text{ or } D_m = \sqrt[3]{D_{ab}\, D_{bc}\, D_{ca}} = \sqrt[3]{6.994 \times 6.994 \times 13.997} = 8.814 \text{ m}$$

Taking the centre to centre distances between the bundles as D_{ab}, D_{bc} and D_{ca},

$D_{eq} = \sqrt[3]{7 \times 7 \times 14} = 8.819$ m which is almost the same, as the accurate value of D_{eq}.

Inductance of the bundled conductor line,

$$L = 0.2 \log_e \frac{D_m}{D_s} \text{ mH/km} = 0.2 \log_e \frac{8.814}{0.0738} = 0.9565 \text{ mH/km}$$

Inductive reactance of the bundled conductor line per phase

$X_L = 2\pi f L = 2\pi \times 50 \times 0.9565 \times 10^{-3} = 0.3$ Ω/km **Ans.**

Radius of conductor in the equivalent line with single conductor per phase,

$r_1 = \sqrt{\frac{n \pi r^2}{\pi}}$ where n is the number of sub-conductors in a bundle

$= \sqrt{2 \times (1.75)^2} = 2.4745 \times 10^{-2}\ m$ $\quad \because n = 2$ and $r = 1.75$ cm

GMR of each conductor,

$$r_1' = 0.7788\, r_1 = 0.7788 \times 2.4745 \times 10^{-2} = 1.927 \times 10^{-2} \text{ m}$$

$$D_{m_1} = \sqrt[3]{D_{ab}\, D_{bc}\, D_{ca}} = \sqrt{7 \times 7 \times 14} = 8.819 \text{ m}$$

$$\text{Inductance, } L_1 = 0.2 \log_e \frac{D_{m1}}{r_1'} \text{ mH/km} = 0.2 \log_e \frac{8.819}{1.927 \times 10^{-2}} = 1.225 \text{ mH/km}$$

Inductive reactance, $X_{L_1} = 2\pi f L_1 = 2\pi \times 50 \times 1.225 \times 10^{-3} = 0.385$ Ω/km

Comments. The inductance of the bundled conductor line is significantly less than that of the line with one solid conductor per phase. The lower inductance leads to a greater transmission capacity of the bundled conductor line.

4.6. CAPACITANCE OF A TRANSMISSION LINE

We know that any two conductors separated by an insulating medium constitute a *condenser* or a *capacitor*. In case of an overhead line two conductors form the two plates of a capacitor and the air between the conductors behaves as the dieletric medium. Thus an overhead line can be assumed to have capacitance between conductors throughout the length of the line. The capacitance is uniformly distributed over the total length of the line and may be regarded as a uniform series of condensers connected between the conductors, as illustrated in Fig. 4.25.

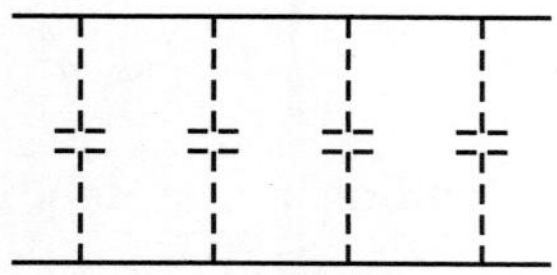

Fig. 4.25

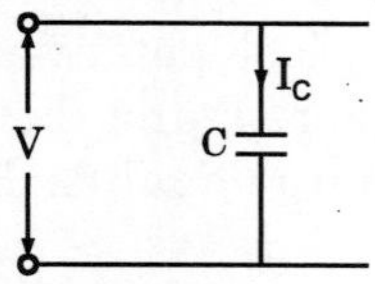

Fig. 4.26

When an alternating *pd* is applied across a transmission line, as shown in Fig. 4.26, it draws a leading current, even when supplying no load. This leading current is in quadrature with the applied voltage and is termed as the *charging current.* It must be noted that the charging current is due to the capacitive effect between the conductors of the line and is not in any way dependent on the load. The strength of the charging current depends upon the voltage of transmission, the capacitance of the line and the frequency of ac supply and is given by the expression.

$$\text{Charging current, } I_C = 2\pi f C V \qquad ...(4.34)$$

where f is the frequency of supply, C is capacitance of the line and V is the voltage.

If the capacitance of an overhead line is high, the line draws more charging current, which compensates or cancels the lagging component (reactive component) of load current (under normal load conditions the load is inductive one). Hence the resultant current flowing in the line is reduced. The reduction in resultant current flowing in the line results in (*i*) reduction of line losses and so increase of transmission efficiency, (*ii*) reduction in voltage drop or improvement of voltage regulation. The other advantages of a transmission line having high capacitance are increased load capacity and improved power factor.

4.6.1. Electric Field and Potential Difference

As we already know, an electric charge causes an electric field around it which theoretically extends up to infinity. If any charge is introduced in this electric field, it will be attracted or repelled according to the nature of the charge and when it will be moved, the work will be done against or by the force acting on the charge due to electric field. Hence potential in an electric field is exactly the same as potential in the gravitational field.

In general electric potential or potential at any point in an electric field is defined as the work done in moving a unit positive charge from infinity to that point.

The concept of electric potential is very important for the determination of capacitance as the latter is defined as the charge per unit potential. Now we will discuss in detail the electric potential owing to various important conductor arrangements.

1. Potential at a Charged Single Conductor

Consider a long straight cylindrical conductor A of radius r metres and having a charge of q coulombs per metre of its length.

The electric field intensity at a distance x from the centre of conductor,

$$E = \frac{q}{2\pi \epsilon_0 \epsilon_r x} \text{ volts/m}$$

Taking air as medium *i.e.* $e_r = 1$

$$E = \frac{q}{2\pi \epsilon_0 x} \text{ volts/m}$$

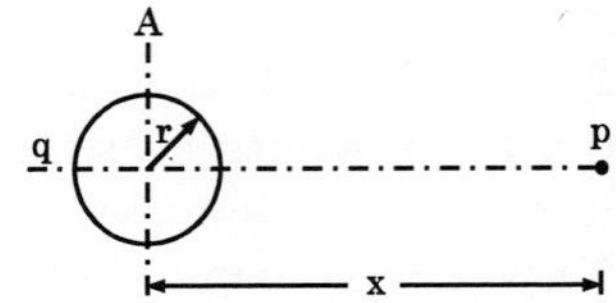

Fig. 4.27

The potential difference between conductor A and infinity distant neutral plane (a plane where E and therefore, potential is zero) will be equal to work done in bringing a unit positive charge against E from infinity to conductor surface and is given as

$$V_A = \int_r^\infty \frac{q}{2\pi \epsilon_0 x} dx = \frac{q}{2\pi \epsilon_0} \int_r^\infty \frac{dx}{x} \quad \text{...(4.35)}$$

2. Potential at a Charged Conductor in a Group of Charged Conductors

Consider a group of long straight conductors A, B, C, D, E ... N having charges of q_1, q_2, q_3, q_4, q_5 ... q_n coulombs per metre length respectively (Fig. 4.28).

Potential of conductor A due to its own charge q_1

$$= \int_r^\infty \frac{q_1}{2\pi \epsilon_0} \frac{dx}{x} \text{ volts}$$

Potential of conductor A due to charge q_2

$$= \int_{d_1}^\infty \frac{q_2}{2\pi \epsilon_0} \frac{dx}{x} \text{ volts}$$

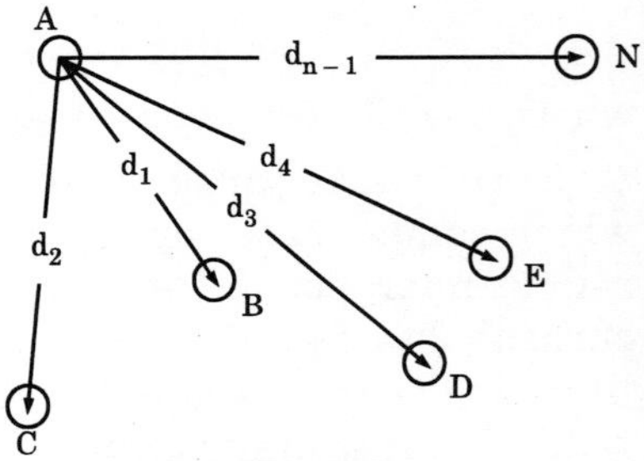

Fig. 4.28

since the field due to the charge q_2 extends from infinity up to a distance d_1 from conductor A.

Similarly potential of conductor A, due to charge q_3 of conductor C placed at a distance of d_2 metres from conductor A

$$= \int_{d_2}^\infty \frac{q_3}{2\pi \epsilon_0} \frac{dx}{x} \text{ volts}$$

Potential of conductor A due to charge q_n of conductor N placed at a distance of d_{n-1} metres from conductor A

$$= \int_{d_{n-1}}^\infty \frac{q_n}{2\pi \epsilon_0} \frac{dx}{x} \text{ volts}$$

So overall potential difference between conductor A and infinite distant neutral plane

$$V_{AN} = \int_r^\infty \frac{q_1}{2\pi \epsilon_0} \frac{dx}{x} + \int_{d_1}^\infty \frac{q_2}{2\pi \epsilon_0} \frac{dx}{x} + \int_{d_2}^\infty \frac{q_3}{2\pi \epsilon_0} \frac{dx}{x} + \cdots + \int_{d_{n-1}}^\infty \frac{q_n}{2\pi \epsilon_0} \frac{dx}{x}$$

$$= \frac{1}{2\pi \in_0}\left[q_1 (\log_e \infty - \log_e r) + q_2 (\log_e \infty - \log_e d_1) + q_3 (\log_e \infty - \log_e d_3) + ... + q_n (\log_e \infty - \log_e d_{n-1})\right]$$

$$= \frac{1}{2\pi \in_0}\left[q_1 \log_e \frac{1}{r} + q_2 \log_e \frac{1}{d_1} + q_3 \log_e \frac{1}{d_2} + ... + q_n \log_e \frac{1}{d_{n-1}} + \log_e \infty (q_1 + q_2 + q_3 + ... + q_n)\right]$$

Assuming balanced load *i.e.* $q_1 + q_2 + q_3 ..... q_n = 0$

$$\text{or } V_{AN} = \frac{1}{2\pi \in_0}\left[q_1 \log_e \frac{1}{r} + q_2 \log_e \frac{1}{d_1} + q_3 \log_e \frac{1}{d_2} + ... + q_n \log_e \frac{1}{d_{n-1}}\right] \text{ volts} \quad ...(4.36)$$

4.7. CAPACITANCE OF A SINGLE PHASE OVERHEAD LINE

Consider a single phase overhead line with two parallel conductors, each of radius r metres placed at a distance of d metres in air. It is assumed that the distance d between the conductors is large in comparison to the radii of the conductors. Therefore the density of charge on either conductor will be practically unaffected by the charge on the other conductor and will, therefore, be uniform throughout the length. A uniformly distributed charge on a conductor acts as though it is concentrated on the conductor axis. Therefore, for the purpose of our present analysis it is assumed that the charge + q coulombs on conductor A and – q coulombs on conductor B are concentrated at the centres of the two conductors which are separated from each other by d metres

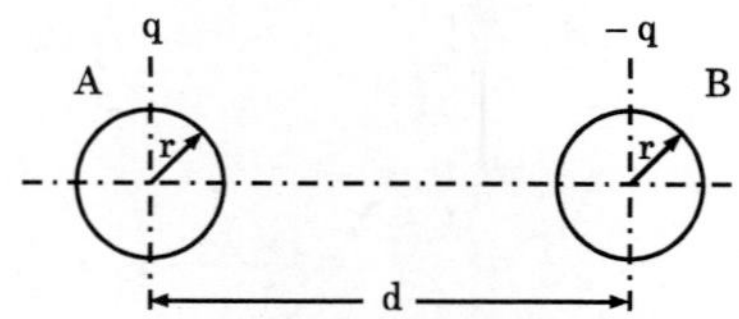

Fig. 4.29

PD between conductor A and neutral 'infinite' plane,

$$V_{AN} = \int_r^\infty \frac{q}{2\pi \in_0} \cdot \frac{d}{dx} + \int_d^\infty \frac{-q}{2\pi \in_0} \frac{dx}{x}$$

$$= \frac{q}{2\pi \in_0}\left[\log_e \infty + \log_e \frac{1}{r} - \log_e \infty + \log_e d\right]$$

$$= \frac{q}{2\pi \in_0} \log_e \frac{d}{r}$$

Similarly $p\ d$ between conductor B and neutral 'infinite' plane,

$$V_{BN} = \int_r^\infty \frac{-q}{2\pi \in_0} \frac{dx}{x} + \int_d^\infty \frac{q}{2\pi \in_0} \frac{dx}{x} = \frac{-q}{2\pi \in_0} \log_e \frac{d}{r} \text{ volts}$$

PD between conductors A and B,

$$V_{AB} = V_{AN} - V_{BN} = \frac{q}{2\pi \in_0} \log_e \frac{d}{r} - \frac{-q}{2\pi \in_0} \log_e \frac{d}{r} = \frac{q}{\pi \in_0} \log_e \frac{d}{r} \text{ volts}$$

Capacitance of the line,

$$C = \frac{q}{V_{AB}} = \frac{q}{\frac{q}{\pi \in_0} \log_e \frac{d}{r}} = \frac{\pi \in_0}{\log_e \frac{d}{r}} \text{ F/m} \quad ...(4.37)$$

The above Eq. (4.37) is for capacitance between two conductors. The capacitance for each conductor, C_n (or phase to neutral) will be double of this value.

Example 4.16. Find out capacitance of a single phase line 30 km long consisting of two parallel wires each 15 mm diameter and 1.5 m apart.

Solution : Radius of each conductor, $r = \frac{15}{2} = 7.5$ mm

Spacing between conductors, $d = 1.5$ m $= 1{,}500$ mm

Capacitance of line, $C = \frac{\pi \in_0}{\log_e \frac{d}{r}} \times$ length of line in metres

$$= \frac{\pi \times 8.854 \times 10^{-12}}{\log_e \frac{1{,}500}{7.5}} \times 30{,}000 = 0.1575 \ \mu\text{F} \ \textbf{Ans.}$$

4.8. CAPACITANCE OF 3-PHASE OVERHEAD LINES

4.8.1. Unsymmetrically Spaced Line

For an untransposed unsymmetrical 3-phase line the capacitances between conductor to neutral of the three conductors are different. Suppose that the line is, as shown in Fig. 4.9, and that voltages V_A, V_B, V_C are applied to the conductors with the result that the charges per metre length are q_1, q_2, q_3 respectively.

Potential of conductor A, ($w.r.t$ to neutral infinite plane)

$$V_{AN} = \frac{q_1}{2\pi \in_0} \int_r^{\infty} \frac{dx}{x} + \frac{q_2}{2\pi \in_0} \int_{d_1}^{\infty} \frac{dx}{x} + \frac{q_3}{2\pi \in_0} \int_{d_3}^{\infty} \frac{dx}{x}$$

$$= \frac{1}{2\pi \in_0} [q_1 \log_e \infty - q_1 \log_e r + q_2 \log_e \infty - q_2 \log_e d_1 + q_3 \log_e \infty - q_3 \log_e d_3]$$

$$= \frac{1}{2\pi \in_0} \left[q_1 \log_e \frac{1}{r} + q_2 \log_e \frac{1}{d_1} + q_3 \log_e \frac{1}{d_3} + (q_1 + q_2 + q_3) \log_e \infty \right]$$

$$= \frac{1}{2\pi \in_0} \left[q_1 \log_e \frac{1}{r} + q_2 \log_e \frac{1}{d_1} + q_3 \log_e \frac{1}{d_3} \right] \text{volts} \quad \because q_1 + q_2 + q_3 = 0 \quad \text{...(4.38)}$$

Similarly $$V_{BN} = \frac{1}{2\pi \in_0} \left[q_2 \log_e \frac{1}{r} + q_1 \log_e \frac{1}{d_1} + q_3 \log_e \frac{1}{d_2} \right] \text{volts} \quad \text{...(4.39)}$$

and $$V_{CN} = \frac{1}{2\pi \in_0} \left[q_3 \log \frac{1}{r} + q_1 \log_e \frac{1}{d_3} + q_2 \log_e \frac{1}{d_2} \right] \text{volts} \quad \text{...(4.40)}$$

Substituting $q_3 = -(q_1 + q_2)$ in Eqs. (4.38) and (4.39) we have

$$V_{AN} = \frac{1}{2\pi \in_0} \left[q_1 \log_e \frac{1}{r} + q_2 \log_e \frac{1}{d_1} - (q_1 + q_2) \log_e \frac{1}{d_3} \right]$$

$$= \frac{1}{2\pi \in_0} \left[q_1 \log_e \frac{d_3}{r} + q_2 \log_e \frac{d_3}{d_1} \right] \text{volts} \quad \text{...(4.41)}$$

$$\text{and } V_{BN} = \frac{1}{2\pi \in_0}\left[q_2 \log_e \frac{1}{r} + q_1 \log_e \frac{1}{d_1} - (q_1 + q_2) \log_e \frac{1}{d_2}\right] \text{ volts}$$

$$= \frac{1}{2\pi \in_0}\left[q_1 \log_e \frac{d_2}{d_1} + q_2 \log_e \frac{d_2}{r}\right] \text{ volts} \quad \text{...(4.42)}$$

Multiplying Eq. (4.41) by $\log_e \frac{d_2}{r}$ and Eq. (4.42) by $\log_e \frac{d_3}{d_1}$ we have

$$V_{AN} \log_e \frac{d_2}{r} = \frac{1}{2\pi\in_0}\left[q_1 \log_e \frac{d_3}{r} \cdot \log_e \frac{d_2}{r} + q_2 \log_e \frac{d_3}{d_1} \log_e \frac{d_2}{r}\right] \quad \text{...(4.43)}$$

$$\text{and } V_{BN} \log_e \frac{d_3}{d_1} = \frac{1}{2\pi\in_0}\left[q_1 \log_e \frac{d_2}{d_1} \cdot \log_e \frac{d_3}{d_1} + q_2 \log_e \frac{d_2}{r} \cdot \log_e \frac{d_3}{d_1}\right] \quad \text{...(4.44)}$$

Subtracting Eq. (4.44) from Eq. (4.43) we have

$$V_{AN} \log_e \frac{d_2}{r} - V_{BN} \log_e \frac{d_3}{d_1} = \frac{q_1}{2\pi\in_0}\left[\log_e \frac{d_3}{r} \cdot \log_e \frac{d_2}{r} - \log_e \frac{d_2}{d_1} \cdot \log_e \frac{d_3}{d_1}\right]$$

$$\text{or } q_1 = 2\pi\in_0 \frac{V_{AN} \log_e \frac{d_2}{r} - V_{BN} \log_e \frac{d_3}{d_1}}{\log_e \frac{d_3}{r} \log_e \frac{d_2}{r} - \log_e \frac{d_2}{d_1} \cdot \log_e \frac{d_3}{d_1}} \quad \text{...(4.45)}$$

Capacitance of conductor A to neutral,

$$C_{AN} = \frac{q_1}{V_{AN}} = 2\pi\in_0 \frac{\log_e \frac{d_2}{r} - \frac{V_{BN}}{V_{AN}} \log_e \frac{d_3}{d_1}}{\log_e \frac{d_3}{r} \log_e \frac{d_2}{r} - \log_e \frac{d_2}{d_1} \cdot \log_e \frac{d_3}{d_1}} \text{ F/m} \quad \text{...(4.46)}$$

Similarly capacitance of conductor B to neutral,

$$C_{BN} = 2\pi\in_0 \frac{\log_e \frac{d_3}{r} - \frac{V_{CN}}{V_{BN}} \log_e \frac{d_1}{d_2}}{\log_e \frac{d_1}{r} \cdot \log_e \frac{d_3}{r} - \log_e \frac{d_3}{d_2} \cdot \log_e \frac{d_1}{d_2}} \text{ F/m} \quad \text{...(4.47)}$$

$$\text{and } C_{CN} = 2\pi\in_0 \frac{\log_e \frac{d_1}{r} - \frac{V_{AN}}{V_{CN}} \log_e \frac{d_2}{d_3}}{\log_e \frac{d_2}{r} \cdot \log_e \frac{d_1}{r} - \log_e \frac{d_1}{d_3} \cdot \log_e \frac{d_2}{d_3}} \text{ F/m} \quad \text{...(4.48)}$$

4.8.2. Unsymmetrical Line With Transposed Conductors

Assuming that the charge per unit length is same in every part of the transposed cycle [Figs. 4.30 (1), (2) and (3)], average value of voltage of conductor A, will be

$$\frac{V_{AN_1} + V_{AN_2} + V_{AN_3}}{3} \quad \text{...(4.49)}$$

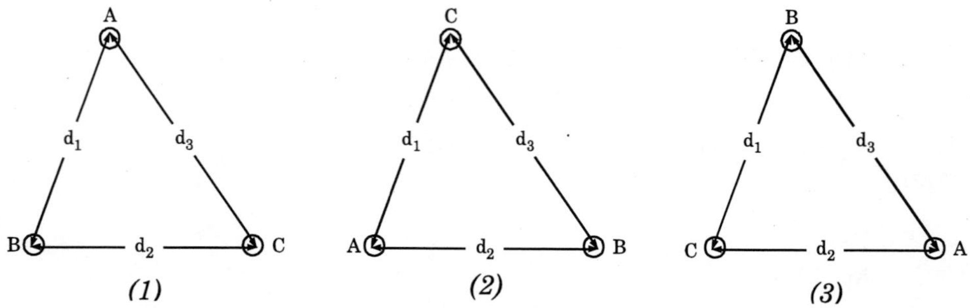

Fig. 4.30

Voltage of conductor A in positions (1), (2), and (3) we have

$$V_{AN_1} = \frac{1}{2\pi \epsilon_0}\left[q_1 \log_e \frac{1}{r} + q_2 \log_e \frac{1}{d_1} + q_3 \log_e \frac{1}{d_3}\right] \text{ volts}$$

$$V_{AN_2} = \frac{1}{2\pi \epsilon_0}\left[q_1 \log_e \frac{1}{r} + q_2 \log_e \frac{1}{d_2} + q_3 \log_e \frac{1}{d_1}\right] \text{ volts}$$

and $$V_{AN_3} = \frac{1}{2\pi \epsilon_0}\left[q_1 \log_e \frac{1}{r} + q_2 \log_e \frac{1}{d_3} + q_3 \log_e \frac{1}{d_2}\right] \text{ volts}$$

So average value of voltage of conductor A,

$$V_{AN} = \frac{V_{AN_1} + V_{AN_2} + V_{AN_3}}{3} = \frac{1}{6\pi \epsilon_0}\left[q_1 \log_e \frac{1}{r^3} + q_2 \log_e \frac{1}{d_1 d_2 d_3} + q_3 \log_e \frac{1}{d_1 d_2 d_3}\right]$$

$$= \frac{1}{6\pi \epsilon_0}\left[q_1 \log_e \frac{1}{r^3} + (q_2 + q_3) \log_e \frac{1}{d_1 d_2 d_3}\right] \text{ volts}$$

Substituting $q_2 + q_3 = -q_1$ in above equation we have

$$V_{AN} = \frac{1}{6\pi \epsilon_0}\left[q_1 \log_e \frac{1}{r^3} - \log_e \frac{1}{d_1 d_2 d_3}\right] \text{ volts}$$

or $$V_{AN} = \frac{q_1}{2\pi \epsilon_0}\left[\log_e \frac{1}{r} + \log_e \sqrt[3]{d_1 d_2 d_3}\right] = \frac{q_1}{2\pi \epsilon_0} \log_e \frac{\sqrt[3]{d_1 d_2 d_3}}{r} \text{ volts}$$

Capacitance of conductor A to neutral,

$$C_{AN} = \frac{q_1}{V_{AN}} = \frac{q_1}{\frac{q_1}{2\pi \epsilon_0} \log_e \frac{\sqrt[3]{d_1 d_2 d_3}}{r}} = \frac{2\pi \epsilon_0}{\log_e \frac{\sqrt[3]{d_1 d_2 d_3}}{r}} \text{ F/m}$$

Similarly, expressions for C_{BN} and C_{CN} can be obtained and we have

$$C_{AN} = C_{BN} = C_{CN} = C_N = \frac{2\pi \epsilon_0}{\log_e \frac{\sqrt[3]{d_1 d_2 d_3}}{r}} \text{ F/m} \quad ...(4.50)$$

4.8.3. Equilaterally Spaced Line

For the equilateral spacing $d_1 = d_2 = d_3 = d$ (say)

$$\text{and } C_{AN} = C_{BN} = C_{CN} = C_N = \frac{2\pi \epsilon_0}{\log_e \frac{\sqrt[3]{d.d.d}}{r}} = \frac{2\pi \epsilon_0}{\log_e \frac{d}{r}} \quad ...(4.51)$$

Example 4.17. Calculate the capacitance of a 100 km long 3-phase, 50 Hz overhead transmission line consisting of 3-conductors, each of diameter 2 cm and spaced 2.5 m at the corners of equilateral triangle. *[U.P. Technical Univ. Elements of Power System 2006-07]*

Solution : Spacing of conductors, $d = 2.5 \ m = 250$ cm

$$\text{Radius of each conductor, } r = \frac{2.0}{2} = 1.0 \text{ cm}$$

$$\text{Capacitance of the line} = \frac{2\pi \epsilon_0}{\log_e \frac{d}{r}} \text{ F/m} = \frac{2\pi \times 8.854 \times 10^{-12}}{\log_e \frac{250}{1}} = 10.075 \times 10^{-12} \text{ F/m}$$

Capacitance of 100 km long line $= 10.075 \times 100 \times 10^3 \times 10^{-12} = 1.0075$ μF **Ans.**

Example 4.18. The 2 cm diameter conductors of a 3-phase, 3-wire transmission line are situated at the corners of a triangle of sides 3.5 cm, 5 m and 8 m. Find the capacitance/conductor/km, if the line is transposed.

Solution : Conductor radius, $r = \frac{2}{2} = 1.0 \text{ cm} = 0.01 \text{ m}$

Spacings, $d_1 = 3.5$ m ; $d_2 = 5$ m and $d_3 = 8$ m

Capacitance per conductor per km,

$$C_N = \frac{2\pi \epsilon_0}{\log_e \frac{\sqrt[3]{d_1 d_2 d_3}}{r}} \times 1{,}000 = \frac{2\pi \times 8.854 \times 10^{-12} \times 1{,}000}{\log_e \frac{\sqrt[3]{3.5 \times 5 \times 8}}{0.01}} = 0.0089 \ \mu\text{F Ans.}$$

Example 4.19. Determine the capacitance and charging current per unit length of the line for the following arrangement of conductors of diameter 20 mm shown in Fig. 4.31.

Solution : $r = \frac{20}{2} = 10 \text{ mm} = 0.01 \text{ m}$

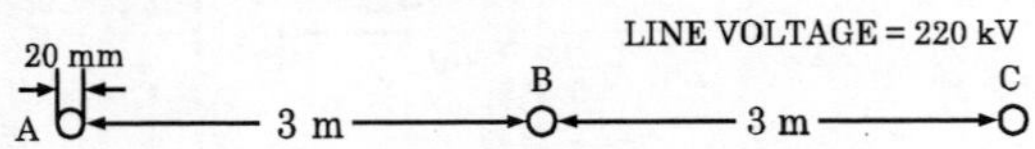

Fig. 4.31

Spacing between conductors

$d_1 = AB = 3$ m; $d_2 = BC = 3$ m; $d_3 = CA = 6$ m

Capacitance per phase per m length,

$$C_N = \frac{2\pi \epsilon_0}{\log_e \frac{\sqrt[3]{d_1 d_2 d_3}}{r}} = \frac{2\pi \times 8.854 \times 10^{-12}}{\log_e \frac{\sqrt[3]{3 \times 3 \times 6}}{0.01}} = 9.3737 \times 10^{-12} \text{ F Ans.}$$

Charging current per phase, $I_C = 2\pi f C_N V_{ph}$

$$= 2\pi \times 50 \times 9.3737 \times 10^{-12} \times \frac{220 \times 1{,}000}{\sqrt{3}} = 0.374 \text{ mA Ans.}$$

Example 4.20. A 3-phase, 50 Hz transmission line has flat horizontal configuration with 3.5 m between adjacent conductors. The conductors are hard drawn seven strand copper wire (outside conductor dia = 1.05 cm). The voltage of the line is 110 kV. Find the capacitnace to neutral and the charging per km. *[U.P. Technical Univ. Elements of Power Systems 2005-06]*

Solution : Conductor radius, $r = \frac{1.05}{2} = 0.5025$ cm

Spacings, $d_1 = 3.5$ m ; $d_2 = 3.5$ m and $d_3 = 7$ m

$$\text{Capacitance to neutral, } C_N = \frac{2\pi \epsilon_0}{\log_e \frac{\sqrt[3]{d_1 d_2 d_3}}{r}} = \frac{2\pi \times 8.854 \times 10^{-12}}{\log_e \frac{\sqrt[3]{3.5 \times 3.5 \times 7} \times 100}{0.5025}} = 8.21 \times 10^{-12} \text{ F/m}$$

or $8.21 \times 10^{-12} \times 1{,}000 = 0.00821$ μF/km **Ans.**

$$\text{Charging current per phase, } I_C = 2\pi f C_N V_{ph} = 2\pi \times 50 \times 0.00821 \times 10^{-6} \times \frac{110 \times 1000}{\sqrt{3}} = 0.164 \text{ A/km}$$ **Ans.**

4.9. CAPACITANCE OF DOUBLE CIRCUIT THREE PHASE OVERHEAD LINES

Normally used conductor configurations are of hexagonal spacing (Fig. 4.17) and flat vertical spacing (Fig. 4.18). It has been found that modified GMD method holds good for determination of capacitance of transposed double circuit 3-phase overhead lines with equilateral spacing (conductors at the vertices of a regular hexagon) and with flat vertical spacing. It is reasonable to assume that the modified GMD method can be used for determination of capacitance of a line with any configuration intermediate between these two configurations.

In the case of calculations of inductance, determination of self GMD (or GMR) of conductor is necessary because of internal flux linkages of the conductor. But in case of calculations of capacitance, since all charges reside on the surface of the conductor, actual radius of the conductor is used.

4.9.1. Symmetrically Spaced Line

Consider a 3-phase double circuit connected in parallel—conductors A, B and C forming one circuit and conductors A′, B′ and C′ forming another circuit (conductors symmetrically spaced).

Let the charge over conductors A, B and C be q_1, q_2 and q_3 coulombs per metre length. Then charge over conductors A′, B′ and C′ will obviously be q_1, q_2 and q_3 coulombs per metre length and $q_1 + q_2 + q_3 = 0$.

Potential of conductor A w.r.t neutral infinite plane (Fig. 4.17),

$$V_{AN} = \frac{1}{2\pi \epsilon_0}\left[q_1 \int_r^\infty \frac{dx}{x} + q_2 \int_d^\infty \frac{dx}{x} + q_3 \int_{\sqrt{3}d}^\infty \frac{dx}{x} + q_1 \int_{2d}^\infty \frac{dx}{x} + q_2 \int_{\sqrt{3}d}^\infty \frac{dx}{x} + q_3 \int_d^\infty \frac{dx}{x}\right]$$

$$= \frac{1}{2\pi \epsilon_0}[q_1 \log_e \infty - q_1 \log_e r + q_2 \log_e \infty - q_2 \log_e d + q_3 \log_e \infty - q_3 \log_e \sqrt{3}\, d$$

$$+ q_1 \log_e \infty - q_1 \log_e 2d + q_2 \log_e \infty - q_2 \log_e \sqrt{3}\, d + q_3 \log_e \infty - q_3 \log_e d]$$

$$= \frac{1}{2\pi \epsilon_0}\left[2(q_1 + q_2 + q_3) \log_e \infty + q_1 \log_e \frac{1}{2dr} + q_2 \log_e \frac{1}{\sqrt{3}\, d^2} + q_3 \log_e \frac{1}{\sqrt{3}\, d^2}\right]$$

$$= \frac{1}{2\pi \epsilon_0}\left[q_1 \log_e \frac{1}{2dr} + q_2 \log_e \frac{1}{\sqrt{3}\, d^2} + q_3 \log_e \frac{1}{\sqrt{3}\, d^2}\right] \qquad \because q_1 + q_2 + q_3 = 0$$

$$= \frac{1}{2\pi \epsilon_0}\left[q_1 \log_e \frac{1}{2dr} + (q_2 + q_3) \log_e \frac{1}{\sqrt{3}\, d^2}\right]$$

$$= \frac{1}{2\pi \epsilon_0}\left[q_1 \log_e \frac{1}{2dr} - q_1 \log_e \frac{1}{\sqrt{3}\, d^2}\right] \qquad \because q_2 + q_3 = -q_1$$

$$= \frac{q_1}{2\pi \epsilon_0} \log_e \frac{\sqrt{3}\, d^2}{2dr} = \frac{q_1}{2\pi \epsilon_0} \log_e \frac{\sqrt{3}\, d}{2r} \text{ volts}$$

Capacitance of conductor A to neutral,

$$C_{AN} = \frac{q_1}{V_{AN}} = \frac{q_1}{\frac{q_1}{2\pi \epsilon_0} \log_e \frac{\sqrt{3}\, d}{2r}} = \frac{2\pi \epsilon_0}{\log_e \frac{\sqrt{3}\, d}{2r}} \text{ F/m}$$

Similarly, expressions for C_{BN} and C_{CN} can be obtained and we have

$$C_{AN} = C_{BN} = C_{CN} = C_N = \frac{2\pi \epsilon_0}{\log_e \frac{\sqrt{3}\, d}{2r}} \text{ F/m} \qquad ...(4.52)$$

This is because the conductors of different phases are symmetrically placed.

Equation (4.52) gives the capacitance of conductor A alone, whereas there are two conductors per phase A and A'. Therefore the capacitance of the system per phase will be twice of the capacitance of one conductor to neutral, *i.e.*,

$$\text{Capacitance per phase, } C = 2C_N = \frac{4\pi \epsilon_0}{\log_e \frac{\sqrt{3}\, d}{2r}} \text{ F/m} \qquad ...(4.53)$$

4.9.2. Flat Vertically Spaced Line

Consider conductors arranged, as shown in Figs. 4.18 (1), (2) and (3), corresponding to different positions in the transposition positions.

Potential of conductor A w.r.t 'infinite' neutral plane [Fig. 4.18 (1)]

$$V_{AN_1} = \frac{1}{2\pi \epsilon_0}\left[q_1 \int_r^\infty \frac{dx}{x} + q_2 \int_{d_1}^\infty \frac{dx}{x} + q_3 \int_{2d_1}^\infty \frac{dx}{x} + q_1 \int_{\sqrt{4d_1^2 + d_2^2}}^\infty \frac{dx}{x} + q_2 \int_{\sqrt{d_1^2 + d_2^2}}^\infty \frac{dx}{x} + q_3 \int_{d_2}^\infty \frac{dx}{x}\right] \text{volts}$$

$$= \frac{1}{2\pi \epsilon_0}\left[q_1 \log_e \frac{1}{r\sqrt{4d_1^2 + d_2^2}} + q_2 \log_e \frac{1}{d_1\sqrt{d_1^2 + d_2^2}} + q_3 \log_e \frac{1}{2d_1 d_2}\right] \text{ volts}$$

Similarly,

$$V_{AN_2} = \frac{1}{2\pi \epsilon_0}\left[q_1 \log_e \frac{1}{r d_2} + q_2 \log_e \frac{1}{d_1\sqrt{d_1^2 + d_2^2}} + q_3 \log_e \frac{1}{d_1\sqrt{d_1^2 + d_2^2}}\right] \text{ volts and}$$

$$V_{AN_3} = \frac{1}{2\pi \epsilon_0}\left[q_1 \log_e \frac{1}{r\sqrt{4d_1^2 + d_2^2}} + q_2 \log_e \frac{1}{2d_1 d_2} + q_3 \log_e \frac{1}{d_1\sqrt{d_1^2 + d_2^2}}\right] \text{ volts}$$

$$V_{AN} = \frac{V_{AN1} + V_{AN2} + V_{AN3}}{3}$$

$$= \frac{1}{6\pi \epsilon_0}\left[q_1 \log_e \frac{1}{r^3 d_2\left(4d_1^2 + d_2^2\right)} + q_2 \log_e \frac{1}{2d_1^3 d_2\left(d_1^2 + d_2^2\right)} + q_3 \log_e \frac{1}{2d_1^3 d_2\left(d_1^2 + d_2^2\right)}\right]$$

$$= \frac{1}{6\pi \epsilon_0}\left[q_1 \log_e \frac{1}{r^3 d_2\left(4d_1^2 + d_2^2\right)} + (q_2 + q_3) \log_e \frac{1}{2d_1^3 d_2\left(d_1^2 + d_2^2\right)}\right]$$

$$= \frac{1}{6\pi \epsilon_0}\left[q_1 \log_e \frac{1}{r^3 d_2\left(4d_1^2 + d_2^2\right)} - q_1 \log_e \frac{1}{2d_1^3 d_2\left(d_1^2 + d_2^2\right)}\right] \qquad \because q_2 + q_3 = -q_1$$

$$= \frac{q_1}{6\pi \epsilon_0} \log_e \frac{2d_1^3 d_2\left(d_1^2 + d_2^2\right)}{r^3 d_2\left(4d_1^2 + d_2^2\right)}$$

$$= \frac{q_1}{6\pi \epsilon_0} \log_e \frac{2\,d_1^3}{r^3} \cdot \frac{d_1^2 + d_2^2}{4\,d_1^2 + d_2^2} = \frac{q_1}{2\pi \epsilon_0} \log_e \frac{2^{1/3} d_1}{r}\left(\frac{d_1^2 + d_2^2}{4\,d_1^2 + d_2^2}\right)^{1/3} \text{ volts}$$

Capacitance of conductor A, $C_{AN} = \dfrac{q_1}{V_{AN}} = \dfrac{q}{\dfrac{q}{2\pi \epsilon_0} \log_e 2^{1/3} \dfrac{d_1}{r}\left(\dfrac{d_1^2 + d_2^2}{4d_1^2 + d_2^2}\right)^{1/3}}$

$$= \frac{2\pi \epsilon_0}{\log_e 2^{1/3} \dfrac{d_1}{r}\left(\dfrac{d_1^2 + d_2^2}{4d_1^2 + d^2}\right)^{1/3}} \text{ F/m} \qquad \text{...(4.54)}$$

Similarly, expressions for capacitances C_{BN} and C_{CN} can be obtained which are same as above

Capacitance per phase will be double of C_{AN} *i.e.*

$$\text{Capacitance per phase, } C = \frac{4\pi \epsilon_0}{\log_e 2^{1/3} \dfrac{d_1}{r}\left(\dfrac{d_1^2 + d_2^2}{4\,d_1^2 + d_2^2}\right)^{1/3}} \text{ F/m} \qquad \text{...(4.55)}$$

Example 4.21. Six conductors of a double-circuit transmission line are arranged as shown in Fig. 4.32 :

The diameter of each conductor is 2.5 cm. Find the capacitive reactance to neutral and the charging current per km per phase at 132 kV and 50 Hz, assuming that the line is regularly transposed. Neglect the effect of earth.

[*Agra Univ. Transmission and Distribution of Electrical Power 1979 Supp.*]

Solution : Radius of conductor, $r = \dfrac{2.5}{5} = 1.25$ cm

The mutual GMD of the circuit is calculated as in case of determination of inductance. The self GMD of the circuit is also determined as in case of determination of inductance except that for r', r is used, as the electric charge, resides on the surface of the conductor unlike the magnetic flux which lies inside the conductor.

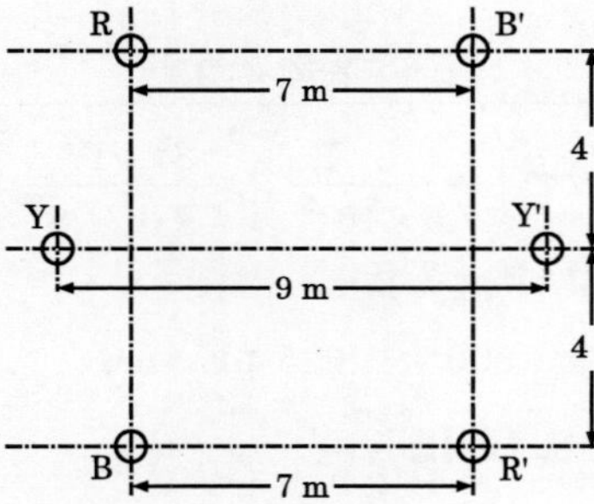

Fig. 4.32

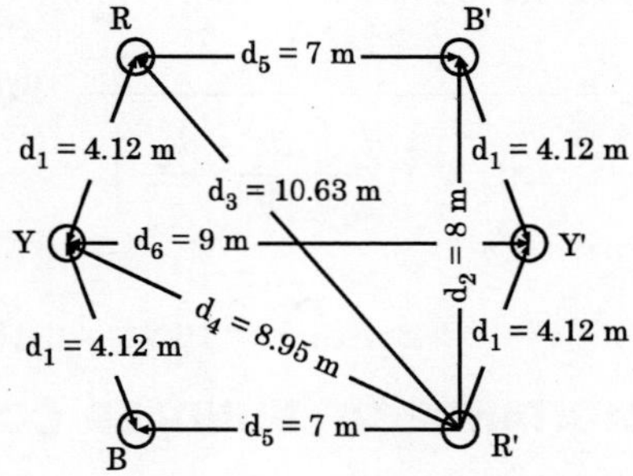

Fig. 4.33

Distance between conductors R and R′, $d_{RR'} = \sqrt{7^2 + (4+4)^2} = 10.63$ m

$$d_{BB'} = d_{RR'} = 10.63 \text{ m}$$

GMR of conductors of phase R,

$$D_{SR} = \sqrt[4]{d_{RR}\, d_{RR'}\, d_{R'R'}\, d_{R'R}} = \sqrt{r \times d_{RR'}} = \sqrt{1.25 \times 10^{-2} \times 10.63} = 0.3645 \text{ m}$$

Similarly, $D_{SY} = \sqrt{r \times d_{YY'}} = \sqrt{1.25 \times 10^{-2} \times 9} = 0.3354$ m

and $D_{SB} = \sqrt{r \times d_{BB'}} = \sqrt{1.25 \times 10^{-2} \times 10.63} = 0.3645$ m

Self, GMD, $D_S = \sqrt[3]{D_{SR}\, D_{SY}\, D_{SB}} = \sqrt[3]{0.3645 \times 0.3354 \times 0.3645} = 0.3545$ m

Distance between conductors R and Y, $D_{RY} = \sqrt{4^2 + (4.5 - 3.5)^2} = 4.12$ m

Distance between conductors R and Y′, $D_{RY'} = \sqrt{(9-1)^2 + 4^2} = 8.95$ m

So $D_{RY} = \sqrt[4]{d_{RY}\, d_{RY'}\, d_{R'Y}\, d_{R'Y'}} = \sqrt[4]{4.12 \times 8.95 \times 8.95 \times 4.12} = 6.072$ m

Similarly, $D_{YB} = \sqrt[4]{(4.12 \times 8.95)^2} = 6.072$ m

and $D_{BR} = \sqrt[4]{(8 \times 7)^2} = 7.483$ m

D_{eq} or $D_m = \sqrt[3]{D_{RY}\, D_{YB}\, D_{BR}} = \sqrt[3]{6.072 \times 6.072 \times 7.483} = 6.51$ m

Capacitance per phase, $C = \dfrac{2\pi \epsilon_0}{\log_e \dfrac{D_m}{D_s}} = \dfrac{2\pi \times 8.854 \times 10^{-12}}{\log_e \dfrac{6.51}{0.3545}} = 19.11 \times 10^{-12}$ F/m or 19.11×10^{-9} F/km

Capacitance reactance to neutral $= \dfrac{1}{2\pi f C} = \dfrac{1}{2\pi \times 50 \times 19.11 \times 10^{-9}} = 166.526$ kΩ **Ans.**

Charging current, $I_C = 2\pi f C V_{ph} = 2\pi \times 50 \times 19.11 \times 10^{-9} \times \dfrac{132 \times 1,000}{\sqrt{3}} = 0.4575$ A/km **Ans.**

Example 4.22. Determine the capacitance per phase of the line of example 4.13.

Solution : Distances $d_1 = 8$ m and $d_2 = 6$ m

Radius of conductor, $r = 3 \times$ radius of each strand of conductor

$$= 3 \times \frac{5}{2} = 7.5 \text{ mm or } 7.5 \times 10^{-3} \text{ m}$$

Capacitance per phase, from Eq. (4.55),

$$C = \frac{4\pi \epsilon_0}{\log_e 2^{1/3} \frac{d_1}{r}\left(\frac{d_1^2 + d_2^2}{4d_1^2 + d_2^2}\right)^{1/3}} \text{ F/m} = \frac{4\pi \times 8.854 \times 10^{-12}}{\log_e 2^{1/3}\left(\frac{8}{7.5 \times 10^{-3}}\right)\left(\frac{8^2 + 6^2}{4 \times 8^2 + 6^2}\right)^{1/3}}$$

$$= 16.25 \times 10^{-12} \text{ F/m}$$

Capacitance of 100 km line = $16.25 \times 10^{-12} \times 100 \times 1{,}000 = 1.625\ \mu\text{F}$ **Ans.**

4.10. CAPACITANCE OF BUNDLED CONDUCTOR LINE

A bundled conductor line is shown in Fig. 4.34. The conductors of any one bundle are in parallel and it is assumed that the charge per bundle divides equally among the conductors of the bundle as $D_{ab} >> s$. Also $D_{ab} - s \simeq D_{ab} + s \simeq D_{ab}$ for the same reason. The results obtained with these assumptions are fairly accurate for usual spacings. Thus if charge on phase A is q_A, then a and a' have a charge of $q_A/2$ each. Similarly the charge is equally divided for phases B and C.

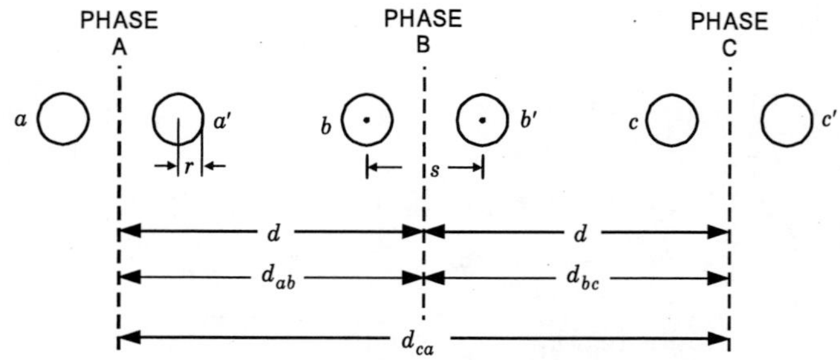

Fig. 4.34. ***Cross-Section of a Bundled Conductor Three-Phase Transmission Line***

Now, writing an equation for the voltage from conductor a to conductor b, we have

$$V_{ab} = \frac{1}{2\pi\varepsilon_0}\left[0.5q_A\left(\log_e \frac{d_{ab}}{r} + \log_e \frac{d_{ab}}{s}\right) + 0.5q_B\left(\log_e \frac{r}{d_{ab}} + \log_e \frac{s}{d_{ab}}\right) + 0.5q_c\left(\log_e \frac{d_{bc}}{d_{ca}} + \log_e \frac{d_{bc}}{d_{ca}}\right)\right]$$

$$= \frac{1}{2\pi\varepsilon_0}\left(q_A \log_e \frac{d_{ab}}{\sqrt{rs}} + q_B \log_e \frac{\sqrt{rs}}{d_{ab}} + q_C \log_e \frac{d_{bc}}{d_{ca}}\right) \quad ...(4.56)$$

Considering the line to be transposed and proceeding in the usual manner, the final result will be

$$C_N = \frac{2\pi\varepsilon_0}{\log_e \frac{\sqrt[3]{d_{ab}d_{bc}d_{ca}}}{\sqrt{rs}}} = \frac{2\pi\varepsilon_0}{\log_e \frac{D_m}{D_s}} \quad ...(4.57)$$

Mutual GMD of the circuit will be determined as mentioned in Art. 4.5.

i.e. $D_m = \sqrt[3]{d_{ab}d_{bc}d_{ca}}$

Self GMD, D_s is determined in the same way as mentioned in Art. 4.5 except that for r', r is used.

Example 4.23. Determine the capacitance and charging current per km of the line of Example 4.14 if the line operates at 132 kV.

Solution :

Self GMD, $D_s = \sqrt{rs} = \sqrt{2.5 \times 10^{-2} \times 0.4} = 0.1$ m

Mutual GMD, $D_m = 8.189$ m

Capacitance per km per phase $C = \frac{2\pi\varepsilon_0}{\log_e \frac{D_m}{D_s}} = \frac{2\pi \times 8.854 \times 10^{-12}}{\log_e \frac{8.189}{0.1}}$ F/m

$= 12.628 \times 10^{-12}$ F/m or 12.628×10^{-9} F/km

$$\text{Charging current per km} = 2\pi f C V_{ph} = 2\pi \times 50 \times 12.628 \times 10^{-9} \times \frac{132{,}000}{\sqrt{3}}$$

assuming supply frequency $f = 50$ Hz

$$= 0.3 \text{ A/km} \quad \textbf{Ans.}$$

Example 4.24. A 50 Hz, 3-phase line (transposed) of an ACSR moose conductor (overall dia = 31.8 mm) per phase has flat horizontal spacing of 10 m between adjacent conductors. Compare the inductive and capacitance reactances in ohm per km per phase of this line with that of a line (transposed) using a three conductor bundle of ACSR hyn *x* conductors (each having overall diameter = 19.6 mm) having 10 m spacing measured from the centre of the bundles. The bundle conductor in each phase are arranged in an equilateral triangle formation with spacing between the conductors in the bundle as 40 cm. *[U.P. Technical Univ. Elements of Power System 2007-08]*

Solution : One Conductor ACSR Moose Conductor Line :

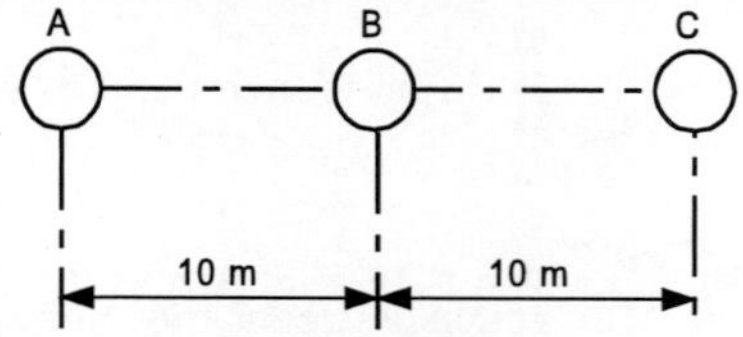

Fig. 4.35

Radius of each conductor, $r = \dfrac{31.8}{2} = 15.9$ mm

GMR, $r' = 0.7788 \times 15.9 = 12.383$ mm

From Eqs. (4.20), (4.21) and (4.22) respectively, we have

$$L_A = 0.2\left[\log_e \frac{d}{r'} + \frac{1}{2}\log_e 2 - j0.866 \log_e 2\right] \text{ mH/km}$$

$$= 0.2\left[\log_e \frac{10{,}000}{12.383} + \frac{1}{2}\log_e 2 + j0.866 \log_e 2\right]$$

$$= 0.2[6.694 + 0.3466 - j0.6] = (1.408 - j0.12) \text{ mH/km}$$

$$L_B = 0.2 \log_e \frac{d}{r'} \text{ mH/km} = 0.2 \log_e \frac{10{,}000}{12.383} = 1.3388 \text{ mH/km}$$

$$L_C = 0.2\left[\log_e \frac{d}{r'} + \frac{1}{2}\log_e 2 + j0.866 \log_e 2\right] \text{ mH/km} = (1.408 + j0.12) \text{ mH/km}$$

The effect of transposition of line is that each conductor has the same average inductance which is given as

$$L_{av} = \frac{L_A + L_B + L_C}{3} = \frac{(1.408 - j0.12) + 1.3388 + (1.408 - j0.12)}{3} \text{ mH/km}$$

$$= 1.385 \text{ mH/km}$$

Inductive reactance per km per phase, $X_L = 2\pi f L_{av} = 2\pi \times 50 \times 1.385 \times 10^{-3} = 0.435\ \Omega$ **Ans.**

$$\text{Capacitance per phase, } C_N = \frac{2\pi\varepsilon_0}{\log_e \dfrac{\sqrt[3]{d_1 d_2 d_3}}{r}} = \frac{2\pi \times 8.854 \times 10^{-12}}{\log_e \dfrac{\sqrt[3]{10 \times 10 \times 20}}{12.383 \times 10^{-3}}}$$

$$= 8.0333 \times 10^{-12} \text{ F/m or } 8.0333 \times 10^{-9} \text{ F/km}$$

Capacitive reactance per phase per km,

$$X_C = \frac{1}{2\pi f C} = \frac{1}{2\pi \times 50 \times 8.0333 \times 10^{-9}}\ \Omega = 0.396 \text{ M}\Omega$$

Three Conductor Bundled Line

$$\text{GMR, } D_s = \sqrt[3]{r' \times s^2} = \sqrt[3]{0.7788 \times r \times s^2} = \sqrt[3]{0.7788 \times 19.6 \times 10^{-3} \times 0.4^2}$$

$$= 0.1347 \text{ m}$$

Taking the centre to centre distances between the bundles as D_{ab}, D_{bc} and D_{ca}.

$$D_{eq} = \sqrt[3]{10 \times 10 \times 20} = \sqrt[3]{2,000} = 12.6 \text{ m}$$

Inductance of bundled conductor line,

$$L' = 0.2 \log_e \frac{D_m}{D_s} \text{ mH/km} = 0.2 \log_e \frac{12.6}{0.1347} \text{ mH/km} = 0.908 \text{ mH/km}$$

Inductive reactance per km per phase,

$$X_L' = 2\pi f L' = 2\pi \times 50 \times 0.908 \times 10^{-3} = 0.285 \ \Omega \textbf{ Ans.}$$

$$\text{GMR, } D_s = \sqrt[3]{r \times s^2} = \sqrt[3]{19.6 \times 10^{-3} \times 0.4^2} = 0.1464 \text{ m}$$

$$\text{Capacitance per km per phase, } C' = \frac{2\pi\varepsilon_0}{\log_e \frac{D_{eq}}{D_s}} = \frac{2\pi \times 8.854 \times 10^{-9}}{\log_e \frac{12.6}{0.1464}} = 0.01249 \ \mu\text{F}$$

$$\text{Capacitive reactance per km per phase} = \frac{1}{2\pi \times 50 \times 0.01249 \times 10^{-6}} = 0.255 \text{ M}\Omega$$

Inductive reactance of three conductor bundled line is 0.285Ω/km/phase, while the inductive reactance of one conductor ACSR Moose conductor line is 0.435 Ω per km per phase **Ans.**

Capacitive reactance of three conductor bundled line is 0.255 MΩ per km per phase while for one conductor ACSR Moose conductor line it is 0.396 MΩ per km per phase.

4.11. EFFECT OF EARTH ON CAPACITANCE OF TRANSMISSION LINE

So far, in determination of capacitance of transmission lines, the presence of earth was ignored. But it is not true; the presence of earth affects the electric field of a line and so the capacitance. The effect of earth on capacitance can be modelled by the method of images.

The electric field of too long, parallel conductors having charge $+q$ and $-q$ per unit is such that it has a zero potential between the conductors, as depicted in Fig. 4.36. If a conducting sheet of infinite dimensions is placed at the zero potential plane, the electric field is not disturbed. Further, if the conductor carrying charge $-q$ is now removed, the electric field above the conducting sheet stays intact, while that below it disappears. Using these well known facts in reverse, the presence of ground below a charged conductor can be replaced by a fictitious conductor having equal and opposite charge and located as far below the ground surface as the overhead conductor above it—such a fictitious conductor is the *mirror image* of the overhead conductor. This method of producing the same electric field as in the presence of earth is called the *method of images* originally suggested by Lord Kelvin.

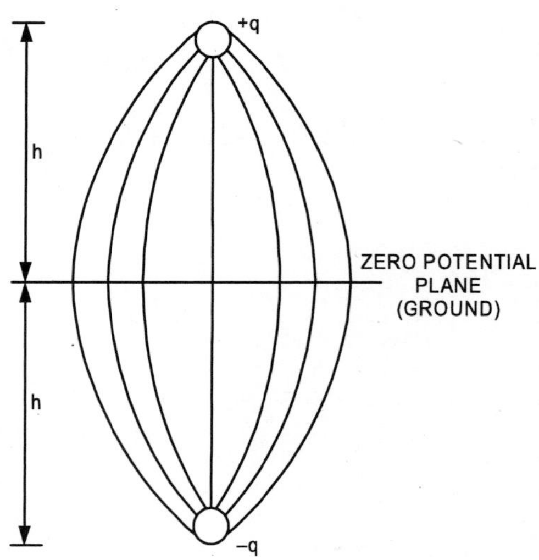

Fig. 4.36. ***Electric Field of Two Long Parallel, Oppositely Charged Conductors***

4.11.1. Effect of Earth on the Capacitance of a Single Phase Transmission Line

Considering the case of a single phase overhead line, assume conductors A′ and B′ as image conductors of conductors A and B respectively, as shown in Fig. 4.37. Let the height of conductors be h metres above the earth and charge of $+q$ coulombs per metre length and $-q$ coulombs per metre length on the conductors A and B respectively.

The equation for the voltage drop V_{AB} as determined by two charged conductors A and B, and their images A′ and B′ may be written as follows :

$$V_{AB} = \frac{1}{2\pi\varepsilon_0}\left[q_A \log_e \frac{d}{r} + q_B \log_e \frac{r}{d} + q'_A \log_e \frac{\sqrt{4h^2+d^2}}{2h} + q'_B \log_e \frac{2h}{\sqrt{4h^2+d^2}}\right] \quad ...(4.58)$$

Substituting $q_A = q'_B = +q$ and $q'_A = q_B = -q$ we have

$$V_{AB} = \frac{q}{\pi\varepsilon_0} \log_e \frac{2hd}{r\sqrt{(4h^2+d^2)}}$$

Capacitance between conductors A and B

$$C_{AB} = \frac{q}{V_{AB}} = \frac{\pi\varepsilon_0}{\log_e \dfrac{d}{r\sqrt{1+\dfrac{d^2}{4h^2}}}} \quad ...(4.59)$$

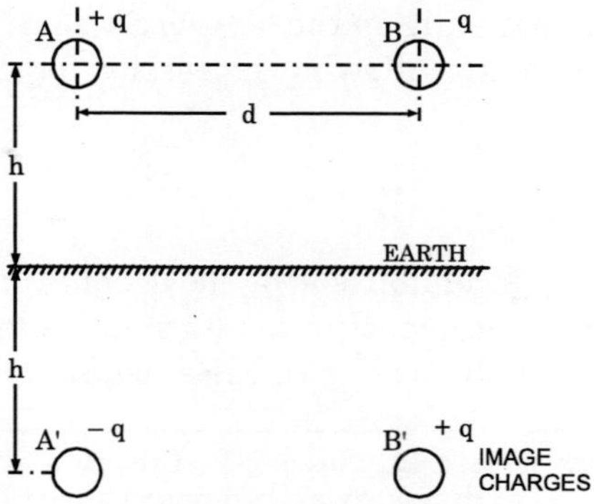

Fig. 4.37. *Single Phase Transmission Lines With Images*

The above expression for capacitance reveals that the presence of earth modifies the radius of conductor r to $r\sqrt{\left(1+\dfrac{d^2}{4h^2}\right)}$. The effect of earth on the capacitance of the system is to increase it. However, normally the distance of separation between the conductors is much smaller than the height of the conductor from the ground, therefore, $r\sqrt{1+\dfrac{d^2}{4h^2}} \simeq r$ and for all practical purposes the effect of earth on line capacitance can be neglected.

4.11.2. Effect of Earth on the Capacitance of a Three Phase Line

Figure 4.38 shows the conductors of a 3-phase line along with image conductors. The line is assumed to be transposed and in the first part of the transposition cycle conductor A is in position 1, B is in position 2 and C is in position 3. Let the charges on line conductors be q_A, q_B and q_C while those on image conductors be $-q_A$, $-q_B$ and $-q_C$ respectively, as shown in the figure.

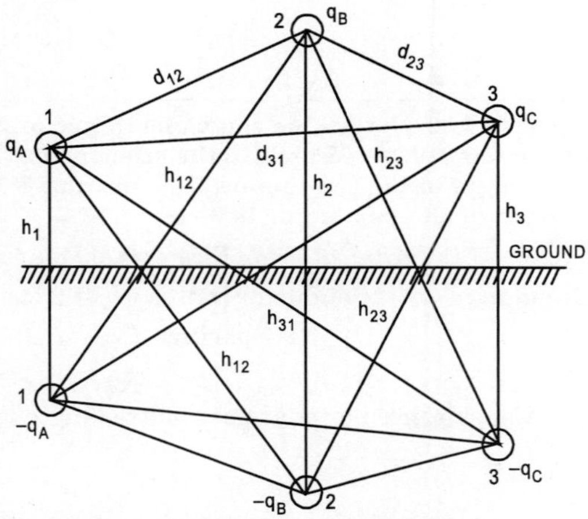

Fig. 4.38. *Conductors of a 3-Phase Line With Image Charges*

The equation for the three sections of transposition cycle can be written for the voltage drop V_{AB} as determined by three charged conductors and their images. With conductor in position 1, B in position 2 and C in position 3, we have

$$V_{AB} = \frac{1}{2\pi\varepsilon_0}\left[q_A\left(\log_e \frac{d_{12}}{r} - \log_e \frac{h_{12}}{h_1}\right) + q_B\left(\log_e \frac{r}{d_{12}} - \log_e \frac{h_2}{h_{12}}\right) + q_C\left(\log_e \frac{d_{23}}{d_{31}} - \log \frac{h_{23}}{h_{31}}\right)\right] \quad ...(4.60)$$

Similarly, equations for V_{AB} can be written for the second and third sections of the transposition cycle. If the fairly accurate assumption of constant charge per unit length of the conductor throughout the transmission cycle is made, the average value of three sections of the cycle is given by

$$V_{AB} = \frac{1}{2\pi\varepsilon_0}\left[q_A\left(\log_e \frac{D_{eq}}{r}\right) - \log_e\left(\frac{h_{12}h_{23}h_{31}}{h_1h_2h_3}\right)^{1/3} + q_B\left(\log_e \frac{r}{D_{eq}} - \log_e\left(\frac{h_1h_2h_3}{h_{12}h_{23}h_{31}}\right)^{1/3}\right)\right] \quad ...(4.61)$$

where $D_{eq} = \sqrt[3]{d_{12}d_{23}d_{31}}$

The equation for the average value of the voltage V_{AC} can be determined in the same way. Proceeding on the lines of Sec. 4.8 and using $V_{AB} + V_{AC} = 3V_{AN}$ and $q_A + q_B + q_C = 0$, we ultimately have the following expression for the capacitance to neutral

$$C_N = \frac{2\pi\varepsilon_0}{\log_e \frac{D_{eq}}{r} - \log_e \sqrt[3]{\frac{h_{12}h_{23}h_{31}}{h_1h_2h_3}}} \text{ F/m to neutral} \quad ...(4.62)$$

A comparison of above equation with Eq. (4.51) reveals that the presence of earth increases the line capacitance by a small amount. However, the effect is very-very small because the heights of conductors are large as compared to the distance between them.

Example 4.25. The conductors in a single phase line are 6 m above the ground. Each conductor is of 1.5 cm diameter and they are spaced 3 m apart. Taking the effect of earth into account, calculate the capacitance per km. *[U.P. Technical Univ. Elements of Power System 2002-03]*

Solution : Radius of each conductor, $r = \frac{1.5}{2} = 0.75$ cm

Spacing between conductors, $d = 3$ m $= 300$ cm

Height of conductors above earth, $h = 6$ m $= 600$ cm

Capacitance per km of line, $C = \frac{\pi\varepsilon_0}{\log_e \frac{d}{r\sqrt{1 + \frac{d^2}{4h^2}}}} \times 1{,}000 \text{ F} = \frac{\pi \times 8.854 \times 10^{-12} \times 1{,}000}{\log_e \frac{300}{0.75\sqrt{\left(1 + \frac{300^2}{4 \times 600^2}\right)}}}$ F

$= 4.666 \times 10^{-9}$ F **Ans.**

Example 4.26. Determine the capacitance of the arrangement shown in the figure when (*i*) the effect of earth is neglected and (*ii*) the effect of earth is considered. The height of the conductors from ground is 10 m and radius of each of the conductor is 2 cm.

[U.P. Technical Univ. Elements of Power System 2008-09]

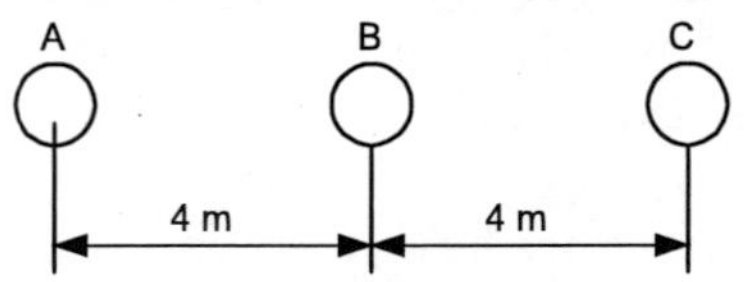

Fig. 4.39

Solution : Conductor radius, $r = 2$ cm $= 0.02$ m

Spacings d_1 = AB = 4 m;

d_2 = BC = 4 m; d_3 = CA = 8 m

Capacitance per phase per metre length (neglecting earth effect)

$$C_N = \frac{2\pi\varepsilon_0}{\log_e \frac{\sqrt[3]{d_1d_2d_3}}{r}} = \frac{2\pi \times 8.854 \times 10^{-12}}{\log_e \frac{\sqrt[3]{4 \times 4 \times 8}}{0.02}} = 10.06 \times 10^{-12} \text{ F}$$

From Fig. 4.40, we have

$$h_1 = h_2 = h_3 = 20 \text{ m}$$

$$h_{12} = h_{23} = \sqrt{20^2 + 4^2} = 20.396 \text{ m}$$

$$h_{31} = \sqrt{20^2 + 8^2} = 21.54 \text{ m}$$

$$r = 0.02 \text{ m}$$

$$D_{eq} = \sqrt[3]{d_1 d_2 d_3} = \sqrt[3]{4 \times 4 \times 8} = 5.0397 \text{ m}$$

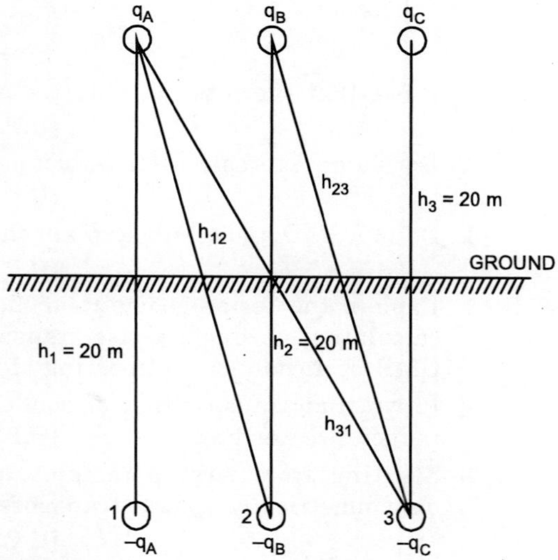

Fig. 4.40

Capacitance per phase per metre length (considering effect of ground)

$$C_N = \frac{2\pi\varepsilon_0}{\log_e \frac{D_{eq}}{r} - \log_e \sqrt[3]{\frac{h_{12}h_{23}h_{31}}{h_1 h_2 h_3}}} \text{ F}$$

$$= \frac{2\pi \times 8.854 \times 10^{-12}}{\log_e \frac{5.0397}{0.02} - \log_e \sqrt[3]{\frac{20.396 \times 20.396 \times 21.54}{20 \times 20 \times 20}}}$$

$$= \frac{55.63 \times 10^{-12}}{5.529 - 0.0378} \text{ F} = 10.13 \times 10^{-12} \text{ F } \textbf{Ans.}$$

4.12. TRANSPOSITION OF THREE PHASE LINES

As already mentioned, the inductance and capacitance of each phase will be different in case of conductors of three phase line being spaced irregularly. The apparent resistance of the conductors is also affected on account of transfer of power between the phases, which occur due to mutual inductance. Thus all the three line constants are affected by irregular spacing of the conductors in a 3-phase line.

Also, due to unsymmetrical spacing, the magnetic field external to the conductors is not zero, thereby causing induced voltages in adjacent electrical circuits, particularly the telephone lines, that may cause disturbances in the telephone lines.

The unbalancing effect on account of irregular spacing of line conductors can be avoided by transposition of line conductors. Transposition of line conductors means changing the positions of the three phases on the line supports twice over the total length of the line. In practice the line conductors should be so transposed that each of the three possible arrangements of conductors exist for one-third of the total length of line. This is illustrated in Figs. 4.10 and 4.41.

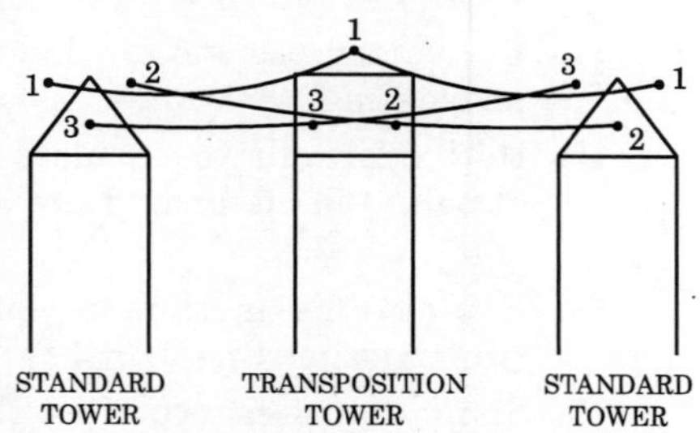

Fig. 4.41. ***Transposition of a 3-Phase Line***

An unbalanced system, may be of conductors symmetrically placed, cannot be represented by three simple inductances, without making transposition. The effect of unbalanced currents is neutralised in case of conductors being transposed regularly at intervals.

The transposition of conductors also reduces the disturbances to the nearby communication circuits. In composite line, the line carrying telephone line conductors below the power line conductors, it is also necessary to transpose the telephone line conductors in order to keep down the disturbances.

EXERCISES

1. Prove that the internal inductance of a conductor is independent of its diameter.
[U.P. Technical Univ. Elements of Power System 2002-03]
2. Derive an expression for inductance of a single phase transmission line.
[U.P. Technical Univ. Elements of Power System 2005-06]
3. Define GMD and GMR. Explain the role of GMD and GMR in line inductance calculation.
[U.P. Technical Univ. Elements of Power System 2007-08]
4. Explain the terms geometrical mean distance (GMD) and 'self GMD' in the inductance calculation of single phase transmission lines with composite conductors. Find the self GMD of a conductor consisting of seven identical strands each having a radius *r*.
5. Prove that transposition of conductors of a three phase transmission line yields equal inductance per phase. **[U.P. Technical Univ. Elements of Power System 2006-07]**
6. Starting from first principles, derive the expression for inductance of a 3-phase unsymmetrically spaced transposed transmission line.
[U.P. Technical Univ. Elements of Power System 2009-10]
7. Starting with flux linkage concept, find the inductance of 3-phase double circuit line with unsymmetrical spacing. **[U.P. Technical Univ. Elements of Power System 2007-08]**
8. Differentiate composite and stranded conductors.
[U.P. Technical Univ. Elements of Power System 2007-08]
9. Mention the contribution of bundle conductors in performance of transmission network.
[U.P. Technical Univ. Elements of Power System 2005-06]
10. Differentiate between stranded conductors and bundled conductors.
[U.P. Technical Univ. Elements of Power System 2007-08]
11. What is a bundle conductor and why is it used ?
[U.P. Technical Univ. Elements of Power System 2006-07]
12. What advantages you get by using "bundled conductors" ?
[U.P. Technical Univ. Elements of Power System 2008-09]
13. Explain the advantages of using bundled conductors in extra high voltage transmission lines and the different configurations.
[U.P. Technical Univ. Elements of Power System 2003-04]
14. Derive expressions for line-to-line capacitance and line to neutral capacitance of a single phase line. **[U.P. Technical Univ. Elements of Power System 2003-04]**
15. Derive formula to calculate the capacitance of a double circuit line. All the conductors are of equal diameter and are spaced hexagonally.
[U.P. Technical Univ. Elements of Power System 2002-03]
16. Explain the method of images for the calculation of capacitance of a three phase transmission line. **[U.P. Technical Univ. Elements of Power System 2002-03]**
17. Show that presence of earth increases capacitance of transmission lines.
[U.P. Technical Univ. Elements of Power System 2004-05]
18. What are the advantages of transposition of conductors ?
[U.P. Technical Univ. Elements of Power System 2004-05]
19. Discuss the transposition of conductors of a three-phase transmission line.
[U.P. Technical Univ. Elements of Power System 2005-06]
20. Explain briefly about the following :
(*i*) GMD. (*ii*) Bundled conductors.
(*iii*) Double circuit lines. (*iv*) Transposition.
[U.P. Technical Univ. Elements of Power System 2009-10]

SHORT ANSWER TYPE QUESTIONS WITH ANSWERS

Q. 1. For determination of inductance of a circuit, determination of flux linkages is essential. Why ?

Ans. It is well known fact that a current carrying conductor is surrounded by concentric circles of magnetic lines. In case of ac system this field set up around the conductor is not constant but changing and links with the same conductor as well as with other conductors. Due to these flux linkages the line possesses inductance, defined as the flux linkages per unit current. Thus for determination of inductance of a circuit, determination of flux linkages is essential.

Q. 2. What is meant by GMR of a conductor ?

Ans. The product $r\, e^{-1/4}$ is known as GMR of a conductor and is equal to 0.7788 times the conductor radius.

Q. 3. What is self GMD of a conductor consisting of seven identical strands, each having a radius r ?

Ans. 2.176 r

Q. 4. What is the effect of transposition of conductors for a three phase transmission line on its inductance ?

Ans. The transposition of conductors of a three phase transmission line yields equal inductance per phase.

Q. 5. What is bundling of conductors ?

Ans. The use of more than one conductor per phase is called the bundling of conductors.

Q. 6. Differentiate composite and bundled conductors.

Ans. The basic difference between a composite conductor and a bundled conductor is that the sub-conductors of a bundled conductor are separated from each other by a constant distance varying from 0.2 m to 0.6 m depending upon designed voltage and surrounding conditions throughout the length of the line with the help of spacers whereas the wires of a composite conductor touch each other. The bundled conductors have filter material or air space inside so that the overall diameter is increased.

Q. 7. For what purposes bundled conductors are primarily used in transmission lines ?

Ans. The bundled conductors are used in EHV transmission lines primarily to reduce corona loss and radio interference.

Q. 8. What are the advantages of bundled conductors. **[U.P.T.U. 2007-08]**

Ans. The advantages of using bundled conductors in EHT transmission lines are : Reduced possibilities of the corona discharge; reduced corona loss; reduced radio interference; reduced losses resulting in increased transmission efficiency; higher charging current resulting in improved pf; reduced inductive reactance per phase; lower surge impedance resulting in increased power transfer capability.

Q. 9. What is the effect of ground on the capacitance of a transmission line ?

Ans. The presence of ground increases the capacitance of a transmission line by a small amount.

Q. 10. Why the effect of ground on the line capacitance can be neglected ?

Ans. The effect of ground on the line capacitance is very-very small because the height of conductors is large as compared to the distance between them.

Q. 11. What is meant by transposition of line conductors ?

Ans. Transposition of line conductors means changing the positions of the three phases on the line supports twice over the total length of the line. The line conductors in practice, are so transposed that each of the three possible arrangements of conductors exist for one-third of the total length of the line.

Q. 12. Why transposition of conductors in a three phase transmission lines is essential ?

Ans. The inductance and capacitance of each phase will be different in case of conductors of three phase line being spaced irregularly. The apparent resistance of the conductors is also affected on account of transfer of power between the phases, which occur due to mutual inductance. Thus all the three line constants are affected by irregular spacing of the conductors in a 3-phase line.

Also, due to unsymmetrical spacing, the magnetic field external to the conductors is not zero, thereby causing induced voltages in adjacent electrical circuits, particularly the telephone lines, that may cause disturbances in the telephone lines.

Transposition of line conductors in transmission lines is essential to counter the unbalancing effect, mentioned above, of irregular spacing of line conductors.

PROBLEMS

1. A single phase line of 230 V has conductor spacing of 1.5 m. The diameter of each conductor is 1.5 cm. Calculate the loop inductance in mH of the line per km. [**Ans.** 2.22 mH]
2. A conductor is composed of 7 identical copper strands, each having a radius *r*. Find the self GMD of the conductor. [**Ans.** 2.176 *r*]
3. A 3-phase transmission line 100 km long has its conductors of 0.5 cm diameter spaced at the corners of the equilateral triangle of 120 cm side. Find the inductance per phase of the system. [**Ans.** 0.129 H]
4. A single circuit 3-phase line operating at 50 Hz has a conductor diameter of 5.18 mm. The spacing between the line conductors is given in metres. D_{ry} = 3.2 ; D_{yb} = 4.0; D_{br} = 5.0. Calculate the inductance and inductive reactance per phase per km. [**U.P.S.C. I.E.S. 1972**]

   [**Ans.** 1.52 mH; 0.477 Ω]
5. The three conductors of a 3-phase transmission line are arranged in a horizontal plane and are 4 metres apart. The diameter of each conductor is 25 mm. Determine the inductance per km of each conductor (line-to-neutral). Assume balanced load and R, Y, B phase sequence.

   Determine also the average inductance per phase for regularly transposed line.

   [**Ans.** (1.273 – *j* 0.12) mH; 1.2 mH; (1.273 + *j* 0.12) mH; 1.249 mH]
6. A part of the transposition cycle of a 3-phase double circuit line is shown in Fig. 1. Radius of each conductor is 12.5 mm. The conductors are solid copper. Determine GMD and GMR and calculate the inductance per km per phase of the line.

   [**Ans.** GMD = 6.61 m; GMR = 0.318 m and H = 0.606 mH/km/phase]

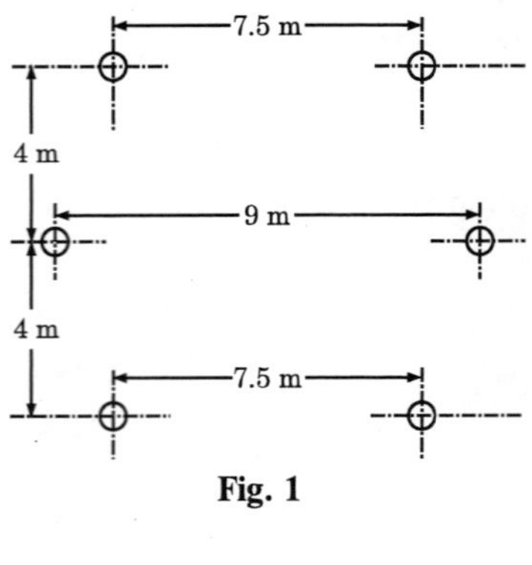

Fig. 1

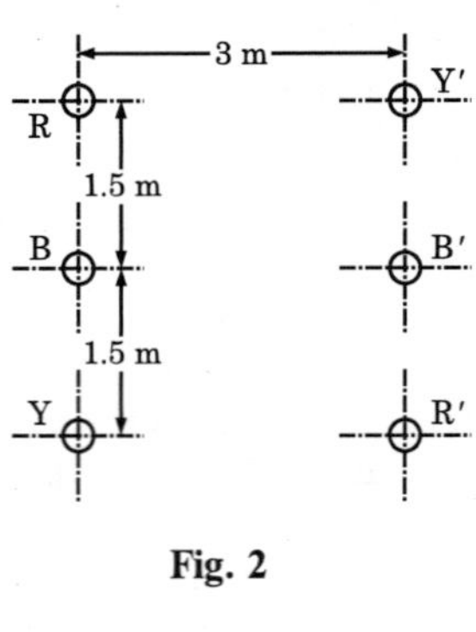

Fig. 2

7. A symmetrical double circuit, 3-phase line is to have its 6 conductors arranged, as shown in Fig. 2. Calculate the inductive reactance per phase in ohms/km. The line is completely transposed and the radius of each conductor is 1.25 cm. [**Agra Univ. 1976**] [**Ans.** 0.16 Ω/km]

8. A 460 kV bundled conductor transmission line has the conductor spacing as shown below. ACSR conductors are used. Determine the inductance per phase per km if the self GMD of each conductor is 0.9 cm. Assume line regularly transposed. [**Ans.** 1.023 mH]

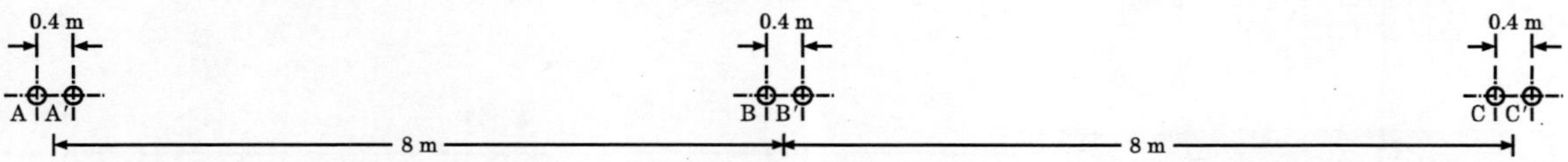

Fig. 3

9. Calculate the capacitance of a single phase overhead line 40 km long consisting of two parallel conductors 1.5 m apart. The conductors are 1.5 cm in diameter. [**Ans.** 0.21 μF]

10. A single phase, 30 km long overhead transmission line consists of two parallel long conductors each 5 mm in diameter and spaced 1.5 m apart. If the line voltage is 50 kV, 50 Hz, determine the charging current of the open-circuited line. [**Ans.** 2.04 A]

11. Calculate the capacitance of a single phase 50 Hz overhead line 40 km long consisting of two parallel wires each 5 mm in diameter, 1.5 metres apart, taking the effect of ground into consideration. The height of the conductors above ground is 7 metres. Derive any formula used. [**A.P.S. Univ. Electric Power Generation, Transmission and Distribution 1978**]

[**Ans.** 0.174 μF]

12. A three phase 50 Hz fully transposed line has conductors equilaterally spaced at 3 metres. The conductor diameter is 3 cm. Determine the capacitance to neutral per km per phase.

[**Gorakhpur Univ. Transmission of Electric Power 1975**]

[**Ans.** 0.0105 μF]

13. The three conductors of a 3-phase, 3-wire line are arranged at the corners of a triangle, the sides of which are 1 m, 1.3 m and 2 m. Calculate the inductance and capacitance per km of the line, when the conductors are transposed. The conductors are 1.5 cm diameter.

[**Ans.** 1.0923 mH; 0.0107 μF]

14. A 220 kV, 50 Hz, 200 km line has its conductors on the corners of a triangle with sides 6 m, 6 m and 12 m. The conductor radius is 1.81 cm. Find the capacitance per phase per km.

[**Ans.** 1.844 μF]

15. A 3-phase, 132 kV, 50 Hz overhead transmission line has steel cored aluminium conductors of equivalent copper area of 1.5 cm^2 and effective diameter of 39.8 mm spaced equilaterally 8 m apart. Calculate the line constants. The resistivity of copper is 1.73 μΩ-cm.

[**Ans.** R = 0.11533 Ω/km/conductor; L = 1.25 mH/km/phase; C = 0.00928 μF/km/phase]

❑❑❑

5

CHAPTER

REPRESENTATION AND PERFORMANCE OF SHORT AND MEDIUM TRANSMISSION LINES

INSIDE THIS CHAPTER

5.1. Introduction **5.2.** Classification of Transmission Lines **5.3.** Regulation and Efficiency of a Transmission Line **5.4.** Short Transmission Lines **5.5.** Medium Transmission Lines **5.6.** Ferranti Effect **5.7.** Charging Current and Line Losses in Open-Circuited Line **5.8.** Effect of Capacitance on Performance of Loaded Line **5.9.** Generalised Circuit Constants.

5.1. INTRODUCTION

The performance of a power system under normal balanced steady state conditions is of primary importance in power system engineering. The transmission line is the main energy corridor in a power system. The performance of a power system depends mainly on the performance of transmission lines in the system. The important considerations in the operation of the transmission lines are voltage drop and power losses occurring in the line and efficiency of transmission. The performance of a transmission line is governed by its four parameters—series resistance R and inductance L, shunt capacitance C and conductance G. The resistance R is due to the fact that every conductor offers opposition to the flow of current. The inductance L is due to the fact that the current carrying conductor is surrounded by the magnetic lines of force. The capacitance of the line is due to the fact that the conductor carrying current forms a capacitor with the earth which is always at lower potential than the conductor and the air between them forms a dielectric medium. The shunt conductance is mainly due to flow of leakage currents over the surface of the insulators especially during bad weather. The effect of line resistance is to cause voltage drop (IR volts) and power loss ($I^2 R$ watts) in the line. The effect of line inductance is to cause voltage drop in the line in quadrature with the current flowing in the conductor, which is equal to $2\pi f LI$ volts where f is supply frequency in Hz, L is the inductance per conductor in Henries and I is the current flowing through the conductor in amperes. The effect of line capacitance is to produce a current; called the *charging current,* which is in quadrature with the voltage. This component of current in the conductor has its maximum value at the sending end of the line and decreases continuously as the receiving end of the line is approached at which its value is zero. The magnitude of charging current I_C at the sending end is given by $2\pi fCV_S$ where f is the supply frequency in Hz, C is the capacitor per conductor in farads and V_S is the voltage at the sending end to neutral in volts. The shunt conductance is in parallel with the system. Generally the leakage currents are small and the shunt conductance G is ignored in calculations.

It is worthnoting that both the reactive drop ($2\pi fLI$) and the charging current ($2\pi fCV_S$) of the line are proportional to the supply frequency and thus have a far greater influence on the performance of a 50 Hz line than on a 25 Hz line. Also that with overhead lines reactance effects are relatively more important owing to wide spacing of conductors which must be employed, whilst in the case of transmission by underground cables reactance effects are small and the capacitance effect predominates.

We are aware that in 3-phase circuit problems it is sufficient to compute results for one phase and subsequently predict results for the remaining two phases by exploiting the 3-phase symmetry. Although the transmission lines are not spaced equilaterally and not transposed, the resulting asymmetry is slight and the phases are considered to be balanced. A 3-phase transmission line carrying an equal load on each phase can, therefore be represented by its single phase equivalent in which inductance is computed for one phase of a balanced 3-phase line and capacitance is computed from line to neutral. While working with a 3-phase transmission line carrying an equal load on each phase, it is usual to assume that all the given voltages are line-to-line values, that all the currents are line currents, that all volt-amperes, power and reactive volt-amperes are 3-phase and that all the power factors are phase power factors (cosine of the angle between the phase voltage and the corresponding phase current). No doubt, all calculations are done in per phase values.

5.2. CLASSIFICATION OF TRANSMISSION LINES

For discussing the performance of the overhead transmission lines they are classified as short, medium and long transmission lines. The overhead transmission lines are classified depending upon the manner in which capacitance is taken into account.

Transmission lines having length lesser than 80 km and operating voltage lower than 20 kV fall in the first category *i.e. short transmission lines*. Due to smaller distance and lower line voltage, the capacitance effects are extremely small and, therefore, can be neglected. Hence the performance of short transmission lines depend upon the resistance and inductance of the line. Though in an actual line, the resistance and inductance are distributed over the whole length, but in case of short lines, the total resistance and inductance are assumed to be lumped at one place. In case of single phase circuits, the total loop resistance and inductance is to be taken into account, whereas, in case of 3-phase circuits only resistance and inductance to neutral *i.e* per phase is required to be taken into account. The effect of generators and transformers can be taken into account by adding their impedances to the impedance of the line.

Transmission lines having length between 80 km and 200 km and line voltage between 20 kV and 100 kV fall into second category *i.e. medium transmission lines*. Owing to appreciable length and voltage of the line the charging current is appreciable and, therefore, capacitance effect cannot be ignored. Though the capacitance is uniformly distributed over the entire length of the line, yet the capacitance may be assumed to be concentrated at one or more points.

The transmission lines having length above 200 km and line voltage above 100 kV fall into third class *i.e. long transmission lines*. In these lines impedance and admittance are to be considered uniformly distributed (not lumped) and, therefore, for their solution rigorous methods are employed.

It is to be emphasised here that exact solution of any transmission line must be based on the fact that the parameters or constants of any line are not lumped but are distributed uniformly throughout the length of the line. However, results obtained by assuming the constants as lumped for short and medium transmission lines are reasonably accurate.

5.3. REGULATION AND EFFICIENCY OF A TRANSMISSION LINE

By performance of a transmission line is meant the determination of regulation and efficiency of the line.

1. *Regulation.* When the load is supplied, there is a voltage drop in the line due to resistance and inductance of the line and, therefore, receiving-end voltage V_R is usually less than sending-end voltage V_S. The voltage drop *i.e.* difference of sending-end voltage and receiving-end voltage expressed as a percentage of receiving-end voltage is called the *regulation*.

When the load is thrown off *i.e* when the line is supplying no load, the receiving-end voltage becomes equal to sending-end voltage and therefore, regulation can be defined as below :

Regulation is defined as the change in voltage at the receiving (or load) end when the full-load is thrown off, the sending-end (or supply) voltage and supply frequency remaining unchanged. It is usually expressed as a percentage of receiving-end voltage.

Mathematically percentage voltage regulation of a transmission line is given by the expression

$$\% \text{ voltage regulation} = \frac{V_S - V_R}{V_R} \times 100 \quad ...(5.1)$$

where V_S is the voltage at sending or supply end and V_R is the voltage at the receiving or load end. *It is to be noted here that* V_S *and* V_R *are the magnitudes of voltages.*

The *lower the voltage regulation, better it is*, because low voltage regulation means little variation in receiving-end voltage due to variation in load current.

Knowledge of voltage regulation helps in maintaining the voltage at the load terminals within prescribed limits* by employing suitable voltage control equipment.

2. *Efficiency.* When the load is supplied there are line losses due to ohmic resistance of line conductors and power delivered at the load end of a transmission line is less than the power supplied at the sending end. Efficiency of a transmission line is defined as below :

Efficiency of a transmission line is defined as the ratio of power delivered at the receiving end to the power sent from the sending end.

Mathematically transmission efficiency is given by the expression

$$\eta_T = \frac{V_R I_R \cos\phi_R}{V_S I_S \cos\phi_S} \times 100 \quad ...(5.2)$$

$$\text{or } \eta_T = \frac{V_R I_R \cos\phi_R}{V_R I_R \cos\phi_R + \text{line losses}} \times 100 \quad ...(5.3)$$

---

* According to IE Rule 54, the voltage at the point of commencement of supply shall not vary by more than

1. ± 6% in the case of low voltage (*i.e.* up to 250 V) and medium voltage (*i.e.* up to 650 V)
2. + 6% to – 9% in the case of high voltage (*i.e.* up to 33 kV)
3. ± $12\frac{1}{2}$% in the case of extra high voltage (*i.e.* exceeding 33 kV)

where V_R, I_R and cos ϕ_R are the receiving-end voltage, current and power factor (all phase values) while V_S, I_S and cos ϕ_S are the sending-end voltage, current and power factor (all phase values)

5.4. SHORT TRANSMISSION LINES

In a short transmission line the shunt conductance and shunt capacitance are neglected and so only the series resistance and inductive reactance are to be considered. The equivalent circuit of a short transmission line is shown in Fig. 5.1 (*a*) where R and X represent the total resistance and inductive reactance of both the conductors (*i.e.* go and return) of a single phase transmission line. Since the effect of shunt capacitance is neglected, the series resistance and reactance are taken as lumped. Thus the equivalent circuit of a short transmission line is a simple ac series circuit.

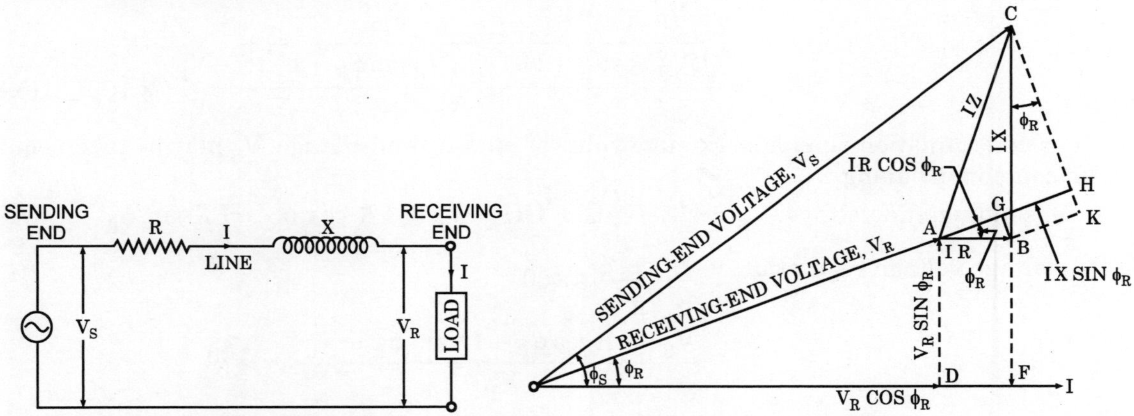

(*a*) *Equivalent Circuit Diagram* (*b*) *Phasor Diagram*

Fig. 5.1. ***Short Transmission Line***

The equivalent circuit shown in Fig. 5.1 (*a*), will represent a 3-phase short transmission line if R and X represent the resistance and inductive reactance to neutral, I is the current in one conductor, V_R is the receiving-end voltage to neutral, V_S is the sending-end voltage to neutral, and the load is that in one phase; the voltage drop found in this way is the drop between a phase and neutral.

From the equivalent circuit shown in Fig. 5.1 (*a*) it is clear that

Receiving-end voltage, $\mathbf{V_R} = \mathbf{V_S} - \mathbf{I}\,(R + jX) = \mathbf{V_S} - \mathbf{IZ}$ ...(5.4)

The quantity I Z is the voltage drop along the line.

Phasor diagram, taking phasor I as the reference phasor, for an inductive load (lagging power factor load) is shown in Fig. 5.1 (*b*) where phasor OI, OA AB, BC, AC and OC represent the load current I, receiving-end voltage V_R, resistive drop I R in line, reactive drop I X in line, line impedance drop I Z and sending-end voltage V_S respectively.

From phasor diagram shown in Fig. 5.1 (*b*),

$$\text{Sending-end voltage, } V_S = OC = \sqrt{OF^2 + FC^2}$$

$$= \sqrt{(OD + DF)^2 + (FB + BC)^2}$$

$$= \sqrt{(V_R \cos\phi_R + IR)^2 + (V_R \sin\phi_R + IX)^2} \quad ...(5.5)$$

$$\text{Sending-end phase angle, } \phi_S = \text{Tan}^{-1}\frac{FC}{OF} = \frac{V_R \sin\phi_R + IX}{V_R \cos\phi_R + IR} \qquad ...(5.6)$$

$$\text{and sending-end power factor, } \cos\phi_S = \frac{OF}{OC} = \frac{V_R \cos\phi_R + IR}{V_S} \qquad ...(5.7)$$

Hence, knowing the conditions existing at one end of the line, those at the other end can easily be determined by means of above relations. Problems may also be solved graphically by drawing the phasor diagram to scale, but owing to very small size of impedance triangle ABC compared with rest of the diagram, accurate results cannot be obtained. So it is advisable to calculate all the quantities trigonometrically, using the graphical method as a rough check,

$$\text{Percentage voltage regulation} = \frac{V_S - V_R}{V_R} \times 100$$

$$= \frac{\sqrt{(V_R \cos\phi_R + IR)^2 + (V_R \sin\phi_R + IX)^2} - V_R}{V_R} \times 100 \qquad ...(5.8)$$

For determination of an approximate value of sending-end voltage, V_S may be taken equal to its component along V_R.

So sending-end voltage, $V_S \simeq OA + AG + GH \simeq V_R + I\,R \cos\phi_R + I\,X \sin\phi_R$ ...(5.9)

$$\text{Percentage voltage regulation} \simeq \frac{V_S - V_R}{V_R} \times 100$$

$$\simeq \frac{V_R + IR\cos\phi_R + IX\sin\phi_R - V_R}{V_R} \times 100$$

$$\simeq \frac{IR\cos\phi_R + IX\sin\phi_R}{V_R} \times 100 \qquad ...(5.10)$$

From the above expression it is obvious that voltage regulation of the line depends upon the resistance and reactance of the line for supplying a given load at a given voltage—larger the value of line resistance and line reactance, greater the percentage regulation.

$$\text{Regulation per unit} = \frac{IR\cos\phi_R}{V_R} + \frac{IX\sin\phi_R}{V_R} = \upsilon_r \cos\phi_R + \upsilon_x \sin\phi_R \qquad ...(5.11)$$

where υ_r and υ_x are the per unit values of resistance and reactance of the line.

Power delivered to load $= V_R I \cos\phi_R$

Line losses $= I^2 R$

Power sent from supply end $= V_R I \cos\phi_R + I^2 R$

$$\text{Efficiency of transmission line, } \eta_T = \frac{\text{Power delivered to load}}{\text{Power supplied from supply end}} \times 100$$

$$= \frac{V_R I \cos\phi_R}{V_R I \cos\phi_R + I^2 R} \qquad ...(5.12)$$

Note: The approximate formula for sending-end voltage V_S [Eq. (5.9)] gives fairly accurate results for lagging power factors (*i.e.* inductive loads) but an appreciable error is caused for leading power factors (*i.e.* capacitive loads).

5.4.1. Effect of Load Power Factor on Transmission Efficiency and Voltage Regulation of Line

To deliver given power at a given voltage at consumer end, line current is inversely proportional to the power factor of the load. The current, hence, the power loss in the line goes on increasing as the power factor of the load goes on decreasing. Thus efficiency of the transmission line decreases with the fall in power factor and vice-versa.

Since regulation of the line is given by

$$= I R \cos \phi_R + I X \sin \phi_R = I Z \left[\frac{R}{Z} \cos \phi_R + \frac{X}{Z} \sin \phi_R\right]$$

$$= I Z [\cos \theta \, . \cos \phi_R + \sin \theta \, . \sin \phi_R] = I Z \cos (\phi_R - \theta) \qquad ...(5.13)$$

where Z is line impedance and is equal to $\sqrt{R^2 + X^2}$ and θ is the line impedance angle and is equal to $\tan^{-1} \frac{X}{R}$

The regulation is zero when $(\phi_R - \theta) = \frac{-\pi}{2}$ and then $\cos \phi_R = \sin \theta = \frac{X}{Z}$. If $(R + jX)$ is the impedance of the line, generator or transformer, θ is positive since X is positive. It is then necessary that ϕ_R be negative for zero regulation. If it is the impedance of a synchronous condenser, θ may be negative, so that ϕ_R may be positive for zero regulation.

The formula given above for zero regulation *viz.* $\phi_R = \theta - \frac{\pi}{2}$ is approximate. The exact relation is determined as follows :

$$V_S^2 = (V_R \cos \phi_R + I R)^2 + (V_R \sin \phi_R + I X)^2$$

For zero regulation $V_S = V_R$ [Refer to Eq. (5.5)]

$$\therefore \; V_R^2 = V_R^2 + 2 I V_R (R \cos \phi_R + X \sin \phi_R) + I^2 (R^2 + X^2)$$

$$\therefore \; R \cos \phi_R + X \sin \phi_R = \frac{-IZ^2}{2 V_R} \qquad \because \; R^2 + X^2 = Z^2$$

$$\text{or} \quad \cos (\phi_R - \theta) = \frac{-IZ}{2 V_R}$$

$$\text{or} \quad \phi_R = \theta - \frac{\pi}{2} - \sin^{-1} \frac{IZ}{2V_R} \qquad ...(5.14)$$

The regulation is negative *i.e.,* the receiving-end voltage is greater than the sending-end voltage, if ϕ_R is negative and numerically greater than the value given in this equation

From the expression for regulation *viz.* $I R \cos \phi_R + I X \sin \phi_R$ the following conclusions can be drawn :

(*i*) When the load power factor, $\cos \phi_R$ is lagging or unity or such leading that $I R \cos \phi_R$ exceeds $I X \sin \phi_R$, the voltage regulation is positive (receiving-end voltage is lesser than the sending-end voltage) and increases with the decrease in power factor for a given load current to be delivered at a given voltage at receiving-end.

(*ii*) When the load power factor, $\cos \phi_R$ is leading to the extent that $I X \sin \phi_R > I R \cos \phi_R$, the voltage regulation is negative (the receiving-end voltage, is more than sending-end voltage) and decreases with the decrease in power factor for a given load current to be delivered at a given voltage at receiving-end.

Note : For leading power factor loads the Eq. (5.10) for voltage regulation of a transmission line is slightly modified. This modification is due to the fact that phase angle ϕ_R is considered positive for lagging power factor loads and negative for leading power factor loads and $\sin(-\phi_R) = -\sin\phi_R$. The modified expression for voltage regulation for leading power factor loads is given below :

$$\%\text{ voltage regulation} = \frac{IR\cos\phi_R - IX\sin\phi_R}{V_R} \times 100$$

5.4.2. Mixed Sending-and Receiving-end Conditions

Sometimes the mixed conditions arise for example when the sending-end voltage and the receiving-end power and power factor are given and the sending-end power and power factor and the receiving-end voltage are to be determined. In such circumstances the solution becomes more complicated. Such problems can be dealt with as follows:

We know that

$$\text{Power delivered, } P = V_R I\cos\phi_R \text{ watts} \qquad ...(5.15)$$

$$\text{Reactive volt-amperes delivered, } Q = V_R I\sin\phi_R \qquad ...(5.16)$$

$$\text{Sending-end voltage, } V_S = \sqrt{(V_R\cos\phi_R + IR)^2 + (V_R\sin\phi_R + IX)^2}$$

or $V_S^2 = V_R^2\cos^2\phi_R + 2V_R IR\cos\phi_R + I^2R^2 + V_R^2\sin^2\phi_R + 2V_R IX\sin\phi_R + I^2X^2$

or $V_S^2 = V_R^2(\cos^2\phi_R + \sin^2\phi_R) + 2V_R(IR\cos\phi_R + IX\sin\phi_R) + I^2(R^2 + X^2)$

or $V_S^2 = V_R^2 + 2V_R I(R\cos\phi_R + X\sin\phi_R) + I^2(R^2 + X^2)$ ...(5.17)

Substituting $V_R I\cos\phi_R = P$, $V_R I\sin\phi_R = Q$ and $I = \dfrac{P}{V_R\cos\phi_R}$ from Eqs. (5.15) and (5.16)

in above Eq. (5.17) we have

$$V_S^2 = V_R^2 + 2(PR + QX) + \frac{(R^2 + X^2)P^2}{V_R^2\cos^2\phi_R}$$

or $$V_S^2 V_R^2 = V_R^4 + 2V_R^2(PR + QX) + \frac{(R^2 + X^2)P^2}{\cos^2\phi_R}$$

or $$V_R^4 - V_R^2[V_S^2 - 2(PR + QX)] + \frac{(R^2 + X^2)P^2}{\cos^2\phi_R} = 0$$

or $$V_R^4 - AV_R^2 + B = 0$$

where $A = V_S^2 - 2(PR + QX)$ and $B = \dfrac{(R^2 + X^2)P^2}{\cos^2\phi_R}$ or $V_R = \sqrt{\dfrac{A \pm \sqrt{(A^2 - 4B)}}{2}}$ ...(5.18)

From the above expression we will get two values for V_R depending on whether the + ve or – ve sign is used. The solution giving the higher value of V_R and consequently higher transmission efficiency is the one required in practice, and therefore the + ve sign is usually taken.

After determining the value of V_R, other quantities can be evaluated.

Graphical Solution. The value of receiving-end voltage V_R can also be determined graphically as explained below:

OA is drawn representing the voltage to neutral at sending end of the line and AO′ is drawn making an angle $\frac{\pi}{2} + \phi_R - \theta$ with AO where $\theta = \tan^{-1}\frac{X}{R}$ (Fig. 5.2).

BO′ is drawn perpendicular to OA and an arc of circle, OCA is drawn with O′ as centre and O′A as radius.

Now with B, mid-point of OA, as centre and BC as radius an arc cutting the arc OCA at C is drawn where $BC = \sqrt{\frac{V_S^2}{4} - (PR + QX)}$. Then OC represents the unknown voltage at the receiving end, V_R.

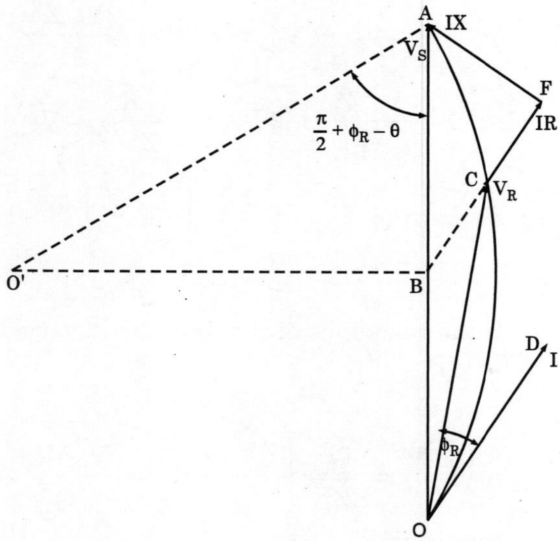

Fig. 5.2

5.4.3. Maximum Power With a Given Regulation

The maximum power that can be transmitted with a given regulation can be determined as follows:

From Eq. (5.17)

$$V_S^2 = V_R^2 + 2\,V_R I\,(R\cos\phi_R + X\sin\phi_R) + I^2\,(R^2 + X^2)$$

Substituting $V_R I\cos\phi_R = P$, $V_R I\sin\phi_R = Q$ and $I^2 = \frac{P^2 + Q^2}{V_R^2}$ from Eqs. (5.15) and (5.16) in above equation we have

$$-V_S^2 + V_R^2 + 2\,PR + 2\,QX + \frac{(P^2 + Q^2)(R^2 + X^2)}{V_R^2} = 0 \qquad \text{...(5.19)}$$

For a given line, *i.e.* given values of R and X, and for given receiving-and sending-end voltages, the above Eq. (5.19) gives relation between P and Q. Active power P delivered, varies with the variation in reactive power Q delivered *i.e.* the above equation gives P as an implicit function of Q. For determining the condition for a maximum value of power P, let us differentiate the above Eq. (5.19) w.r.t Q and put $\frac{dP}{dQ} = 0$

$$\text{So } 2X + \frac{2Q\,(R^2 + X^2)}{V_R^2} = 0$$

$$\text{or } Q = \frac{-V_R^2 X}{R^2 + X^2} \qquad \text{...(5.20)}$$

i.e. for maximum value of P, the value of Q is a constant negative value.

The maximum value of P is determined by substituting this value of Q in Eq. (5.19) which then becomes

$$-V_S^2 + V_R^2 + 2\,PR - \frac{2V_R^2 X^2}{R^2 + X^2} + \frac{P^2\,(R^2 + X^2)}{V_R^2} + \frac{V_R^4 X^2}{(R^2 + X^2)^2} \times \frac{(R^2 + X^2)}{V_R^2} = 0$$

or $$-V_S^2 + V_R^2 + 2\,PR - \frac{2V_R^2 X^2}{Z^2} + \frac{P^2 Z^2}{V_R^2} + \frac{V_R^2 X^2}{Z^2} = 0 \qquad \because R^2 + X^2 = Z^2$$

or $$\frac{P^2 Z^2}{V_R^2} + 2\,PR + \frac{V_R^2 R^2}{Z^2} - V_S^2 = 0$$

So we find $$P_{max} = \frac{V_R^2}{Z^2}\left[-R \pm \sqrt{R^2 - R^2 + \frac{V_S^2 Z^2}{V_R^2}}\right] = \frac{V_R^2}{Z^2}\left[\frac{ZV_S}{V_R} - R\right] \qquad ...(5.21)$$

taking +ve sign of the square root.

If the reactance of the line can be varied, P_{max} will vary, and will have a final maximum when $\frac{dP_{max}}{dX} = 0.$

As $$P_{max} = \frac{V_R^2}{Z^2}\left[\frac{ZV_S}{V_R} - R\right] = \frac{V_R V_S}{Z} - \frac{V_R^2 R}{Z^2} = \frac{V_R V_S}{\sqrt{R^2 + X^2}} - \frac{V_R^2 R}{R^2 + X^2}$$

the condition is

$$\frac{V_R V_S X}{(R^2 + X^2)^{3/2}} = \frac{2V_R^2 RX}{(R^2 + X^2)^2} \quad \text{which yields } X = 0$$

$$\text{or } (R^2 + X^2)^{1/2} = \frac{2RV_R}{V_S} \quad i.e.\ X = R\sqrt{\left(\frac{4V_R^2}{V_S^2} - 1\right)}$$

The value of P_{max} is then $$P_{max} = \frac{V_S^2}{4R} \qquad ...(5.22)$$

Example 5.1. A single-phase line is transmitting 1,100 kW power to a factory at 11 kV and at 0.8 pf lagging. It has a total resistance of 2 Ω and a loop reactance of 3 Ω. Determine *(i)* the voltage at the sending end, *(ii)* percentage regulation and *(iii)* transmission efficiency.

Solution : Load current, $$I = \frac{P \times 1{,}000}{V_R \cos\phi_R} = \frac{1{,}100 \times 1{,}000}{11{,}000 \times 0.8} = 125 \text{ A}$$

(i) Voltage at sending end, $$V_S = \sqrt{(V_R\cos\phi_R + IR)^2 + (V_R\sin\phi_R + IX)^2}$$
$$= \sqrt{(11{,}000 \times 0.8 + 125 \times 2)^2 + (11{,}000 \times 0.6 + 125 \times 3)^2}$$
$$= \sqrt{(9{,}050)^2 + (6{,}975)^2} = 11{,}426 \text{ V } \textbf{Ans.}$$

(ii) Percentage regulation $$= \frac{V_S - V_R}{V_R} \times 100 = \frac{11{,}426 - 11{,}000}{11{,}000} \times 100 = 3.873\% \textbf{ Ans.}$$

Line losses $= I^2 R = (125)^2 \times 2 = 31{,}250$ W or 31.25 kW

(iii) Transmission efficiency, $$\eta_T = \frac{\text{Power delivered}}{\text{Power delivered + line loss}} \times 100$$
$$= \frac{1{,}100}{1{,}100 + 31.25} \times 100 = 97.24\% \textbf{ Ans.}$$

Example 5.2. A short three phase transmission line of parameters R = 0.4 Ω and X = 0.4 Ω is delivering 2,000 kVA to a load at a pf of 0.8 lagging at the receiving end of the line. If the load voltage is 3,000 V, determine the voltage regulation and efficiency of the line.

[*U.P. Technical Univ. Elements of Power System 2004-05*]

Solution : Phase voltage at sending end, $V_R = \dfrac{3{,}000}{\sqrt{3}} = 1{,}732$ V

Load delivered, $P = 2{,}000$ kVA

Load power factor, $\cos\phi_L = 0.8$ lagging

$$\text{Line current, } I = \frac{P \text{ in VA}}{3 \times V_R} = \frac{2{,}000 \times 1{,}000}{3 \times 1{,}732} = 384.9 \text{ A}$$

$$\begin{aligned}\text{Sending end voltage per phase, } V_S &= V_R + IR\cos\phi_R + IX\sin\phi_R \\ &= V_R + I(R\cos\phi_R + X\sin\phi_R) \\ &= 1{,}732 + 384.9(0.4 \times 0.8 + 0.4 \times 0.6) \\ &= 1{,}732 + 384.9 \times 0.56 = 1{,}947.5 \text{ V}\end{aligned}$$

$$\text{Percentage regulation} = \frac{V_S - V_R}{V_R} \times 100 = \frac{1{,}947.5 - 1{,}732}{1{,}732} \times 100 = 12.44\% \textbf{ Ans.}$$

$$\text{Total line losses} = \frac{3I^2R}{1{,}000} = \frac{3 \times (384.9)^2 \times 0.4}{1{,}000} = 177.8 \text{ kW}$$

$$\text{Power output} = \text{kVA} \times \cos\phi = 2{,}000 \times 0.8 = 1{,}600 \text{ kW}$$

$$\text{Transmission efficiency, } \eta_T = \frac{\text{Output power}}{\text{Output power + line losses}} \times 100$$

$$= \frac{1{,}600}{1{,}600 + 177.8} \times 100 = 90\% \textbf{ Ans.}$$

Example 5.3. A 15 km long 3-phase overhead line delivers 5 MW at 11 kV at 0.8 lagging p f. Line loss is 12% of power delivered. Line inductance is 1.1 mH per km per phase. Find sending-end voltage and voltage regulation.

Solution : Power delivered, $P = 5$ MW $= 5{,}000$ kW

$$\text{Phase voltage at receiving end, } V_R = \frac{V_{RL}}{\sqrt{3}} = \frac{11{,}000}{\sqrt{3}} = 6{,}351 \text{ V}$$

$$\text{Line current, } I = \frac{P \times 1{,}000}{\sqrt{3}\, V_{RL} \cos\phi} = \frac{5{,}000 \times 1{,}000}{\sqrt{3} \times 11{,}000 \times 0.8} = 328 \text{ A}$$

$$\text{Line losses, } 3I^2R = 12\% \text{ of } P = \frac{12}{100} \times 5{,}000 \times 1{,}000 = 600{,}000 \text{ W}$$

$$\text{Line resistance, } R = \frac{\text{Line losses}}{3I^2} = \frac{600{,}000}{3 \times 328^2} = 1.859\ \Omega/\text{phase}$$

Inductive reactance per phase, $X = 2\pi f L = 2\pi \times 50 \times 15 \times 1.1 \times 10^{-3} = 5.1836\ \Omega$/phase

Phase voltage at sending end, $V_S = V_R + IR\cos\phi_R + IX\sin\phi_R$

$= 6{,}351 + 328 \times 1.859 \times 0.8 + 328 \times 5.1836 \times 0.6 = 7{,}858.93$ V

Line voltage at sending end, $V_{SL} = \sqrt{3}\ V_S = \sqrt{3} \times 7{,}858.93 = 13{,}612$ V or 13.612 kV **Ans.**

$$\text{Voltage regulation} = \frac{V_{SL} - V_{RL}}{V_{RL}} \times 100 = \frac{13.612 - 11}{11} \times 100 = 23.746\ \% \textbf{ Ans.}$$

Example 5.4. A 3-phase transmission line 50 km long consists of three hard drawn copper conductors in a 1.2 m delta. Load conditions at receiving end are: 10,000 kVA at 0.8 pf lagging 33,000 volts, 50 Hz. Line is designed so that transmission loss is approximately 10%. Find (*i*) the sending-end voltage and power factor; (*ii*) efficiency; and (*iii*) regulation.

Solution : Power supplied, S = 10,000 kVA

$$\text{Line current, } I = \frac{\text{S in kVA} \times 1{,}000}{\sqrt{3} \times V_{RL}} = \frac{10{,}000 \times 1{,}000}{\sqrt{3} \times 33{,}000} = 174.95 \text{ A}$$

$$\text{Line losses, } 3\,I^2 R = 10\ \%\text{ of power delivered} = \frac{10}{100} \times 10{,}000 \times 1{,}000 \times 0.8$$

$$= 800{,}000 \text{ watts}$$

$$\text{Line resistance, } R = \frac{800{,}000}{3 \times (174.95)^2} = 8.7125\ \Omega$$

$$\text{Area of x-section of conductor, } a = \frac{\rho l}{R} = \frac{1.73 \times 10^{-6} \times 50 \times 1{,}000 \times 100}{8.7125} = 0.9928 \text{ cm}^2$$

$$\text{Radius of conductor, } r = \sqrt{\frac{a}{\pi}} = \sqrt{\frac{0.9928}{\pi}} = 0.562 \text{ cm}$$

$$\text{Inductance of line, } L = 0.2 \log_e \frac{d}{r'} \text{ mH/km} \times \text{length of line in km}$$

$$= \left[0.2 \log_e \frac{120}{0.7788 \times 0.562}\right] \times 10^{-3} \times 50 = 56.14 \text{ mH}$$

$$\text{Inductive reactance, } X = 2\pi f L = 2\pi \times 50 \times 56.14 \times 10^{-3} = 17.637\ \Omega$$

$$\text{Sending-end phase voltage, } V_S = V_R + I R \cos\phi_R + I X \sin\phi_R$$

$$= \frac{33{,}000}{\sqrt{3}} + 174.95 \times 8.7125 \times 0.8 + 174.95 \times 17.637 \times 0.6$$

$$= 19{,}052.6 + 3{,}070.8 = 22{,}123.36 \text{ V}$$

$$\text{Sending-end line voltage, } V_{SL} = \sqrt{3} \times 22{,}123.36 = 38{,}320 \text{ V or } 38.32 \text{ kV}$$ **Ans.**

$$\text{Sending-end pf, } \cos\phi_R = \frac{V_R \cos\phi_R + IR}{V_S} = \frac{19{,}052.6 \times 0.8 + 174.95 \times 8.7125}{22{,}123.36}$$

$$= 0.7578 \text{ lagging}$$ **Ans.**

$$\text{Transmission efficiency, } \eta_T = \frac{\text{Output}}{\text{Output} + \text{losses}} \times 100 = \frac{10{,}000 \times 0.8}{10{,}000 \times 0.8 + 800} \times 100 = 90.91\ \%$$ **Ans.**

$$\text{Voltage regulation} = \frac{V_{SL} - V_{RL}}{V_{RL}} \times 100 = \frac{38.32 - 33}{33} \times 100 = 16.12\ \%.$$

Example 5.5. A short 3-phase transmission line connected to a 33 kV, 50 Hz generating station at the sending end is required to supply a load of 10 MW at 0.8 lagging power factor at 30 kV at the receiving end. If the minimum transmission efficiency is to be limited to 96% estimate the per phase values of resistance and inductance the line.

Solution : Power delivered, P = 10 MW = 10,000 kW

$$\text{Phase voltage at receiving end, } V_R = \frac{V_{RL}}{\sqrt{3}} = \frac{30{,}000}{\sqrt{3}} = 17{,}320 \text{ V}$$

$$\text{Line current, } I = \frac{P \times 1{,}000}{\sqrt{3}\, V_{RL} \cos\phi_R} = \frac{10{,}000 \times 1{,}000}{\sqrt{3} \times 30{,}000 \times 0.8} = 240.56 \text{ A}$$

$$\text{Line losses, } \frac{3I^2 R}{1{,}000} = \text{Power delivered} \times \left(\frac{1}{\eta_T} - 1\right) = 10{,}000 \left(\frac{1}{0.96} - 1\right) = 416.67 \text{ kW}$$

$$\text{or } R = \frac{416.67 \times 1{,}000}{3 \times (240.56)^2} = 2.4\ \Omega/\text{phase} \textbf{ Ans.}$$

$$\text{Phase voltage at sending end, } V_S = \frac{V_{SL}}{\sqrt{3}} = \frac{33{,}000}{\sqrt{3}} = 19.052 \text{ V}$$

$$\text{Since } V_S = V_R + I R \cos\phi_R + I X \sin\phi_R$$

$$\therefore\ 19{,}052 = 17{,}320 + 240.56 \times 2.4 \times 0.8 + 240.56 \times X \times 0.6$$

$$\text{or } X = \frac{19{,}052 - 17{,}320 - 240.56 \times 2.4 \times 0.8}{240.56 \times 0.6} = 8.8\ \Omega/\text{phase}$$

$$\text{Inductance per phase, } L = \frac{X}{2\pi f} = \frac{8.8}{2\pi \times 50} = 28 \text{ mH/phase} \textbf{ Ans.}$$

Example 5.6. A 3-phase line, 3 km long delivers 3,000 kW at a power factor of 0.8 (lagging) to a load. If the voltage at supply end is 11 kV, determine the voltage at the load end and the efficiency of transmission. The resistance and reactance per km of each conductor are 0.4 Ω and 0.8 respectively.

[*U.P. Technical Univ. Elements of Power System 2006-07*]

Solution : Resistance of each conductor, $R = 3 \times 0.4 = 1.2\ \Omega$

Reactance of each conductor, $X = 3 \times 0.8 = 2.4\ \Omega$

Let the phase voltage at the load end be V_R and load current be I amperes

$$\text{Phase voltage at supply end, } V_S = \frac{V_{SL}}{\sqrt{3}} = \frac{11{,}000}{\sqrt{3}} = 6{,}351 \text{ V}$$

$$\text{Now phase voltage at sending end, } V_S = V_R + I(R\cos\phi_R + X\sin\phi_R)$$

$$\text{or } 6{,}351 = V_R + I(1.2 \times 0.8 + 2.4 \times 0.6)$$

$$\text{or } V_R = 6{,}351 - 2.4\, I \qquad ...(i)$$

$$\text{also } I = \frac{\text{kW output} \times 1{,}000}{3 \times V_R \times \cos\phi_R} = \frac{3{,}000 \times 1{,}000}{3 \times V_R \times 0.8} = \frac{12{,}50{,}000}{V_R} \qquad ...(ii)$$

Solving Eqs. (*i*) and (*ii*) we get $V_R = 5{,}837$ V and I = 214 A

$$\text{Line voltage at load end, } V_{RL} = \sqrt{3} \times V_R = \sqrt{3} \times 5{,}837 = 10{,}110 \text{ V} \textbf{ Ans.}$$

$$\text{Transmission efficiency, } \eta_T = \frac{\text{Output}}{\text{Output + line losses}} \times 100$$

$$= \frac{3{,}000 \times 1{,}000}{3{,}000 \times 1{,}000 + 3 \times (214)^2 \times 1.2} \times 100 = 94.8\% \textbf{ Ans.}$$

Example 5.7. The sending-end voltage of a 3-phase overhead line with a resistance and reactance per phase of 5 Ω and 20 Ω respectively is 46.85 kV while the receiving-end voltage is 33 kV. Determine the power output at 0.8 pf (lag) and the sending-end power factor.

Solution : Phase voltage at receiving end, $V_R = \frac{33{,}000}{\sqrt{3}} = 19{,}050$ V

Let load current be I amperes

$$\text{Now phase voltage at sending end, } V_S = V_{RL} + I(R\cos\phi_R + X\sin\phi_R)$$

$$\text{or} \quad \frac{46.85 \times 1,000}{\sqrt{3}} = 19,050 + I\,(5 \times 0.8 + 20 \times 0.6)$$

$$\text{or } I = 500 \text{ amperes}$$

$$\text{Power output} = \frac{\sqrt{3}\, V_{RL}\, I \cos\phi_R}{1,000} \text{ kW} = \frac{\sqrt{3} \times 33,000 \times 500 \times 0.8}{1,000} = 22,853 \text{ kW } \textbf{Ans.}$$

$$\text{Power factor at sending end, } \cos\phi_S = \frac{V_R \cos\phi_R + IR}{V_S} = \frac{19,050 \times 0.8 + 500 \times 5}{\frac{46,850}{\sqrt{3}}} = 0.656 \text{ (lag) } \textbf{Ans.}$$

Graphical Method. Draw $OA = \frac{46,850}{\sqrt{3}}$ *i.e.* 27,048 V (on scale 1 cm = 2,000 V) to represent phase voltage at supply end and BO′ perpendicular to OA and AO′ making an angle $\left(\frac{\pi}{2} + \cos^{-1} 0.8 - \tan^{-1} \frac{20}{5}\right)$ *i.e.* (90° + 36.87° – 75.97° = 50.9°) with OA. Draw an arc of circle with O′ as centre and O′A as radius. Now with O as centre and OC representing 19,050 V, the phase voltage at load end as radius draw an arc cutting the arc ACO at C.

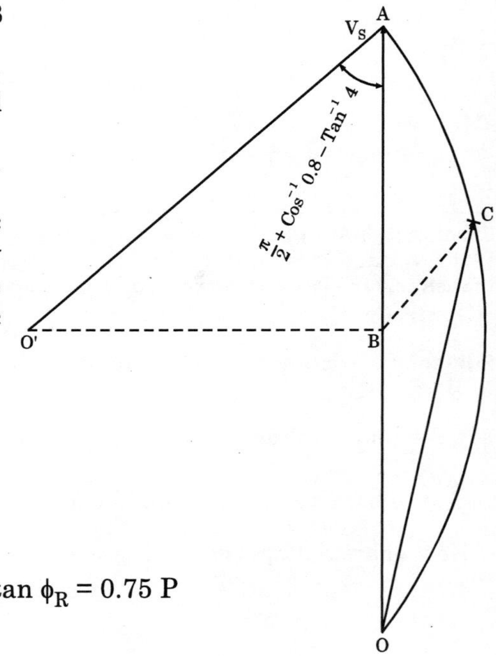

Fig. 5.3

On measurement BC = 3.3 × 2,000 = 6,600 V

$$BC^2 = \frac{V_S^2}{4} - (PR + QX)$$

$$(6,600)^2 = \frac{1}{4}\left(\frac{46,850}{\sqrt{3}}\right)^2 - [P \times 5 + P \times 0.75 \times 20]$$

$$\because \; Q = P \tan\phi_R = 0.75\, P$$

$$\text{or } P = \left[\frac{(46,850)^2}{12} - (6,600)^2\right] \times \frac{1}{20}$$

$$= 69,67,510 \text{ W or } 6,967.51 \text{ kW}$$

Total power output = 3 P = 3 × 6,967.51 = 20,902.5 kW **Ans.**

$$\text{Load current, } I = \frac{20,902.5 \times 1,000}{\sqrt{3} \times 33,000 \times 0.8} = 457 \text{ A}$$

Power factor at sending end,

$$\cos\phi_S = \frac{V_R \cos\phi_R + IR}{V_S} = \frac{19,050 \times 0.8 + 457 \times 5}{\frac{46,850}{\sqrt{3}}} = 0.648 \text{ (lag) } \textbf{Ans.}$$

5.5. MEDIUM TRANSMISSION LINES

As already discussed, for transmission lines of length up to 80 km and transmitting power at relatively low voltage; (voltage below 20 kV), the capacitance is too small that its effects can be neglected. However, the effects of shunt capacitance become more and more pronounced with the increase in length and operating voltage of the line. Since medium transmission lines have sufficient length exceeding 80 km and usually operate at voltages exceeding 20 kV, the capacitive current is appreciable and, therefore, line capacitance is to be taken into account. The capacitive current is always flowing in the line while the supply end switches are closed even though the receiving end of the line may be open-circuited. The magnitude of capacitive current flowing at any point along the line is that required to charge the section of the line between the given point and the receiving end, hence it has a maximum value at the sending end and diminishes at a practically uniform rate down to zero at the receiving end.

Actually the capacitance of the line is uniformly distributed over its entire length, as illustrated in Fig. 5.4. However, in order to make the calculations simple, the capacitance of the system is assumed to be divided up, and 'lumped' in the form of capacitors shunted across the line at one or more points—the more the points the closer the approximation.

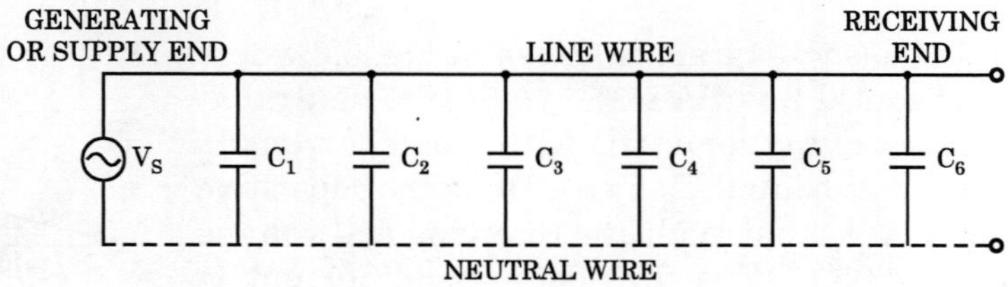

Fig. 5.4

The most common methods of representation, called the *localised capacitance methods*, are (*i*) end condenser method, nominal-T method (or middle condenser method) and nominal-π method (or split-condenser method).

The above methods of solution of transmission lines are only of academic interest as far as long lines are concerned because these involve considerable error owing to the actual distributed capacitance having been assumed lumped. However, for lines up to 200 km at 50 Hz, the error involved by the use of these methods will be of the order of 10 per cent for sending-end voltage calculations which is an acceptable approximation. It is interesting to note that the end condenser method and the nominal T- method over compensates (giving sending-end voltages which are too low), while the nominal-π method under compensates. Because of the approximations involved in T and π-methods, they are often referred to as nominal-T and nominal-π methods.

5.5.1. End Condenser Method

In this method the capacitance of the line is assumed to be lumped at the load-end, as shown in Fig. 5.5. This method overestimates the effect of capacitance.

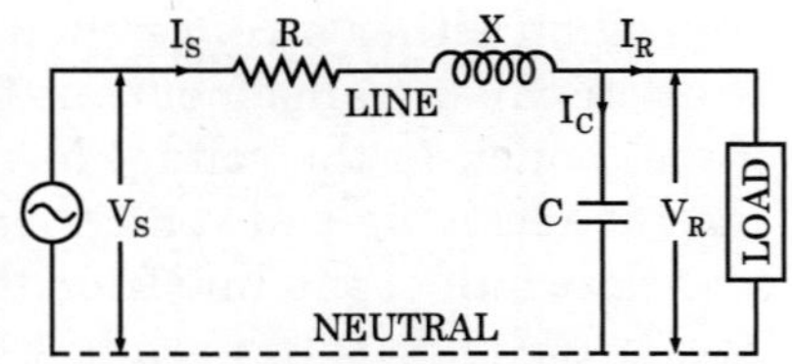

Fig. 5.5. ***Equivalent Circuit For End Condenser Method***

Let the load current per phase be $\mathbf{I}_R$ lagging behind the load end phase voltage $\mathbf{V}_R$ by and angle ϕ_R. Let resistance, and reactance per phase of the line be R and X ohms respectively. Capacitance per phase of line be C farads.

Taking receiving-end phase voltage as reference phasor we have $\mathbf{V}_R = V_R\,(1 + j\,0)$

Receiving-end or load-end current, $\mathbf{I}_R = I_R\,(\cos\phi_R - j\sin\phi_R)$

Capacitive current, $\mathbf{I}_C = j\,\mathbf{V}_R\,\omega\,C = j\,2\,\pi f\,C\,\mathbf{V}_R$

Sending-end current, $\mathbf{I}_S = \mathbf{I}_R + \mathbf{I}_C = I_R\,(\cos\phi_R - j\sin\phi_R) + j\,2\,\pi f\,C\,\mathbf{V}_R$

Voltage drop in line (per phase) $= \mathbf{I}_S\,\mathbf{Z} = \mathbf{I}_S\,(R + jX)$

$= [I_R\,(\cos\phi_R - j\sin\phi_R) + j\,2\,\pi f\,C\,\mathbf{V}_R]\,(R + jX)$

$= I_R\,R\cos\phi_R + I_R\,X\sin\phi_R - 2\,\pi f\,C\,V_R\,X + j\,[I_R\,X\cos\phi_R - I_R\,R\sin\phi_R + 2\,\pi f\,C\,V_R\,R]$

$= I_R\,[R\cos\phi_R + X\sin\phi_R) + j\,(X\cos\phi_R - R\sin\phi_R)] - 2\,\pi f\,C\,V_R\,(X - jR)$

Phase voltage at the sending end, $\mathbf{V_S} = \mathbf{V_R} + \mathbf{I_S Z}$ ...(5.23)

$$\text{Percentage regulation} = \frac{V_S - V_R}{V_S} \times 100$$

$$\text{Transmission efficiency, } \eta_T = \frac{\text{Power delivered per phase}}{\text{Power delivered per phase + line losses / phase}} \times 100$$

$$= \frac{V_R\, I_R \cos\phi_R}{V_R\, I_R \cos\phi_R + I_S^2 R} \times 100 \quad ...(5.24)$$

Phasor diagram is shown in Fig. 5.6 where horizontal line OA is the voltage to neutral at the receiving end; OD is the load current I_R that lags behind V_R by ϕ_R; DF is the capacitive current I_C that leads the receiving-end voltage V_R by 90°; OF is the sending-end current I_S that flows through the line and is the phasor sum of receiving-end current I_R and capacitive current I_C; AB is the resistive voltage drop I_SR that is in phase with I_S; BC is the reactive voltage drop $I_S X$ that leads the I_S by 90°; OC is the sending-end voltage V_S that is the phasor sum of V_R, $I_S R$ and $I_S X$ (or V_R and $I_S Z$); and ϕ_S, the phase angle between the sending-end voltage V_S and sending-end current I_S is the sending-end phase angle that determines the sending-end power factor.

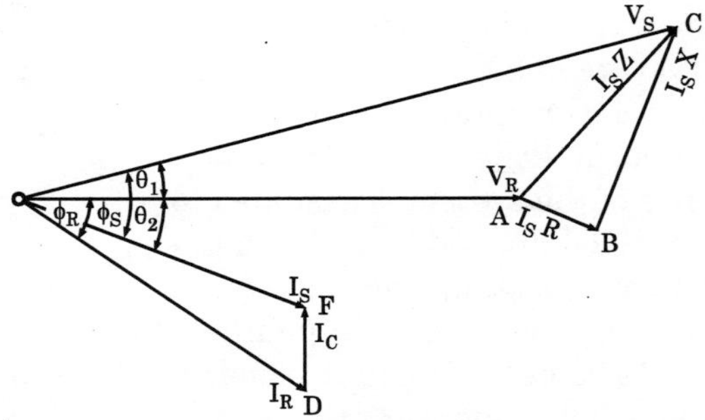

Fig. 5.6. *Phasor Diagram For End Condenser Method*

5.5.2. Nominal T-Method (or Middle Condenser Method)

In this method the whole of the line capacitance is assumed to be concentrated at the middle point of the line and half the line resistance and reactance (*i.e.* $\frac{R}{2}$ and $\frac{X}{2}$) to be lumped on either side, as shown in Fig. 5.7. Thus in this arrangement, at the terminals of the condenser the voltage has a value V′ intermediate between the sending-end voltage V_S and the receiving-end voltage V_R. A capacitive current I_C leading V′ by 90° accordingly flows over first half of the line from the sending-end. The current in the receiving-end half of the line is I_R, and in the sending-end half of the line is I_S which is the phasor sum of the receiving-end (or load) current I_R and charging or capacitance current I_C.

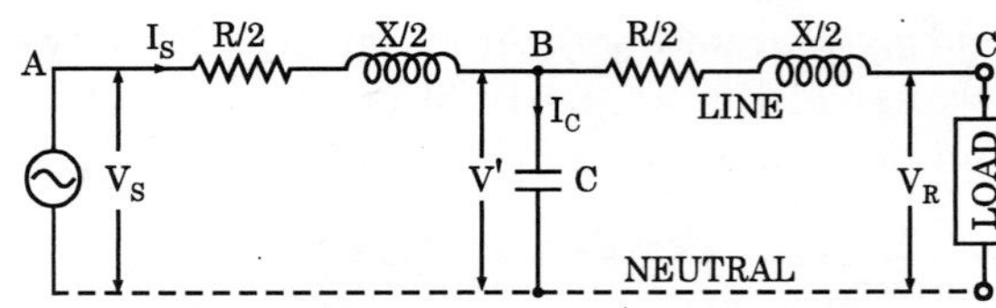

Fig. 5.7. *Equivalent Circuit For Nominal-T Method*

Taking receiving-end phase voltage (voltage to neutral) as the reference phasor we have

$$\mathbf{V_R} = V_R\,(1 + j\,0)$$

$$\text{Receiving-end or load current, } \mathbf{I_R} = I_R\,(\cos\phi_R - j\sin\phi_R)$$

$$\text{Voltage across the condenser, } \mathbf{V} = \mathbf{V_R} + \mathbf{I_R}\frac{\mathbf{Z}}{2} = \mathbf{V_R} + \frac{\mathbf{I_R}}{2}(R + jX)$$

$$= V_R\,(1 + j\,0) + \frac{1}{2} I_R\,(\cos\phi_R - j\sin\phi_R)\,(R + jX) \quad ...(5.25)$$

$$\text{Capacitive current, } I_C = j\,\omega\,C\,\mathbf{V'} = j\,2\,\pi f\,C\,\mathbf{V'} \quad ...(5.26)$$

$$\text{Sending-end current, } \mathbf{I_S} = \mathbf{I_R} + \mathbf{I_C} = I_R\,(\cos\phi_R - j\sin\phi_R) + 2\,\pi f\,C\,\mathbf{V'} \quad ...(5.27)$$

$$\text{Sending-end voltage, } \mathbf{V_S} = \mathbf{V'} + \mathbf{I_S}\frac{(R + jX)}{2} \qquad ...(5.28)$$

$$\text{Transmission efficiency, } \eta_T = \frac{V_R I_R \cos\phi_R}{V_R I_R \cos\phi_R + I_R^2 \frac{R}{2} + I_S^2 \frac{R}{2}} \qquad ...(5.29)$$

In the phasor diagram shown in Fig. 5.8, horizontal line OA is the voltage to neutral at the receiving-end; OB is the current in the receiving-end half of the line (*i.e.*I_R) drawn ϕ_R degrees behind OA; ϕ_R is the load power factor angle; AC drawn parallel to OB is the ohmic voltage drop in the receiving-end half of the line, CD drawn 90° ahead of OB is the reactive voltage drop in the receiving-end half of the line; OD is the voltage across the capacitor; BF drawn 90° ahead of OD is the charging or capacitive current I_C; OF (the phasor sum of OB and BF) is the sending-end current I_S; DG drawn parallel to OF is the ohmic voltage drop in the sending-end half of the line; GH drawn 90° ahead of OF is the reactive voltage drop in the sending end half of the line; OH is the sending-end voltage; and ϕ_S (the angle between OH and OF) is the phase angle at the sending-end.

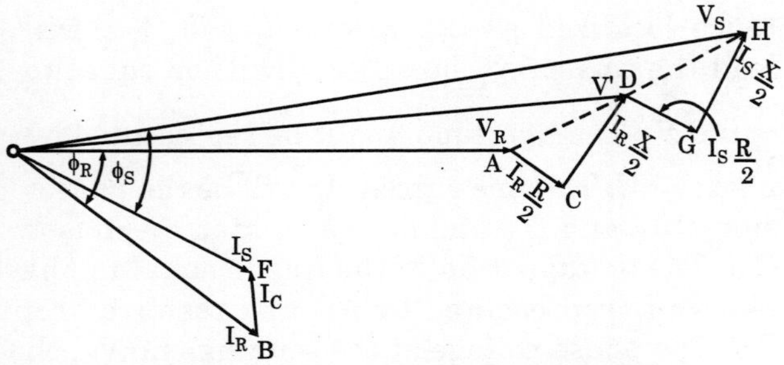

Fig. 5.8. *Phasor Diagram For Nominal-T Method*

5.5.3. Nominal-π Method (or Split-Condenser Method)

In this method the capacitance of each line conductor is assumed to be divided into two halves, one half being shunted between line conductor and neutral at the receiving-end and the other half at the sending end, as shown in Fig. 5.9. The current flowing in the line at any point in between the two capacitors is I_L which is phasor sum of the load current I_R and the current I_{CR} drawn by the receiving-end capacitor. Knowing the line current I_L, the voltage at the sending-end can easily be determined by the fundamental impedance method used for short lines. The sending-end voltage is, therefore, more readily calculated by this method than by nominal-T method which needs the calculation of two separate halves of the line. Though the capacitance at the sending-end has no effect on the voltage drop in the line but if the sending-end current I_S and power factor cos ϕ_S are to be determined, the current drawn by the sending-end capacitor, I_{CS} is to be added vectorially to that in the line.

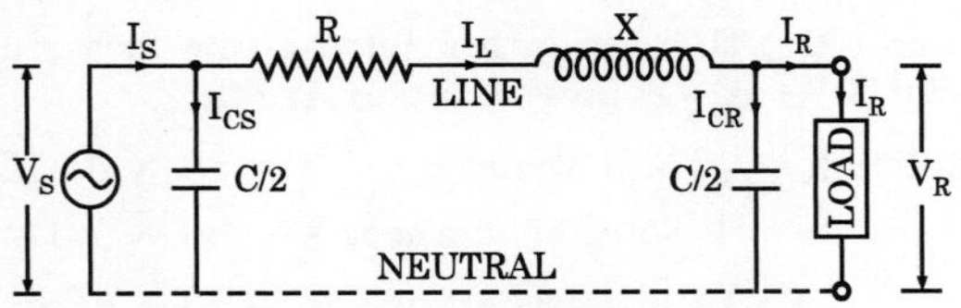

Fig. 5.9. *Equivalent Circuit For Nominal-π Method*

Taking receiving-end phase voltage (voltage to neutral) as the reference phasor we have

$$\mathbf{V_R} = V_R (1 + j\,0)$$

$$\text{Load current, } \mathbf{I_R} = I_R (\cos\phi_R - j \sin\phi_R)$$

$$\text{Charging current at the receiving end, } \mathbf{I_{CR}} = j\,\omega \frac{C}{2} \mathbf{V_R}$$

$$\text{Line current, } \mathbf{I_L} = \mathbf{I_R} + \mathbf{I_C}$$

$$\text{Sending-end voltage, } \mathbf{V_S} = \mathbf{V_R} + \mathbf{I_L Z} = \mathbf{V_R} + \mathbf{I_L}\,(R + jX) \qquad ...(5.30)$$

$$\text{Charging current at the sending end, } \mathbf{I_{CS}} = j\,\omega \frac{C}{2}\, \mathbf{V_S}$$

$$\text{Sending-end current, } \mathbf{I_S} = \mathbf{I_L} + \mathbf{I_{CS}} \quad \text{...(5.31)}$$

In phasor diagram shown in Fig. 5.10 horizontal line OA is the voltage to neutral at the receiving end V_R; OB is the load current I_R drawn ϕ_R degrees behind OA; ϕ_R is the load power factor angle. The capacitive current due to capacitance localised at the receiving end, I_{CR} will lead the phasor V_R by 90° and will be equal to $\frac{1}{2}\omega C V_R$ in magnitude. Let it be represented by phasor BC. The line current I_L will be the phasor sum of I_R and I_{CR} and is represented by phasor OC. The voltage drop in the line is equal to phasor sum of resistive drop $I_L R$ in phase with phasor I_L represented by AD and reactive drop $I_L X$ in quadrature with I_L represented by DF. The phase voltage at the sending-end V_S, the phasor sum of receiving-end phase voltage V_R and the line drop is represented by phasor OF. The sending end current I_S, which is phasor sum of I_L and I_{CS} (I_{CS} is equal to $\frac{1}{2}\omega C V_S$, in magnitude and leads V_S by 90°), is represented by phasor OG.

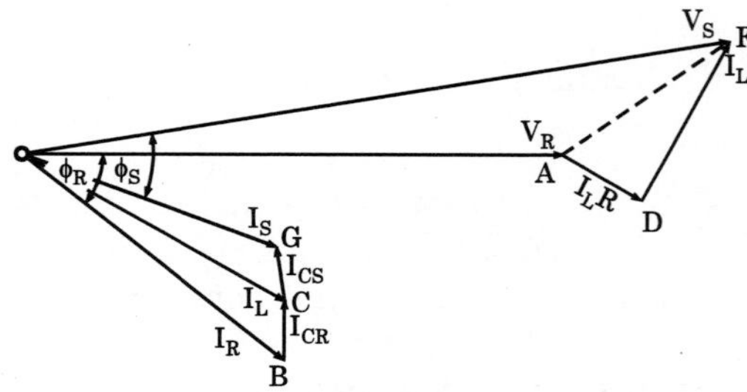

Fig. 5.10. ***Phasor Diagram For Nominal-π Method***

Example 5.8. A single phase medium transmission line 80 km long has the following constants.

Resistance per km = 0.3125 Ω; Reactance per km = 1.0 Ω; Susceptance per km = 17.5 × 10⁻⁶ S. Receiving-end voltage = 66 kV.

Assuming that the total capacitance of the line is localized at the receiving-end alone, determine *(i)* the sending-end current *(ii)* the sending-end voltage *(iii)* the sending-end power factor *(iv)* the regulation *(v)* the line losses and *(vi)* transmission efficiency. The line is delivering 15.0 MW at 0.8 power factor lagging.

Solution : Total resistance, $R = 80 \times 0.3125 = 25\ \Omega$

Total reactance, $X = 80 \times 1.0 = 80\ \Omega$

Total susceptance, $Y = 80 \times 17.5 \times 10^{-6}$
$= 14 \times 10^{-4}$ S

Receiving-end voltage, $V_R = 66$ kV $= 66{,}000$ V

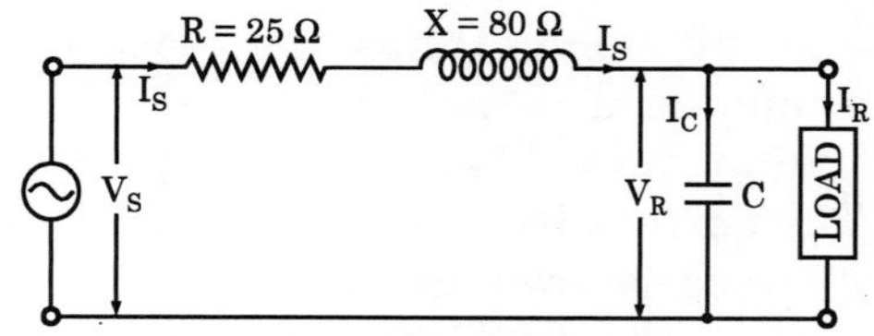

Fig. 5.11

$$\text{Load current, } I_R = \frac{15 \times 10^6}{66{,}000 \times 0.8} = 284 \text{ A}$$

Power factor at receiving end, $\cos\phi_R = 0.8$ and $\sin\phi_R = 0.6$

Taking receiving-end voltage V_R as reference phasor we have

$$\mathbf{V_R} = 66{,}000\ (1 + j\,0) \text{ volts}$$

$$\text{Load current, } \mathbf{I_R} = 284\ (0.8 - j\,0.6) \text{ A} = (227.2 - j\,170.4) \text{ A}$$

$$\text{Charging current, } \mathbf{I_C} = j\,Y\,V_R = j\,0.0014 \times 66{,}000 = j\,92.4 \text{ A}$$

(i) Sending-end current,

$$\mathbf{I_S} = \mathbf{I_R} + \mathbf{I_C} = (227.2 - j\,170.4) + j\,92.4$$
$$= (227.2 - j\,78)$$
$$= 240.2\ \angle{-18.94°} \text{ A} \quad \textbf{Ans.}$$

(ii) Sending-end voltage,

$$\mathbf{V_S} = \mathbf{V_R} + \mathbf{I_S}\,(R + j\,X)$$
$$= (66{,}000 + j\,0) + (227.2 - j\,78)\,(25 + j\,80)$$
$$= (77{,}920 + j\,16{,}226) = 79{,}592\ \angle 11.76° \text{ volts } \textbf{Ans.}$$

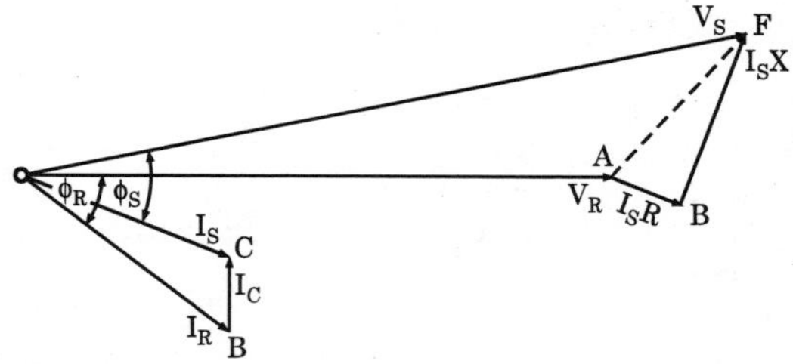

Fig. 5.12

Sending-end power factor angle, $\phi_S = 11.76° - (-18.94°) = 30.7°$

(*iii*) Sending-end power factor, $\cos \phi_S = \text{Cos } 30.7° = 0.86$ (lag) **Ans.**

$$(iv)\ \text{Regulation} = \frac{V_S - V_R}{V_{R_2}} = \frac{79{,}592 - 66{,}000}{66{,}000} = 0.206 \text{ or } 20.6\ \%\ \textbf{Ans.}$$

$$(v)\ \text{Line losses} = \frac{I_S^2 R}{1{,}000} = \frac{(240.2)^2 \times 25}{1{,}000}\ \text{kW} = 1{,}442.4\ \text{kW}\ \textbf{Ans.}$$

$$(vi)\quad \text{Transmission efficiency, } \eta_T = \frac{\text{Power output}}{\text{Power output + line losses}} \times 100$$

$$= \frac{15{,}000}{15{,}000 + 1{,}442.4} \times 100 = 91.23\%\ \textbf{Ans.}$$

Example 5.9. Calculate the sending end voltage, efficiency and regulation of a three phase, 50 Hz, 100 km long transmission line, supplying a load of 20 MW at 0.8 power factor lag, at 66 kV. The resistance per phase is 10 ohms; the inductance per phase is 111.7 mH; and the capacitance per phase is 0.9954 μF. Use nominal T Method. [*U.P. Technical Univ. Elements of Power System 2002-03*]

Solution : Line resistance per phase, $R = 10\ \Omega$

Line reactance per phase, $X = 2\pi f L = 2 \times \pi \times 50 \times 111.7 \times 10^{-3} = 35.1\ \Omega$

Line susceptance per phase, $Y = 2\pi f C = 2\pi \times 50 \times 0.9954 \times 10^{-6} = 3.127 \times 10^{-4}$ S

Receiving-end line-to-line voltage, $V_{RL} = 66$ kV $= 66{,}000$ V

$$\text{Receiving-end line voltage per phase, } V_R = \frac{66{,}000}{\sqrt{3}} = 38{,}105\ \text{V}$$

$$\text{Load current, } I_R = \frac{\text{MW} \times 10^6}{\sqrt{3} \times V_{RL} \times \cos\phi} = \frac{20 \times 10^6}{\sqrt{3} \times 66 \times 1{,}000 \times 0.8} = 218.7\ \text{A}$$

Taking receiving-end phase voltage as reference phasor, we have

$$\mathbf{V_R} = 38{,}105 \angle 0°\ \text{V} = (38{,}105 + j0)\ \text{V}$$

$$\text{Load current, } \mathbf{I_R} = 218.7(\cos\phi - j\sin\phi) = 218.7(0.8 - j0.6) = (175 - j131.22)\ \text{A}$$

$$= 218.7 \angle -36.87°\ \text{A}$$

$$\text{Line impedance per phase, } \mathbf{Z} = (10 + j35.1) = 36.5 \angle 74.1°\ \Omega$$

$$\text{Phase voltage across condenser, } \mathbf{V'} = V_R + \frac{1}{2}\mathbf{I_R Z_R} = (38{,}105) + \frac{1}{2} \times 218.7 \angle -36.87° \times 36.5 \angle 74.1°$$

$$= (41{,}283 + j2{,}415)\ \text{V}$$

$$\text{Charging current, } \mathbf{I_C} = \mathbf{V'} \times \mathbf{Y} = (41{,}283 + j2{,}415) \times j3.127 \times 10^{-4} = (-0.75 + j12.91)\ \text{A}$$

$$\text{Sending-end current, } \mathbf{I_S} = \mathbf{I_R} + \mathbf{I_C} = (175 - j131.22) + (-0.75 + j12.91)$$

$$= (174.25 - 118.31)\ \text{A} = 210.62 \angle -34.175°\ \text{A}$$

Sending-end voltage per phase,

$$\mathbf{V_S} = \mathbf{V} + \frac{1}{2}\mathbf{I_S Z} = (41{,}283 + j2{,}415) + \frac{1}{2} \times 210.62 \angle -34.175° \times 36.5 \angle 74.1°$$

$$= (44{,}231 + j4{,}882) = 44{,}500\ \text{V} \angle 6.3°\ \text{V}$$

$$\text{Sending-end line volts, } \mathbf{V_{SL}} = \sqrt{3}\, V_S = \sqrt{3} \times 44{,}500 = 77{,}075\ \text{V}\ \textbf{Ans.}$$

$$\text{Percentage regulation} = \frac{V_{SL} - V_{RL}}{V_{RL}} = \frac{77{,}075 - 66{,}000}{66{,}000} \times 100 = 16.78\%\ \textbf{Ans.}$$

$$\text{Transmission efficiency, } \eta = \frac{\sqrt{3} V_{RL} I_{RL} \cos \phi_R}{\sqrt{3} V_{SL} \times I_{SL} \times \cos \phi_S} \times 100$$

$$= \frac{\sqrt{3} \times 66{,}000 \times 218.7 \times 0.8}{\sqrt{3} \times 77{,}075 \times 210.62 \cos \angle 6.3 - (-34.175°)} \times 100$$

$$= \frac{66{,}000 \times 218.7 \times 0.8}{77{,}075 \times 210.62 \times 0.7607} \times 100 = 93.5\% \textbf{ Ans.}$$

Example 5.10. A transmission line 200 km long has the following constants

Resistance per km = 0.15 Ω
Reactance per km = 0.50 Ω
Susceptance per km = 2.0×10^{-6} mho

Voltage at the receiving end is 132 kV. The transmission line is delivering 50 MVA at 0.85 pf lagging at the receiving end. Calculate (*i*) sending-end voltage (*ii*) sending-end current (*iii*) voltage regulation, (*iv*) line eficiency.

Solution : Resistance of line per phase, R = 200 × 0.15 = 30 Ω

Reactance of line per phase, X = 200 × 0.50 = 100 Ω

Capacitive susceptance per phase, Y = $200 \times 2.0 \times 10^{-6} = 4 \times 10^{-4}$ S

Receiving-end line voltage, V_{RL} = 132 kV or 132,000 V

$$\text{Receiveing-end phase voltage, } V_R = \frac{132{,}000}{\sqrt{3}} = 76{,}210 \text{ V}$$

$$\text{Load current, } I_R = \frac{50 \times 10^6}{\sqrt{3} \times 132{,}000} = 218.7 \text{ A}$$

Line impedance, **Z** = (R + *j* X) = (30 + *j* 100) Ω

Using nominal T-method (Fig. 5.7) and taking receiving-end voltage as reference phasor we have

$$\mathbf{V_R} = (76{,}210 + j\,0) \text{ volts}$$

$$\text{Power factor, } \cos \phi = 0.85 \text{ and } \sin \phi = \sqrt{1 - 0.85^2} = 0.5268$$

$$\text{So } \mathbf{I_R} = 218.7\,(0.85 - j\,0.5268) = (185.9 - j\,115.2) \text{ A}$$

$$\text{Phase voltage across condenser, } \mathbf{V'} = \mathbf{V_R} + \mathbf{I_R} \frac{\mathbf{Z}}{2}$$

$$= (76{,}210 + j\,0) + (185.9 - j\,115.2) \times \left(\frac{30 + j\,100}{2}\right)$$

$$= (84{,}758.5 + j\,7{,}567) \text{ V}$$

$$\text{Charging current, } \mathbf{I_C} = \mathbf{V'} \times \mathbf{Y} = (84{,}758.5 + j\,7{,}567) \times j \times 4 \times 10^{-4} \text{ A}$$

$$= (-\,3.0268 + j\,33.9) \text{ A}$$

$$(i) \text{ Sending-end current, } \mathbf{I_S} = \mathbf{I_R} + \mathbf{I_C} = (185.9 - j\,115.2) + (-\,3.0268 + j\,33.9)$$

$$= (182.8732 - j\,81.3) = 200.13 \angle -23.97° \text{ A } \textbf{Ans.}$$

$$\text{Sending-end voltage, } \mathbf{V_S} = \mathbf{V'} + \mathbf{I_S} \frac{Z}{2} = (84{,}758.5 + j\,7{,}567) + \frac{1}{2}(182.8732 - j\,81.3) \times (30 + j\,100)$$

$$= (91{,}566.6 + j\,7{,}924.16) = 91{,}910 \angle 4.95° \text{ V}$$

$$(ii) \text{ Sending-end line voltage, } V_{LS} = \sqrt{3} \times 91{,}910 = 159{,}191 \text{ V or } 159.2 \text{ kV } \textbf{Ans.}$$

$$(iii) \text{ Voltage regulation} = \frac{V_{SL} - V_{RL}}{V_{RL}} \times 100 = \frac{159.2 - 132}{132} \times 100 = 20.6\% \textbf{ Ans.}$$

$$(iv)\ \text{Line efficiency, } \eta = \frac{\sqrt{3}\, V_{RL}\, I_{RL} \cos\phi_R}{\sqrt{3}\, V_{SL}\, I_{SL} \cos\phi_S} \times 100$$

$$= \frac{\sqrt{3} \times 1{,}32{,}000 \times 218.7 \times 0.85}{\sqrt{3} \times 1{,}59{,}200 \times 200.13 \times \cos[4.95° - (-23.97°)]} \times 100$$

$= 87.99\%$ **Ans.**

Example 5.11. A single phase transmission line delivers 1 MVA at a power factor of 0.71 lagging, 22 kV, 50 Hz. The loop resistance is 15 Ω, the loop inductance 0.2 H and the capacitance 0.5 μF. Using nominal π method find the sending end voltage, current and power factor.

[U. P. Technical Univ. Elements of Power System 2005-06]

Solution : Load delivered, S = 1 MVA = 10^3 kVA

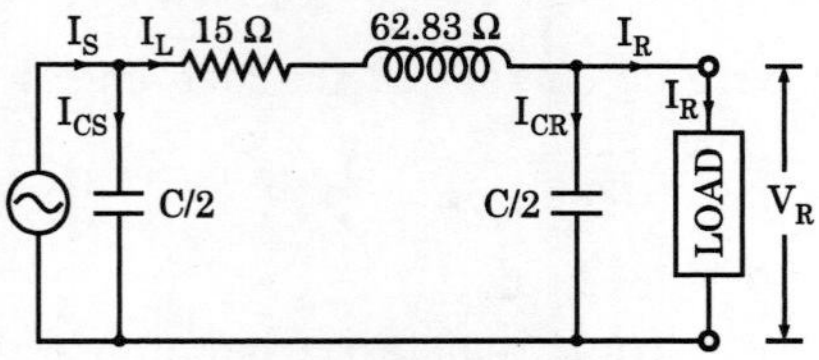

Fig. 5.13

$$\text{Receiving-end current, } I_R = \frac{S}{V_R} = \frac{10^3 \times 1{,}000}{22 \times 1{,}000} = 45.455 \text{ A}$$

Taking receiving-end voltage V_R as reference phasor we have

$$\mathbf{V_R} = 22{,}000\ (1 + j\,0) \text{ V}$$

$$\mathbf{I_R} = 45.455\ (0.71 - j\,0.7042) = (32.273 - j\,32) \text{ A}$$

Inductive reactance of line, $X = 2\pi f L = 2\pi \times 50 \times 0.2 = 62.83\ \Omega$

Line impedance, $Z = \sqrt{R^2 + X^2} = \sqrt{15^2 + 62.83^2} = 64.6° \angle 76.57°\ \Omega$

Capacitive susceptance of line, $Y = 2\pi f C = 2\pi \times 50 \times 0.5 \times 10^{-6} = 1.5708 \times 10^{-4}$ S

Charging current at the receiving end,

$$\mathbf{I_{CR}} = \frac{1}{2}\mathbf{Y}\,\mathbf{V_R} = \frac{1}{2}(0 + j\,1.5708 \times 10^{-4}) \times (22{,}000 + j\,0) = j\,1.728 \text{ A}$$

Line current, $\mathbf{I_L} = \mathbf{I_R} = \mathbf{I_{CR}} = (32.273 - j\,32) + j\,1.728 = (32.273 - j\,30.272)$ A

Impedance drop in line $= \mathbf{I_L}\mathbf{Z} = (32.273 - j\,30.272)\ (15 + j\,62.83) = (2{,}386.1 + j1{,}573.6)$ V

(*a*) Sending-end voltage, $\mathbf{V_S} = \mathbf{V_R} + \mathbf{I_L}\,Z = (22{,}000 + j\,0) + (2{,}386.1 + j\,1{,}573.6)$

$= (24{,}386.1 + j\,1{,}573.6) = 24{,}437 \angle 3.69°$ V **Ans.**

Charging current at the sending end,

$$\mathbf{I_{CS}} = \frac{1}{2}\mathbf{Y}\,\mathbf{V_S} = \frac{1}{2}(0 + j\,1.5708 \times 10^{-4})\ (24{,}386.1 + j\,1{,}573.6)$$

$= (-\,0.124 + j\,1.915)$ A

(*b*) Sending-end current, $\mathbf{I_S} = \mathbf{I_L} + \mathbf{I_{CS}} = (32.273 - j\,30.272) + (-\,0.124 + j\,1.915)$

$= (32.149 - j28.357) = 42.868 \angle -41.4°$ A **Ans.**

Sending-end power factor angle, $\phi_S = 3.69° - (-\,41.4°) = 45.09°$

(*c*) Sending-end power factor $= \cos\phi_S = \cos 45.09° = 0.706$ (lag) **Ans.**

Example 5.12. A 3-phase, 50 Hz, 150 km long line has resistance, inductive reactance and capacitive shunt admittance of 0.1 Ω, 0.5 Ω and 3 × 10^{-6} ℧ per km per phase. If the line delivers 50 MW at 110 kV at 0.8 pf lagging, determine the sending end voltage. Assume a nominal π circuit for the line.

[U. P. Technical Univ. Elements of Power System 2003-04]

Solution : Load delivered, P = 50 MW

$$\text{Phase voltage at receiving end, } V_R = \frac{110}{\sqrt{3}} = 63.508 \text{ kV} = 63{,}508 \text{ V}$$

Power factor, $\cos \phi_R = 0.8$ (lag)

$$\sin \phi_R = \sin \cos^{-1} 0.8 = 0.6$$

Receiving end current, $I_R = \dfrac{\text{Load supplied in MW} \times 10^6}{\sqrt{3} V_{RL} \cos \phi_R} = \dfrac{50 \times 10^6}{\sqrt{3} \times 110 \times 10^3 \times 0.8} = 328$ A

Taking receiving-end phase voltage as the reference phasor, we have

$$\mathbf{V_R} = 63{,}508 + j0 = 63{,}508 \angle 0^\circ \text{ V}$$

$$\mathbf{I_R} = 328 \angle -36.87^\circ \text{ A} = (262.4 - j196.8) \text{ A}$$

Capacitive admittance per phase, $Y = 3 \times 10^{-6} \times 150 = 4.5 \times 10^{-4}$ S

Charging current at the load end, $\mathbf{I_{CR}} = \dfrac{1}{2}\mathbf{YV_R} = \dfrac{1}{2} \times j4.5 \times 10^{-4} \times (63{,}508 + j0) = j14.3$ A

Line current, $\mathbf{I_L} = \mathbf{I_R} + \mathbf{I_{CR}} = (262.4 - j196.8) + j14.3 = (262.4 - j182.5)$ A

$$= 319.625 \angle -34.82^\circ \text{ A}$$

Total resistance per phase, $R = 0.1 \times 150 = 15\ \Omega$

Total reactance per phase, $X_L = 0.5 \times 150 = 75\ \Omega$

Line impedance per phase, $\mathbf{Z} = (R + jX_L) = (15 + j75)\ \Omega = 76.485 \angle 78.69^\circ\ \Omega$

Sending-end phase voltage, $\mathbf{V_S} = \mathbf{V_R} + \mathbf{I_L Z}$

$$= (63{,}508 + j0) + 319.625 \angle -34.82^\circ \times 76.485 \angle 78.69^\circ$$

$$= (63{,}508 + j0) + 24{,}446.5 \angle 43.87^\circ$$

$$= (81{,}132 + j16{,}942) \text{ V} = 82{,}882 \angle 11.795^\circ \text{ V}$$

Sending end voltage (line-to-line), $\mathbf{V_{SL}} = \sqrt{3} \times 82{,}882 = 143{,}556$ V or 143.556 kV **Ans.**

Example 5.13. A 50 Hz, 3-phase transmission line 30 km has a total series impedance of (40 + *j* 125) ohms and shunt admittance of 10^{-3} mho. The load is 50 MW at 220 kV with 0.8 lagging power factor. Find the sending-end voltage, current and power factor (use nominal-π representation).

Solution : Phase voltage at the receiving end, $V_R = \dfrac{220}{\sqrt{3}} = 127$ kV or 127,000 V

Power factor, $\cos \phi = 0.8$ (lag) and $\sin \phi = 0.6$

Receiving-end current, $I_R = \dfrac{\text{Load supplied in MW} \times 10^6}{\sqrt{3} \times V_{RL} \text{ in volts} \times \cos \phi} = \dfrac{50 \times 10^6}{\sqrt{3} \times 220 \times 1{,}000 \times 0.8} = 164$ A

Taking phase voltage at receiving end, V_R as the reference phasor we have

$$\mathbf{V_R} = (12{,}700 + j\,0) \text{ volts}$$

$$\mathbf{I_R} = 164\,(0.8 - j\,0.6) = (131.2 - j\,98.4) \text{ A}$$

Charging current at receiving end, $\mathbf{I_{CR}} = \dfrac{1}{2}\ \mathbf{Y\ V_R} = \dfrac{1}{2}\ (0 + j\,10^{-3}) \times (127{,}000 + j\,0) = (0 + j\,63.5)$

Line current, $\mathbf{I_L} = \mathbf{I_R} + \mathbf{I_{CR}} = (131.2 - j\,98.4) + (0 + j\,63.5) = (131.2 - j\,34.9)$

Sending-end voltage, $\mathbf{V_S} = \mathbf{V_R} + \mathbf{I_L\ Z}$

$$= (127{,}000 + j\,0) + (131.2 - j\,34.9)\ (40 + j\,125)$$

$$= 136{,}610 + j\,15{,}004 = 1{,}37{,}432 \angle 6.27^\circ \text{ V}$$

Sending-end line voltage, $V_{SL} = \sqrt{3}\ V_S = \sqrt{3} \times 1{,}37{,}432$

$$= 238{,}038 \text{ V or } 238.04 \text{ kV } \textbf{Ans.}$$

Sending end current, $\mathbf{I_S} = \mathbf{I_L} + \frac{1}{2}\,\mathbf{Y\,V_S}$

$= (131.2 - j\,34.9) + \frac{1}{2}\,(0 + j\,10^{-3})\,(1{,}36{,}610 + j\,15{,}004)$

$= (123.82 + j\,33.4) = 128.25\ \angle 15.1°$ A **Ans.**

Sending-end power factor angle, $\phi_S = 15.1° - (6.27°) = 8.83°$

Sending-end power factor $= \text{Cos}\ \phi_S = \text{Cos}\ 8.83° = 0.988$ (lag) **Ans.**

Example 5.14. A 50 Hz, 3-phase transmission line of length 30 km has a total series impedance of $(40 + j125)$ ohms and shunt admittance of 10^{-3} mho. The load is 50 MW at 220 kV with 0.8 lagging power. Find the sending-end voltage. (Use Nominal-π representation).

[*U.P. Technical Univ. Elements of Power System 2005-06*]

Solution : Refer to Solution of Example 5.13.

Example 5.15. A 3-phase, 100 km transmission line is delivering 50 MW, 0.8 pf lagging at 132 kV. Each conductor is having resistance of 0.1 Ω/km, reactance of 0.3 Ω/km and shunt admittance of 3×10^{-6} ℧/km. If the load is balanced and leakage is neglected, calculate the sending end voltage, efficiency and sending end power factor of line using nominal π representation.

[*U.P. Technical Univ. Elements of Power System 2005-06*]

Solution : Line resistance per phase, $R = 0.1 \times 100 = 10\ \Omega$

Line reactance per phase, $X_L = 0.3 \times 100 = 30\ \Omega$

Line impedance per phase, $\mathbf{Z} = (10 + j30) = 31.62\ \angle 71.565°\ \Omega$

Capacitive admittance per phase, $\mathbf{Y} = 3 \times 10^{-6} \times 100 = 3 \times 10^{-4}$ S

Load supplied, $P = 50$ MW

Phase voltage at receiving end, $V_R = \frac{132}{\sqrt{3}} = 76.21 \text{ kV} = 76{,}210 \text{ V}$

Power factor, $\cos\phi_R = 0.8$ lagging

$\sin\phi_R = \sqrt{1 - 0.8^2} = 0.6$

Receiving-end current, $I_R = \frac{\text{Load supplied in MW} \times 10^6}{\sqrt{3}V_{RL}\cos\phi_R}$

$= \frac{50 \times 10^6}{\sqrt{3} \times 132 \times 10^3 \times 0.8} = 273.4$ A

Taking receiving-end phase voltage as the reference phasor, we have

$$\mathbf{V_R} = (76{,}210 + j0)\text{ V} = 76{,}210\ \angle 0°\text{ V}$$

$$\mathbf{I_R} = 273.4\,(0.8 - j0.6) = (218.7 - j164) = 273.4\ \angle -36.87°\text{ A}$$

Charging current at load end, $\mathbf{I_{CR}} = \frac{1}{2}\mathbf{Y\,V_R} = \frac{1}{2} \times j3 \times 10^{-4} \times (76{,}210 + j0) = j11.4$ A

Line current, $\mathbf{I_L} = \mathbf{I_R} + \mathbf{I_{CR}} = (218.7 - j164) + j11.4$

$= (218.7 - j152.6)$ A $= 266.68\ \angle -34.9°$ A

Sending-end phase voltage, $\mathbf{V_S} = \mathbf{V_R} + \mathbf{I_L\,Z} = (76{,}210 + j0) + 266.68\ \angle -34.9° \times 31.62\ \angle 71.565°$

$= (76{,}210 + j0) + 8{,}432.4\ \angle 36.665°$

$= (82{,}974 + j5{,}035) = 83{,}127\ \angle 3.47°$ V

Sending-end line voltage, $\mathbf{V_{SL}} = \sqrt{3} \times 83{,}127 = 143{,}980 \text{ V} = 143.98 \text{ kV}$ **Ans.**

Charging current at the sending end, $\mathbf{I_{CS}} = \frac{1}{2}\mathbf{Y}\mathbf{V_S} = \frac{1}{2} \times j3 \times 10^{-4} \times (82{,}974 + j5{,}035)$

$= (-0.755 + j12.45) \text{ A}$

Sending-end current, $\mathbf{I_S} = \mathbf{I_L} + \mathbf{I_{CS}} = (218.7 - j152.6) + (-0.755 + j12.45)$

$= (217.945 - j140.15) \text{ A} = 259.12 \angle -32.74° \text{ A}$

Sending-end power factor angle, $\phi_S = 3.47° - (-32.74°) = 36.21°$

Sending-end power factor $= \cos\phi_S = \cos 36.21° = 0.8069$ (lagging) **Ans.**

$$\text{Transmission efficiency, } \eta = \frac{\sqrt{3}V_{RL}\, I_R \cos\phi_R}{\sqrt{3}\, V_{SL} I_S \cos\phi_S} \times 100$$

$$= \frac{50 \times 10^6}{\sqrt{3} \times 143{,}980 \times 259.12 \times 0.8069} \times 100 = 95.893\% \quad \textbf{Ans.}$$

Example 5.16. A 20 km 3-phase transmission line delivers at the furthest end a 5,000 kVA load at 33 kV and 0.8 p f lagging and at the mid-point, a 10,000 kVA load at 0.7071 p f lagging. Neglecting the changes in line power factor and using approx formula, find the voltage at the mid-point and at the sending-end and also the total current. The resistance and reactance per phase per km are 0.6 Ω and 1.5 Ω respectively.

Solution : Phase voltage at receiving end, $V_R = \frac{33}{\sqrt{3}} \times 1{,}000 = 19{,}052 \text{ V}$

Power factor, $\cos\phi = 0.8$

$\sin\phi = \sqrt{1 - 0.8^2} = 0.6$

Receiving-end current, $I_1 = \frac{S_1}{\sqrt{3}\, V_{RL}}$

$= \frac{5{,}000 \times 1{,}000}{\sqrt{3} \times 33 \times 1{,}000}$

$= 87.5 \text{ A}$

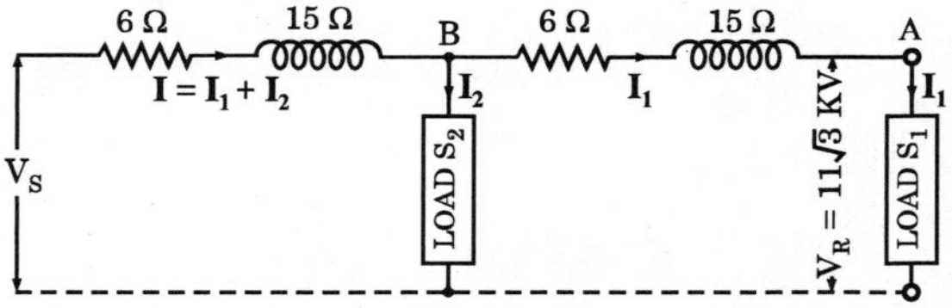

Fig. 5.14

Total resistance of line from receiving end to mid-point (*i.e.* of 10 km length)

$R_1 = 0.6 \times 10 = 6 \ \Omega$

Total reactance of line from receiving end to mid-point,

$X_1 = 1.5 \times 10 = 15 \ \Omega$

Taking phase voltage at receiving end A, V_R as the reference phasor, we have

$\mathbf{V_R} = (19{,}052 + j\,0)$ volts

$\mathbf{I_1} = 87.5\,(0.8 - j\,0.6) = (70 - j\,52.5) \text{ A}$

Voltage at the mid-point B, $V_B = \mathbf{V_R} + \mathbf{I_1}\mathbf{Z_1} = (19{,}052 + j\,0) + (70 - j\,52.5)\,(6 + j\,15)$

$= (20{,}259.5 + j\,735) = 20{,}273 \angle 2.1°$ volts

Line voltage at mid-point, $V_{BL} = \sqrt{3}\, V_B = \sqrt{3} \times 20{,}273 = 35{,}114 \text{ V}$ or 35.114 kV **Ans.**

Current, $I_2 = \frac{S_2}{\sqrt{3}\, V_{BL}} = \frac{10{,}000 \times 1{,}000}{\sqrt{3} \times 35{,}114} = 164.4 \text{ A}$

Neglecting changes in line power factor,

$\mathbf{I_2} = 164\,(\cos\phi_2 - j\sin\phi_2) = 164.4\,(0.7071 - j\,0.7071)$

$= (116.26 - j\,116.26) \text{ A}$

Total current flowing in the line from mid-point to sending end,

$$\mathbf{I} = \mathbf{I_1} + \mathbf{I_2} = (70 - j\,52.5) + (116.26 - j\,116.26)$$

$$= (186.26 - j\,168.76)\text{ A} = 251.35 \angle -42.2^\circ \text{ A} \quad \textbf{Ans.}$$

Impedance of line from mid-point to sending end,

$$\mathbf{Z_2} = (R_2 + j\,X_2) = (10 \times 0.6 + j\,10 \times 1.5) = (6 + j\,15)\ \Omega$$

Voltage at the sending end, $\mathbf{V_S} = \mathbf{V_B} + \mathbf{I\,Z_2}$

$$= (20{,}259.5 + j\,735) + (186.26 - j\,168.76)(6 + j\,15)$$

$$= 23{,}908.5 + j\,2{,}516.3 = 24{,}040 \angle 6^\circ \text{ V}$$

Line voltage at sending end, $V_{SL} = \sqrt{3}\,V_S = \sqrt{3} \times 24{,}040 = 41{,}640$ V or 41.64 kV **Ans.**

Example 5.17. Two overhead lines are connected in parallel to supply a load of 10 MW at 0.8 p f (lagging) and 30 kV. The resistance and reactance of one line (A) are 5.5 Ω and 13.5 Ω, respectively; those of the other line (B) are 6 Ω and 11 Ω respectively. Calculate (*a*) the kVA supplied by each line (*b*) the power supplied by each line.

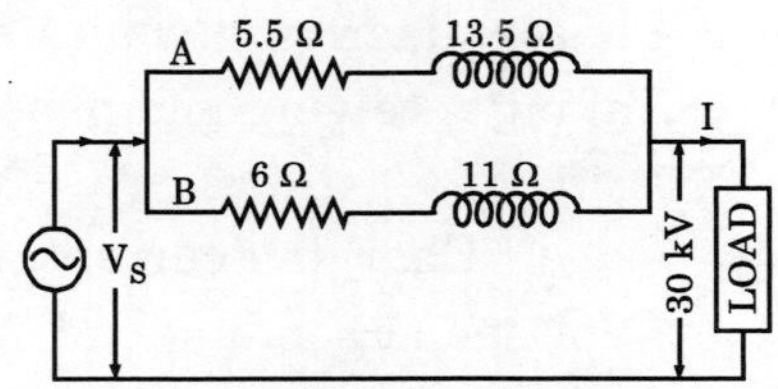

Fig. 5.15

Solution : Impedance of line A, $\mathbf{Z_A} = 5.5 + j\,13.5$

$$= 14.58 \angle 67^\circ 50' \text{ ohms}$$

Impedance of line B, $\mathbf{Z_B} = 6 + j\,11.0$

$$= 12.53 \angle 61^\circ 24' \text{ ohms}$$

$$\mathbf{Z_A} + \mathbf{Z_B} = (5.5 + j\,13.5) + (6 + j\,11.0)$$

$$= (11.5 + j\,24.5) = 27.06 \angle 64^\circ 51' \text{ ohms}$$

Total load supplied in kVA, $\mathbf{S} = \dfrac{10{,}000}{0.8} \angle -36^\circ 52' = 12{,}500 \angle -36^\circ 52'$

(*a*) Load supplied by line A, $\mathbf{S_A} = \mathbf{S} \cdot \dfrac{\mathbf{Z_B}}{\mathbf{Z_A} + \mathbf{Z_B}}$

$$= 12{,}500 \angle -36^\circ 52' \times \frac{12.53 \angle 61^\circ 24'}{27.06 \angle 64^\circ 51'}$$

$$= \frac{12{,}500 \times 12.53}{27.06} \angle -36^\circ 52' + 61^\circ 24' - 64^\circ 51'$$

$$= 5{,}790 \angle -40^\circ 19' \text{ kVA} = 5{,}790 \text{ kVA at p f } 0.763 \text{ (lag)} \quad \textbf{Ans.}$$

Load supplied by line **B**, $\mathbf{S_B} = \mathbf{S} \cdot \dfrac{\mathbf{Z_A}}{\mathbf{Z_A} + \mathbf{Z_B}}$

$$= 12{,}500 \angle -36^\circ 52' \times \frac{14.58 \angle 67^\circ 50'}{27.06 \angle 64^\circ 51'}$$

$$= \frac{12{,}500 \times 14.58}{27.06} \angle -36^\circ 52' + 67^\circ 50' - 64^\circ 51'$$

$$= 6{,}730 \angle -33^\circ 53' \text{ kVA} = 6{,}730 \text{ kVA at pf } 0.83 \text{ (lag)} \quad \textbf{Ans.}$$

(*b*) Power supplied by line **A** = 5,790 × 0.763 = 4,415 kW **Ans.**

Power supplied by line **B** = 6,730 × 0.83 = 5,585 kW **Ans.**

5.6. FERRANTI EFFECT

In a medium or long transmission line when open-circuited or lightly loaded the receiving -end voltage is found more than the sending-end voltage. *This phenomenon of rise in voltage at the receiving end of the open-circuited or lightly loaded line is called the Ferranti effect* owing to its first being observed on the Deptford mains laid down by S.Z. de Ferranti in 1890. This effect is due to voltage drop across the line inductance, due to charging current, being in phase with the applied voltage at the sending end of the line. Thus both capacitance and inductance are necessary to cause this phenomenon. The capacitance, and therefore, charging current is negligible in short lines but significant in medium and long transmission lines. On a 300 km line operating on 50 Hz supply, the receiving-end voltage on open-circuit is usually found to be roughly 5% higher than the sending-end voltage.

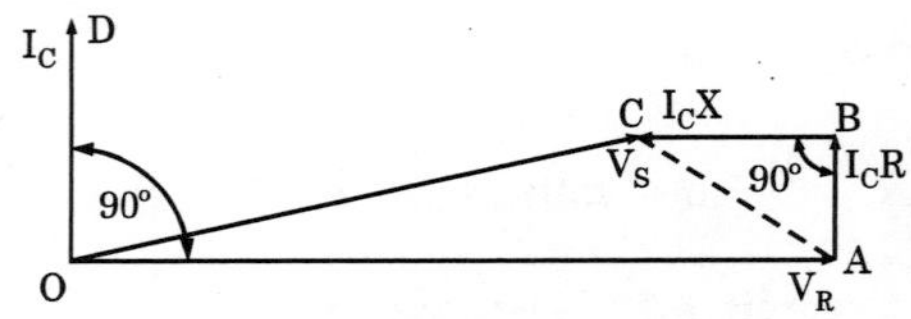

Fig. 5.16

Phasor diagram given in Fig. 5.16 illustrates the Ferranti effect.

Taking receiving-end phase voltage as reference phasor we have

$\mathbf{V_R} = V_R\,(1 + j\,0)$ represented by phasor OA

Charging current, $\mathbf{I_C} = j\,\omega\,C\,V_R$ represented by phasor OD.

Voltage at sending end, $\mathbf{V_S} = \mathbf{V_R}$ + resistive drop in line + reactive drop in line

$= \mathbf{V_R} + \mathbf{I_C}\,R$ (represented by phasor AB) $+ j\,\mathbf{I_C}\,X$ (represented by phasor BC)

$= \mathbf{V_R} + \mathbf{I_C}\,(R + j\,X) = \mathbf{V_R} + j\,\omega\,C\,V_R\,(R + j\,\omega\,L)$

or $\mathbf{V_S} = \mathbf{V_R} - \omega^2\,C\,L\,V_R + j\,\omega\,C\,R\,\mathbf{V_R}$ represented by phasor OC.

The overhead transmission lines having long length to bring the effect into prominence usually have a fairly large reactance compared with resistance, and by taking advantage of this fact a simple approximate formula for determining the magnitude of the effect can be derived.

Neglecting resistive drop $\mathbf{I_C}\,\mathbf{R}$ we get

Rise in voltage $= \mathbf{OA} - \mathbf{OC}$ = Inductive drop in line (refer to Fig. 5.16).

If $\mathbf{C_0}$ and $\mathbf{L_0}$ are the capacitance and inductance of the transmission line per km length and l is the length of line in km, the capacitive reactance, $\mathbf{X_C} = \dfrac{1}{\omega\,l\,C_0}$.

As the capacitance of the line is uniformly distributed over the entire length of the line, therefore, the average current flowing throughout the line

$$I_C = \frac{1}{2}\,\frac{V_R}{X_C} = \frac{1}{2}\,V_R\,C_0\,\omega\,l$$

The reactance of the transmission line, $X = \omega\,L_0\,l$

$$\text{Rise in voltage} = \mathbf{I_C X} = \frac{1}{2}\,V_R\,C_0\,\omega\,l \times \omega\,L_0\,l = \frac{1}{2}\,\omega^2\,l^2\,V_R\,C_0\,L_0 \text{ volts}$$

The quantity $\dfrac{1}{\sqrt{L_0\,C_0}}$ is constant for all overhead lines and is equal to the velocity of propagation of electromagnetic waves. It is nearly equal to the speed of light *i.e.* 3×10^5 km/s. So

Rise in voltage at the receiving end $= \frac{1}{2}\omega^2 l^2 V_R \times \frac{1}{(3 \times 10^5)^2} = \frac{\omega^2 l^2 V_R \times 10^{-10}}{18}$ V ...(5.32)

From Eq. (5.32) it is obvious that the excess of voltage at the receiving end of an open-circuited line is proportional to the square of the length of the line. It is found that quite a small load eliminates this effect, and where there are transformers in the circuit at the receiving end, the magnetising current acts in such a way.

For long high voltage and extra high voltage lines shunt reactors are provided to absorb part of charging current or shunt capacitive VA of the line on no-load and light load, in order to prevent over-voltages on the line.

Example 5.18. Determine the percentage rise in voltage at the receiving end of a transmission line of length 200 km operating at 50 Hz.

Solution : Rise in voltage $= \frac{\omega^2 l^2 V_R \times 10^{-10}}{18}$ volts

Percentage rise in voltage $= \frac{\omega^2 l^2 V_R \times 10^{-10}}{18 \times V_R} \times 100$ %

$$= \frac{\omega^2 l^2 \times 10^{-8}}{18} = \frac{(2\pi \times 50)^2 \times (200)^2 \times 10^{-8}}{18} = 2.19\% \quad \textbf{Ans.}$$

5.7. CHARGING CURRENT AND LINE LOSSES IN OPEN-CIRCUITED LINE

In transmission line operation the voltage at the receiving end is usually maintained constant irrespective of the magnitude of the load of the system. The charging current of an open-circuited line is the amount of current flowing into the line at the sending end with normal voltage and no load at the receiving end. The calculation of charging current becomes much simplified if it is assumed that voltage throughout the length of the line is same, *i.e.* there is no voltage drop in the line.

Making the assumption of voltage equality along the line, the actual value of charging current will fall uniformly from its maximum value at the sending end to zero at the receiving-end.

Charging current at a distance x km from the sending end of a l km long line is given as $I_C \frac{l - x}{l}$ where I_C is the maximum value of charging current.

The average value of the square of this quantity

$$= \frac{1}{l}\int_0^l \frac{I_C^2 (l-x)^2}{l^2} = \frac{I_C^2}{l^3}\int_0^l (l-x)^2\, dx = \frac{I_C^2}{l^3}\left[\frac{(l-x)^3}{3 \times (-1)}\right]_0^l = \frac{I_C^2}{3} \text{ A}^2$$

RMS value of charging current, $I_{Crms} = \sqrt{\frac{I_C^2}{3}} = \frac{I_C}{\sqrt{3}}$ A ...(5.33)

Power loss in each of the line conductor $= \left(\frac{I_C}{\sqrt{3}}\right)^2 R = \frac{1}{3} I_C^2 R$ watts ...(5.34)

5.8. EFFECT OF CAPACITANCE ON PERFORMANCE OF LOADED LINE

On inductive load, as usual in practice, the effect of capacitance is to reduce voltage drop along the line and also to reduce the copper losses of the line. The magnitude of this effect depends on the phase angle at the receiving end.

In any line, the current flowing through it can be split up into two components, the load current and the charging current. The effect of load current with lagging or unity power factor is to make the receiving-end voltage lower than sending-end voltage whereas the charging current tends to raise the receiving-end voltage. As a result, the voltage drop along the line is diminished by the amount given by the Eq. (5.32).

In case of short transmission lines the regulation and voltage drop are numerically equal. But in medium and long transmission lines, where capacitance effects have to be taken into account, the regulation is greater than voltage drop by the rise in voltage due to Ferranti effect.

The presence of capacitance also influences the power losses of the system by an amount depending on the relative values of the charging current and the quadrature component of the load current. The charging current tends to neutralize the quadrature component (lagging) of the load current, consequently the line current is reduced and, therefore, line losses are reduced. It can be easily seen that reduction in line current and line losses takes place so long as the receiving-end charging current is less than twice the quadrature component of the load current.

At unity power factor of the load there will be no quadrature component of the load current and so the effect of charging current in a line supplying a non-inductive load is always to increase the line losses. With small highly inductive load, the charging current may exceed twice the quadrature component of load current and give increased line losses.

5.9. GENERALISED CIRCUIT CONSTANTS

In any passive, bilateral and linear network with two input and two output terminals, the input voltage and the input current can be expressed in terms of output voltage and output current. Incidently, a transmission line is a 4-terminal network; two input terminals where power enters the network, and two output terminals where the power leaves the network. Such a circuit is passive as it does not contain any source of emf, linear as its impedance is independent of current flowing and bilateral as its impedance is independent of direction of current flowing.

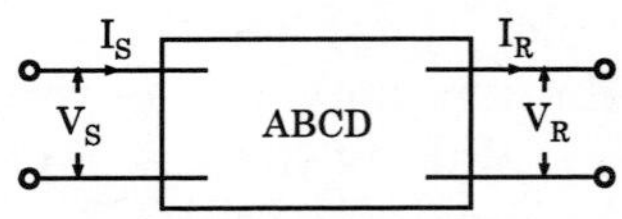

Fig. 5.17

The input voltage per phase (voltage at sending end) and the input current (current at sending end) of a transmission line can be expressed as

$$\mathbf{V_S} = \mathbf{A}\,\mathbf{V_R} + \mathbf{B}\,\mathbf{I_R} \qquad ...(5.35)$$

$$\mathbf{I_S} = \mathbf{C}\,\mathbf{V_R} + \mathbf{D}\,\mathbf{I_R} \qquad ...(5.36)$$

where $\mathbf{V_S}$ is the sending-end voltage per phase, $\mathbf{I_S}$ is the sending-end current, $\mathbf{V_R}$ is the receiving-end voltage per phase, $\mathbf{I_R}$ is the receiving-end current, and **A**, **B**, **C** and **D** (generally complex numbers) are the constants, called the *generalised circuit constants* of a transmission line. The values of these constants A, B, C and D depend upon the particular method of solution employed. Once the values of these constants for a particular transmission line are known, the performance of the line can be determined easily.

The values of A, B, C and D can be determined as follows :

With receiving-end open-circuited

Receiving-end current $\mathbf{I_R}$ is zero, therefore, from Eqs. (5.35) and (5.36)

$$\mathbf{A} = \mathbf{V_S}/\mathbf{V_R} \qquad ...(5.37)$$

$$\text{and } \mathbf{C} = \mathbf{I_S}/\mathbf{V_R} \qquad ...(5.38)$$

With receiving-end short-circuited

Receiving-end voltage $\mathbf{V_R}$ is zero, therefore, from Eqs. (5.35) and (5.36)

$$\mathbf{B} = \mathbf{V_S}/\mathbf{I_R} \quad ...(5.39)$$

$$\text{and } \mathbf{D} = \mathbf{I_S}/\mathbf{I_R} \quad ...(5.40)$$

From the above expressions **A**, **B**, **C** and **D** can be defined as follows :

A can be defined as the ratio of sending end voltage $\mathbf{V_S}$ to receiving-end voltage $\mathbf{V_R}$ when the line is open-circuited on receiving-end. **A** is dimensionless and usually a complex number.

B is defined as the ratio of sending-end voltage $\mathbf{V_S}$ to receiving-end current $\mathbf{I_R}$ when the line is short-circuited on receiving end. **B** is also the complex number and has the unit of ohm.

C is defined as the ratio of sending-end current $\mathbf{I_S}$ to receiving-end voltage $\mathbf{V_R}$ when the line is open-circuited on the receiving end. **C** is also the complex number and has the unit of *siemen*.

D is defined as the ratio of sending end current $\mathbf{I_S}$ to receiving-end current $\mathbf{I_R}$ when the line is short-circuited on the receiving-end. **D** is dimensionless and usually a complex number

For passive network $\quad \mathbf{AD} - \mathbf{BC} = 1 \quad ...(5.41)$

and in symmetrical network $\quad \mathbf{A} = \mathbf{D} \quad ...(5.42)$

The values of generalised constants **A**, **B**, **C** and **D** for various methods of solution adopted are determined as follows :

(i) Short Lines (Simple Series Impedance Method). In this method the effect of the line capacitance is neglected and, therefore, the line can be considered to have series impedance. The circuit diagram is shown in Fig. 5.1 (*a*). Here we have

$$\mathbf{I_S} = \mathbf{I_R} \quad ...(5.43)$$

$$\text{and } \mathbf{V_S} = \mathbf{V_R} + \mathbf{I_R}\,\mathbf{Z} \quad ...(5.44)$$

Comparing Eq. (5.44) with Eq. (5.35) and Eq. (5.43) with Eq. (5.36) we have

$$\mathbf{A} = 1;\ \mathbf{B} = \mathbf{Z};\ \mathrm{C} = 0 \ \text{ and } \ \mathbf{D} = 1$$

The phasor diagram is shown in Fig. 5.1 (*b*)

Incidently $\mathbf{A} = \mathbf{D}$ and $\mathbf{AD} - \mathbf{BC} = 1 \times 1 - \mathbf{Z} \times 0 = 1$

(ii) End Condenser Method. In this method the capacitance of the line is assumed to be lumped at the load end, as shown in Fig. 5.5. This method overestimates the effect of capacitance.

Here we have $\quad \mathbf{I_S} = \mathbf{I_C} + \mathbf{I_R} = \mathbf{Y}\,\mathbf{V_R} + \mathbf{I_R} \quad ...(5.45)$

$$\text{and } \mathbf{V_S} = \mathbf{V_R} + \mathbf{Z}\,(\mathbf{I_C} + \mathbf{I_R})$$

$$= \mathbf{V_R} + \mathbf{Z}\,\mathbf{I_R} + \mathbf{Z}\,\mathbf{Y}\,\mathbf{V_R} = \mathbf{V_R}\,(1 + \mathbf{Z}\,\mathbf{Y}) + \mathbf{Z}\,\mathbf{I_R} \quad ...(5.46)$$

Comparing Eq. (5.46) with Eq. (5.35) and Eq. (5.45) with Eq. (5.36) we have

$$\mathbf{A} = 1 + \mathbf{ZY};\ \mathbf{B} = \mathbf{Z};\ \mathbf{C} = \mathbf{Y} \text{ and } \mathbf{D} = 1$$

The phasor diagram is shown in Fig. 5.6

Incidently $\mathbf{AD} - \mathbf{BC} = (1 + \mathbf{ZY}) \times 1 - \mathbf{ZY} = 1$

(iii) Nominal T-Method. In this method the capacitance of each conductor is assumed to be concentrated at the mid-point, as shown in Fig. 5.7. The phasor diagram is shown in Fig. 5.8

Here we have

$$\mathbf{V_S} = \mathbf{V'} + \mathbf{I_S}\,\frac{\mathbf{Z}}{2} \quad ...(5.47)$$

$$\text{and } \mathbf{V'} = \mathbf{V_R} + \mathbf{I_R}\,\frac{\mathbf{Z}}{2} \quad ...(5.48)$$

$$\text{and } \mathbf{I_S} = \mathbf{I_R} + \mathbf{I_C} = \mathbf{I_R} + \mathbf{Y}\,\mathbf{V'} = \mathbf{I_R} + \mathbf{Y}\left(\mathbf{V_R} + \mathbf{I_R}\,\frac{\mathbf{Z}}{2}\right) = \mathbf{Y}\,\mathbf{V_R} + \mathbf{I_R}\left(1 + \frac{\mathbf{YZ}}{2}\right) \quad ...(5.49)$$

Substituting the value of $\mathbf{V}'$ from Eq. (5.48) in Eq. (5.47) we have

$$\mathbf{V_S} = \mathbf{V_R} + \mathbf{I_R}\frac{\mathbf{Z}}{2} + \mathbf{I_S}\frac{\mathbf{Z}}{2}$$

$$= \mathbf{V_R} + \mathbf{I_R}\frac{\mathbf{Z}}{2} + \frac{\mathbf{Z}}{2}\left[\mathbf{Y}\mathbf{V_R} + \mathbf{I_R}\left(1+\frac{\mathbf{YZ}}{2}\right)\right] = \left(1+\frac{\mathbf{YZ}}{2}\right)\mathbf{V_R} + \left(\mathbf{Z}+\frac{\mathbf{YZ}^2}{4}\right)\mathbf{I_R} \quad ...(5.50)$$

Comparing Eq. (5.50) with Eq. (5.35) and Eq. (5.49) with Eq. (5.36) we have

$$\mathbf{A} = \mathbf{D} = 1+\frac{\mathbf{YZ}}{2};\ \mathbf{B} = \mathbf{Z}\left(1+\frac{\mathbf{YZ}}{4}\right) \text{ and } \mathbf{C} = \mathbf{Y}$$

$$\text{Incidently } \mathbf{AD} - \mathbf{BC} = \left(1+\frac{\mathbf{YZ}}{2}\right)^2 - \mathbf{ZY}\left(1+\frac{\mathbf{YZ}}{4}\right) = 1$$

(iv) Nominal π-Method. In this method the capacitance of each conductor is assumed to be divided into two halves, one half being shunted between conductor and neutral at the receiving end and the other half at the sending end, as shown in Fig. 5.11. The phasor diagram is shown in Fig. 5.12.

Here we have $$\mathbf{I_L} = \mathbf{I_R} + \frac{\mathbf{Y}}{2}\mathbf{V_R} \quad ...(5.51)$$

$$\text{and } \mathbf{I_S} = \mathbf{I_L} + \frac{\mathbf{Y}}{2}\mathbf{V_S} \quad ...(5.52)$$

$$\text{and } \mathbf{V_S} = \mathbf{V_R} + \mathbf{I_L}\mathbf{Z} \quad ...(5.53)$$

Substituting $\mathbf{I_L} = \mathbf{I_R} + \frac{1}{2}\mathbf{Y}\mathbf{V_R}$ from Eq. (5.51) we have

$$\mathbf{V_S} = \mathbf{V_R} + \mathbf{Z}\left(\mathbf{I_R} + \frac{1}{2}\mathbf{Y}\mathbf{V_R}\right) = \mathbf{V_R}\left(1+\frac{1}{2}\mathbf{YZ}\right) + \mathbf{I_R}\mathbf{Z} \quad ...(5.54)$$

From Eqs. (5.51) and (5.52) we have

$$\mathbf{I_S} = \mathbf{I_R} + \frac{1}{2}\mathbf{Y}\mathbf{V_R} + \frac{1}{2}\mathbf{Y}\mathbf{V_S} \quad ...(5.55)$$

Substituting the value of $\mathbf{V_S}$ from Eq. (5.54) in Eq. (5.55) we have

$$\mathbf{I_S} = \mathbf{I_R} + \frac{1}{2}\mathbf{Y}\mathbf{V_R} + \frac{1}{2}\mathbf{Y}\left[\left(1+\frac{1}{2}\mathbf{YZ}\right)\mathbf{V_R} + \mathbf{I_R}\mathbf{Z}\right]$$

$$= \mathbf{V_R}\mathbf{Y}\left(1+\frac{1}{4}\mathbf{YZ}\right) + \mathbf{I_R}\left(1+\frac{1}{2}\mathbf{YZ}\right) \quad ...(5.56)$$

Comparing Eq. (5.54) with Eq. (5.35) and Eq. (5.56) with Eq. (5.36) we have

$$\mathbf{A} = \mathbf{D} = 1 + \frac{1}{2}\mathbf{YZ};\ \mathbf{B} = \mathbf{Z} \text{ and } \mathbf{C} = \mathbf{Y}\left(1+\frac{1}{4}\mathbf{YZ}\right)$$

$$\text{Incidently } \mathbf{AD} - \mathbf{BC} = \left(1+\frac{1}{2}\mathrm{YZ}\right)^2 - \mathbf{ZY}\left(1+\frac{1}{4}\mathbf{YZ}\right) = 1$$

5.9.1. Proof For Relation AD – BC = 1

Consider a two terminal pair network with parameters **A, B, C** and **D** connected to an ideal voltage source* at one end and short-circuited at other end (Fig. 5.18).

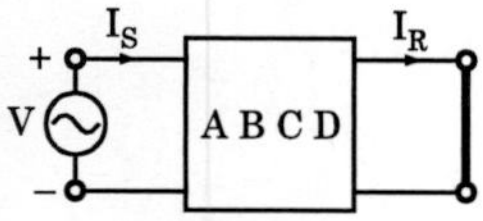

Fig. 5.18

Receiving-end voltage, $V_R = 0$

Now from Eq. (5.35) we have

$$\mathbf{V} = \mathbf{A} \times 0 + \mathbf{BI_R}$$

$$\text{or } \mathbf{I_R} = \frac{\mathbf{V}}{\mathbf{B}} \quad \text{...(5.57)}$$

Now let the sending-end terminals be short-circuited and the voltage source be connected to receiving-end terminals, as shown in Fig. 5.19.

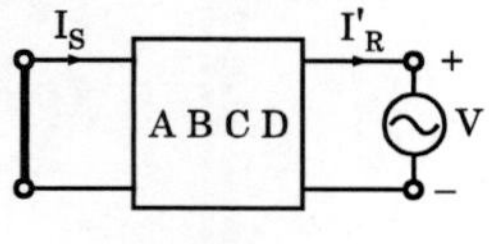

Fig. 5.19

Now from Eq. (5.35) we have $0 = \mathbf{AV} + \mathbf{BI'_R}$ ...(5.58)

and from Eq. (5.36)

$$\mathbf{I_S} = \mathbf{CV} + \mathbf{DI'_R}$$

$$\text{or } -\mathbf{I_R} = \mathbf{CV} + \mathbf{DI'_R} \quad \text{...(5.59)}$$

because $\mathbf{I_S} = -\mathbf{I_R}$ as transmission line is a linear bilateral network.

Substituting the values of $\mathbf{I_R}$ and $\mathbf{I'_R}$ from Eqs. (5.57) and (5.58) respectively in above Eq. (5.59) we have

$$\frac{-\mathbf{V}}{\mathbf{B}} = \mathbf{C\,V} + \mathbf{D}\,\frac{(-\mathbf{AV})}{\mathbf{B}}$$

$$\text{or } -1 = \mathbf{BC} - \mathbf{AD}$$

$$\text{or } \mathbf{AD} - \mathbf{BC} = 1 \quad \text{...(5.60)}$$

5.9.2. Two Networks in Tandem

Two networks** are said to be connected in tandem when the output of one network is connected to the input of the other network, as illustrated in Fig. 5.20. Let the two networks having constants $\mathbf{A_1}$, $\mathbf{B_1}$, $\mathbf{C_1}$, $\mathbf{D_1}$ and $\mathbf{A_2}$, $\mathbf{B_2}$, $\mathbf{C_2}$, $\mathbf{D_2}$ be connected in series

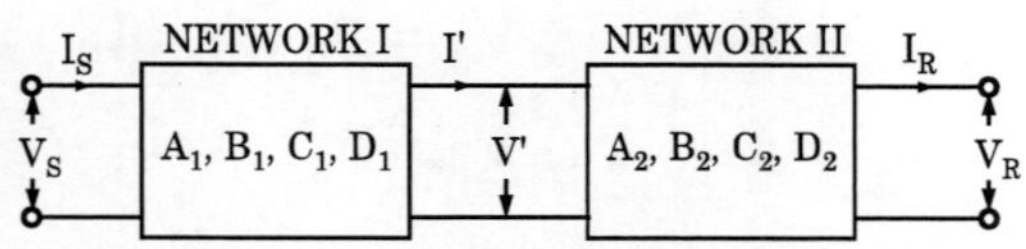

Fig. 5.20. ***Two Networks in Tandem***

For network I,
$$\mathbf{V_S} = \mathbf{A_1V'} + \mathbf{B_1I'} \quad \text{...(5.61)}$$
$$\text{and } \mathbf{I_S} = \mathbf{C_1V'} + \mathbf{D_1I'} \quad \text{...(5.62)}$$

For network II,
$$\mathbf{V'} = \mathbf{A_2V_R} + \mathbf{B_2I_R} \quad \text{...(5.63)}$$
$$\text{and } \mathbf{I'} = \mathbf{C_2V_R} + \mathbf{D_2I_R} \quad \text{...(5.64)}$$

Substituting the values of **V′** and **I′** from Eqs. (5.63) and (5.64) in Eqs. (5.61) and (5.62) respectively we have

$$\mathbf{V_S} = \mathbf{A_1}\,(\mathbf{A_2V_R} + \mathbf{B_2I_R}) + \mathbf{B_1}\,(\mathbf{C_2V_R} + \mathbf{D_2I_R})$$
$$= (\mathbf{A_1A_2} + \mathbf{B_1C_2})\,\mathbf{V_R} + (\mathbf{A_1B_2} + \mathbf{B_1D_2})\,\mathbf{I_R} \quad \text{...(5.65)}$$

$$\text{and } \mathbf{I_S} = \mathbf{C_1}\,(\mathbf{A_2V_R} + \mathbf{B_2I_R}) + \mathbf{D_1}\,(\mathbf{C_2V_R} + \mathbf{D_2I_R})$$
$$= (\mathbf{C_1A_2} + \mathbf{D_1C_2})\,\mathbf{V_R} + (\mathbf{C_1B_2} + \mathbf{D_1D_2})\,\mathbf{I_R} \quad \text{...(5.66)}$$

* Voltage source with zero internal impedance is called the ideal voltage source.

** Two networks can be two transmission lines or a transformer connected to a transmission line etc.

Therefore, for the two networks in tandem, the generalised constants of the equivalent network are

$$\left.\begin{aligned} A &= A_1A_2 + B_1C_2 \\ B &= A_1B_2 + B_1D_2 \\ C &= A_2C_1 + C_2D_1 \\ D &= B_2C_1 + D_1D_2 \end{aligned}\right\} \quad \text{...(5.67)}$$

5.9.3. Two Networks in Parallel

Let the two networks be connected in parallel, as illustrated in Fig. 5.21, and the currents at the sending end and receiving end be I_{S_1} and I_{R_1} for network **I** and I_{S_2} and I_{R_2} for network **II**

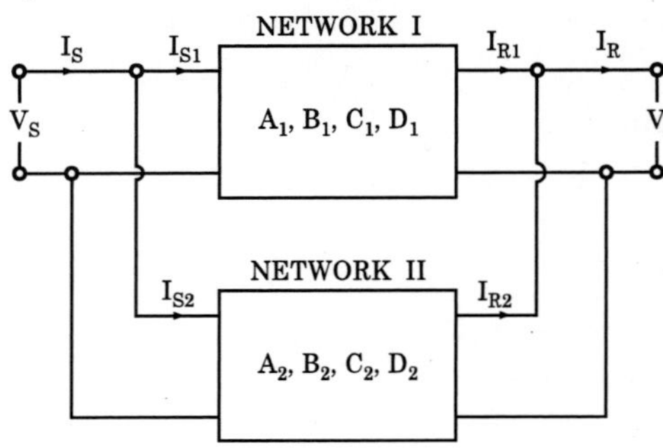

Fig. 5.21. *Two Networks in Parallel*

Here $I_S = I_{S_1} + I_{S_2}$ ...(5.68)

$I_R = I_{R_1} + I_{R_2}$ ...(5.69)

For network I, $V_S = A_1V_R + B_1I_{R_1}$ ...(5.70)

and $I_{S_1} = C_1V_R + D_1I_{R_1}$ ...(5.71)

For network II, $V_S = A_2V_R + BI_{R_2}$ ...(5.72)

$I_{S_2} = C_2V_R + D_2I_{R_2}$ ...(5.73)

Comparing Eqs. (5.70) and (5.72) we have

$$A_1V_R + B_1I_{R_1} = A_2V_R + B_2I_{R_2}$$

or $(A_1 - A_2)\, V_R = B_2I_{R_2} - B_1I_{R_1}$

or $(A_1 - A_2)\, V_R = B_2\,(I_R - I_{R_1}) - B_1I_{R_1}$ $\qquad \because I_{R_1} + I_{R_2} = I_R$

or $(A_1 - A_2)\, V_R = B_2\,I_R - I_{R_1}\,(B_1 + B_2)$

$$\text{or } I_{R_1} = \frac{B_2\,I_R - (A_1 - A_2)\,V_R}{B_1 + B_2} \quad \text{...(5.74)}$$

Substituting the value of I_{R_1} from above Eq. (5.74) in Eq. (5.70) we have

$$V_S = A_1V_R + \frac{B_1}{B_1 + B_2}[B_2\,I_R - (A_1 - A_2)\,V_R]$$

$$= \left[A_1 - \frac{B_1\,(A_1 - A_2)}{B_1 + B_2}\right]V_R + \frac{B_1\,B_2}{B_1 + B_2}\,I_R = \frac{A_1\,B_2 + B_1\,A_2}{B_1 + B_2}\,V_R + \frac{B_1\,B_2}{B_1 + B_2}\,I_R \quad \text{...(5.75)}$$

Substituting the values of I_{S_1} and I_{S_2} from Eqs. (5.71) and (5.73) in Eq. (5.68) we have

$$I_S = C_IV_R + D_1I_{R_1} + C_2V_R + D_2I_{R_2} = (C_1 + C_2)\,V_R + D_1I_{R_1} + D_2\,(I_R - I_{R_1})$$

$$= (C_1 + C_2)\,V_R + (D_1 - D_2)\,I_{R_1} + D_2I_R$$

Substituting the value of I_{R_1} from Eq. (5.74) we have

$$I_S = (C_1 + C_2) V_R + (D_1 - D_2) \frac{B_2 I_R - (A_1 - A_2) V_R}{B_1 + B_2} + D_2 I_R$$

$$= \left[C_1 + C_2 - \frac{(A_1 - A_2)(D_1 - D_2)}{B_1 + B_2} \right] V_R + \frac{B_1 D_2 + D_1 B_2}{B_1 + B_2} I_R \quad ...(5.76)$$

Therefore, for the two networks connected in parallel, the generalised constants of the equivalent network are

$$\left. \begin{aligned} A &= \frac{A_1 B_2 + A_2 B_1}{B_1 + B_2} \\ B &= \frac{B_1 B_2}{B_1 + B_2} \\ C &= C_1 + C_2 - \frac{(A_1 - A_2)(D_1 - D_2)}{B_1 + B_2} \\ D &= \frac{B_1 D_2 + B_2 D_1}{B_1 + B_2} \end{aligned} \right\} \quad ...(5.77)$$

Example 5.19. Find the A, B, C, D parameters of a 3-phase 80 km 50 Hz transmission line with series impedance of 0.15 + *j* 0.78 ohm per km and a shunt capacitance of *j* 5 × 10^{-6} mho per km.

Solution : Total series impedance of the line, $\mathbf{Z} = (0.15 + j\,0.78) \times 80 = (12 + j\,62.4)\ \Omega$

Total shunt capacitance of the line $\mathbf{Y} = j\,5 \times 10^{-6} \times 80 = j\,4 \times 10^{-4}$ mho

In nominal T-configuration

$$A = D = 1 + \frac{1}{2} YZ = 1 + \frac{1}{2} (12 + j\,62.4) \times j4 \times 10^{-4} = (0.98752 + j\,0.0024) \quad \textbf{Ans.}$$

$$B = Z \times \left(1 + \frac{YZ}{4}\right) = (12 + j\,62.4) \left[1 + \frac{1}{4} \times j4 \times 10^{-4} \times (12 + j62.4)\right]$$

$$= (12 + j\,62.4)(0.99376 + j\,0.0012)$$

$$= (11.85024 + j\,11.91072) \text{ ohms} \quad \textbf{Ans.}$$

$$C = Y = j\,4 \times 10^{-4} = 0.0004 \text{ S} \quad \textbf{Ans.}$$

Example 5.20. A 3-phase overhead transmission line has a total series impedance per phase of 200 $\angle 80°$ ohms and a total shunt admittance of 0.0013 $\angle 90°$ mho per phase. The line delivers a load of 80 M W at 0.8 pf lagging and 220 kV between the lines. Calculate :

(*a*) The ABCD constants of the line

(*b*) The sending-end voltage, current and power factor of the line, and

(*c*) The efficiency of transmission.

Solution : Receiving-end phase voltage, $V_R = \dfrac{220{,}000}{\sqrt{3}} = 127{,}000$ V

Receiving-end current, $I_R = \dfrac{80 \times 10^6}{\sqrt{3} \times 220 \times 1{,}000 \times 0.8} = 262.43$ A

Power factor, $\cos \phi = 0.8$

and $\sin \phi = \sqrt{1 - 0.8^2} = 0.6$

Taking receiving-end phase voltage as the reference phasor we have

$$\mathbf{V_R} = 127{,}000\ (1 + j\,0) = 127{,}000\angle 0^\circ \text{ volts}$$

$$\text{and } \mathbf{I_R} = 262.43\ (0.8 - j\,0.6) = (210 - j\,157.46) = 262.43\angle -36.87^\circ \text{ A}$$

$$\text{Series impedance, } \mathbf{Z} = 200\angle 80^\circ\ \Omega$$

$$\text{Shunt admittance, } \mathbf{Y} = 0.0013\angle 90^\circ \text{ siemens}$$

Using nominal T-method

$$\mathbf{A} = \mathbf{D} = 1 + \frac{1}{2}\mathbf{YZ} = 1 + \frac{1}{2} \times 0.0013\angle 90^\circ \times 200\angle 80^\circ = 1 + 0.13\angle 170^\circ$$

$$= 0.872 + j\,0.02257 = 0.872\angle 1.5^\circ \ \textbf{Ans.}$$

$$\mathbf{B} = \mathbf{Z}\left(1+\frac{1}{4}\mathbf{YZ}\right) = 200\angle 80^\circ \left(1+\frac{1}{4} \times 0.0013\angle 90^\circ \times 200\angle 80^\circ\right)$$

$$= 200\angle 80^\circ\ (1 + 0.065\angle 170^\circ) = 200\angle 80^\circ\ (0.936 + j\,0.0113)$$

$$= 200\angle 80^\circ \times 0.936\angle 0.69^\circ = 187.2\angle 80.69^\circ\ \Omega \ \textbf{Ans.}$$

$$\mathbf{C} = \mathbf{Y} = 0.0013\angle 90^\circ \text{ siemens } \textbf{Ans.}$$

Sending-end phase voltage,

$$\mathbf{V_S} = \mathbf{AV_R} + \mathbf{B\,I_R} = 0.872\angle 1.5^\circ \times 127{,}000\angle 0^\circ + 187.2\angle 80.69^\circ \times 262.43\angle -36.87^\circ$$

$$= 110{,}744\angle 1.5^\circ + 49{,}127\angle 43.82^\circ$$

$$= 110{,}700 + j\,2{,}900 + 35{,}446 + j\,34{,}015 = 146{,}146 + j\,36{,}915 = 150{,}736\angle 14.18^\circ \text{ volts}$$

Sending-end line voltage, $V_{SL} = \sqrt{3}\,V_S = \sqrt{3} \times 150{,}736 = 261{,}000$ V or 261 kV **Ans.**

Sending-end current,

$$\mathbf{I_S} = \mathbf{CV_R} + \mathbf{DI_R} = 0.0013\angle 90^\circ \times 127{,}000\angle 0^\circ + 0.872\angle 1.5^\circ \times 262.43\angle -36.87^\circ$$

$$= 165.1\angle 90^\circ + 228.84\angle -35.37^\circ$$

$$= (0 + j165.1) + 186.6 - j\,132.46 = 186.6 + j\,32.64 = 189.43\angle 9.92^\circ \text{ A } \textbf{Ans.}$$

Sending-end power factor $= \text{Cos}\ \phi_S = \text{Cos}\ (14.18^\circ - 9.92^\circ) = 0.9972$ (lag) **Ans.**

$$\text{Power input} = \sqrt{3}\ V_{SL}\ I_S \cos\phi_S \times 10^{-6} \text{ MW}$$

$$= \sqrt{3} \times 2{,}61{,}000 \times 189.43 \times 0.9972 \times 10^{-6} = 85.4 \text{ MW } \textbf{Ans.}$$

(*c*) Transmission efficiency, $\eta = \dfrac{\text{Power output}}{\text{Power input}} \times 100 = \dfrac{80}{85.4} \times 100 = 93.68\ \%$ **Ans.**

Example 5.21. Find the following for a single circuit transmission line delivering a load of 50 MVA at 110 kV and p f 0.8 lagging.

(*i*) Sending-end voltage (*ii*) sending-end current (*iii*) sending-end power (*iv*) efficiency of transmission.

Given A = D = 0.98 $\angle 3^\circ$; B = 110 $\angle 75^\circ$ Ω ; C = 0.0005 $\angle 80^\circ$ ℧

Solution : Receiving-end phase voltage, $V_R = \dfrac{110 \times 1{,}000}{\sqrt{3}} = 63{,}500$ V

$$\text{Receiving-end current, } I_R = \frac{50 \times 10^6}{\sqrt{3} \times 110 \times 10^3} = 262.43 \text{ A}$$

$$\text{Cos } \phi = 0.8 \text{ and } \sin \theta = 0.6$$

Taking receiving-end phase voltage as reference phasor, we have

$$\mathbf{V_R} = 63{,}500\ (1 + j\,0) = 63{,}500 \angle 0^\circ \text{ V}$$

$$\mathbf{I_R} = 262.43\ (0.8 - j\,0.6) = 262.43 \angle -36.87^\circ \text{ A}$$

$$\mathbf{A} = \mathbf{D} = 0.98 \angle 3^\circ;\ \mathbf{B} = 110 \angle 75^\circ\ \Omega;\ \mathbf{C} = 0.0005 \angle 80^\circ \text{ siemen}$$

Sending-end phase voltage,

$$\mathbf{V_S} = \mathbf{AV_R} + \mathbf{BI_R} = 0.98 \angle 3^\circ \times 63{,}500 \angle 0^\circ + 110 \angle 75^\circ \times 262.43 \angle -36.87^\circ$$

$$= 62{,}230 \angle 3^\circ + 28{,}867 \angle 38.13^\circ$$

$$= (62{,}144 + j\,3{,}257) + (22{,}707 + j\,17{,}824) = 84{,}851 + j\,21{,}081 = 87{,}430 \angle 13.95^\circ \text{ V}$$

(*i*) Sending-end line voltage,

$$V_{SL} = \sqrt{3} \times 87{,}430 = 151{,}433 \text{ V or } 151.433 \text{ kV } \textbf{Ans.}$$

(*ii*) Sending-end current, $\mathbf{I_S} = \mathbf{CV_R} + \mathbf{DI_R}$

$$= 0.0005 \angle 80^\circ \times 63{,}500 \angle 0^\circ + 0.98 \angle 3^\circ \times 262.43 \angle -36.87^\circ$$

$$= 31.75 \angle 80^\circ + 257.18 \angle -33.87^\circ$$

$$= 5.513 + j\,31.27 + 213.53 - j\,143.33 = 219.043 - j\,112.06$$

$$= 246 \angle -27.1^\circ \text{ A } \textbf{Ans.}$$

Sending-end phase angle, $\phi_S = 13.95^\circ - (-27.1^\circ) = 41.05^\circ$

Sending-end power factor $= \text{Cos } \phi_S = \text{Cos } 41.05^\circ = 0.7541$ (lagging)

(*iii*) Sending-end power $= \sqrt{3}\, V_{SL}\, I_S \cos \phi_S \times 10^{-6}$

$$= \sqrt{3} \times 1{,}51{,}433 \times 246 \times 0.7541 \times 10^{-6} = 48.66 \text{ MW } \textbf{Ans.}$$

(*iv*) Transmission efficiency, $\eta = \dfrac{\text{Power output}}{\text{Power input}} \times 100 = \dfrac{50 \times 0.8}{48.66} \times 100 = 82.2\%$ **Ans.**

Example 5.22. A 3-phase, 50 Hz, 300 km long transmission line has the following parameters: $r = 0.15$ Ω/km ; $x = 0.5$ Ω/km ; $y = 3 \times 10^{-6}$ ℧/km.

If the line is represented by nominal-π, determine

(*i*) the ABCD constants of the line (derive the formula you might have used);

(*ii*) the power at unity power factor that can be received if the voltage at each end is maintained at 220 kV;

(*iii*) the sending-end voltage required if a load of 200 MW at 0.85 lagging power factor is to be delivered with the receiving-end voltage remaining at 220 kV.

Solution : Total line parameters are :

$$\text{Resistance, } R = r \times l = 0.15 \times 300 = 45\ \Omega$$

$$\text{Reactance, } X = x \times l = 0.5 \times 300 = 150\ \Omega$$

$$\text{Susceptance, } Y = y \times l = 3 \times 10^{-6} \times 300 = 9 \times 10^{-4} \text{ S}$$

For nominal π-equivalent

$$(i)\ \mathbf{A} = \mathbf{D} = 1 + \frac{1}{2}\mathbf{YZ} = 1 + \frac{1}{2} \times j\,9 \times 10^{-4} \times (45 + j\,150)$$

$$= (0.9325 + j\,0.02025) \text{ or } 0.9327 \angle 1.24^\circ \textbf{ Ans.}$$

$$\mathbf{B} = \mathbf{Z} = (45 + j\,150)\ \Omega = 156.6 \angle 73.3^\circ\ \Omega \textbf{ Ans.}$$

$$\mathbf{C} = \mathbf{Y}\left(1+\frac{1}{4}\mathbf{YZ}\right) = j\,9 \times 10^{-4}\left[1+\frac{1}{4}\times j9\times 10^{-4}\,(45+j150)\right]$$

$$= j\,9 \times 10^{-4}\,(0.96625 + j\,0.010125) = 8.67 \times 10^{-4}\,\angle 90.6^\circ \text{ S } \textbf{Ans.}$$

(*ii*) Taking receiving-end phase voltage as the reference phasor, we have

$$\mathbf{V_R} = \frac{220}{\sqrt{3}} \times 1{,}000\,(1 + j\,0) = 127{,}000\ \angle 0^\circ \text{ V}$$

Load current, $\mathbf{I_R} = I_R\ \angle 0^\circ$ A ∵ power factor is unity

Sending-end phase voltage, $\mathbf{V_S} = \mathbf{AV_R} + \mathbf{BI_R}$

$$= 0.9327\ \angle 1.24^\circ \times 127{,}000\ \angle 0^\circ + 156.6\ \angle 73.3^\circ \times I_R\ \angle 0^\circ$$

$$= 118{,}425 + j\,2{,}563 + 45\,I_R + j\,150\,I_R = (118{,}425 + 45\,I_R) + j\,(2{,}563 + 150\,I_R)$$

$$\text{or } V_S = \sqrt{(118{,}425 + 45\,I_R)^2 + (2{,}563 + 150\,I_R)^2}$$

$$16{,}133 \times 10^6 = 14{,}024.5 \times 10^6 + 6.57 \times 10^6 + 10{,}658 \times 10^6\,I_R + 0.7689 \times 10^6\,I_R + 0.024525 \times 10^6\,I_R^2$$

$$\text{or } 0.024525\,I_R^2 + 11.427\,I_R - 21{,}02 = 0$$

$$\text{or } I_R = \frac{-11.427 \pm \sqrt{(11.427)^2 + 4 \times 0.024525 \times 2{,}102}}{2 \times 0.024525} = 141.175 \text{ A}$$

Power received, $P = \sqrt{3}\ V_{RL}\ I_R \cos\theta \times 10^{-6}$ MW

$$= \sqrt{3} \times 220{,}000 \times 141.175 \times 1.0 \times 10^{-6} = 53.8 \text{ Mw } \textbf{Ans.}$$

(*iii*) $V_R = 127{,}000$ V

$$I_R = \frac{200 \times 10^6}{\sqrt{3} \times 220{,}000 \times 0.85} = 617.5 \text{ A}$$

Phase angle, $\phi = \text{Cos}^{-1}\,0.85 = 31.79^\circ$ (lagging)

Taking receiving-end phase voltage as the reference phasor we have

$$\mathbf{V_R} = 127{,}000\ \angle 0^\circ \text{ V and } \mathbf{I_R} = 617.5\ \angle -31.79^\circ \text{ A}$$

Sending-end phase voltage,

$$\mathbf{V_S} = \mathbf{AV_R} + \mathbf{BI_R}$$

$$= 0.9327\ \angle 1.24^\circ \times 1{,}27{,}000\ \angle 0^\circ + 156.6\ \angle 73.3^\circ \times 617.5\ \angle -31.79^\circ$$

$$= 118{,}469\ \angle 1.24^\circ + 96{,}700\ \angle 41.51^\circ$$

$$= 118{,}441 + j\ 2{,}563.72 + 72{,}413 + j\ 64{,}088$$

$$= 190{,}854 + j\ 66{,}652 = 202{,}158\ \angle 19.25^\circ \text{ V}$$

Sending-end line voltage, $V_{SL} = \sqrt{3} \times 202{,}158 = 350{,}150$ V or 350.15 kV **Ans.**

Example 5.23. The A, B, C, D constants of a three-phase transmission line are A = D = 0.936 + *j*0.016, B = 33.5 + *j*138 Ω and C = (–0.9280 + *j*901.223) × 10^{-6} mho.

The load at the receiving end is 40 MW at 200 kV with pf of 0.86 lagging. Find the magnitude of the sending-end voltage, current, power, and voltage regulation. Assume that the magnitude of the sending-end voltage remains constant. [*U.P. Technical Univ. Elements of Power System 2008-09*]

Solution : Receiving-end phase voltage, $V_R = \dfrac{200 \times 1{,}000}{\sqrt{3}} = 115{,}470$ V

$$\text{Receiving-end current, } I_R = \frac{40 \times 10^6}{\sqrt{3} \times 200 \times 10^3 \times 0.86} = 134.27 \text{ A}$$

$$\text{Power factor, } \cos\phi = 0.86$$

$$\text{Phase angle, } \phi = \text{Cos}^{-1}\, 0.86 = -30.68°$$

$$\sin\phi = 0.51$$

$$\mathbf{A} = \mathbf{D} = 0.936 + j0.016 = 0.93614 \angle 0.98°$$

$$\mathbf{B} = 33.5 + j138 = 142 \angle 76.355° \,\Omega$$

$$\mathbf{C} = (-0.9280 + j901.223) \times 10^{-6} = 901.2235 \angle 90.06° \times 10^{-6} \,\mho$$

Taking receiving-end phase voltage as reference phasor, we have

$$\mathbf{V_R} = 115{,}470(1 + j0) = 115{,}470 \angle 0° \text{ V}$$

$$\mathbf{I_R} = 134.27 \angle -30.68° \text{ A}$$

Sending-end phase voltage,

$$\mathbf{V_S} = \mathbf{A}\mathbf{V_R} + \mathbf{B}\mathbf{I_R} = 0.93614 \angle 0.98° \times 115{,}470 \angle 0° + 142 \angle 76.355° \times 134.27 \angle -30.68°$$

$$= 108{,}096 \angle 0.98° + 19{,}066 \angle 45.675°$$

$$= (108{,}080 + j1848.8) + (13{,}322 + j13{,}639.6)$$

$$= (121{,}402 + j15{,}488.4) = 122{,}386 \angle 7.27° \text{ V}$$

Sending-end line voltage, $V_{SL} = \sqrt{3} \times V_S = \sqrt{3} \times 122{,}386 = 211{,}979$ V or 211.98 kV **Ans.**

Sending-end current, $\mathbf{I_S} = \mathbf{C}\mathbf{V_R} + \mathbf{D}\mathbf{I_R}$

$$= 901.2235 \times 10^{-6} \angle 90.06° \times 115{,}470 \angle 0° + 0.93614 \angle 0.98° \times 134.27 \angle -30.68°$$

$$= 104.064 \angle 90.06° + 125.696 \angle -29.7°$$

$$= (-0.109 + j104.064) + (109.18 - j62.28)$$

$$= (109.071 + j41.784) = 116.8 \angle 20.96° \text{ A}$$ **Ans.**

Sending-end phase angle, $\phi_S = 20.96° - 7.27° = 13.69° = 0.9716$ (leading) **Ans.**

Sending-end power $= \sqrt{3} V_{SL} I_S \times \cos\phi_S \times 10^{-6}$ MW

$$= \sqrt{3} \times 211.98 \times 10^3 \times 116.8 \times 0.9716 \times 10^{-6} = 41.666 \text{ MW}$$ **Ans.**

Voltage regulation $= \dfrac{V_{SL} - V_{RL}}{V_{RL}} \times 100 = \dfrac{211.98 - 200}{200} \times 100 = 5.99\%$ **Ans.**

Example 5.24. Two 3-phase transmission lines have the generalised constants :

$$\mathbf{A_1} = \mathbf{D_1} = 0.98 \angle 2°;\ \mathbf{B_1} = 28 \angle 69° \,\Omega;\ \mathbf{C_1} = 0.0002 \angle 88° \text{ S}$$

$$\mathbf{A_2} = \mathbf{D_2} = 0.95 \angle 3°;\ \mathbf{B_2} = 40 \angle 85° \,\Omega;\ \mathbf{C_2} = 0.0004 \angle 90° \text{ S}$$

They are connected in series and delivers a load current of 200 A at 0.95 p f (lagging) at 110 kV. Determine the sending-end voltage and sending-end current.

Solution : Generalised constants of two lines connected in series

$$\mathbf{A} = \mathbf{A_1 A_2 + B_1 C_2} = 0.98 \angle 2^\circ \times 0.95 \angle 3^\circ + 28 \angle 69^\circ \times 0.0004 \angle 90^\circ$$
$$= 0.931 \angle 5^\circ + 0.0112 \angle 159^\circ$$
$$= 0.9275 + j\,0.08114 + 0.0105 + j\,0.004 = 0.917 + j\,0.08514 = 0.92 \angle 5.3^\circ$$

$$\mathbf{B} = \mathbf{A_1 B_2 + B_1 D_2} = 0.98 \angle 2^\circ \times 40 \angle 85^\circ + 28 \angle 69^\circ \times 0.95 \angle 3^\circ$$
$$= 39.2 \angle 87^\circ + 26.6 \angle 72^\circ = 2.0516 + j\,39.1463 + 8.2199 + j\,25.2981$$
$$= 10.2715 + j\,64.4444 = 65.26 \angle 80.94^\circ\ \Omega$$

$$\mathbf{C} = \mathbf{C_1 A_2 + D_1 C_2} = 0.0002 \angle 88^\circ \times 0.95 \angle 3^\circ + 0.98 \angle 2^\circ \times 0.0004 \angle 90^\circ$$
$$= 0.00019 \angle 91^\circ + 0.000392 \angle 92^\circ$$
$$= -\,0.000003 + j\,0.00019 - 0.00001368 + j\,0.000392$$
$$= -\,0.00001668 + j\,0.000582 = 0.000582 \angle 91.67^\circ\ \text{S}$$

$$\mathbf{D} = \mathbf{C_1 B_2 + D_1 D_2} = 0.0002 \angle 88^\circ \times 40 \angle 85^\circ + 0.98 \angle 2^\circ + 0.95 \angle 3^\circ$$
$$= -\,00794 + j\,0.000975 + 0.9275 + j\,0.08114 = 0.91956 + j\,0.082115 = 0.923 \angle 5.1^\circ$$

Receiving-end phase voltage, $V_R = \dfrac{110{,}000}{\sqrt{3}} = 63{,}500$ V

Receiving-end current, $I_R = 200$ **A**

Receiving-end power factor angle, $\phi_R = \text{Cos}^{-1}\, 0.95 = 18.195^\circ$ (lagging)

Taking receiving-end phase voltage as reference phasor, we have

$$\mathbf{V_R} = 63{,}500 \angle 0^\circ \ \text{V}$$

$$\text{and } \mathbf{I_R} = 200 \angle -18.195^\circ \ \text{A}$$

Sending-end phase voltage,

$$\mathbf{V_S} = \mathbf{AV_R + B\,I_R} = 0.92 \angle 5.3^\circ \times 63{,}500 \angle 0^\circ + 65.26 \angle 80.94^\circ \times 200 \angle -18.195^\circ$$
$$= 58{,}420 \angle 5.3^\circ + 13{,}052 \angle 62.745^\circ = 58{,}170 + j\,5{,}396 + 5{,}977 + j\,11{,}602$$
$$= 64{,}147 + j\,16{,}998 = 66{,}361 \angle 14.84^\circ \ \text{V}$$

Sending-end line voltage,

$V_{SL} = \sqrt{3} \times 66{,}361 = 114{,}940$ V or 114.94 kV **Ans.**

Sending-end current,

$$\mathbf{I_S} = \mathbf{C\,V_R + D\,I_R} = 0.000582 \angle 91.67^\circ \times 63{,}500 \angle 0^\circ + 0.923 \angle 5.1^\circ \times 200 \angle -18.195^\circ$$
$$= 36.957 \angle 91.67^\circ + 184.6 \angle -13.095^\circ$$
$$= -\,1.077 + j\,36.9413 + 179.8 - j\,41.8241$$
$$= 178.723 - j\,4.8828 = 178.8 \angle -1.56^\circ \ \text{A}$$ **Ans.**

Example 5.25. Two identical three-phase transmission lines are connected in parallel to supply a load of 100 MW at 132 kV and 0.8 p f lagging at the receiving-end. The ABCD constants of each transmission line are as follows :

A = D = 0.98 $\angle 1^\circ$; B = 100 $\angle 75^\circ$ Ω per phase ; C = 0.0005 $\angle 90^\circ$ siemens per phase.

Determine (*a*) the ABCD constants of the combined network (*b*) the sending-end power factor. [*GATE 1995*]

Solution : Generalised constants of two lines combined in parallel are :

$$\mathbf{A} = \frac{\mathbf{A_1 B_2 + A_2 B_1}}{\mathbf{B_1 + B_2}} = \frac{0.98 \angle 1^\circ \times 100 \angle 75^\circ + 0.98 \angle 1^\circ \times 100 \angle 75^\circ}{100 \angle 75^\circ + 100 \angle 75^\circ} = 0.98 \angle 1^\circ \text{ Ans.}$$

$$\mathbf{B} = \frac{\mathbf{B_1 \cdot B_2}}{\mathbf{B_1 + B_2}} = \frac{100 \angle 75^\circ \times 100 \angle 75^\circ}{100 \angle 75^\circ + 100 \angle 75^\circ} = 50 \angle 75^\circ\ \Omega \text{ Ans.}$$

$$\mathbf{C} = \mathbf{C_1 + C_2} - \frac{\mathbf{(A_1 - A_2)(D_1 - D_2)}}{\mathbf{B_1 + B_2}} = \mathbf{C_1 + C_2}$$

$\because \mathbf{A_1 = A_2}$ and $\mathbf{D_1 = D_2}$ So, $\mathbf{(A_1 - A_2)} = 0$ and $\mathbf{(D_1 - D_2)} = 0$

$= 0.0005 \angle 90^\circ + 0.0005 \angle 90^\circ = 0.001 \angle 90^\circ$ S **Ans.**

$$\mathbf{D} = \frac{\mathbf{B_1 D_2 + B_2 D_1}}{\mathbf{B_1 + B_2}} = \mathbf{A} = 0.98 \angle 1^\circ \text{ Ans.}$$

Receiving-end phase voltage, $\mathbf{V_R} = \frac{132 \times 1,000}{\sqrt{3}} = 76,210$ V

Receiving-end current, $\mathbf{I_R} = \frac{100 \times 10^6}{\sqrt{3} \times 132 \times 10^3 \times 0.8} = 546.7$ A

Taking receiving-end phase voltage as the reference phasor, we have

$\mathbf{V_R} = 76,210 \angle 0^\circ$ V and $\mathbf{I_R} = 546.7 \angle -36.87^\circ$ A

$\because$ Phase angle, $\phi_R = \text{Cos}^{-1}\, 0.8 = -36.87^\circ$

Sending-end phase voltage, $\mathbf{V_S = AV_R + BI_R}$

$= 0.98 \angle 1^\circ \times 76,210 \angle 0^\circ + 50 \angle 75^\circ \times 546.7 \angle -36.87^\circ$

$= 74,685.8 \angle 1^\circ + 27,335 \angle 38.13^\circ = 74,674.4 + j\, 1,303.4 + 21,502 + j\, 16,877.9$

$= 96,176.4 + j\, 18,181.3 = 97,879.8 \angle 10.7^\circ$ V

Sending-end current, $\mathbf{I_S = CV_R + DI_R}$

$= 0.001 \angle 90^\circ \times 76,210 \angle 0^\circ + 0.98 \angle 1^\circ \times 546.7 \angle -36.87^\circ$

$= 76.21 \angle 90^\circ + 535.766 \angle -35.87^\circ$

$= j\, 76.21 + 434.15 - j\, 313.93 = 434.15 - j\, 237.72 = 495 \angle -28.7^\circ$ A

Sending-end phase angle, ϕ_S = Phase angle between $\mathbf{V_S}$ and $\mathbf{I_S} = 10.7^\circ - (-28.7^\circ) = 39.4^\circ$ (lagging)

Sending-end power factor = Cos ϕ_S = Cos $39.4^\circ = 0.7727$ (lagging) **Ans.**

EXERCISES

1. Along with circuit representations, explain the basis of categorization of short, medium and long transmission lines. **[U.P. Technical Univ. Elements of Power System 2003-04]**
2. Identify the criterion for classification of transmission lines. Which transmission line representation is close to actual system dynamics ?
[U.P. Technical Univ. Elements of Power System 2005-06]
3. Draw the phasor diagram of a short transmission line and derive an expression for voltage regulation. **[U.P. Technical Univ. Elements of Power System 2006-07]**
4. Obtain the expression of 'voltage regulation' and 'efficiency' of a short transmission line in terms of line parameters. **[U.P. Technical Univ. Elements of Power System 2007-08]**
5. Define regulation, efficiency, losses and power factor at each end of the line and explain how these characteristics are affected by the constants of the line.

6. Define precisely the 'regulation' of a transmission line and discuss qualitatively its dependence on the load power factor. Show that the regulation is zero when $\phi_R = \theta - \pi/2$ where ϕ_R is the load power factor angle and θ is the line impedance angle. Can the regulation be negative ? If so, under what conditions ? State clearly the assumptions made.
7. Show a diagrammatic representation of a medium line in terms of its nominal-π model. Indicate also the values of series impedance and shunt admittance of the model in terms of the resistance, inductance and capacitance per unit length of the line.
8. What is Ferranti effect ? Explain it with the help of phasor diagram.
9. Explain by means of phasor diagram the 'Ferranti effect' in transmission lines. **[U.P. Technical Univ. Elements of Power System 2007-08]**
10. What is Ferranti Effect ? Derive an expression for the voltage rise of an unloaded line.
11. A long transmission line is not loaded ? What will be the voltage at the receiving end compared to the sending-end voltage ? Give reasons.
12. Explain the Ferranti effect and derive an expression for power loss in an open-circuited line.
13. Explain the physical significance of the generalised circuit constants A, B, C and D of a transmission line. Find the values of A, B, C and D in the following approximate methods in terms of Z and Y.
    (*i*) Simple series impedance method.
    (*ii*) Nominal π-method.
    (*iii*) Nominal T-method.
14. Obtain A, B, C, D constants for a short transmission line. **[U.P. Technical Univ. Elements of Power System 2007-08]**
15. Derive the A, B, C, D constants for the transmission line represented by nominal-T section. **[U.P. Technical Univ. Elements of Power System 2005-06]**
16. Derive A, B, C, D constants of a medium length transmission line and hence prove that AD – BC = 1. **[U.P. Technical Univ. Elements of Power System 2005-06]**
17. The ABCD parameters of two transmission networks are A_1, B_1, C_1, D_1 and A_2, B_2, C_2, D_2 respectively. Determine the overall ABCD parameters (*a*) when the two networks are connected in series and (*b*) in parallel
18. Write short notes on the following :
    (*i*) Classification of transmission lines.
    (*ii*) Regulation and efficiency of transmission line.
    (*iii*) Ferranti effect.
    (*iv*) Generalised circuit constants for lines in series.

SHORT ANSWER TYPE QUESTIONS WITH ANSWERS

Q. 1. What are the factors which govern the performance of a transmission line ?

Ans. The factors governing the performance of a transmission line are series resistance R and inductance L, shunt capacitance C and conductance G. However, the effect of conductance is negligible.

Q. 2. How do you classify transmission lines ?

Ans. Transmission lines are classified depending upon the manner in which capacitance is taken into account.

Q. 3. What are the causes of voltage drop and line loss in a transmission line ?

Ans. The voltage drop in a transmission line is due to its resistance and inductance whereas the line loss is due to its resistance.

Q. 4. What is Ferranti effect ?

Ans. The phenomenon of rise in voltage at the receiving end of the open-circuited or lightly loaded line is called the Ferranti effect.

Q. 5. How does rise in voltage at the receiving end of an open-circuited line depend upon the length of the line and system operating voltage ?

Ans. Rise in voltage at the receiving end of an open-circuited line is proportional to the square of the length of the line and varies directly in proportion to the system operating voltage.

Q. 6. How do the terms 'impedance drop', 'voltage drop' and 'voltage regulation', in connection with transmission lines differ ?

Ans. The impedance drop in a line is the voltage drop in the line impedance and is equal to the phasor difference between the sending-ending voltage and the receiving-end voltage.

The voltage drop in a line is the arithmetical difference between the sending-end and receiving-end voltages.

The voltage regulation is defined as the change in voltage at the receiving-end when full-load is thrown off, the sending-end voltage being held constant.

Q. 7. Why in medium and long transmission lines, regulation is greater than voltage drop ?

Ans. The regulation is greater than voltage drop in a medium/long transmission line due to rise in voltage owing to Ferranti effect.

Q. 8. What are the units of generalised constants of a transmission line ?

Ans. Generalised constants A and D are dimensionless whereas the units of generalised constants B and C are ohms and siemens (mho) respectively.

PROBLEMS

1. A single phase transmission line has a resistance of 0.22 Ω and inductive reactance of 0.36 Ω. Find the voltage at the sending end to give 500 kVA at 2,000 V at the receiving end at a load power factor of 0.707 lagging. Also find the efficiency of transmission at this load.
[**Ans.** 2,102.37 V ; 96.26 %]
2. A 1-phase transmission line has a resistance of 0.20 Ω and an inductive reactance of 0.40 Ω. Find the voltage at the sending end to give 500 kVA at 2 kV at receiving end at load power factors of (*i*) unity (*ii*) 0.707 lagging. Illustrate with suitable phasor diagrams.
[**Ans.** (*i*) 2,052.44 V (*ii*) 2,106 V]
3. A power of 250 kW is delivered at the end of a single phase line 15 km long at 1,000 V, and 0.8 p f lag, 50 Hz. The resistance of the conductor is 0.03 Ω/km and the inductance/conductor/km is 0.15 mH. Find the sending-end voltage, sending-end power factor and % regulation. Draw the phasor diagram. [**Ans.** 1,501.5 V ; 0.72 (lag) ; 73.99 %]
4. An overhead, single phase transmission line delivers 1,100 kW at 33 kV at 0.8 pf lagging. The total resistance of the line is 10 Ω and the total inductive reactance is 15 Ω. Determine (*i*) sending-end voltage (*ii*) sending-end power factor and (*iii*) transmission efficiency.
[**Ans.** 33,710 V (*ii*) 0.794 (lag) (*iii*) 98.5 %]
5. A single phase line has an impedance of $5 \angle 60°$ ohm. It is supplying a load of 120 A, 3.3 kV at 0.8 pf lagging. Calculate its regulation. [**Ans.** 16.94 %]
6. What is the maximum length in km for a single phase transmission line having copper conductors of 0.775 cm^2 cross-section over which 200 kW at unity power factor and at 3,300 V are to be delivered ? The efficiency of transmission is 90 percent. Take specific resistance as 1.725 μΩ-cm. [**Ans.** 13.6 km]
7. A short three phase transmission line having resistance and reactance 5 Ω and 12 Ω respectively supplies 10 MVA load at 0.8 p f (lagging). When receiving-end voltage is maintained at 33 kV, find the sending -end voltage. Draw a phasor diagram to explain.
[**Ans.** 36.394 kV]

8. A 3-phase transmission line delivers 10,000 kW at 33 kV and 0.8 lagging power factor. The resistance and inductive reactance per conductor is 3 Ω and 6 Ω respectively. Calculate the sending-end voltage and the efficiency of transmission.
[U.P.S.C. I.E.S. (Elec. Engg.) 1972] [**Ans.** 35.2 kV; 96.5%]

9. Find the sending-end voltage, percentage voltage drop and efficiency of a 20 km, 20 kV, 50 Hz, 3-phase transmission line delivering 5 MW at a power factor of 0.8. The conductors have an effective diameter of 12 mm and a x-sectional area of 95 mm^2. These are arranged in an equilateral triangular formation, the distance between their centres being one metre.
[**Ans.** 22.2 kV : 12 % : 93.3%]

10. A 3-phase transmission line delivers a load of 5 MW at 0.8 p f (lagging). Resistance of each conductor is 0.5 Ω/km. Receiving-end voltage is 33 kV. If the line loss is not to exceed 10%, determine the length of the line. [**Ans.** 27.9 km]

11. A 3-phase line has a resistance of 5.31 Ω and an inductance of 0.0176 H. Power is transmitted at 33 kV, 50 Hz from one end and the load at the receiving-end is 3,600 kW at 80 percent power factor. Find the line current, receiving-end voltage, sending-end power factor and efficiency of transmission. **[A.P.S. Univ. 1978]** [**Ans.** 81.5 A, 31.9 kV, 0.796, 97.2%]

12. A 3-phase 20 km long line delivers 3 MW at 11 kV. Line loss is equal to 10% of power delivered. Line inductance is 0.9156 mH/phase/km. Find sending-end voltage and regulation if the load power factor is 0.8 lagging. [**Ans.** 12,863.66 V ; 16.94%]

13. An 11 kV, 3-phase, 50 Hz ac power line supplies power to a load of 1 MW at 0.8 p f lag situated at a distance of 10 km. The resistance of each conductor is 0.5 Ω per km and reactance is 0.56 Ω per km. Determine (*i*) line current (*ii*) line losses (*iii*) receiving-end voltage and (*iv*) efficiency of transmission.
[**Ans.** (*i*) 71.5 A (*ii*) 76.68 kW (*iii*) 10.09 kV (*iv*) 92.88%]

14. A 3-phase transmission line has the following constants (line to neutral) :
R = 10 Ω, inductive reactance = 20 Ω, capacitive susceptance = 4×10^{-4} S
Using the nominal-T method, calculate the sending-end voltage, line current, power factor and the efficiency of transmission when supplying a balance load of 10 MW at 66 kV, power factor 0.8 lagging. [**Ans.** 69.544 kV; 100.53 A; 0.853 (lag); 96.79%]

15. A 3-phase transmission line has the following data :
Resistance of each conductor = 20 Ω
Inductive reactance of each conductor = 40 Ω
Capacitive susceptance to neutral = 0.0004 ℧
Load at the receiving end = 12,500 kVA at 0.8 p f lagging
Voltage at the receiving end = 66 kV
Find (*a*) sending-end voltage and current (*b*) sending-end power factor (*c*) efficiency (*d*) regulation of the line. Use nominal T-method and draw the phasor diagram.
[A.P.S. Univ. 1977] [**Ans.** (*a*) 73.122 kV; 100.125 A (*b*) 0.84 (lag) (*c*) 93.88% (*d*) 10.8%]

16. A 1,000 km long, 3-phase, 50 Hz transmission line has resistance per phase per km = 0.1 Ω, reactance per phase per km = 0.5 Ω; susceptance per phase per km = 10×10^{-6} ℧. If the line supplies a load of 20 MW at 0.9 pf lagging at 65 kV at the receiving end, calculate by the nominal- π method the regulation and efficiency of the line. Neglect leakance.
[Bombay Univ. Elec. Power-I, March 1971] [**Ans.** 15.15% ; 94.56%]

17. A 3-ϕ, 50 Hz, 150 km long overhead line has the following constants :
Resistance/phase/km = 0.15 Ω
Inductance/phase/km = 1.2 mH
Capacitance/phase/km = 0.0096 μF
The line supplies a load of 30 MW at 0.9 power factor lagging at a line voltage of 132 kV at the receiving end.
Determine voltage, current and power factor at the sending end and also percentage regulation of the line. [**Ans.** 142.08 kV; 132.73 A; 0.9588 (lag); 9.037%]

18. A total load of 4 MW at 0.8 pf (lagging) is transmitted to a 33/11 kV substation by two 33 kV OH lines drawn in parallel. One line has a conductor resistance 2 Ω per phase and reactance 15 Ω per phase and corresponding values for the other line being 1.5 Ω and 12 Ω respectively. Calculate power transmitted by each OH line. [**Ans.** 1.784 MW; 2.216 MW]

19. The parameters of a 132 kV, 50 Hz, 3-ϕ transmission line are:

$$R = 12.84\ \Omega\ ;\ L = 0.162\ H\ ;\ C = 1.014 \times 10^{-6}\ F$$

Calculate the line constants, **A**, **B**, **C** and **D** for (i) nominal T-circuit and (ii) nominal π-circuit.

[**Ans.** (*i*) $A = D = 0.992\ \angle 0.1°$; $B = 52.2\ \angle 75.95°$; Ω; $C = 0.00032\ \angle 90°$ S

(*ii*) $A = D = 0.992\ \angle 0.1°$; $B = 52.4\ \angle 75.9°$; Ω; $C = 0.000318\ \angle 90.05°$ S

20. A three-phase, 220 kV, 50 Hz transmission line 200 km long transmits load of 75 MW at 0.8 pf lagging at the receiving end. Resistance of the line per phase per km is 0.8 Ω. Susceptance per phase per km = 6×10^{-6} ℧. Reactance per phase per km = 0.8 Ω. Determine (*a*) A, B, C, D constants of the line. (*b*) the sending-end voltage and current (*c*) sending-end power factor. (*d*) sending-end active and reactive powers (*e*) regulation and efficiency of the line.

[**Ans.** (*a*) $A = D = 0.909\ \angle 6.06°$, $B = 215.66\ \angle 47.89°$ Ω $C = 1.2 \times 10^{-3}\ \angle 90°$ S (by nominal T-method) (*b*) 291.7 kV, 195.78 A (*c*) 0.9981 (leading) (*d*) 98.71 MW; 6.094 MVAR (*e*) 32.57%, 75.98%]

21. A balanced load of 50 MW is supplied at 132 kV, 50 Hz and 0.8 lagging power factor by means of a transmission line. The series impedance of a single conductor is (20 + *j* 50) Ω and the total phase-neutral admittance is 300×10^{-6} S. Shunt leakance may be neglected. Using the nominal T-approximation, calculate the voltage at the sending-end of the line.

If the load is removed and the sending-end voltage remains constant, find the percentage rise in voltage at the receiving end. [**Ans.** 153.3 kV; 17%]

22. A 220 kV, 50 Hz, 3-phase overhead transmission line delivers a load of 75,000 kW at 0.8 pf lagging at the receiving end and has the following constants :

$A = D = 0.9\ \angle 0.6°$; $B = 153.2\ \angle 84.6°$ Ω and $C = 0.0012\ \angle 90°$ S

Calculate the sending-end parameters. [**Ans.** 247 kV; 180 A; 4.92°; 0.996 (lag); 76,700 kW]

23. A 132 kV overhead transmission line has the following line constants :

$$A = D = 0.99\ \angle 0.3°;\ B = 70\ \angle 69°\ \Omega;\ C = 4 \times 10^{-4}\ \angle 90°\ ℧$$

Calculate the sending-end voltage of transmission line if the load delivered at the receiving end is 45 MVA at 0.8 pf lag. [**Ans.** 167.215 kV]

24. A transmission line consists of two lines I and II connected in series, the line I being at the sending end and line II at the receiving end. The circuits have the following constants :

Constants	Line I	Line II
A = D	$0.895\ \angle 1.4°$	$0.987\ \angle 0.3°$
B	$182.5\ \angle 78.6°$ Ω	$76\ \angle 67.1°$ Ω
C	$0.00111\ \angle 90.47°$ ℧	$0.00037\ \angle 90.0°$ ℧

Determine the **A**, **B**, **C** and **D** constants of the whole line which is equivalent to the two lines in series. [**Ans.** $A = 0.816\ \angle 2.7°$; $B = 247\ \angle 76.1°$ Ω; $C = 0.00142\ \angle 90.97°$ S ; $D = 0.8071\ \angle 47.0°$]

□□□

6

CHAPTER

REPRESENTATION AND PERFORMANCE OF LONG TRANSMISSION LINES

INSIDE THIS CHAPTER

6.1. Introduction **6.2.** Analysis of Long Transmission Lines (Rigorous Method of Solution of Long Transmission Lines) **6.3.** Evaluation of Constants ABCD **6.4.** Surge Impedance **6.5.** Surge Impedance Loading **6.6.** Transmission Line With Series Impedance at The Receiving End **6.7.** Transmission Line With Series Impedance at The Sending End **6.8.** Transmission Line With Series Impedances at Both Ends **6.9.** Transmission Line With Transformers at Both Ends **6.10.** Equivalent T-Network of a Long Transmission Line **6.11.** Equivalent π-Network of a Long Transmission Line **6.12.** Constant Voltage Transmission **6.13.** Synchronous Phase Modifiers or Compensators **6.14.** Tuned Power Lines

6.1. INTRODUCTION

It is well known that the line constants (resistance, inductance, capacitance and conductance) of a transmission line are uniformly distributed over the entire length of the line. So far it has been assumed that the line has lumped constants and the line calculations made with such assumptions gave results with reasonable accuracy. If such an assumption of lumped line constants is applied to long transmission lines (lines having length more than 200 km), serious errors are introduced in the line performance calculations. Therefore, the performance calculations of long transmission lines are made with line constants uniformly distributed over the entire length of the line so that the results with fair degree of accuracy are obtained. Rigorous mathematical treatment is required for the solution of such transmission lines. The equivalent circuit of a 3-phase long transmission line is represented schematically by Fig. 6.1.

The whole line is divided into n-sections, each section having line constants $\frac{1}{n}$th of those for the whole line.

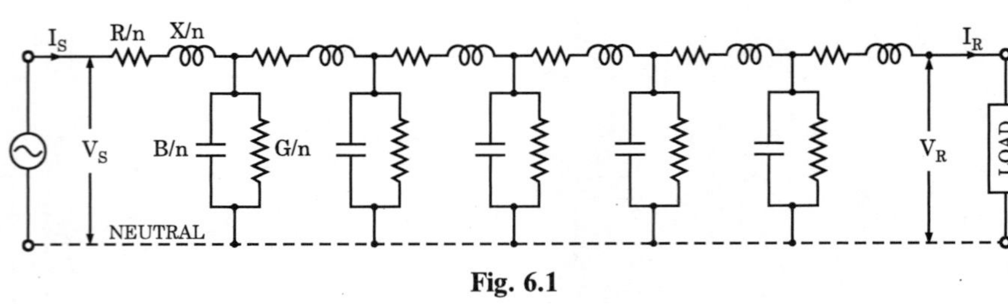

Fig. 6.1

The following points are worthnoting:

(*i*) The line constants (resistance, inductive reactance, capacitive susceptance and conductance) are uniformly distributed over the entire length of the line as is actually the case.

(*ii*) The resistance (R) and inductive reactance (X) are the series elements.

(*iii*) The capacitive susceptance (B) and leakage conductance (G) are shunt elements. The leakage susceptance is due to the fact the capacitance exists between line and neutral. The leakage conductance (G) takes into account the energy losses occurring through leakage over the insulators, or due to corona effect between the conductors.

Thus shunt admittance, $Y = \sqrt{G^2 + B^2}$

(*iv*) The leakage current through the shunt admittance is maximum at the sending end of the transmission line and decreases continuously as we move towards the receiving end of the line and becomes zero at the receiving end.

6.2. ANALYSIS OF LONG TRANSMISSION LINES (RIGOROUS METHOD OF SOLUTION OF LONG TRANSMISSION LINES)

The exact equivalent circuit of a transmission line is shown in Fig. 6.2.

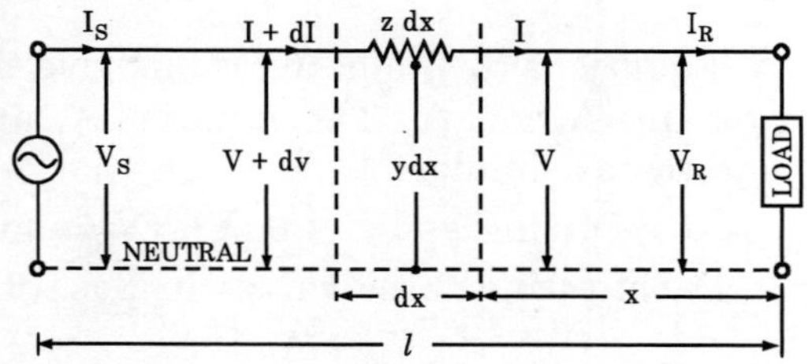

Fig. 6.2

Consider an infinitely small length *dx* of the line at a distance *x* from the receiving end.

Let r = Resistance per unit length of line

x = Reactance per unit length of line

b = Susceptance per unit length of line

g = Conductance per unit length of line

$z = \sqrt{r^2 + x^2}$, impedance per unit length of line

$y = \sqrt{g^2 + b^2}$, admittance per unit length of line

$\mathbf{V}$ = Voltage per phase at the end of element towards receiving end

$\mathbf{V} + d\mathbf{V}$ = Voltage per phase at the end of element towards sending end

$\mathbf{V_R}$ = Voltage per phase at receiving end

$\mathbf{V_S}$ = Voltage per phase at sending end

$\mathbf{I_R}$ = Current per phase at receiving end

$\mathbf{I_S}$ = Current per phase at sending end

$\mathbf{I} + d\mathbf{I}$ = Current entering the element dx

and $\mathbf{I}$ = Current leaving the element dx

Now the series impedance of element *dx* of the line = $z\,dx$

The shunt admittance of element *dx* of the line = $y\,dx$

Obviously the rise in voltage over the element length in the direction of increasing x

$d\mathbf{V} = \mathbf{I}\,z\,dx$, voltage drop in the element dx

$$\text{or} \quad \frac{d\mathbf{V}}{dx} = z\mathbf{I} \qquad \text{...(6.1)}$$

Similarly, the difference of current entering the element and that of leaving the element,

$d\mathbf{I} = \mathbf{V}\,y\,dx$, the current drawn by this element or $\dfrac{d\mathbf{I}}{dx} = \mathrm{V}y$ ...(6.2)

Differentiating expression (6.1) w.r.t. x we get

$$\frac{d^2\,\mathbf{V}}{d\,x^2} = z\,\frac{d\mathbf{I}}{dx} = z\,\mathbf{V}\,y \quad \text{...(6.3)} \quad \because \text{ From Eq. (6.2) } \frac{d\mathrm{I}}{dx} = \mathrm{V}y$$

The solution of above differential equation is

$$\mathbf{V} = \mathrm{A}_1 e^{\sqrt{yz}\,x} + \mathrm{A}_2 e^{-\sqrt{yz}\,x} \quad \text{...(6.4)}$$

where A_1 and A_2 are unknown constants. Differentiating Eq. (6.4) w.r.t. x we get

$$\frac{d\mathbf{V}}{dx} = \sqrt{yz}\,\left\{\mathrm{A}_1\,e^{\sqrt{yz}\,x} - \mathrm{A}_2\,e^{-\sqrt{yz}\,x}\right\} \quad \text{...(6.5)}$$

and from Eq. (6.1) we have

$$\mathbf{I} = \frac{1}{z}\frac{d\mathbf{V}}{dx}$$

$$\therefore\ \mathbf{I} = \frac{1}{z}\sqrt{yz}\,\left\{\mathrm{A}_1\,e^{\sqrt{yz}\,x} - \mathrm{A}_2\,e^{-\sqrt{yz}\,x}\right\} = \sqrt{y/z}\,\left\{\mathrm{A}_1\,e^{\sqrt{yz}\,x} - \mathrm{A}_2\,e^{-\sqrt{yz}\,x}\right\} \quad \text{...(6.6)}$$

Equations (6.4) and (6.5) thus give the expressions for **V** and **I** in the form of unknown constants A_1 and A_2. The values of A_1 and A_2 can be determined by applying receiving-end conditions as under

At receiving end $x = 0$, $\mathbf{V} = \mathbf{V}_\mathrm{R}$ and $\mathbf{I} = \mathbf{I}_\mathrm{R}$

Substituting these values in Eqs. (6.4) and (6.6) we get

$$\mathbf{V}_\mathrm{R} = \mathrm{A}_1 + \mathrm{A}_2 \quad \text{...(6.7)}$$

$$\text{and } \mathbf{I}_\mathrm{R} = \sqrt{\frac{y}{z}}(\mathrm{A}_1 - \mathrm{A}_2) \quad \text{...(6.8)}$$

For a transmission line $\sqrt{\dfrac{z}{y}}$ is a constant, called the *characteristic constant,* Z_C. and $\sqrt{yz}$ is another constant, called the *propagation constant,* γ, both complex quantities.

From Eqs. (6.7) and (6.8) we have

$$\mathrm{A}_1 = \frac{1}{2}\mathbf{V}_\mathrm{R} + \frac{1}{2}\mathbf{I}_\mathrm{R}\sqrt{\frac{z}{y}} = \frac{1}{2}[\mathbf{V}_\mathrm{R} + \mathbf{I}_\mathrm{R}\,\mathrm{Z}_\mathrm{C}]$$

$$\text{and } \mathrm{A}_2 = \frac{1}{2}\mathbf{V}_\mathrm{R} - \frac{1}{2}\mathbf{I}_\mathrm{R}\sqrt{\frac{z}{y}} = \frac{1}{2}[\mathbf{V}_\mathrm{R} - \mathbf{I}_\mathrm{R}\,\mathrm{Z}_\mathrm{C}]$$

Thus the expressions for **V** and **I** become

$$\mathbf{V} = \frac{1}{2}[\mathbf{V}_\mathrm{R} + \mathbf{I}_\mathrm{R}\,\mathrm{Z}_\mathrm{C}]\,e^{\gamma x} + \frac{1}{2}[\mathbf{V}_\mathrm{R} - \mathbf{I}_\mathrm{R}\,\mathrm{Z}_\mathrm{C}]\,e^{-\gamma x} \quad \text{...(6.9)}$$

$$\text{and } \mathbf{I} = \frac{1}{\mathrm{Z}_\mathrm{C}}\left[\frac{1}{2}\{\mathbf{V}_\mathrm{R} + \mathbf{I}_\mathrm{R}\,\mathrm{Z}_\mathrm{C}\}\,e^{\gamma x} - \frac{1}{2}\{\mathbf{V}_\mathrm{R} - \mathbf{I}_\mathrm{R}\,\mathrm{Z}_\mathrm{C}\}\,e^{-\gamma x}\right]$$

$$\text{or } \mathbf{I} = \frac{1}{2}\left[\left(\frac{\mathbf{V}_\mathrm{R}}{\mathrm{Z}_\mathrm{C}} + \mathbf{I}_\mathrm{R}\right)\right]e^{\gamma x} - \frac{1}{2}\left[\left(\frac{\mathbf{V}_\mathrm{R}}{\mathrm{Z}_\mathrm{C}} - \mathbf{I}_\mathrm{R}\right)\right]e^{-\gamma x} \quad \text{...(6.10)}$$

Expanding and re-arranging Eqs. (6.9) and (6.10) we get

$$\mathbf{V} = \mathbf{V}_\mathrm{R}\cosh\gamma x + \mathbf{I}_\mathrm{R}\,\mathrm{Z}_\mathrm{C}\sinh\gamma x \quad \text{...(6.11)}$$

$$\text{and } \mathbf{I} = \mathbf{I}_R \cosh \gamma x + \frac{\mathbf{V}_R}{Z_C} \sinh \gamma x \quad ...(6.12)$$

The sending-end voltage, $\mathbf{V}_S$ and sending-end current, $\mathbf{I}_S$ can be obtained by substituting $x = l$ in the above Eqs. (6.11) and (6.12)

$$\mathbf{V}_S = \mathbf{V}_R \cosh \gamma l + \mathbf{I}_R Z_C \sinh \gamma l \quad ...(6.13)$$

$$\text{and } \mathbf{I}_S = \mathbf{I}_R \cosh \gamma l + \frac{\mathbf{V}_R}{Z_C} \sinh \gamma l \quad ...(6.14)$$

$$\text{Now, } \gamma l = \sqrt{yz}\, l = \sqrt{yl \times zl} = \sqrt{\mathbf{Y Z}}$$

where **Z** is the total impedance of the line and **Y** is the total admittance of the line.

The expressions for sending-end voltage and sending-end currents are

$$\mathbf{V}_S = \mathbf{V}_R \cosh \sqrt{\mathbf{Y Z}} + \mathbf{I}_R Z_C \sinh \sqrt{\mathbf{Y Z}} \quad ...(6.15)$$

$$\text{and } \mathbf{I}_S = \mathbf{I}_R \cosh \sqrt{\mathbf{Y Z}} + \frac{\mathbf{V}_R}{Z_C} \sinh \sqrt{\mathbf{Y Z}} \quad ...(6.16)$$

Comparing above equations with the general voltage and current equations of the line we have

$$\mathbf{A} = \mathbf{D} = \text{Cosh} \sqrt{\mathbf{Y Z}}$$

$$\mathbf{B} = Z_C \sinh \sqrt{\mathbf{Y Z}}$$

$$\text{and } \mathbf{C} = \frac{1}{Z_C} \sinh \sqrt{\mathbf{Y Z}}$$

The relation $\mathbf{A D} - \mathbf{B C} = 1$ holds for $\mathbf{A D} - \mathbf{B C} = \cosh^2 \sqrt{\mathbf{Y Z}} - \sinh^2 \sqrt{\mathbf{Y Z}} = 1$

Equations (6.13) and (6.14) can be written in matrix form as :

$$\begin{bmatrix} \mathbf{V}_S \\ \mathbf{I}_S \end{bmatrix} = \begin{bmatrix} \cosh \gamma l & Z_C \sinh \gamma l \\ \dfrac{1}{Z_C} \sinh \gamma l & \cosh \gamma l \end{bmatrix} \begin{bmatrix} \mathbf{V}_R \\ \mathbf{I}_R \end{bmatrix} \quad ...(6.17)$$

Thus sending-end values can be obtained if $\mathbf{V}_R$, $\mathbf{I}_R$ and line parameters are known.

In case $\mathbf{V}_S$, $\mathbf{I}_S$ are known, $\mathbf{V}_R$ and $\mathbf{I}_R$ can be easily found by inverting the above Eq. (6.17)

$$\begin{bmatrix} \mathbf{V}_R \\ \mathbf{I}_R \end{bmatrix} = \begin{bmatrix} \cosh \gamma l & -Z_C \sinh \gamma l \\ \dfrac{-1}{Z_C} \sinh \gamma l & \cosh \gamma l \end{bmatrix} \begin{bmatrix} \mathbf{V}_S \\ \mathbf{I}_S \end{bmatrix} \quad ...(6.18)$$

6.3. EVALUATION OF CONSTANTS ABCD

ABCD constants can be evaluated by three methods given below:

1. *Convergent Series (Complex Angle) Method.* In this method the hyperbolic sine and cosine are expressed in terms of their power series. The expansions are

$$\text{Cosh } x = 1 + \frac{x^2}{2!} + \frac{x^4}{4!} + \frac{x^6}{6!} + ...$$

$$\text{and } \sinh x = x + \frac{x^3}{3!} + \frac{x^5}{5!} + \frac{x^7}{7!} + ...$$

It is never worthwhile going beyond the seventh power. We then get

$$\mathbf{A} = \mathbf{D} = \text{Cosh}\,\sqrt{\mathbf{YZ}} = 1 + \frac{\mathbf{YZ}}{2} + \frac{\mathbf{Y}^2\mathbf{Z}^2}{24} + \frac{\mathbf{Y}^3\mathbf{Z}^3}{720} + \ldots.. \quad \ldots(6.19)$$

$$\mathbf{B} = Z_C \sinh\sqrt{\mathbf{YZ}} = Z_C\sqrt{\mathbf{YZ}}\left\{1 + \frac{\mathbf{YZ}}{6} + \frac{\mathbf{Y}^2\mathbf{Z}^2}{120} + \frac{\mathbf{Y}^3\mathbf{Z}^3}{5{,}040} + \ldots\right\}$$

$$= \sqrt{\frac{\mathbf{Z}}{\mathbf{Y}}} \times \sqrt{\mathbf{YZ}}\left\{1 + \frac{\mathbf{YZ}}{6} + \frac{\mathbf{Y}^2\mathbf{Z}^2}{120} + \frac{\mathbf{Y}^3\mathbf{Z}^3}{5{,}040} + \ldots..\right\}$$

$$= \mathbf{Z}\left\{1 + \frac{\mathbf{YZ}}{6} + \frac{\mathbf{Y}^2\mathbf{Z}^2}{120} + \frac{\mathbf{Y}^3\mathbf{Z}^3}{5{,}040} + \ldots..\right\} \quad \ldots(6.20)$$

and $$\mathbf{C} = \frac{1}{Z_C}\sinh\sqrt{\mathbf{YZ}} = \sqrt{\frac{\mathbf{Y}}{\mathbf{Z}}} \times \sqrt{\mathbf{YZ}}\left\{1 + \frac{\mathbf{YZ}}{6} + \frac{\mathbf{Y}^2\mathbf{Z}^2}{120} + \frac{\mathbf{Y}^3\mathbf{Z}^3}{5{,}040} + \ldots..\right\}$$

$$= \mathbf{Y}\left[1 + \frac{\mathbf{YZ}}{6} + \frac{\mathbf{Y}^2\mathbf{Z}^2}{120} + \frac{\mathbf{Y}^3\mathbf{Z}^3}{5{,}040} + \ldots..\right] \quad \ldots(6.21)$$

2. *Convergent Series (Real Angle) Method.* In this method the hyperbolic sines and cosines are expanded by the trigonometrical formulae given below, and tables of the trigonometric functions and hyperbolic functions of real numbers are used. Thus we express $\sqrt{\mathbf{YZ}}$ in the form

$$\sqrt{\mathbf{YZ}} = \alpha + j\beta \text{ where both } \alpha \text{ and } \beta \text{ are real}$$

Then $$\cosh\sqrt{\mathbf{YZ}} = \text{Cosh}\,(\alpha + j\beta)$$

$$= \text{Cosh}\,\alpha\,\cosh j\beta + \sinh\alpha\,\sinh j\beta = \text{Cosh}\,\alpha\cos\beta + j\sinh\alpha\sin\beta$$

and $$\sinh\sqrt{\mathbf{YZ}} = \text{Sinh}\,\alpha\,\cosh j\beta + \cosh\alpha\,\sinh j\beta = \text{Sinh}\,\alpha\cos\beta + j\cosh\alpha\sin\beta.$$

3. *By Use of Hyperbolic Charts.* Woodruff has prepared a convenient set of charts for evaluating the hyperbolic functions of complex variables, usually required in transmission line calculations. But such charts are usually not available.

In general, method 1 is less laborious to use than method 2; moreover the tables of the sinh and cosh of complex numbers are not easy to use, and are not given for small differences.

6.4. SURGE IMPEDANCE

The square root of the ratio of line impedance (**Z**) and shunt admittance (**Y**) is called the *'surge impedance'* (Z_0) of the line

i.e. $$Z_0 = \sqrt{\frac{\mathbf{Z}}{\mathbf{Y}}}$$

where $\mathbf{Z} = R + jX$ and $\mathbf{Y} = G + jB$

Characteristic impedance (Z_C) has also been defined in the same way. Actually characteristic impedance is a more correct nomenclature. Surge impedance is the characteristic impedance of a loss-free line.

For a heavy copper conductor and well insulated line the resistance (R) and leakage conductance (G) can be taken as zero and therefore, surge impedance, $Z_0 = \sqrt{\frac{L}{C}}$ which is a pure resistance. Its value varies between 400 Ω and 600 Ω in case of overhead transmission lines and between 40 Ω and 60 Ω incase of underground cables.

The surge impedance can also be evaluated by measuring the line impedance at the sending end when

(*i*) the line at the receiving end is open-circuited.

(*ii*) the line at the receiving end is short-circuited.

When the line is open-circuited ($\mathbf{I_R}$) is zero so, the sending-end voltage ($\mathbf{V_S}$) and sending -end current ($\mathbf{I_S}$) are given as :

$$\mathbf{V_S = A\,V_R} \text{ and } \mathbf{I_S = C\,V_R}$$

$$\therefore\ \mathbf{Z_{OC}} = \frac{\mathbf{V_S}}{\mathbf{I_S}} = \frac{\mathbf{A\,V_R}}{\mathbf{C\,V_R}} = \frac{\mathbf{A}}{\mathbf{C}} \quad \text{...(6.22)}$$

Similarly, when the line is short-circuited at the receiving end, the receiving-end voltage($\mathbf{V_R}$) reduces to zero and the sending-end voltage ($\mathbf{V_S}$) and the sending-end current ($\mathbf{I_S}$) are given as :

$$\mathbf{V_S = B\,I_R} \text{ and } \mathbf{I_S = D\,I_R}$$

$$\therefore\ \mathbf{Z_{SC}} = \frac{\mathbf{V_S}}{\mathbf{I_S}} = \frac{\mathbf{BI_R}}{\mathbf{DI_R}} = \frac{\mathbf{B}}{\mathbf{D}} \quad \text{...(6.23)}$$

Multiplying Eqs. (6.22) and (6.23) we have

$$\mathbf{Z_{OC}\cdot Z_{SC}} = \frac{\mathbf{A}}{\mathbf{C}} \times \frac{\mathbf{B}}{\mathbf{D}} = \frac{\mathbf{A}}{\mathbf{C}} \times \frac{\mathbf{B}}{\mathbf{A}} = \frac{\mathbf{B}}{\mathbf{C}} \quad \text{...(6.24)}$$

∵ For a transmission line (which happens to be bilateral network) $\mathbf{D = A}$

Substituting $\mathbf{B} = \sqrt{\frac{\mathbf{Z}}{\mathbf{Y}}}\ \sinh\sqrt{\mathbf{YZ}}$ and $\mathbf{C} = \sqrt{\frac{\mathbf{Y}}{\mathbf{Z}}}\ \sinh\sqrt{\mathbf{YZ}}$ in Eq. (6.24) we have

$$\mathbf{Z_{OC}\cdot Z_{SC}} = \frac{\mathbf{B}}{\mathbf{C}} = \frac{\sqrt{\mathbf{Z/Y}}\sinh\sqrt{\mathbf{YZ}}}{\sqrt{\mathbf{Y/Z}}\sinh\sqrt{\mathbf{YZ}}} = \frac{\mathbf{Z}}{\mathbf{Y}} = Z_0^2 \quad \text{...(6.25)}$$

$$\therefore \text{ Surge impedance } Z_0 = \sqrt{\mathbf{Z_{OC}\,Z_{SC}}}$$

6.5. SURGE IMPEDANCE LOADING

This is defined as the load (of unity power factor) that can be delivered by the line of negligible resistance.

$$\therefore \text{ Power transmitted, } P_R = \frac{V_{RL}^2}{Z_0} \text{ MW} \quad \text{...(6.26)}$$

where V_{RL} is the receiving-end voltage in kV and Z_0 is the surge impedance in ohms.

P_R is called the '*surge impedance loading*' or '*natural power*' of the line.

The above expression gives a limit of the maximum power* that can be delivered by a line and is useful in the design of transmission lines. This can be used for the comparison of loads that can be carried on the transmission lines at different voltages.

The power transmitted through a long transmission line can be increased either by increasing the value of receiving-end line voltage (V_{RL}) or by reducing surge impedance (Z_0).

* This may not necessarily be the maximum loading on the transmission lines as discussed else where.

Nowadays the trend is of employing higher and higher voltages for transmission, therefore, this is the most commonly adopted method for increasing the power limit of heavily loaded long transmission lines. But there is a limit beyond which it is neither economical nor practicable to increase the value of receiving-end line voltage.

Since the spacing between the conductors, which depends upon the line voltage employed, cannot be reduced much, so the value of surge impedance (Z_0) cannot be varied as such. However some artificial means, such as series capacitors (capacitors in series with the transmission lines) or shunt capacitors (capacitors in parallel with the transmission lines), can be used to reduce the value of surge impedance (Z_0).

For a loss free transmission line surge impedance, $Z_0 = \sqrt{\frac{L}{C}}$ and propagation constant, $\gamma = j\omega\sqrt{LC} = j\beta$, where β is the phase shift, which determines the torque angle δ between V_S and V_R and hence the system stability.

By use of series capacitors surge impedance (Z_0) and the phase shift (β) get reduced due to decrease of line inductance (L). This improves the system stability limit also. But it causes difficulties under short-circuit conditions of the system as no satisfactory protection of capacitors has yet been devised. These capacitors are also helpful in reducing line drops and so voltage variations.

By use of shunt capacitors though the surge impedance (Z_0) is reduced but the phase shift β is increased due to increase in the value of C. Hence stability conditions become worse specially when synchronous machines are used as loads. So this method is not used in case of long transmission lines specially where stability limits are present.

Example 6.1. A 3-phase, 50 Hz, 1,000 km long transmission line has the following line constants per phase per km uniformly distributed :

$$r = 0.22\ \Omega;\ x = 0.45\ \Omega;\ g = 4 \times 10^{-9}\ S \text{ and } b = 2.53 \times 10^{-6}\ S$$

Determine the auxiliary constants (*i*) by using convergent series of complex angles (*ii*) by using convergent series of real angles.

Solution : Impedance per km per phase, $\mathbf{z} = (r + jx) = (0.22 + j\,0.45)\ \Omega$

Admittance per km per phase, $\mathbf{y} = (0.004 + j\,2.53) \times 10^{-6}\ S$

Total series impedance, $\mathbf{Z} = (0.22 + j\,0.45) \times 1{,}000$

$$= (220 + j\,450)\ \Omega = 500.9\ \angle 63.95^\circ\ \Omega$$

Total shunt admittance, $\mathbf{Y} = (0.004 + j\,2.53) \times 10^{-6} \times 1{,}000$

$$= (0.004 + j\,2.53) \times 10^{-3}\ S = 2.53 \times 10^{-3}\ \angle 89.91^\circ\ S$$

Hence $\mathbf{YZ} = 2.53\ 10^{-3}\ \angle 89.91^\circ \times 500.9\ \angle 63.95^\circ$

$$= 1.267277\ \angle 153.86^\circ = (-1.1376 + j\ 0.5583)$$

$$\mathbf{Y^2Z^2} = (1.267277)^2\ \angle 307.72^\circ = 1.606\ \angle 307.72^\circ = (0.9826 - j\,1.27)$$

$$\mathbf{Y^3Z^3} = (1.267277)^3\ \angle 461.58^\circ = 2.0352354\ \angle 101.58^\circ = (-0.408 + j\,1.9938)$$

(*i*) Now auxiliary line constants

$$\mathbf{A} = \mathbf{D} = 1 + \frac{\mathbf{YZ}}{2} + \frac{\mathbf{Y^2Z^2}}{24} + \frac{\mathbf{Y^3Z^3}}{720} + \ldots$$

$$= 1 + \frac{-1.1376 + j0.5583}{2} + \frac{0.9826 - j1.27}{24} + \frac{-0.408 + j1.9938}{720} + \ldots$$

$$= (1 - 0.5688 + 0.0409 - 0.0005) + j\,(0.2792 - 0.0529 + 0.0028) = 0.4716 + j\,0.2291$$

$$\mathbf{B} = \mathbf{Z}\left\{1+\frac{\mathbf{YZ}}{6}+\frac{\mathbf{Y}^2\mathbf{Z}^2}{120}+\frac{\mathbf{Y}^3\mathbf{Z}^3}{5,040}+\cdots\right\}\Omega$$

$$= (220 + j\ 450)\left\{1+\frac{-1.1376+j\,0.5583}{6}+\frac{0.9826-j\,1.27}{120}+\frac{-0.408+j\,1.9938}{5,040}+\cdots\right\}$$

$$= (220 + j\ 450)\ [(1 - 0.1896 + 0.0081 - 0.00008) + j\,(0.09305 - 0.01058 + 0.0004)]\ \Omega$$

$$= (220 + j\ 450)\ (0.81842 + j\,0.08287)\ \Omega$$

$$= 500.9\ \angle 63.95^\circ \times 0.8226\ \angle 5.78^\circ\ \Omega = 412.04\ \angle 69.73^\circ = (142.75 + j\,386.52)\ \Omega$$

$$\mathbf{C} = \mathbf{Y}\left\{1+\frac{\mathbf{YZ}}{6}+\frac{\mathbf{Y}^2\mathbf{Z}^2}{120}+\frac{\mathbf{Y}^3\mathbf{Z}^3}{5,040}+\cdots\right\}\mathrm{S}$$

$$= 2.53 \times 10^{-3}\ \angle 89.91^\circ \times 0.8226\ \angle 5.78^\circ\ \mathrm{S}$$

$$= 2.08 \times 10^{-3}\ \angle 95.69^\circ = (-\,0.0002 + j\ 0.00207)\ \mathrm{S}$$

(*ii*) Using convergent series of real angles

We have $\mathbf{Z} = 500.9\ \angle 63.59^\circ\ \Omega$ and $\mathbf{Y} = 2.53 \times 10^{-3}\ \angle 89.91^\circ$ S

$$\therefore\ \mathbf{YZ} = 2.53 \times 10^{-3}\ \angle 89.91^\circ \times 500.9\ \angle 63.95^\circ = 1.26728\ \angle 153.86^\circ$$

and $\sqrt{\mathbf{YZ}} = 1.125735\ \angle 76.93^\circ = 0.25458 + j\ 1.09658$

Since $\alpha + j\,\beta = 0.25458 + j\ 1.09658$

$\therefore\ \alpha = 0.25458$ radians or 14.586° and $\beta = 1.09658$ radians or 62.83°

From tables

Cosh α = Cosh $14.586^\circ = 1.03258$ and sinh α = sinh $14.586^\circ = 0.25734$

Cos β = Cos $62.83^\circ = 0.4566$ and sin β = Sin $62.83^\circ = 0.8896$

$$\text{Cosh}\ \sqrt{\mathbf{YZ}} = \text{Cosh}\ 14.586^\circ \cos 62.83^\circ + j \sinh 14.586^\circ \sin 62.53^\circ$$

$$= 1.03258 \times 0.4566 + j\ 0.25734 \times 0.8896 = 0.4715 + j\ 0.2289 = 0.524\ \angle 25.9^\circ$$

$$\text{Sinh}\ \sqrt{\mathbf{YZ}} = \text{Sinh}\ \alpha \cos \beta + j \cosh \alpha \sin \beta$$

$$= \text{Sinh}\ 14.586^\circ \cos 62.83^\circ + j \cosh 14.586^\circ \sin 62.83^\circ$$

$$= 0.25734 \times 0.4566 + j\ 1.03258 \times 0.8896 = 0.1175 + j\ 0.9186 = 0.9261\ \angle 82.71^\circ$$

Auxiliary line constants

$$\mathbf{A} = \mathbf{D} = \text{Cosh}\ \sqrt{\mathbf{YZ}} = 0.4715 + j\ 0.2289 = 0.524\ \angle 25.9^\circ$$

$$\mathbf{B} = Z_C \sinh \sqrt{\mathbf{YZ}} = \sqrt{\frac{\mathbf{Z}}{\mathbf{Y}}} \times 0.9261\ \angle 82.71^\circ$$

$$= \sqrt{\frac{500.9\ \angle 63.95^\circ}{2.53 \times 10^{-3}\ \angle 89.91^\circ}} \times 0.9261\ \angle 82.71^\circ = 412.07\ \angle 69.73^\circ\,\Omega$$

$$\mathbf{C} = \frac{1}{Z_C} \sinh \sqrt{\mathbf{YZ}} = \sqrt{\frac{2.53 \times 10^{-3}\ \angle 89.91^\circ}{500.9\ \angle 63.95^\circ}} \times 0.9261\ \angle 82.71^\circ$$

$$= 2.08 \times 10^{-3}\ \angle 95.69^\circ\ \mathrm{S}$$

Note : *Results obtained by both methods are almost same.*

Example 6.2. A 3-phase overhead transmission line has a total series impedance per phase of 200 $\angle 80^\circ$ ohms and a total shunt admittance of 0.0013 $\angle 90^\circ$ siemen per phase. The line delivers a load of 80 MW at 0.8 pf lagging and 220 kV between the lines. Determine the sending-end voltage and current by rigorous method.

Solution : Receiving-end phase voltage,

$$V_R = \frac{220,000}{\sqrt{3}} = 127,000 \text{ V}$$

$$\text{Receiving-end current, } I_R = \frac{80 \times 10^6}{\sqrt{3} \times 220,000 \times 0.8} = 262.43 \text{ A}$$

Power factor angle, $\phi = \text{Cos}^{-1}\ 0.8 = 36.87°$ lagging

Taking receiving-end phase voltage as reference phasor we have

$$\mathbf{V_R} = 127,000 \angle 0° \text{ V and } \mathbf{I_R} = 262.43 \angle -36.87° \text{ A}$$

Series impedance, $\mathbf{Z} = 200 \angle 80°\ \Omega$; Shunt admittance, $\mathbf{Y} = 0.0013 \angle 90°$ S

$$\mathbf{Y\,Z} = 0.0013 \angle 90° \times 200 \angle 80° = 0.26 \angle 170° = -0.25605 + j\,0.04515$$

$$\mathbf{Y^2\,Z^2} = 0.26^2 \angle 340° = 0.0676 \angle 340° = 0.06352 - j\,0.02312$$

$$\mathbf{Y^3Z^3} = 0.26^3 \angle 510° = 0.017576 \angle 510° = -0.01522 + j\,0.008788$$

Now auxiliary line constants,

$$\mathbf{A} = \mathbf{D} = 1 + \frac{\mathbf{YZ}}{2} + \frac{\mathbf{Y^2Z^2}}{24} + \frac{\mathbf{Y^3Z^3}}{720} + ...$$

$$= 1 + \frac{-0.25605 + j\,0.04515}{2} + \frac{0.06352 - j\,0.02312}{24} + \frac{-0.01522 + j\,0.008788}{720} + ...$$

$$= (1 - 0.12802 + 0.00265 - 0.00002) + j\,(0.02257 - 0.00096 + 0.00001)$$

$$= (0.8746 + j\ 0.02162) = 0.8749 \angle 1.42°$$

$$\mathbf{B} = \mathbf{Z}\left\{1 + \frac{\mathbf{YZ}}{6} + \frac{\mathbf{Y^2\,Z^2}}{120} + \frac{\mathbf{Y^3\,Z^3}}{5,040} + ...\right\}\Omega$$

$$= 200 \angle 80° + \left\{1 + \frac{-0.25605 + j\,0.04515}{6} + \frac{0.06352 - j\,0.02312}{120} + \frac{-0.01522 + j\,0.008788}{5,040}\right\}\Omega$$

$$= 200 \angle 80° \{(1 - 0.042675 + 0.00053 - 0.000003) + j\,(0.007525 - 0.00019 + 0.000002)\}$$

$$= 200 \angle 80° (0.957852 + j\,0.007337) = 200 \angle 80° \times 0.95788 \angle 0.43°\ \Omega = 191.57 \angle 80.43°\ \Omega$$

$$\mathbf{C} = \mathbf{Y}\left\{1 + \frac{\mathbf{YZ}}{6} + \frac{\mathbf{Y^2\,Z^2}}{120} + \frac{\mathbf{Y^3\,Z^3}}{5,040} + ...\right\}\text{S}$$

$$= 0.0013 \angle 90° \times 0.95788 \angle 0.43° = 0.001245 \angle 90.43° \text{ S}$$

Sending-end phase voltage,

$$\mathbf{V_S} = \mathbf{AV_R} + \mathbf{BI_R}$$

$$= 0.8749 \angle 1.42° \times 127,000 \angle 0° + 191.57 \angle 80.43° \times 262.43 \angle -36.87°$$

$$= 111,112 \angle 1.42° + 50,273 \angle 43.56° = 111,078 + j\,2,753 + 36,430 + j\,34,644$$

$$= 147,508 + j\,37,397 = 152,174 \angle 14.23° \text{ volts}$$

Sending end line voltage,

$$V_{SL} = \sqrt{3}\ V_S = \sqrt{3} \times 152,174 = 263,574 \text{ volts} = 263.574 \text{ kV}$$ **Ans.**

$$\text{Sending-end current, } \mathbf{I_S} = \mathbf{C\,V_R} + \mathbf{D\,I_R}$$

$$= 0.001245 \angle 90.43° \times 127,000 \angle 0° + 0.8749 \angle 1.42° \times 262.43 \angle -36.87°$$

$$= 158.12 \angle 90.43° + 229.6 \angle -35.45°$$
$$= -1.2115 + j\,158.1154 + 187.0372 - j\,133.1662$$
$$= 185.8257 + j\,24.95 = 187.5 \angle 7.65° \text{ A } \textbf{Ans.}$$

6.6. TRANSMISSION LINE WITH SERIES IMPEDANCE AT THE RECEIVING END

Line auxiliary constants, as determined before, are changed due to insertion of series impedance at the receiving end of a transmission line. Normally an impedance is supposed to consist of resistance and inductance and the current flowing through it is the same as the receiving-end current, $\mathbf{I_R}$. The circuit is shown in Fig. 6.3.

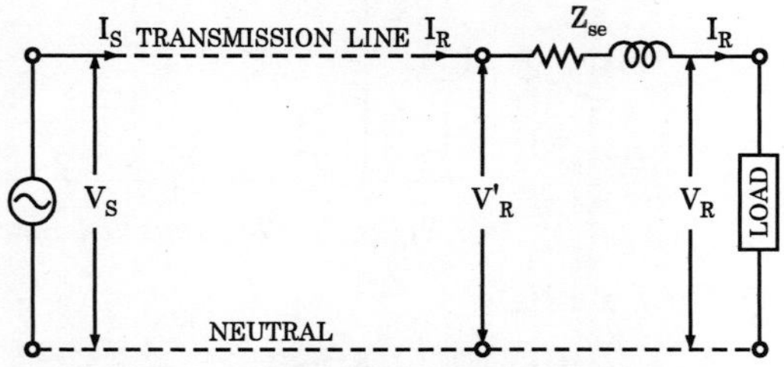

Fig. 6.3. ***Transmission Line With Series Impedance At The Receiving End***

Let the auxiliary constants of the transmission line be **A, B, C** and **D** and $\mathbf{Z_{se}}$ by the value of series impedance of the line at the receiving-end. If $\mathbf{V_R}$ is the phase voltage at the load end and $\mathbf{I_R}$ is the load current, then the phase voltage at the end of transmission line will be

$$\mathbf{V'_R = V_R + I_R Z_{se}} \quad ...(6.27)$$

and $\mathbf{V_S = A\,V'_R + B\,I_R = A\,(V_R + I_R Z_{se}) + B\,I_R = A\,V_R + (A\,Z_{se} + B)\,I_R}$ ...(6.28)

and $\mathbf{I_S = C\,V'_R + D\,I_R}$

Substituting the value of $\mathbf{V'_R}$ from Eq. (6.27) we have

$$\mathbf{I_S = C\,(V_R + I_R Z_{se}) + D\,I_R = C\,V_R + (C\,Z_{se} + D)\,I_R} \quad ...(6.29)$$

Comparing with the general voltage and current expressions of the transmission line we have

$$\mathbf{A_0 = A\ ;\ B_0 = A\,Z_{Se} + B\ ;\ C_0 = C} \text{ and } \mathbf{D_0 = C\,Z_{se} + D} \quad ...(6.30)$$

6.7. TRANSMISSION LINE WITH SERIES IMPEDANCE AT THE SENDING END

Figure 6.4. represents a transmission line with a series impedance $\mathbf{Z_{se}}$ at the sending end. In this case the current passing through the series impedance $\mathbf{Z_{se}}$ is the same as the sending end current $\mathbf{I_s}$

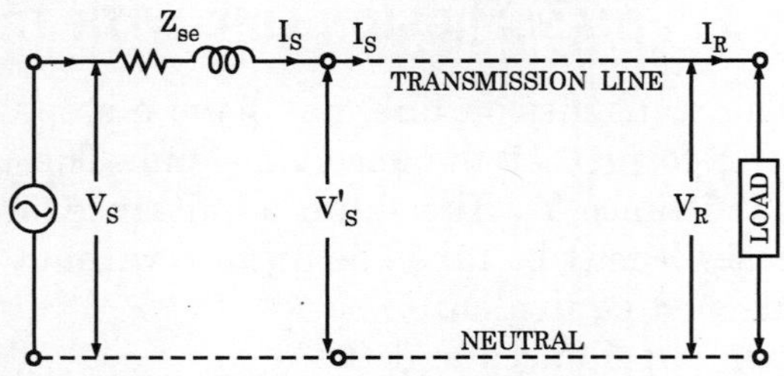

Fig. 6.4. ***Transmission Line With Series Impedance At The Sending End***

If **A, B, C** and **D** are the auxiliary constants of the transmission line, $\mathbf{V_R}$ and $\mathbf{V_S}$ are the phase voltages at the load and supply ends respectively, $\mathbf{I_R}$ is the load current and $\mathbf{V'_s}$ is the phase voltage at the starting end of the transmission line.

then $\mathbf{V'_S = AV_R + B\,I_R}$ ...(6.31)

and $\mathbf{I_S = C\,V_R + D\,I_R}$ ...(6.32)

Also $\mathbf{V_S = V'_S + I_S Z_{se}}$ ...(6.33)

Substituting the values of $\mathbf{V'_S}$ and $\mathbf{I_S}$ from Eqs. (6.31) and (6.32) in Eqs. (6.33) we have

$$\mathbf{V_S = A\,V_R + BI_R + (C\,V_R + D\,I_R)\,Z_{se} = (A + C\,Z_{se})\,V_R + (B + D\,Z_{se})\,I_R} \quad ...(6.34)$$

Comparing Eqs. (6.34) and (6.32) with the general expressions for voltage and current we have

$$\mathbf{A_0 = A + C\,Z_{se};\ B_0 = B + D\,Z_{se};\ C_0 = C} \text{ and } \mathbf{D_0 = D = A} \quad ...(6.35)$$

6.8. TRANSMISSION LINE WITH SERIES IMPEDANCES AT BOTH ENDS

In this case there are two series impedances, one at the receiving end and the other at the sending end, as shown in Fig. 6.5.

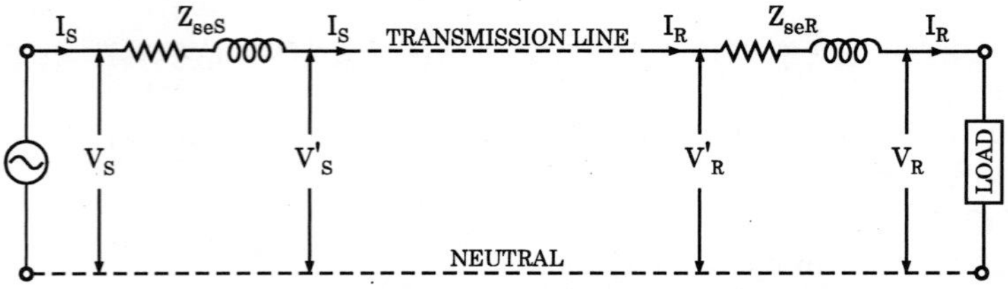

Fig. 6.5. *Transmission Line With Series Impedances At Both Ends*

Now we have

$$\mathbf{V'_R = V_R + I_R Z_{seR}} \quad ...(6.36)$$

$$\text{and } \mathbf{V'_S = A\,V'_R + BI_R = A\,(V_R + I_R Z_{seR}) + BI_R = AV_R + (B + A\,Z_{seR})I_R} \quad ...(6.37)$$

$$\mathbf{I_S = CV'_R + DI_R = C\,(V_R + I_R Z_{seR}) + DI_R = C\,V_R + (D + C\,Z_{seR})I_R} \quad ...(6.38)$$

$$\text{and } \mathbf{V_S = V'_S + I_S Z_{seS}}$$

$$\mathbf{= AV_R + (B + A\,Z_{seR})I_R + \{C\,V_R + (D + C\,Z_{seR})\,I_R\}\,Z_{seS}}$$

$$\mathbf{= (A + C\,Z_{seS})V_R + \{B + A\,Z_{seR} + (D + C\,Z_{seR})\,Z_{seS}\}\,I_R} \quad ...(6.39)$$

Comparing Eqs. (6.39) and (6.38) with the general expressions for voltage and current of the transmission line we have

$$\mathbf{A_0 = A + C\,Z_{seS}}$$

$$\mathbf{B_0 = B + A\,Z_{seR} + (D + C\,Z_{seR})\,Z_{seS}} \quad ...[6.40\ (a)]$$

$$\mathbf{= B + A\,Z_{seR} + (A + C\,Z_{seR})\,Z_{seS}} \quad \because\ \mathbf{A = D}$$

$$\mathbf{= B + A\,(Z_{seR} + Z_{seS}) + C\,Z_{seR}\,Z_{seS}} \quad ...[6.40\ (b)]$$

$$\mathbf{C_0 = C} \quad ...[6.40\ (c)]$$

$$\mathbf{D_0 = D + C\,Z_{seR}} \quad ...[6.40\ (d)]$$

6.9. TRANSMISSION LINE WITH TRANSFORMERS AT BOTH ENDS

The transmission line may have a step-up transformer at the sending end and step-down transformer at the receiving end. The transformer has series impedance $\mathbf{Z_T}$ and shunt admittance $\mathbf{Y_T}$. The shunt admittance of transformer is usually so high that in most of the cases it may be taken as infinite without much loss of accuracy in which case transformers are just equivalent to series impedances and the equivalent circuit is, as shown in Fig. 6.5.

In case the shunt admittance of transformers are to be taken into account, the equivalent circuit will be as shown in Fig. 6.6.

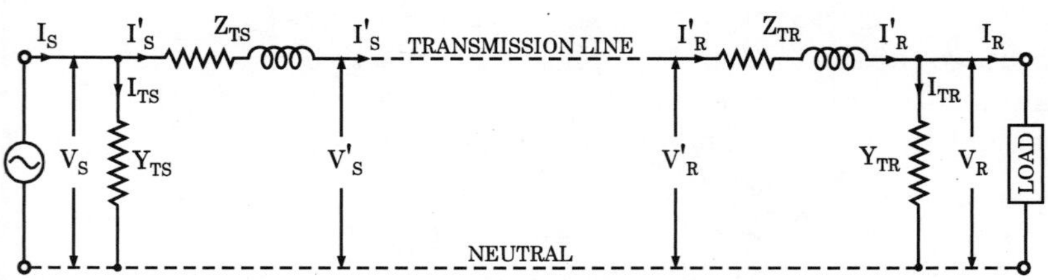

Fig. 6.6. *Transmission Line With Transformers At Both Ends*

In this case the sending-end voltage and current for the transmission line are $\mathbf{V'_S}$ and $\mathbf{I'_S}$ respectively.

Also the receiving-end voltage and current for the transmission line are $\mathbf{V'_R}$ and $\mathbf{I'_R}$ respectively.

$\mathbf{I_{TS}}$ and $\mathbf{I_{TR}}$ are the magnetising currents taken by the transformers at the sending end and receiving end of the transmission line respectively.

Now applying the transmission line treatment to the line we have

$$\mathbf{V'_S = A\,V'_R + B\,I'_R} \quad ...(6.41)$$

and $\mathbf{I_{TR} = Y_{TR}\,V_R}$

$$\mathbf{I'_R = I_R + I_{TR} = I_R + Y_{TR}\,V_R} \quad ...(6.42)$$

$$\therefore\ \mathbf{V'_R = V_R + I'_R\,Z_{TR} = V_R + (I_R + I_{TR})\,Z_{TR} = V_R + (I_R + Y_{TR}V_R)\,Z_{TR}} \quad ...(6.43)$$

Substituting the value of $\mathbf{V'_R}$ from Eq. (6.43) and value of $\mathbf{I'_R}$ from Eq. (6.42) in Eq. (6.41) we get

$$\mathbf{V'_S = A\{V_R + (I_R + Y_{TR}V_R)\,Z_{TR}\} + B\,(I_R + Y_{TR}\,V_R)}$$

Considering the sending-end current for the line

$$\mathbf{I'_S = C\,V'_R + D\,I'_R = C\,\{V_R + (I_R + Y_{TR}\,V_R)\,Z_{TR}\} + D\,(I_R + Y_{TR}V_R)}$$

$$\begin{aligned}
\text{Also } \mathbf{V_S} &= \mathbf{V'_S + I'_S\,Z_{TS}} \\
&= \mathbf{A\,\{V_R + (I_R + Y_{TR}\,V_R)\,Z_{TR}\} + B\,(I_R + Y_{TR}\,V_R)} \\
&\qquad \mathbf{+\,C\,Z_{TS}\,\{V_R + (I_R + Y_{TR}\,V_R)\,Z_{TR}\} + D\,Z_{TS}\,(I_R + Y_{TR}\,V_R)} \\
&= \mathbf{[A\,(1 + Y_{TR}\,Z_{TR} + Z_{TS}\,Y_{TR}) + B\,Y_{TR} + C\,Z_{TS}\,(1 + Y_{TR}\,Z_{TR})]\,V_R} \\
&\qquad \mathbf{+\,[A\,Z_{TR} + B + C\,Z_{TS}\,Z_{TR} + A\,Z_{TS}]\,I_R} \quad ...(6.44)
\end{aligned}$$

$$\because A = D$$

$$\begin{aligned}
\text{and } \mathbf{I_S} &= \mathbf{I'_S + I_{TS}} \\
&= \mathbf{C\,\{V_R + (I_R + Y_{TR}\,V_R)\,Z_{TR}\} + D\,(I_R + Y_{TR}\,V_R) + Y_{TS}\,V_S} \\
&= \mathbf{C\,\{V_R + (I_R + Y_{TR}\,V_R)\,Z_{TR}\} + A\,(I_R + Y_{TR}\,V_R)} \\
&\quad \mathbf{+\,[\{A\,(1 + Y_{TR}\,Z_{TR} + Y_{TR}\,Z_{TS}) + B\,Y_{TR} + C\,Z_{TS}\,(1 + Y_{TR}\,Z_{TR})\}\,V_R} \\
&\quad \mathbf{+\,\{A\,Z_{TR} + B + C\,Z_{TS}\,Z_{TR} + A\,Z_{TS}\}\,I_R]\,Y_{TS}} \\
&= \mathbf{[C + C\,Y_{TR}\,Z_{TR} + A\,Y_{TR} + A\,Y_{TS} + A\,Y_{TR}\,Z_{TR}\,Y_{TS} + A\,Y_{TR}\,Z_{TS}\,Y_{TS}} \\
&\quad \mathbf{+\,B\,Y_{TR}\,Y_{TS} + C\,Z_{TS}\,Y_{TS} + C\,Z_{TS}\,Y_{TR}\,Z_{TR}\,Y_{TS}]\,V_R} \\
&\quad \mathbf{+\,I_R\,[C\,Z_{TR} + A + A\,Z_{TR}\,Y_{TS} + B\,Y_{TS} + C\,Z_{TS}\,Z_{TR}\,Y_{TS} + A\,Z_{TS}\,Y_{TS}]} \\
&= \mathbf{[C\,(1 + Y_{TR}\,Z_{TR} + Z_{TS}\,Y_{TS} + Z_{TS}\,Z_{TR}\,Y_{TR}\,Y_{TS}) + B\,Y_{TR}\,Y_{TS}} \\
&\quad \mathbf{+\,A\,(Y_{TR} + Y_{TS} + Y_{TR}\,Y_{TS}\,Z_{TR} + Y_{TR}\,Y_{TS}\,Z_{TS}]\,V_R} \\
&\quad \mathbf{+\,[A\,(1 + Y_{TS}\,Z_{TR} + Z_{TS}\,Y_{TS}) + B\,Y_{TS} + C\,Z_{TR}\,(1 + Z_{TS}\,Y_{TS})]\,I_R} \quad ...(6.45)
\end{aligned}$$

Comparing Eqs. (6.44) and (6.45) with the general expressions for voltage and current of the transmission line we have

$$\left.\begin{aligned}
\mathbf{A_0} &= \mathbf{A\,(1 + Y_{TR}\,Z_{TR} + Y_{TR}\,Z_{TS}) + B\,Y_{TR} + C\,Z_{TS}(1 + Y_{TR}\,Z_{TR})} \\
\mathbf{B_0} &= \mathbf{A\,(Z_{TR} + Z_{TS}) + B + C\,Z_{TS}\,Z_{TR}} \\
\mathbf{C_0} &= \mathbf{A\,(Y_{TR} + Y_{TS} + Y_{TR}\,Y_{TS}\,Z_{TR} + Y_{TR}\,Y_{TS}\,Z_{TS}) + B\,Y_{TR}\,Y_{TS}} \\
&\quad \mathbf{+\,C\,(1 + Y_{TR}\,Z_{TR} + Z_{TS}\,Y_{TS} + Z_{TR}\,Z_{TS}\,Y_{TR}\,Y_{TS})} \\
\mathbf{D_0} &= \mathbf{A\,(1 + Y_{TS}\,Z_{TR} + Z_{TS}\,Y_{TS}) + B\,Y_{TS} + C\,Z_{TR}\,(1 + Z_{TS}\,Y_{TS})}
\end{aligned}\right\} \quad ...(6.46)$$

For identical transformers impedances $\mathbf{Z_{TR}}$ and $\mathbf{Z_{TS}}$ will be equal and also shunt admittances $\mathbf{Y_{TR}}$ and $\mathbf{Y_{TS}}$ will be equal.

Let $\mathbf{Z_{TR}} = \mathbf{Z_{TS}} = \mathbf{Z_T}$ (say) and $\mathbf{Y_{TR}} = \mathbf{Y_{TS}} = \mathbf{Y_T}$ (say)

Substituting the above values of $\mathbf{Z_{TR}}$, $\mathbf{Z_{TS}}$, $\mathbf{Y_{TR}}$ and $\mathbf{Y_{TS}}$ in Eq. (6.46) we get

$$\left.\begin{aligned} \mathbf{A_0} &= \mathbf{A}\ (1 + 2\,\mathbf{Y_T Z_T} + \mathbf{B\,Y_T} + \mathbf{C\,Z_T}\ (1 + \mathbf{Y_T Z_T}) \\ \mathbf{B_0} &= 2\,\mathbf{A\,Z_T} + \mathbf{B} + \mathbf{C\,Z_T^2} \\ \mathbf{C_0} &= 2\,\mathbf{A\,Y_T}\ (1 + \mathbf{Y_T Z_T}) + \mathbf{B\,Y_T^2} + \mathbf{C}\ (1 + \mathbf{Y_T Z_T})^2 \\ \mathbf{D_0} &= \mathbf{A}\ (1 + 2\,\mathbf{Y_T Z_T}) + \mathbf{B\,Y_T} + \mathbf{C\,Z_T}\ (1 + \mathbf{Z_T Y_T}) = \mathbf{A_0} \end{aligned}\right\} \quad ...(6.47)$$

6.10. EQUIVALENT T-NETWORK OF A LONG TRANSMISSION LINE

The equivalent T-network of a long transmission line is determined to represent the line accurately by assuming suitable values of lumped constants.

Let the transmission line having auxiliary constants A, B, C and D be represented by a T-network shown in Fig. 6.7.

Now we have

$$\mathbf{I_S} = \mathbf{I_R} + \left(\mathbf{V_R} + \frac{1}{2}\mathbf{I_R Z}\right)\mathbf{Y} = \mathbf{Y\,V_R} + \left(1 + \frac{1}{2}\mathbf{YZ}\right)\mathbf{I_R} \quad ...(6.48)$$

$$\text{and } \mathbf{V_S} = \mathbf{V_R} + \frac{1}{2}\,\mathbf{I_R Z} + \frac{1}{2}\,\mathbf{I_S Z}$$

$$= \mathbf{V_R} + \frac{1}{2}\mathbf{I_R Z} + \frac{1}{2}\mathbf{Z}\left\{\mathbf{YV_R} + \left(1 + \frac{1}{2}\mathbf{YZ}\right)\mathbf{I_R}\right\} = \left(1 + \frac{1}{2}\,\mathbf{YZ}\right)\mathbf{V_R} + \left(\mathbf{Z} + \frac{1}{4}\,\mathbf{YZ^2}\right)\mathbf{I_R} \quad ...(6.49)$$

Comparing Eqs. (6.49) and (6.48) with the general expressions for voltage and current of the transmission line we have

$$\mathbf{A} = \mathbf{D} = 1 + \frac{1}{2}\mathbf{YZ} \quad ...(6.50)$$

$$\mathbf{B} = \mathbf{Z} + \frac{1}{4}\,\mathbf{YZ^2} \quad ...(6.51)$$

$$\mathbf{C} = \mathbf{Y} \quad ...(6.52)$$

From Eqs. (6.50) and (6.52) we have

$$\mathbf{Y} = \mathbf{C} \quad ...(6.53)$$

$$\text{and } \mathbf{Z} = \frac{2(\mathbf{A} - 1)}{\mathbf{C}} \quad ...(6.54)$$

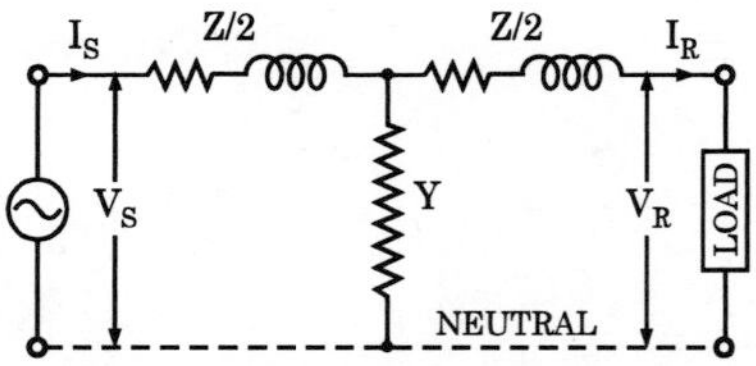

Fig. 6.7. *Equivalent T-Network of a Long Transmission Line*

6.11. EQUIVALENT π-NETWORK OF A LONG TRANSMISSION LINE

Sometimes the equivalent π-network of a long transmission line is required to be determined to represent the line accurately by assuming suitable values of lumped constants.

Let the transmission line having auxiliary constants A, B, C and D be represented by a π-network shown in Fig. 6.8.

The sending-end voltage and current are given as

$$\mathbf{V_S} = \mathbf{V_R} + \mathbf{I\,Z} = \mathbf{V_R} + \left(\mathbf{I_R} + \frac{1}{2}\mathbf{Y\,V_R}\right)\mathbf{Z} = \left(1 + \frac{1}{2}\mathbf{YZ}\right)\mathbf{V_R} + \mathbf{Z\,I_R} \quad ...(6.55)$$

$$\text{and } \mathbf{I_S} = \mathbf{I_R} + \frac{1}{2}\mathbf{Y\,V_R} + \frac{1}{2}\mathbf{Y\,V_S} = \mathbf{I_R} + \frac{1}{2}\,\mathbf{Y\,V_R} + \frac{1}{2}\,\mathbf{Y}\left\{\left(1 + \frac{1}{2}\mathbf{YZ}\right)\mathbf{V_R} + \mathbf{ZI_R}\right\}$$

$$= \left(\mathbf{Y} + \frac{1}{4}\mathbf{Y^2 Z}\right)\mathbf{V_R} + \left(1 + \frac{1}{2}\mathbf{YZ}\right)\mathbf{I_R} \quad ...(6.56)$$

Comparing Eqs. (6.55) and (6.56) with the general expressions for voltage and current of the transmissions line we have

$$\mathbf{A} = \mathbf{D} = 1 + \frac{1}{2}\,\mathbf{Y\,Z} \qquad \text{...(6.57)}$$

$$\mathbf{B} = \mathbf{Z} \qquad \text{...(6.58)}$$

$$\mathbf{C} = \mathbf{Y} + \frac{1}{4}\,\mathbf{Y}^2\mathbf{Z} \qquad \text{...(6.59)}$$

From Eqs. (6.57) and (6.58) we get

$$\mathbf{Z} = \mathbf{B} \qquad \text{...(6.60)}$$

$$\text{and } \mathbf{Y} = \frac{2(\mathbf{A} - 1)}{\mathbf{B}} \qquad \text{...(6.61)}$$

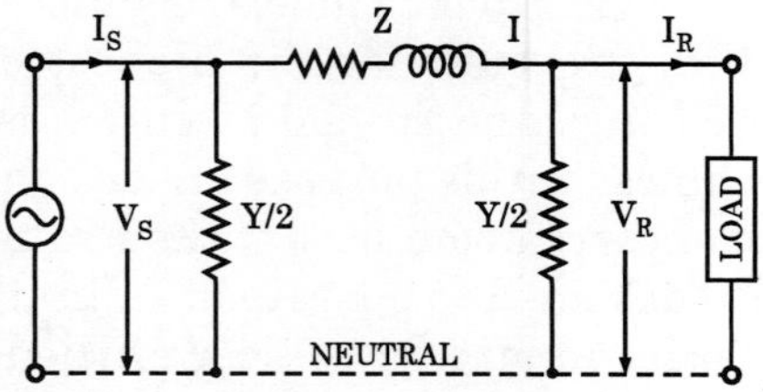

Fig. 6.8. ***Equivalent π-Network of a Long Transmission Line***

6.12. CONSTANT VOLTAGE TRANSMISSION

For constant voltage transmission, specially designed synchronous motors, called the *synchronous phase modifiers,* are installed at the receiving end, which maintain the voltage drop along the line constant. With the change in load, power factor of the system is changed by the synchronous motors and thus voltage drop along the line remains constant. The merits and demerits of the system are given below :

Merits. (*i*) Possibility of carrying increased power for a given conductor size in case of long distance heavy power transmission.

(*ii*) Improvement of power factor at times of moderate and heavy loads.

(*iii*) Possibility of better protection for the line due to possible use of higher terminal reactances.

(*iv*) Availability of steady voltage at all loads at the line terminals.

(*v*) Improvement in system stability due to inertia effect of synchronous phase modifier and reduction in effect of sudden changes in load.

Demerits. (*i*) Increase of short-circuit current of the system and, therefore, increase in the circuit breaker ratings.

(*ii*) Increase risk of interruption of supply due to falling of synchronous motors, called the *synchronous phase modifiers,* out of synchronism.

(*iii*) Lower reserve of lines in case of line trouble.

6.13. SYNCHRONOUS PHASE MODIFIERS OR COMPENSATORS

It is well known that a synchronous motor can be made to take either a lagging or leading current from the line by altering its excitation. Idle-running synchronous motors were first employed in connection with power plants to correct for low power factor of the load, and thus reduce the current and power losses in the feeders and generators. The synchronous machines used for power factor improvement are usually referred to as '*synchronous condensers*' as they are always required to take a leading current.

However, an idle-running synchronous motor can be used for voltage regulating purposes by connecting it in parallel with the load at the receiving end of a line, as shown in Fig. 6.9. Since the machines used for voltage regulation are required to run part of the time with leading current and part of the time with lagging current depending upon the load conditions, therefore, it is more appropriate to refer to these machines as '*synchronous phase modifiers*'. They are usually of salient pole design with 6 or 8 poles

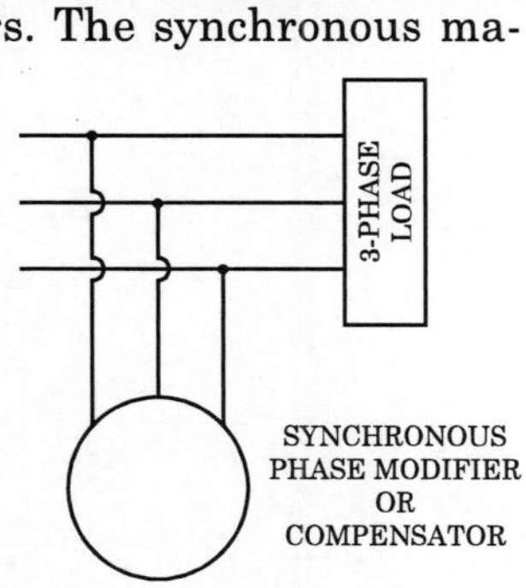

Fig. 6.9.

with ratings up to 60 MVA, 11 kV and connected to high voltage system through transformers.

Synchronous phase modifiers differ from the ordinary synchronous motors in as much as they are built for the highest economical speeds, and provided with smaller shafts and bearings and special attention is paid for securing a high overall efficiency. Standard machines for this purpose are designed to give their full-load output at leading power factor, and can carry about 50% of their rated capacity at lagging power factor. Machines can, however, be designed to operate at full rating on both leading and lagging power factors but they are larger in size, have poor efficiency and are more expensive than standard machines.

A transmission line needs lagging VARs (reactive volt-amperes) at the receiving end during peak load conditions to avoid excessive voltage drop and these lagging VARs are supplied by the synchronous phase modifier when it is over-excited as it absorbs leading VARs when over-excited. During off-peak hours the transmission line needs leading VARs at the receiving end to prevent any voltage rise. These leading VARs are supplied by the synchronous phase modifier when it is under-excited as it absorbs lagging VARs when under-excited. Thus synchronous phase modifier or compensator is a very flexible source of supply of reactive VARs (leading or lagging as per need) and is well suited for voltage regulation (or compensation).

Example 6.3. A 220 kV, 50 Hz, 3-phase transmission line is connected to station bus-bars at the sending end through a 10 MVA; 11/220 kV transformer having impedance of (1 + *j*8)%. The auxiliary constants of the transmission line are :

$$\mathbf{A = D = 0.9 \angle 0.6^\circ \; ; \; B = 153.2 \angle 84.6^\circ \; \Omega \text{ and } C = 0.0012 \angle 90^\circ \; S.}$$

Determine the A_0, B_0, C_0 and D_0 constants of the equivalent circuit of the line and the transformer connected at the supply end.

Solution : Transformer impedance = $(1 + j\,8)\%$

Impedance of transformers in ohms referred to HV side,

$$\mathbf{Z}_{se} = \frac{(1+j8)}{100} \times \frac{(220)^2}{100} = (4.84 + j\,38.72)\ \Omega/\text{phase} = 39.02 \angle 82.87^\circ\ \Omega/\text{phase}$$

The auxiliary constants of the transmission line are :

$$\mathbf{A} = \mathbf{D} = 0.9 \angle 0.6^\circ,\ \mathbf{B} = 153.2 \angle 84.6^\circ\ \Omega \text{ and } \mathbf{C} = 0.0012 \angle 90^\circ\ \text{S}$$

$\mathbf{A}_0$, $\mathbf{B}_0$, $\mathbf{C}_0$ and $\mathbf{D}_0$ constants of the equivalent circuit of the line and the transformer connected at the supply end (Fig. 6.4) are given below [refer to Eq. (6.35)]

$$\begin{aligned}
\mathbf{A}_0 &= \mathbf{A} + \mathbf{C}\,\mathbf{Z}_{se} \\
&= 0.9 \angle 0.6^\circ + 0.0012 \angle 90^\circ \times 39.02 \angle 82.87^\circ \\
&= 0.9 \angle 0.6^\circ + 0.046824 \angle 172.87^\circ \\
&= (0.8535166 + j\,0.0152346) = 0.8536 \angle 1.02^\circ \ \textbf{Ans.}
\end{aligned}$$

$$\begin{aligned}
\mathbf{B}_0 &= \mathbf{B} + \mathbf{D}\,\mathbf{Z}_{se} \\
&= 153.2 \angle 84.6^\circ + 0.9 \angle 0.6^\circ \times 39.02 \angle 82.87^\circ \\
&= 153.2 \angle 84.6^\circ + 35.118 \angle 83.43^\circ \\
&= 18.43549 + \text{j}\ 187.40743 = 188 \angle 84.38^\circ\ \Omega\ \textbf{Ans.}
\end{aligned}$$

$$\mathbf{C}_0 = \mathbf{C} = 0.0012 \angle 90^\circ\ \text{S}\ \textbf{Ans.}$$

$$\mathbf{D}_0 = \mathbf{A} = 0.9 \angle 0.6^\circ\ \textbf{Ans.}$$

Example 6.4. A long transmission line has the following auxiliary constants:

$A = D = 0.98 \angle 2°$ and $B = 28 \angle 69°\ \Omega$

Two similar transformers, each of impedance 12 $\angle 80°$ ohms are interposed at the two ends of the above transmission line. Determine the values of A_0, B_0, C_0 and D_0 to represent the line with the transformers.

Solution : The auxiliary constants of the transmission line are

$$\mathbf{A} = \mathbf{D} = 0.98 \angle 2° \text{ and } \mathbf{B} = 28 \angle 69°\ \Omega$$

$$\mathbf{C} = \frac{\mathbf{AD} - 1}{\mathbf{B}} \qquad \because \mathbf{AD} - \mathbf{BC} = 1$$

$$= \frac{0.98 \angle 2° \cdot 0.98 \angle 2° - 1}{28 \angle 69°}$$

$$= \frac{0.9604 \angle 4° - 1}{28 \angle 69°} = \frac{0.9580605 + j0.067 - 1}{28 \angle 69°} = \frac{0.079 \angle 122°}{28 \angle 69°} = 0.00282 \angle 53°\ \text{S}$$

Series impedance connected at both ends of the transmission line

$$\mathbf{Z}_{se} = 12 \angle 80°\ \Omega$$

$\mathbf{A}_0$, $\mathbf{B}_0$, $\mathbf{C}_0$ and $\mathbf{D}_0$ constants of the equivalent circuit of the line and similar transformers connected at both ends (Fig. 6.5) are given below [refer Eq. (6.40)]

$$\mathbf{A}_0 = \mathbf{A} + \mathbf{C}\,\mathbf{Z}_{se}$$

$$= 0.98 \angle 2° + 0.00282 \angle 53° \times 12 \angle 80° = 0.98 \angle 2° + 0.03384 \angle 133°$$
$$= (0.9794 + j\,0.0342) + (-\,0.0231 + j\,0.0248)$$

$$= (0.9563 + j\,0.059) = 0.958 \angle 3.53° \textbf{ Ans.}$$

$$\mathbf{B}_0 = \mathbf{B} + 2\mathbf{A}\,\mathbf{Z}_{se} + \mathbf{C}\,\mathbf{Z}_{se}^2$$

$$= 28 \angle 69° + 2 \times 0.98 \angle 2° \times 12 \angle 80° + 0.00282 \angle 53° \cdot (12 \angle 80°)^2$$

$$= 28 \angle 69° + 23.52 \angle 82° + 0.40608 \angle 213°$$
$$= (10.0343 + j\,26.1403) + (3.2734 + j\,23.2911) + (-\,0.3406 - j\,0.2212)$$

$$= 12.9671 + j\,49.2102 = 50.89 \angle 75.24°\ \Omega \textbf{ Ans.}$$

$$\mathbf{C}_0 = \mathbf{C} = 0.00282 \angle 53°\ \text{S} \textbf{ Ans.}$$

$$\mathbf{D}_0 = \mathbf{A}_0 = 0.958 \angle 3.53° \textbf{ Ans.}$$

Example 6.5. A long transmission line has the following auxiliary constants :

$A = D = 0.92 \angle 53°$; $B = 65.3 \angle 81°\ \Omega$

Two similar transformers, each of series impedance 100 $\angle 70°\ \Omega$ and admittance 0.0002 $\angle -75°$ S are connected at the two ends of the above transmission line. Determine A_0, B_0, C_0 and D_0 for the combination.

Solution : The auxiliary constants of the transmission lines are

$$\mathbf{A} = \mathbf{D} = 0.92 \angle 5.3° \text{ and } \quad \mathbf{B} = 65.3 \angle 81°\ \Omega$$

$$\mathbf{C} = \frac{\mathbf{AD} - 1}{\mathbf{B}} = \frac{(0.92 \angle 5.3°)^2 - 1}{65.3 \angle 81°} = \frac{0.8464 \angle 10.6° - 1}{65.3 \angle 81°} \qquad \because \mathbf{AD} - \mathbf{BC} = 1$$

$$= \frac{-\,0.1681 + j\,0.1557}{65.3 \angle 81°} = \frac{0.229 \angle 137.2°}{65.3 \angle 81°} = 0.0035 \angle 56.2°\text{S}$$

Impedance of transformers connected at both ends of transmission line,

$$\mathbf{Z_T} = 100 \angle 70^\circ\ \Omega$$

Shunt admittance of transformers connected at both ends of transmission line,

$$\mathbf{Y_T} = 0.0002 \angle -75^\circ\ \text{S}$$

$\mathbf{A_0}$, $\mathbf{B_0}$, $\mathbf{C_0}$ and $\mathbf{D_0}$ constants of the equivalent circuit of the transmission line and similar transformers connected at both ends (Fig. 6.6) are given below [refer to Eq. (6.47)]

$$\mathbf{A_0} = \mathbf{D_0} = \mathbf{A}(1 + 2\mathbf{Y_T Z_T}) + \mathbf{B\,Y_T} + \mathbf{C\,Z_T}(1 + \mathbf{Y_T Z_T})$$

$$= 0.92 \angle 5.3^\circ (1 + 2 \times 0.0002 \angle -75^\circ \times 100 \angle 70^\circ) + 65.3 \angle 81^\circ \times 0.0002 \angle -75^\circ$$
$$+ 0.0035 \angle 56.2^\circ \times 100 \angle 70^\circ (1 + 0.0002 \angle -75^\circ \times 100 \angle 70^\circ)$$

$$= 0.92 \angle 5.3^\circ (1 + 0.04 \angle -5^\circ) + 0.01306 \angle 6^\circ + 0.35 \angle 126.2^\circ (1 + 0.02 \angle -5^\circ)$$

$$= 0.92 \angle 5.3^\circ (1 + 0.03985 - j\,0.0034) + 0.01306 \angle 6^\circ + 0.35 \angle 126.2^\circ (1.02 - j\,0.0017)$$

$$= 0.92 \angle 5.3^\circ \times 1.04 \angle -0.2^\circ + 0.01306 \angle 6^\circ + 0.35 \angle 126.2^\circ \times 1.02 \angle -0.1^\circ$$

$$= 0.9568 \angle 5.1^\circ \times 0.01306 \angle 6^\circ + 0.357 \angle 126.1^\circ$$
$$= (0.9530 + j\,0.0850) + (0.013 + j\,0.0014) + (-0.2103 + j\,0.2884)$$

$$= 0.7557 + \text{j}\ 0.3748 = 0.84354 \angle 26.38^\circ \ \textbf{Ans.}$$

$$\mathbf{B_0} = 2\,\mathbf{A\,Z_T} + \mathbf{B} + \mathbf{C}\ \mathbf{Z_T^2}$$

$$= 2 \times 0.92 \angle 5.3^\circ \times 100 \angle 70^\circ + 65.3 \angle 81^\circ + 0.0035 \angle 56.2^\circ \cdot (100 \angle 70^\circ)^2$$

$$= 184 \angle 75.3^\circ + 65.3 \angle 81^\circ + 35 \angle 196.2^\circ$$
$$= (46.6915 + j\,177.9773) + (10.2152 + j\,64.496) + (-33.6103 - j\,9.7647)$$

$$= 23.2964 + 232.7086 = 233.87 \angle 84.28^\circ\ \Omega\ \textbf{Ans.}$$

$$\mathbf{C_0} = 2\,\mathbf{A\,Y_T}(1 + \mathbf{Y_T\,Z_T}) + \mathbf{B\,Y_T^2} + \text{C}\,(1 + \mathbf{Y_T\,Z_T})^2$$

$$= 2 \times 0.92 \angle 5.3^\circ \times 0.0002 \angle -75^\circ (1 + 0.0002 \angle -75^\circ \times (100 \angle 70^\circ)$$
$$+ 65.3 \angle 81^\circ \times (0.0002 \angle -75^\circ)^2 + 0.0035 \angle 56.2^\circ (1 + 0.0002 \angle -75^\circ \times 100 \angle 70^\circ)^2$$

$$= 0.000368 \angle -69.7^\circ (1 + 0.02 \angle -5^\circ) + 0.0000026 \angle -69^\circ + 0.0035 \angle 56.2^\circ (1 + 0.02 \angle -5^\circ)^2$$

$$= 0.000368 \angle -69.7^\circ (1.01992 - j\,0.00174) + \text{zero} + 0.0035 \angle 56.2^\circ (1.01992 - j\,0.00174)^2$$

$$= 0.000368 \angle -69.7^\circ \times 1.02 \angle -0.1^\circ + 0.0035 \angle 56.2^\circ \times 1.0404 \angle -0.2^\circ$$

$$= 0.000375 + \angle -69.8^\circ + 0.0036414 \angle 56^\circ$$
$$= (0.00013 - j\,0.000352) + (0.00204 + j\,0.003019)$$

$$= (0.00217 + j\,0.002667)\ \text{S} = 0.003438 \angle 50.9^\circ\ \text{S}\ \textbf{Ans.}$$

Example 6.6. A long transmission line has the following auxiliary constants:

$$\mathbf{A = D = 0.945 \angle 1.02^\circ,\ B = 82.3 \angle 73.03^\circ\ \Omega\ \text{and}\ C = 0.001376 \angle 90.4^\circ\ S}$$

Determine (*i*) equivalent T-network and (*ii*) π-network.

Solution : (*i*) Equivalent T-network is shown in Fig. 6.7 in which

Shunt admittance $\mathbf{Y} = \mathbf{C} = 0.001376 \angle 90.4^\circ$ S **Ans.**

and impedance, $\mathbf{Z} = \dfrac{2(\mathbf{A} - 1)}{\mathbf{C}} = \dfrac{2\,(0.945 \angle 1.02^\circ - 1)}{0.001376 \angle 90.4^\circ}$

$$= 2 \times \frac{-0.0552 + j\,0.01682}{0.000688 \angle 90.4^\circ} = 2 \times \frac{0.05766 \angle 163^\circ}{0.000688 \angle 90.4^\circ} = 167.75 \angle 72.6^\circ\ \Omega\ \textbf{Ans.}$$

(*ii*) Equivalent π-network is shown in Fig. 6.8 in which

Series impedance, $\mathbf{Z} = \mathbf{B} = 82.3 \angle 73.03° \ \Omega$ **Ans.**

$$\text{and shunt admittance, } \mathbf{Y} = \frac{2(\mathbf{A}-1)}{\mathbf{B}} = \frac{2(0.945 \angle 1.02° - 1)}{82.3 \angle 73.03°}$$

$$= \frac{2 \times 0.05766 \angle 163°}{82.3 \angle 73.03°} = 0.001401 \angle 89.97° \text{ S } \textbf{Ans.}$$

6.14. TUNED POWER LINES

The behaviour of a transmission line is described by the Eqs. (6.15) and (6.16), as given below :

$$\mathbf{V_S} = \mathbf{V_R} \cosh \sqrt{\mathbf{YZ}} + \mathbf{I_R} Z_C \sinh \sqrt{\mathbf{YZ}}$$

$$\text{and } \mathbf{I_S} = \mathbf{I_R} \cosh \sqrt{\mathbf{YZ}} + \frac{\mathbf{V_R}}{Z_C} \sinh \sqrt{\mathbf{YZ}}$$

For an overhead line shunt conductance G is always negligible and it is sufficiently accurate to neglect line resistance R as well. So with this approximation

$$\mathbf{Y} = j\omega C l \text{ and } \mathbf{Z} = j\omega L l$$

where L and C are the inductance and capacitance per unit length and l the length. Thus

$$\sqrt{\frac{\mathbf{Z}}{\mathbf{Y}}} = \sqrt{\frac{j\omega L l}{j\omega C l}} = \sqrt{\frac{L}{C}}$$

$$\text{and } \sqrt{\mathbf{YZ}} = \sqrt{j\omega C l \times j\omega L l} = j\omega l \sqrt{LC}$$

So the characteristic equations become

$$\mathbf{V_S} = \mathbf{V_R} \cosh (j\omega l \sqrt{LC}) + \mathbf{I_R} Z_C \sinh (j\omega l \sqrt{LC})$$

$$\text{and } \mathbf{I_S} = \mathbf{I_R} \cosh (j\omega l \sqrt{LC}) + \mathbf{V_R} \times \frac{1}{Z_C} \sinh (j\omega l \sqrt{LC})$$

$$\text{or } \mathbf{V_S} = \mathbf{V_R} \cos (\omega l \sqrt{LC}) + j Z_C \mathbf{I_R} \sin (\omega l \sqrt{LC}) \quad \text{...(6.62 a)}$$

$$\text{and } \mathbf{I_S} = \mathbf{I_R} \cos (\omega l \sqrt{LC}) + \frac{j}{Z_C} \mathbf{V_R} \sin (\omega l \sqrt{LC}) \quad \text{...(6.62 b)}$$

$$\because \cosh (j\omega l \sqrt{LC}) = \cos (\omega l \sqrt{LC})$$

$$\text{and } \sinh (j\omega l \sqrt{LC}) = j \sin (\omega l \sqrt{LC})$$

If there is no load, the receiving-end current I_R is zero. The charging current becomes

$$\mathbf{I_S} = \frac{j}{Z_C} \mathbf{V_R} \sin (\omega l \sqrt{LC})$$

$$\text{while } \mathbf{V_S} = \mathbf{V_R} \cos (\omega l \sqrt{LC})$$

Thus sending-end voltage V_S and sending-end current I_S are in quadrature. If the relation

$$\omega l \sqrt{LC} = n\pi \text{ where } n = 1, 2, 3, \ldots.. \quad \text{...(6.63)}$$

is satisfied, there is no charging current on zero load. If the expression $\omega l \sqrt{LC} = \pi$, the $\cos (\omega l \sqrt{LC}) = -1$ and then for load conditions, the Eqs. 6.62 (a) and 6.62 (b) become

$$\mathbf{V_S} = -\mathbf{V_R} \text{ and } \mathbf{I_S} = -\mathbf{I_R}$$

and if $(\omega l \sqrt{LC}) = 2\pi$, the $\cos \omega l (\sqrt{LC}) = 1$ and Eqs. 6.62 (a) and 6.62 (b) become

$$\mathbf{V_S} = \mathbf{V_R} \text{ and } \mathbf{I_S} = \mathbf{I_R}$$

In both the cases the receiving-end voltage and current are numerically equal to the sending-end values, so that there is no voltage drop on load. Such a line is called *a tuned line.*

For 50 Hz, the length of line for tuning is

$$l = \frac{n\pi}{2\pi f\sqrt{LC}} = \frac{n}{2\times 50\sqrt{LC}} = \frac{n}{100} \times \text{velocity of light}$$

$$\because \frac{1}{\sqrt{LC}} = v, \text{ the velocity of light} = 3 \times 10^8 \text{ m/s or } 3 \times 10^5 \text{ km/s}$$

$$\text{or } l = \frac{n}{100} \times 3 \times 10^5 \text{ km} = 3{,}000 \text{ km}, 6{,}000 \text{ km} \ldots..$$

It is too long a distance of transmission from the point of view of cost and efficiency. For a given line (length of line and frequency being fixed), the line may be tuned by increasing L or C *i.e.* by adding series inductances or shunt capacitances at intervals along the line length, as illustrated in Figs. 6.10 (*a*) and 6.10 (*b*). This method is impractical and uneconomical for power transmission and is used in telephony where higher frequencies are employed.

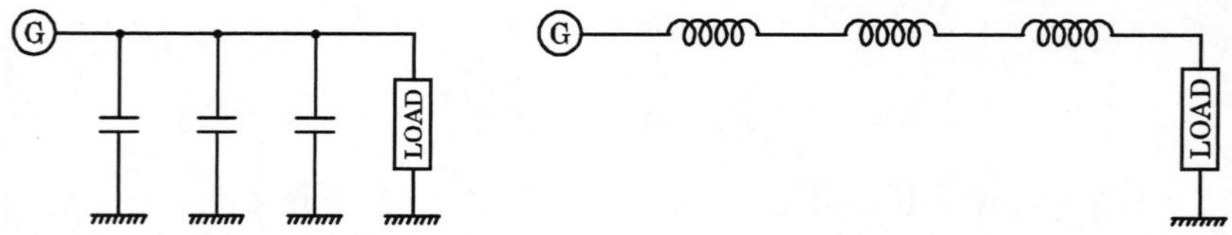

(*a*) *By Shunt Capacitance* (*b*) *By Series Inductance*

Fig. 6.10. ***Timing of Transmission Line***

Another method of tuning power lines, which is being presently experimented with, is shown in Fig. 6.11. The capacitors in series with the line neutralize the voltage drop due to the line inductance, and the inductors across the line neutralize the charging current due to line capacitance. This method is known as the *method of tuning by the use of compensating sections.*

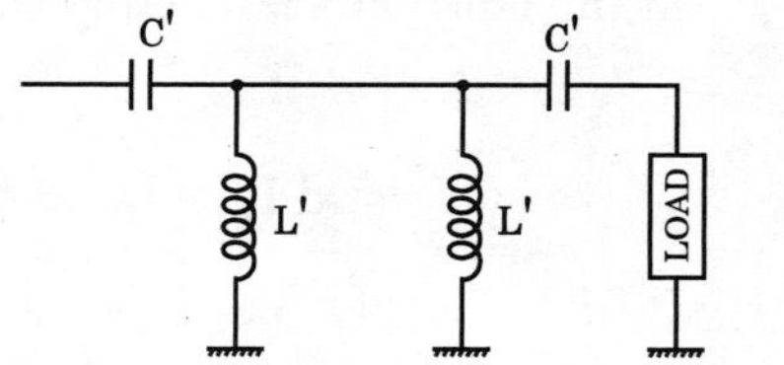

Fig. 6.11. ***Tuning of Power Transmission Line By Compensating Sections***

The values of L′ and C′ may be found with fair accuracy by replacing the line by its nominal-π equivalent (Fig. 5.9). The two capacitances C′ are equivalent to a single capacitance $\frac{1}{2}$ C′ which will neutralize the inductance L of the line impedance Z if $\omega^2 L \left(\frac{1}{2}C'\right) = 1$. The inductance L′ will neutralize the charging current provided $\omega^2 L' \left(\frac{1}{2}C\right) = 1$. The exact values required in long lines, can be obtained from the π-equivalent shown in Fig. 6.8. Thus $\frac{1}{2}$ C′ must neutralize Z, which is given as,

$$Z = \left(\sqrt{\frac{L}{C}}\right), \sinh (j\,\omega\sqrt{LC}) = j\sqrt{\frac{L}{C}} \sin(\omega\sqrt{LC}) \text{ so that}$$

$$\frac{1}{j\,\omega\,C'} + Z = 0 \text{ giving } C' = \frac{2}{\omega\sqrt{\frac{L}{C}} \sin\omega\sqrt{LC}} \quad \ldots(6.64)$$

Similarly, L′ must neutralize, Y/2 giving $L' = \dfrac{1}{\omega \sqrt{C/L} \tan\left(\dfrac{1}{2} \omega \sqrt{LC}\right)}$ ...(6.65)

EXERCISES

1. Obtain equivalent-π model of a long transmission line.

   [U.P. Technical Univ. Elements of Power System 2008-09]

2. Derive the expression for sending end voltage and current in terms of receiving end voltage and current for long transmission lines. What do you understand by the "characteristic impedance" and "propagation constant" in it.

   [U.P. Technical Univ. Elements of Power System 2009-10]

3. A long transmission line is terminated by an impedance equal to its characteristic impedance. Show that

   $$\frac{V_R}{V_S} = e^{-Yl}$$

   where V_R = Receiving end voltage
   V_S = Sending end voltage
   Y = Propagation constant
   l = Length of line

   [U.P. Technical Univ. Elements of Power System 2004-05]

4. Derive the A, B, C, D constants of a long transmission line from the basic principle.

   [U.P. Technical Univ. Elements of Power System 2004-05]

5. What do you understand by the generalised constants of a transmission line ? What is their significance ?
6. Define the generalised A, B, C, D constants of a transmission line and determine their values for a long transmission line. Hence show A = D and AD – BC = 1.
7. Starting from first principles deduce expressions for ABCD constants of a long line in terms of its parameters. Define propagation constant and characteristic impedance.
8. Develop the long line exact equations in hyperbolic terms for sending-end voltage and current. **[U.P.S.C. I.E.S. Elec. Engineering II; 1999]**
9. With reference to long transmission lines, give physical interpretation of the terms 'characteristic empedance' and 'propagation constant.' What is meant by surge impedance ?

   [U.P.S.C. I.E.S. Elec. Engineering II; 1998]

10. Explain surge impedance loading with respect to an overhead transmission line. Show that

    $$Z_C = \sqrt{Z_{OC} Z_{SC}}$$

    [U.P. Technical Univ. Elements of Power System 2006-07]

11. Explain briefly about "SIL". **[U.P. Technical Univ. Elements of Power System 2009-10]**
12. Explain the concept of regulation in long transmission line. If regulation is very high then what can be seen adversely in power network.

    [U.P. Technical Univ. Elements of Power System 2005-06]

13. Draw the equivalent circuit of a long transmission line. Derive from fundamentals the following relationships between sending-and receiving-end voltages and currents.

    $$\mathbf{V_S = AV_R + BI_R \text{ and } I_S = CV_R + DI_R}$$

14. What are the A, B, C, D constants of a long and medium length transmission lines ? In this connection explain 'characteristic impedance' and 'nominal' T and 'nominal' π representations and where these are applicable. What are 'series' and 'shunt compensation' ?

15. Show a diagrammatic representation of a medium long line in terms of its nominal-π model. Indicate also the values of the series impedance and shunt admittance of the model in terms of the resistance, inductance and capacitance per unit length of the line.
16. Explain the function of a synchronous phase modifier placed at the receiving end of the transmission line.
17. Write short notes on the following :
    (*i*) A, B, C, D constants of long transmission line.
    (*ii*) Surge impedance of a transmission line.
    (*iii*) SIL.
    (*iv*) Synchronous phase modifier.
    (*v*) Tuned power lines.

SHORT ANSWER TYPE QUESTIONS WITH ANSWER

Q. 1. What is the range of surge impedance for an overhead transmission line ?

Ans. 400–600 Ω.

Q. 2. What is the range of surge impedance in case of underground cables ?

Ans. 40–60 Ω.

Q. 3. What is the surge impedance loading of a line ?

Ans. The unity power factor load, that can be delivered by a line of negligible resistance, is called the surge impedance loading.

Q. 4. What is meant by 'natural load' as applied in connection with a transmission line ?

Ans. Natural load of a transmission line is defined as the load (of unity power factor) that can be delivered by the line of negligible resistance.

Q. 5. How does synchronous phase modifier differ from a synchronous motor ?

Ans. Synchronous phase modifier differs from an ordinary synchronous motor in as much as they are built for the highest economical speeds, and provided with smaller shafts and bearings and special attention is paid for securing a high overall efficiency.

Q. 6. Why high and extra voltages are adopted for transmission of bulk power over long distances ?

Ans. This is because the maximum power transferred over a given line increases with the increase in V_S and V_R.

PROBLEMS

1. Determine the auxiliary constants of a 3-ϕ, 50 Hz, 150 km long transmission line having resistance, inductance and capacitance per phase per km of 0.2 Ω, 1.5 mH and 0.008 μF respectively. Neglect the conductance of the line.

[**Ans.** **A** = **D** = 0.987 $\angle 0.3°$; **B** = 76 $\angle 67.1°$ Ω ; **C** = 0.37 × 10^{-3} $\angle 90.1°$ S]

2. The constants for a 132 kV line are as under :

A = **D** = 0.98 $\angle 3°$; **B** = 110 $\angle 75°$ Ω ; **C** = 0.0034 $\angle 80°$ **S**

The load at the receiving end is 40 MVA at 132 kV and power factor 0.8 lagging. Draw the circle diagram and find out (*i*) sending-end voltage (*ii*) leading MVAR for this load is at the receiving end, if the sending -end voltage is 140 kV. [**Ans.** (*i*) 175.8 kV (*ii*) 23 MVAR]

❑❑❑

UNIT-III

CORONA AND INTERFERENCE ELECTROSTATIC AND ELECTROMAGNETIC INTERFERENCE WITH COMMUNICATION LINES AND OVERHEAD LINE INSULATORS

Contents in This Unit

CORONA AND INTERFERENCE : Phenomenon of Corona; Corona Formation; Calculation of Potential Gradient, Corona Loss; Factors Affecting Corona, Methods of Reducing Corona and Interference.

ELECTROSTATIC AND ELECTROMAGNETIC INTERFERENCE WITH COMMUNICATION LINES

OVERHEAD LINE INSULATORS : Type of Insulators and Their Applications, Potential Distribution Over a String of Insulators, Methods of Equalizing the Potential, String Efficiency.

CHAPTER 7. Corona.

CHAPTER 8. Electrostatic and Electromagnetic Interference With Communication Lines.

CHAPTER 9. Overhead Line Insulators.

7

CHAPTER

CORONA

INSIDE THIS CHAPTER

7.1. Introduction **7.2.** Factors Affecting Corona **7.3.** Triple-Harmonic Currents Due to Corona **7.4.** Advantages and Disadvantages of Corona **7.5.** Dielectric Strength of Air **7.6.** Disruptive Critical Voltage **7.7.** Visual Critical Voltage **7.8.** Corona Power Loss **7.9.** Factors Affecting Corona Loss **7.10.** Methods of Reducing Corona Effects **7.11.** Voltage limitations of Lines.

7.1. INTRODUCTION

When an alternating potential difference is applied across two conductors, whose spacing is large in comparison with their diameters, then the atmospheric air surrounding the conductors is subjected to electrostatic stresses. At low voltage there is no change in the condition of atmospheric air around the conductors. However, when the potential difference is gradually increased, a stage arrives when a faint luminous glow of violet colour appears together with a hissing noise. This phenomenon is called "*visual corona*" and is accompanied by the production of ozone which is readily detected because of its characteristic odour. The glow is due to the fact that the atmospheric air around the conductor becomes conducting due to electrostatic stresses. If the potential difference is raised still further, the glow and the noise will increase in intensity until eventually a spark-over owing to breakdown of air insulation will take place. The whole phenomenon—the hissing noise, the violet glow, and production of ozone gas is known as the *corona*. This phenomenon is very much evident in transmission lines of 100 kV and above.

If the conductors are perfectly uniform and smooth the glow will be uniform throughout the length of the conductor, otherwise the rough points will appear brighter. If the spacing between the conductors is not very large as compared with their diameters (interaxial difference is less than 15 times the diameter of the conductors) corona may bridge the conductors and cause flash-over before any luminous glow is observed. It happens due to the fact that the distance between the conductors being smaller, there is no time for the glow to occur.

In EHV ac transmission lines the visible glow of corona is mostly uniform about both the conductors but in case of HVDC transmission the glow is uniform and brighter about the positive conductor but spotty about the negative conductor.

Corona is accompanied by a loss of energy which increases very rapidly once the visual critical voltage is exceeded. Power loss, due to corona, heavily depends upon weather condition—during humid and moist climate corona loss is much increased. The energy loss accompanied by the phenomenon, called the corona, is dissipated in the form of light, heat, sound and chemical action. In case of ac system the current due to corona is non-sinusoidal. In practice this non-sinusoidal current and the non-sinusoidal voltage drop caused by it may be more important than the power loss.

The effects of corona are summarized as below.

(*i*) A violet glow is observed around the conductor.

(*ii*) It produces a hissing noise.

(*iii*) It produces ozone which can be readily detected by its characteristic odour.

(*iv*) The glow is maximum over rough and dirty surfaces of the conductor.

(*v*) It is accompanied by a power loss—the wattmeter connected in the electric circuit will show a reading.

(*vi*) The charging current under corona condition increases because the corona induces harmonic currents.

7.2. FACTORS AFFECTING CORONA

Since corona occurs due to the ionization of the air surrounding the line conductors, it is affected by the physical state of the atmosphere as well as by the condition of the line. The corona is affected by the following factors.

(*i*) *Atmosphere* : Corona is caused by the bombardment of molecules, with subsequent dislodging of electrons, by the ionised particles. Corona will thus be affected by the physical state of the atmosphere. The voltage gradient for the breakdown of the air is proportional to its density. In the stormy weather the number of ions may be more than normal, and as such the corona may occur at much less voltage compared with fair weather.

(*ii*) *Conductor* : The corona is considerably affected by the size (diameter), shape (stranded or smooth) and surface condition (dirty or clean) of the conductor. The corona decreases with the increase in diameter of conductor. A stranded conductor gives rise to more corona than a solid conductor. For stranded conductors, the shape of the x-section is a series of arcs of circles each of much smaller diameter than the conductor as a whole. The potential gradient for such a conductor will, therefore, be greater than for the equivalent smooth conductor so the breakdown voltage for such a conductor will be somewhat less than for a smooth conductor, the ratio being between 0.88 and unity. The effect of dirt on the surface is to increase the irregularity and thereby decreasing the breakdown voltage further, an average value for the ratio when both stranding and dirt are taken into account lies between 0.85 and unity.

(*iii*) *Spacing Between Conductors* : With the increase in spacing between the conductors the electrostatic stresses are reduced and therefore, the corona effect is reduced. If the spacing between the conductors is made very large as compared with their diameters, there may not be any corona effect.

(*iv*) *Line Voltage* : Line voltage largely affects the corona. At low voltage, there is no corona effect, but when the line voltage is increased to such a value that electrostatic stresses developed at the conductor surface make the atmospheric air surrounding the conductor conducting, corona effect appears.

7.3. TRIPLE-HARMONIC CURRENTS DUE TO CORONA

Corona forms when the voltage of a conductor passes the disruptive critical voltage and disappears when the voltage descends through the same value. This occurs on each conductor every half-cycle, and develops a pulsation in the voltage wave having three times the generator frequency. In an earthed system this triple-frequency voltage causes a triple-frequency current to flow through the capacitance of the system to earth and back through the earthed neutral. This effect is accentuated by the fact that the effective capacitance of the conductors pulsates at triple frequency due to the increase and decrease of effective diameter of conductors caused by the corona. Because of corona triple frequency currents flow through the ground in case of a non-earthed system, the neutral has a voltage to earth of triple frequency.

7.4. ADVANTAGES AND DISADVANTAGES OF CORONA

In the past, the corona was considered as something to be avoided because of the energy loss associated with it and the distortion of the waveform. But now the corona is considered beneficial because of the following advantages.

1. On the formation of corona the sheath of air surrounding the conductor becomes conductive and there is a virtual increase in conductor diameter and due to this virtual increase in conductor diameter the maximum potential gradient or maximum electrostatic stress is reduced. Thus probability of flash-over is reduced and system performance is improved.
2. Effects of transients produced by lightning and other causes are reduced, since charges induced on the line by lightning or other causes will be partially dissipated as a corona loss. In this way it acts as a safety valve and sometimes lines have been purposely designed to have an operating voltage near to the critical voltage in order to do away with the necessity for, and expense of lightning arrestor gear. An objection to this scheme is that the critical voltage is not fixed for a given line, but may vary considerably with changes in the weather.

Disadvantages of corona are given below.

1. There is a definite loss of power, although this is not important except under abnormal weather conditions.
2. There is a non-sinusoidal voltage drop due to non-sinusoidal corona current and these may cause some interference with neighbouring communication circuits due to electromagnetic and electrostatic induction effects.
3. Owing to the formation of corona, ozone gas is produced which chemically reacts with the conductor and causes corrosion.

7.5. DIELECTRIC STRENGTH OF AIR

The value of potential gradient at which complete disruption of air occurs, is called the *disruptive strength* or *dielectric strength* of air. The breakdown strength of air at a barometric pressure of 760 mm, and temperature of 25° C is 30 kV/cm (maximum) or 21.1. kV/cm (rms) and is denoted by g_0. The value of dielectric strength of air (*i.e.* g_0) depends upon the density of air—g_0 is proportional to density of air over a wide range and thus directly proportional to the barometric pressure, and inversely proportional to the absolute temperature. Thus breakdown strength of air at a barometric pressure of b cm of mercury and temperature of t° C becomes $\delta\, g_0$ where

$$\delta = \frac{b \times 10}{760} \times \frac{273+25}{273+t} = \frac{3.92\, b}{273+t} \qquad ...(7.1)$$

7.6. DISRUPTIVE CRITICAL VOLTAGE

Let us consider the two-wire line shown in Fig. 7.1, where r is the radius of the line conductors and d is the distance between their centres, where d is very large as compared to r. If a positive charge of q coulombs per metre of conductor length is given to conductor A, then a negative charge of q coulombs per metre of conductor length will be induced on conductor B.

Consider point P at a distance of x metres from conductor A.

Electric field intensity at point P due to charge on conductor A $= \dfrac{q}{2\pi\,\varepsilon_0\,x}$ acting towards B as shown.

Electric field intensity at point P due to induced charge on conductor B $= \dfrac{q}{2\pi\,\varepsilon_0\,(d-x)}$ acting towards B as shown

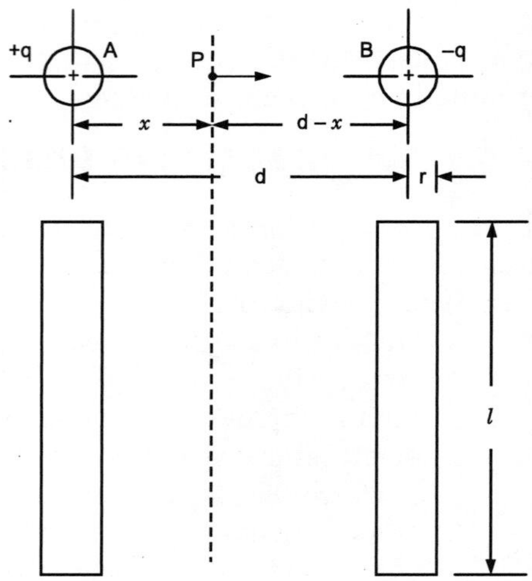

Fig. 7.1

Resultant electric field intensity at point P,

$$E_x = \frac{q}{2\pi\,\varepsilon_0\,x} + \frac{q}{2\pi\,\varepsilon_0\,(d-x)} = \frac{q}{2\pi\,\varepsilon_0}\left(\frac{1}{x}+\frac{1}{d-x}\right)$$

PD between conductors A and B,

$$V = \int_r^{d-r} E_x\,dx = \int_r^{d-r} \frac{q}{2\pi\,\varepsilon_0}\left(\frac{1}{x}+\frac{1}{d-x}\right)dx$$

$$= \frac{q}{2\pi\,\varepsilon_0}\Big[\log x - \log(d-x)\Big]_r^{d-r} = \frac{q}{\pi\varepsilon_o}\log_e\frac{d-r}{r}$$

Now since r is very small as compared to d, $d-r \simeq d$

$$\text{and, therefore, } V = \frac{q}{\pi\varepsilon_0}\log_e\frac{d}{r} \qquad ...(7.2)$$

Now gradient at any point x from the centre of the conductor A is given by

$$E_x = \frac{q}{2\pi\varepsilon_0}\left(\frac{1}{x}+\frac{1}{d-x}\right) = \frac{q}{2\pi\varepsilon_0}\cdot\frac{d}{x(d-x)}$$

Substituting for q from Eq. (7.2) in above equation, we have

$$E_x = \frac{\pi\varepsilon_0 V}{\log_e\frac{d}{r}}\cdot\frac{1}{2\pi\varepsilon_0}\cdot\frac{d}{x(d-x)} = \frac{V}{2\log_e\frac{d}{r}}\cdot\frac{d}{x(d-x)}$$

where V is voltage between two conductors.

$$\text{and } E_x = \frac{V'}{\log_e\frac{d}{r}}\cdot\frac{d}{x(d-x)} \qquad ...(7.3)$$

where V′ is the line to neutral voltage of the system.

$$\text{In case of 3-phase system, } V' = \frac{\text{Line-to-line voltage}}{\sqrt{3}} = \frac{V_L}{\sqrt{3}}$$

From the expression for the potential gradient it is clear that for a given transmission system, the potential gradient increases as x decreases *i.e.* the potential gradient is maximum when $x = r$, the surface of the conductor, and this value is given by

$$g_{max} = E_r = E_{max} = \frac{V'd}{r(d-r)\log_e \frac{d}{r}} \simeq \frac{V'}{r \log_e \frac{d}{r}} \quad ...(7.4)$$

where r is the radius of conductors in cm, d is the spacing in cm and V' is the value of voltage of the conductors to neutral, both g_{max} and V being expressed in rms values.

When the disruptive gradient of air is reached at the conductor surface

$$g_0 = \frac{V_{d_0}}{r \log_e \frac{d}{r}}$$

or disruptive critical voltage, $V_{d_0} = g_0 \, r \log_e \frac{d}{r}$ ...(7.5)

In practice, corrections have to be applied to the above formula for air density and surface conditions of the conductor and thus the complete formula becomes

$$V_{d_0} = g_0 \, \delta \, m_0 \, r \log_e \frac{d}{r} \quad ...(7.6)$$

where δ is the air density factor [Refer to Eq. (7.1)] and m_0 is the conductor surface condition factor.

The value of m_0 may be taken as unity, 0.98 – 0.93 and 0.87 – 0.80 for polished, roughened (or weathered) and stranded conductors respectively.

In the case of stranded conductors the variations in m_0 is due to the number of strands employed, and the varying ratio between the strand diameter and overall diameter in different sizes of conductors. It also depends on the surface conditions of the conductor, often improving after a conductor has been in service for a long time, and become weathered and the roughness and abrasions oxidised away. Any mutilation of the conductor during construction or stringing lowers m_0 considerably.

The value of V_{d_0} given by Eq. (7.6) is known as *disruptive critical voltage*, and is defined as the minimum phase to neutral voltage at which corona occurs and holds good for fair-weather conditions. Its value is considerably reduced during bad atmospheric conditions, such as fog, sleet, rain and snowstorms.

Example 7.1. A 3-phase transmission line consists of 1 cm radius conductors spaced symmetrically 4 metres apart. Dielectric strength of air is 30 kV/cm. Determine the line voltage for commencing of corona.

Solution :

Spacing of conductors, d = 4 m = 400 cm

Radius of conductors, r = 1 cm

Dielectric strength of air, $g_0 = \frac{30}{\sqrt{2}}$ kV/cm = 21.21 kV/cm

Line voltage for commencing of corona $= \sqrt{3}\, V_{d_0} = \sqrt{3} \times g_0 \, r \log_e \frac{d}{r}$

$= \sqrt{3} \times 21.21 \times 1 \times \log_e \frac{400}{1} = 220$ kV **Ans.**

Example 7.2. Define disruptive critical voltage. A 3-phase, 220 kV, 50 Hz transmission line consists of 1.2 cm radius conductors spaced 2 m at the corners of an equilateral triangle. Calculate the disruptive critical voltage between the lines. Irregularity factor = 0.96; temperature 20° C. Barometric pressure 72.2 cm of mercury. Dielectric strength of air = 21.1 kV (rms)/cm.

[*U.P. Technical Univ. Elements of Power System 2005-06*]

Solution : Air density factor, $\delta = \dfrac{3.92\,b}{273+t} = \dfrac{3.92 \times 72.2}{273+20} = 0.966$

Disruptive critical voltage to neutral, $V_{d_0} = g_0\,\delta\,m_0\,r\log_e \dfrac{d}{r}$

$$= 21.1 \times 0.966 \times 0.96 \times 1.2 \log_e \frac{200}{1.2} = 120.1 \text{ kV (rms)}$$

Disruptive critical voltage from line to line $= \sqrt{3}\,V_{d_0} = \sqrt{3} \times 120.1 = 208$ kV (rms) **Ans.**

Example 7.3. A 132 kV line with 2 cm diameter conductors is built so that corona takes place if the line voltage exceeds 210 kV (rms). If the value of the potential gradient at which ionisation occurs can be taken as 30,000 volts/cm, find the spacing between the conductors.

Solution : Conductor radius, $r = \dfrac{\text{Conductor diameter}}{2} = \dfrac{2}{2} = 1$ cm

Dielectric strength of air, $g_0 = \dfrac{30{,}000}{\sqrt{2}} = 21{,}210$ V/cm $= 21.21$ kV/cm

Disruptive critical voltage to neutral, $V_{d_0} = \dfrac{210}{\sqrt{3}} = 70\sqrt{3}$ kV assuming 3-phase line

Assuming smooth conductors *i.e.* irregularity factor $m_0 = 1$, standard pressure and temperature for which air density factor, $\delta = 1$ and spacing of conductors d cm we have

$$70\sqrt{3} = 21.21 \times 1 \times 1 \times 1 \times \log_e \frac{d}{1} \quad \therefore\ V_{d_0} = g_0\,m_0\,\delta\,r\log_e \frac{d}{r}$$

$$\text{or}\quad \log_e d = \frac{70\sqrt{3}}{21.21} = 5.715$$

or Spacing between conductors, $d = e^{5.715} = 304$ cm or 3.04 m **Ans.**

Example 7.4. A 3-phase overhead transmission line operates at 132 kV, 50 Hz. The conductors are arranged in a 3.0 m delta configuration. What is the minimum diameter of conductor that can be used for no corona under fair weather condition ? Assume air density factor of 0.95 and irregularity factor of 0.85. [*U.P. Technical Univ. Elements of Power System 2004-05*]

Solution : Spacing of conductors, d = 3.0 m = 300 cm

Dielectric strength of air, g_0 = 21.21 kV/cm (rms), as assumed

Irregularity factor, m_0 = 0.85

Air density factor, δ = 0.95

Disruptive critical voltage to neutral, $V_{d_0} = \dfrac{132}{\sqrt{3}} = 76.21$ kV

$$76.21 = 21.21 \times 0.85 \times 0.95 \times r \times \log_e \frac{300}{r} \quad \because\ V_{d_0} = g_0 m_0 \delta r \log_e \frac{d}{r}$$

$$\text{or}\quad r\log_e \frac{300}{r} = \frac{76.21}{21.21 \times 0.85 \times 0.95} = 4.45$$

Solving above equation by trial and error method, we have

$$r = 0.75 \text{ cm}$$

or Diameter of conductors (minimum) $= 2 \times 0.75 = 1.5$ cm **Ans.**

Example 7.5. A conductor with 2.5 cm diameter is passed through a porcelain bushing having $\varepsilon_r = 4$ and internal and external diameters of 3 cm and 9 cm respectively. The voltage between the conductor and an earthed clamp surrounding the porcelain is 20 kV rms. Determine whether corona will be present in the air-space round the conductor. [*U.P. Technical Univ. Elements of Power System 2008-09*]

Solution : Maximum gradient on the conductor surface,

$$g_{1\max} = \frac{q}{2\pi\varepsilon_0 r}$$

Maximum gradient on the inner side of the porcelain,

$$g_{2\max} = \frac{q}{2\pi\varepsilon_0\varepsilon_r r_1}$$

$$\text{So, } g_{2\max} = \frac{r}{\varepsilon_r r_1} g_{1\max} = \frac{1.25}{4\times 1.5} g_{1\max} = 0.2083\, g_{1\max}$$

$$\text{Now, } 20 = g_{1\max}\, r \log_e \frac{1.5}{1.25} + g_{2\max} \times 1.5 \log_e \frac{4.5}{1.5}$$

$$= 1.25\, g_{1\max} \log_e \frac{1.5}{1.25} + 0.2083\, g_{1\max} \times 1.5 \log_e 3$$

$$= 0.2279 g_{1\max} + 0.3433\, g_{1\max} = 0.5712 g_{1\max}$$

$$\text{or } g_{1\max} = \frac{20}{0.5712} = 35 \text{ kV/cm.}$$

Since, the gradient exceeds the dielectric strength of air (21.1 kV/cm), corona will be present.

7.7. VISUAL CRITICAL VOLTAGE

In the case of parallel wires it is found that visual corona does not begin at the voltage V_{d_0} at which the disruptive gradient of air g_0 is reached, but at a higher voltage V_{v_0}, called the *visual critical voltage*. Visual critical voltage is defined as the minimum phase to neutral voltage at which glow appears all along the line conductors. Thus when corona begins, the potential gradient g_v at the conductor surface is higher than the disruptive gradient g_0. Contrary to what might be expected, g_v, the apparent strength of air is not constant but depends on the size of the conductors, air being apparently stronger at the surface of small conductors than large ones.

Peek states that the disruptive critical voltage must be so exceeded that the stress is greater than the breakdown value up to a distance of $0.3\sqrt{\delta r}$ cm from the conductor. Thus visual corona will occur when the breakdown value is attained at the distance $r + 0.3\sqrt{\delta r}$ from the axis, instead of at the distance r. This requires that the voltage to neutral be $1+\frac{(0.3)}{\sqrt{\delta r}}$ times the disruptive critical voltage. Thus the visual critical voltage is

$$V_{v_0} = g_0\, \delta\, m_v\, r\left(1+\frac{0.3}{\sqrt{\delta r}}\right)\log_e \frac{d}{r} \text{ kV (rms) to neutral} \quad ...(7.7)$$

where m_v is a roughness factor, which is unity for smooth conductors. When the wire is stranded or rough, m_v is less than unity. m_v may be taken as 0.98 – 0.93 for rough conductor exposed to atmospheric severities and 0.72 for local corona on stranded conductors.

Because of irregular surface of the conductor, the corona does not start simultaneously on the whole surface but it takes place at different points of the conductor which are pointed and this is known as *local corona*. For this $m_v = 0.72$. For decided or general corona along the length of the conductor m_v may be taken as 0.82.

Example 7.6. Calculate the critical disruptive voltage and critical voltages for local and general corona on a three phase overhead transmission line, consisting of three stranded copper conductors spaced 2.5 m apart at the corners of an equilateral triangle. Air temperature and pressure are 21°C and 73.6 cm Hg respectively. The conductor diameter, irregularity factor and surface factors are 10.4 mm, 0.85 and 0.7, 0.8 respectively. *[U.P. Technical Univ. Elements of Power System 2002-03]*

Solution : Conductor radius, $r = \dfrac{10.4}{2} = 5.2 \text{ mm or } 0.52 \text{ cm}$

Spacing of conductors, $d = 2.5 \text{ m} = 250 \text{ cm}$

Dielectric strength of air, $g_0 = 21.1$ kV/cm (rms) as assumed

Air density factor, $\delta = \dfrac{3.92b}{273+t} = \dfrac{3.92 \times 73.6}{273+21} = 0.9813$

Critical disruptive voltage to neutral, $V_{d_0} = g_0\, \delta\, m_0\, r \log_e \dfrac{d}{r}$

$$= 21.1 \times 0.9813 \times 0.85 \times 0.52 \log_e \frac{250}{0.52} = 56.52 \text{ kV (rms)}$$

Critical disruptive line-to-line voltage $= \sqrt{3} \times 56.52 = 97.89$ kV (rms)

Visual critical voltage to neutral, $V_{vo} = g_0 \delta m_v r \left(1 + \dfrac{0.3}{\sqrt{\delta r}}\right) \log_e \dfrac{d}{r}$ kV (rms)

Here $m_v = 0.7$ for local corona

and $m_v = 0.8$ for decided or general corona

Hence V_{vo} for local corona $= 21.1 \times 0.9813 \times 0.7 \times 0.52 \left(1 + \dfrac{0.3}{\sqrt{0.9813 \times 0.52}}\right) \log_e \dfrac{250}{0.52}$

$= 66$ kV (rms)

Line-to-line visual critical voltage for local corona $= \sqrt{3} \times 66 = 114.3$ kV (rms) **Ans.**

Line-to-line visual critical voltage for general corona $= \dfrac{0.8}{0.7} \times 114.3 = 130$ kV (rms) **Ans.**

7.8. CORONA POWER LOSS

The formation of corona is associated with a loss of power, which will have some effect on the efficiency of the transmission line, but will not be of much importance to have any appreciable effect on the voltage regulation.

As in the case of the critical voltage for corona, the power loss is affected both by atmospheric and line conditions.

Under the fair weather conditions, the following empirical formula for the corona power loss has been ascertained due to Peek

$$P_c = \frac{244}{\delta}(f + 25)\sqrt{\frac{r}{d}}\left(V_{ph} - V_{d_0}\right)^2 \times 10^{-5} \text{ kW/km/phase} \quad ...(7.8)$$

where V_{ph} is voltage to neutral in kV, V_{do} is the disruptive critical voltage to neutral in kV and f is the supply frequency in Hz.

Under the stormy weather conditions, V_{d_0} is taken to be 0.8 times its fair weather value and power loss due to corona is given by

$$P'_c = \frac{244}{\delta}(f+25)\sqrt{\frac{r}{d}}\left(V_{ph} - 0.8V_{d_0}\right)^2 \times 10^{-5} \text{kW/km/phase} \quad ...(7.9)$$

As a matter of fact, with perfectly smooth and cylindrical conductors no loss occurs until the visual critical voltage is reached, when the loss suddenly takes a definite value as calculated by the Eq. (7.8) or Eq. (7.9). It then follows the quadratic law for higher voltages. The empirical relation as derived by Peek has certain limitations and holds good only under certain conditions (*i.e* the supply frequency lies between 25 and 120 Hz, the conductor radius is greater than 2.5 mm and ratio $\frac{V_{ph}}{V_{d_0}}$ exceeds 1.8). Also a small error in m_0, the irregularity factor, will lead to wrong results when using the above formula.

When the ratio $\frac{V_{ph}}{V_{d_0}}$ is less than 1.8 Peterson's formula is to be applied for determining the corona losses and is given as

$$P_c = \frac{21 \times 10^{-6} f (V_{ph})^2}{\left(\log_{10}\frac{d}{r}\right)^2} \times \text{K kW/km/phase} \quad ...(7.10)$$

where K is a factor which varies with the ratio of $\frac{V_{ph}}{V_{d_0}}$ as given below :

V_{ph}/V_{do}	0.6	0.8	1.0	1.2	1.4	1.6	1.8	2.0	2.2
K	0.012	0.018	0.05	0.08	0.30	1.0	3.5	6.0	8.0

Example 7.7. A certain 3-φ equilaterally spaced transmission line has a total corona loss of 53 kW at 106 kV and a loss of 98 kW at 110.9 kV. What is the disruptive critical voltage between lines ? What is the corona loss at 113 kV ? [*U.P. Technical Univ. Elements of Power System 2004-05*]

Solution : Power loss due to corona for three phases is given by

$$P_c = 3 \times \frac{244}{\delta}(f+25)\sqrt{\frac{r}{d}}(V_{ph} - V_{d0})^2 \times 10^{-5} \text{ kW / km}$$

Taking air density factor δ, supply frequency f, radius of conductor r and spacing of conductors d constant

$$P_c \propto (V_{ph} - V_{d0})^2$$

$$\therefore \quad 53 \propto \left(\frac{106}{\sqrt{3}} - V_{d_0}\right) \propto \left(61.2 - V_{d_0}\right)^2 \quad ...(i)$$

$$\text{and } 98 \propto \left(\frac{110.9}{\sqrt{3}} - V_{d_0}\right)^2 \propto \left(64 - V_{d_0}\right)^2 \quad ...(ii)$$

Dividing Eq. (*ii*) by Eq. (*i*), we get

$$\frac{98}{53} = \frac{(64 - V_{d_0})^2}{(61.2 - V_{d_0})^2}$$

$$\text{or } V_{d_0} = 53.4 \text{ kV}$$

and total corona loss at 113 kV $P_c \propto \left(\frac{113}{\sqrt{3}} - V_{d_0}\right)^2 \propto \left(65.2 - V_{d_0}\right)^2$ ...(*iii*)

Dividing Eq. (*iii*) by Eq. (*i*), we get

$$\frac{P_c}{53} = \frac{(65.2 - V_{d_0})^2}{(61.2 - V_{d_0})^2} = \frac{(65.2 - 53.4)^2}{(61.2 - 53.4)^2}$$

$$\text{or Corona loss, } P_c = 53 \times \left(\frac{11.8}{7.8}\right)^2 = 121.3 \text{ kW } \textbf{Ans.}$$

Disruptive critical voltage between lines = $\sqrt{3} \times V_{d_0} = \sqrt{3} \times 53.4 = 92.5$ kV **Ans.**

Example 7.8. Determine (*i*) the critical disruptive voltage (*ii*) the visual critical voltage; and (*iii*) the corona loss under foul weather condition for 3-phase line 160 km long, conductor dia. 1.036 cm; 2.44 m delta spacing. Air temperature 26.6 ° C, corresponding to an approximate barometric pressure of 73.15 cm of mercury, operating voltage 110 kV at 50 Hz, surface irregularity factor 0.85. Assume a value of m_v= 0.72.

Disruptive voltage under foul weather = 0.8 × fair weather value

Solution : Conductor radius, $r = \frac{1.036}{2} = 0.518$ cm

Spacings of conductors, d = 2.44 m = 244 cm

Dielectric strength of air, g_0 = 21.1 kV/cm (rms), as assumed

Irregularity factor, m_0 = 0.85

Roughness factor, m_v = 0.72

$$\text{Air density factor, } \delta = \frac{3.92\, b}{273 + t} = \frac{3.92 \times 73.15}{273 + 26.6} = 0.957$$

Critical disruptive voltage to neutral, $V_{d_0} = g_0 \delta\, m_0 r \log_e \frac{d}{r}$

$$= 21.1 \times 0.957 \times 0.85 \times 0.518 \log_e \frac{244}{0.518}$$

$= 54.72$ kV (rms) **Ans.** ...(*i*)

Visual critical voltage to neutral,

$$V_{v_0} = g_0 \delta\, m_v\, r \left(1 + \frac{0.3}{\sqrt{\delta r}}\right) \log_e \frac{d}{r}$$

$$= 21.1 \times 0.957 \times 0.72 \times 0.518 \left(1 + \frac{0.3}{\sqrt{0.957 \times 0.518}}\right) \log_e \frac{244}{0.518} = 66.1 \text{ kV (rms) } \textbf{Ans.} \quad ...(ii)$$

Voltage to which the conductors are subjected, $V_{ph} = \frac{110}{\sqrt{3}} = 63.51$ kV ...(*iii*)

Hence from Eqs. (*i*), (*ii*) and (*iii*) it is obvious that corona will take place but it will not be visual.

Corona loss under foul weather condition (using Peek's formula under stormy weather conditions),

$$P'_c = \frac{244}{\delta}(f+25)\sqrt{\frac{r}{d}}\left(V_{ph} - 0.8V_{d_0}\right)^2 \times 10^{-5} \text{ kW/ km/phase}$$

$$= \frac{244}{0.957}(50+25)\sqrt{\frac{0.518}{244}}(63.51 - 0.8 \times 54.72)^2 \times 10^{-5} \text{ kW/km/phase} = 3.43 \text{ kW/km/phase}$$

Total corona loss for 160 km, 3-phase line = 160 × 3 × 3.43 = 1,647 kW **Ans.**

Corona loss under foul weather condition (using Peterson formula for stormy weather conditions)

In this case $\frac{V_{ph}}{V_{d_0}} = \frac{63.51}{0.8 \times 54.72} = 1.45$

For this value of $\frac{V_{ph}}{V_{d_0}}$, K is 0.46 (see Table given in the text on page 193)

$$\text{Corona loss} = \frac{21 \times 10^{-6} f (V_{ph})^2}{\left(\log_{10}\frac{d}{r}\right)^2} \times \text{K kW/km/phase} = \frac{21 \times 10^{-6} \times 50 \times (63.51)^2 \times 0.46}{\left(\log_{10}\frac{244}{0.518}\right)^2}$$

$$= 2.726 \text{ kW/km/phase}$$

Total corona loss for 160 km, 3-phase line = 160 × 3 × 2.726 = 1,308.5 kW **Ans.**

Example 7.9. Find the corona characteristic of 110 kV, 50 Hz, 3-phase transmission line, 175 km long consisting of three 1 cm diameter stranded copper conductors spaced in 3 metre delta arrangement. Temperature 26 °C and barometric pressure is 74 cm, m_0 = 0.85, m_v for local corona = 0.72 and m_v for general corona = 0.82.

Solution : Conductor radius, $r = \frac{1}{2} = 0.5$ cm

Spacing of conductors, d = 3 m = 300 cm

Dielectric strength of air = 21.1 kV/cm (rms) as assumed

Irregularity factor, m_0 = 0.85

Roughness factor, m_v = 0.72 for local corona

Roughness factor, m'_v = 0.82 for general corona

Voltage to which the conductors are subjected, $V_{ph} = \frac{110}{\sqrt{3}} = 63.51$ kV ...(*i*)

Air density factor, $\delta = \frac{3.92\,b}{273+t} = \frac{3.92 \times 74}{273+26} = 0.97$

Critical disruptive voltage to neutral, $V_{d_0} = g_0 \delta\, m_0 r \log_e \frac{d}{r} = 21.1 \times 0.97 \times 0.85 \times 0.5 \log_e \frac{300}{0.5}$

$= 55.6$ kV (rms) ...(*ii*)

Visual critical voltage to neutral in case of local corona,

$$V_{v_0} = g_0 \delta\, m_v r \left(1 + \frac{0.3}{\sqrt{\delta r}}\right) \log_e \frac{d}{r}$$

$$= 21.1 \times 0.97 \times 0.72 \times 0.5 \left(1 + \frac{0.3}{\sqrt{0.97 \times 0.5}}\right) \log_e \frac{300}{0.5} = 67.44 \text{ kV (rms)} \quad ...(iii)$$

Visual critical voltage to neutral in case of general corona,

$$V_{v_0} = 67.44 \times \frac{m'_v}{m_v} = 67.44 \times \frac{0.82}{0.72} = 76.8 \text{ kV (rms)} \qquad ...(iv)$$

From Eqs. (*i*), (*ii*), (*iii*) and (*iv*) it is obvious that the corona will take place but it will not be visual.

Power loss due to corona for fair weather conditions

According to Peek's formula, $P_c = \frac{244}{\delta}(f+25)\sqrt{\frac{r}{d}}\left(V_{ph} - V_{d_0}\right)^2 \times 10^{-5}$ kW/km/phase

$$= \frac{244}{0.97}(50+25)\sqrt{\frac{0.5}{300}}(63.51 - 55.6)^2 \times 10^{-5} = 0.482 \text{ kW/km/phase}$$

Total corona loss for 175 km, 3-phase line = 175 × 3 × 0.482 = 253 kW **Ans.**

According to Peterson formula

$$\frac{V_{ph}}{V_{d_0}} = \frac{63.51}{55.6} = 1.142$$

For this value of $\frac{V_{ph}}{V_{d_0}}$, K = 0.0713 (see Table given in the text on page 193)

$$\text{and } P_c = \frac{21 \times 10^{-6} \times f\left(V_{ph}\right)^2}{\left(\log_{10}\frac{d}{r}\right)^2} \times K \text{ kW/km/phase}$$

$$= \frac{21 \times 10^{-6} \times 50 \times (63.51)^2}{\left(\log_{10}\frac{300}{0.5}\right)^2} \times 0.0713 = 0.3912 \text{ kW/km/phase}$$

Total corona loss for 175 km, 3-phase line = 175 × 3 × 0.3912 = 205.4 kW **Ans.**

Power loss due to corona for stormy weather conditions

Disruptive critical voltage, $V_{d_0} = 0.8\ V_{d_0} = 0.8 \times 55.6 = 44.48$ kV

Total corona loss for 175 km, 3-phase line, according to Peek's formula,

$$P'_c = 175 \times 3 \times \frac{244}{\delta}(f+25)\sqrt{\frac{r}{d}}(V_{ph} - V_{d_0})^2 \times 10^{-5} \text{ kW}$$

$$= 175 \times 3 \times \frac{244}{0.97}(50+25) \times \sqrt{\frac{0.5}{300}} \times (63.51 - 44.48)^2 \times 10^{-5} \text{ kW} = 1{,}464 \text{ kW } \textbf{Ans.}$$

According to Peterson formula

$$\frac{V_{ph}}{V_{d_0}} = \frac{63.51}{44.48} = 1.4278$$

For this value of $\frac{V_{ph}}{V_{d_0}}$, K = 0.395 (see Table given in the text on page 193)

$$\text{and Corona loss, } P'_c = \frac{21 \times 10^{-6} \times f \times (V_{ph})^2}{\left(\log_{10} \frac{d}{r}\right)^2} \times k \text{ kW/km/phase}$$

$$= \frac{21 \times 10^{-6} \times 50 \times (63.51)^2}{\left(\log_{10} \frac{300}{0.5}\right)^2} \times 0.395 = 2.167 \text{ kW/km/phase}$$

Total corona loss = 175 × 3 × 2.167 = 1,138 kW **Ans.**

7.9. FACTORS AFFECTING CORONA LOSS

There are numerous factors such as system frequency, system voltage, air conductivity, air density, conductor radius, surface of conductor, load current and atmospheric conditions, on which the corona loss depends.

1. Effect of System Frequency. From Eqs. (7.8), (7.9) and (7.10) it is obvious that corona loss varies directly as the system frequency.

2. Effect of System Voltage. The electric field in the space around the conductors depends mainly on the potential difference between the conductors. Greater the potential difference, greater is the electric field and, therefore, greater is the power loss due to corona. In the region near the disruptive critical voltage, the rate of increase of corona power loss with the increase in system voltage is small but when V_{ph} is large as compared with V_{d_0} corona loss increases at a very fast rate with the increase in system voltage. This fact is quite evident from Eqs. (7.8), (7.9) and (7.10).

3. Effect of Conductivity of Air. The conductivity of air depends on the number of ions per unit volume of air, the size and charge per ion. Both of these factors vary with altitude and atmospheric conditions. During rain and thunderstorms, ion content increases and, therefore, atmosphere becomes more conducting. Higher conductivity leads to greater corona loss.

4. Effect of Density of Air. Corona loss increases with the decrease in the density of air. The corona loss of a transmission line passing through a hilly area is higher than that of a similar line in plains due to reduced value of δ at high altitudes.

5. Effect of Conductor Radius. The electric field intensity decreases with the increase in radius of conductor. Hence with conductors of large radius, electric field intensity decreases resulting in lower corona power loss.

6. Effect of Conductor Surface. The potential gradient at the surface of a stranded conductor is greater than that for the equivalent solid conductor. So breakdown voltage is low and corona loss is more for stranded conductors.

Roughness of the surface of the conductor causes a field distortion and gives rise to high potential gradient causing higher corona loss.

7. Effect of Atmospheric Conditions. Corona loss, particularly local corona discharges, increases in rain and bad atmospheric conditions, such as fog, sleet and snowstorms. The effect is more pronounced if the conductor is of larger diameter because on smaller diameter conductors the radius of a rain drop may not be very much smaller than the radius of the conductor itself.

8. Effect of Load Current. The heating of the conductor due to flow of load current through it has an indirect reducing effect on the corona loss. Heating of conductor prevents deposition of dew or snow on the surface of the conductor and reduces corona loss.

During rains, the heating of conductor has no effect on the corona loss but, after the rain it accelerates the drying of the conductor surface. The time duration for which the drops remain on the surface is reduced and so the corona loss is reduced.

For long transmission lines passing through routes of varying altitudes, the average value of corona loss is determined by determining the corona loss per km at a number of points and then taking out an average.

9. Bundling of Conductors. A bundled conductor consists of two or more parallel sub-conductors at a spacing of several diameters.These group of conductors form the phase conductors.Thus the effective diameter of the bundled conductor is much larger than that of the equivalent single conductor. The value of large diameter, as obvious from the equation for corona loss, will reduce the corona loss.

7.10. METHODS OF REDUCING CORONA EFFECTS

The critical voltage can be raised either by increasing the spacing between the conductors or the diameter of conductors. By increasing the spacing between the conductors, the voltage at which corona occurs is raised and hence corona effects can be avoided. However, the spacing cannot be increased to a large extent because it will result in heavier supporting structures (*e.g.*, the bigger cross-arms and supports) and therefore high cost. The diameter of the conductors can be increased by using hollow conductors with a hemp core. Steel-cored aluminium conductors have a large diameter for a given conductivity and weight, and are thus good from the point of view of corona.

By using bundled conductors also corona effects can be avoided because effective diameter of the bundled conductor is much larger than that of the equivalent single conductor.

7.11. VOLTAGE LIMITATIONS OF LINES

The basis for the design of a transmission line is essentially financial, the most economical line being the most acceptable. It is because power loss due to corona is of no great importance. It is considered satisfactory to design a line for operation at a voltage just below the disruptive critical voltage for fair weather (taking $\delta = 1$). It is economical to have a small corona loss in bad weather, *i.e* for a fraction of the year (storms are experienced at intervals) rather than have large conductors and heavy supporting structures to avoid corona entirely.

Moreover, corona is considered beneficial because of its inherent advantages, mentioned in Art 7.4.

EXERCISES

1. What is meant by corona ? What are the various factors which affect corona ? How can the corona effect be minimised ? Discuss the advantages and disadvantages of corona.
2. Explain corona and discuss its effects on performance of transmission lines.
3. Explain the situation under which corona phenomenon starts in high voltage power transmission lines. Identify the factors that affect the corona alongwith reasoning.

   [U.P. Technical Univ. Elements of Power System 2005-06]
4. Explain corona and discuss its effects on performance of transmission lines.

   [U.P. Technical Univ. Elements of Power System 2006-07]
5. Explain the factors which determine the formation of corona on overhead lines. Derive an expression for the disruptive critical voltage between two smooth circular conductors, assuming the breakdown strength of air to be 30 kV/cm (peak).

   [U.P. Technical Univ. Elements of Power System 2003-04]

6. Deduce from first principles expressions for the disruptive critical voltage between two smooth circular wires. Find the ratio of critical voltages when the distance between the conductors is halved, other data remaining the same.

[U.P. Technical Univ. Elements of Power System 2007-08]

7. What is corona ? Show that the maximum critical disruptive voltage occurs when the radius of conductor is $\frac{d}{e}$ where d is the distance between conductors.

[U.P. Technical Univ. Elements of Power System 2007-08]

8. State and explain the Peek's empirical formula for calculation of corona power loss for a power transmission line. **[U.P. Technical Univ. Elements of Power System 2005-06]**

9. Write and explain the expression for power loss due to corona. What factors affect the corona loss ? **[U.P. Technical Univ. Elements of Power System 2009-10]**

10. What is meant by corona ? How would you reduce power loss due to corona ? Discuss the advantages and disadvantages of corona.

[U.P. Technical Univ. Elements of Power System 2005-06]

11. Discuss advantages and disadvantages of corona. How can corona loss be minimized ?

[U.P. Technical Univ. Elements of Power System 2008-09]

12. Discuss the factors on which corona loss depends. Suggest different means of reducing corona loss. **[U.P. Technical Univ. Elements of Power System 2007-08]**

13. Explain corona loss. How is disruptive critical voltage estimated ? What are the factors that affect the corona loss ? Give advantages and disadvantages of it.

[U.P. Technical Univ. Elements of Power System 2006-07]

14. Explain the role of bundled conductors in reducing corona loss.

[U.P. Technical Univ. Elements of Power System 2004-05]

15. Explain 'corona loss' at extra high voltages and its disadvantages. How is disruptive critical voltage estimated ?
    What are the effects of 'air density', 'temperature' and 'bundling' of conductors on corona loss ?

16. Write short notes on the following :
    (*i*) Corona.
    (*ii*) Effects of corona.
    (*iii*) Corona loss.
    (*iv*) Methods of reducing corona loss.

SHORT ANSWER TYPE QUESTIONS WITH ANSWERS

Q. 1. What is corona ?

Ans. When the voltage between conductors of an overhead line exceeds the disruptive critical voltage value, a hissing noise accompanied by a violet glow appears. This phenomenon is called the corona.

Q. 2. What are the various factors which affect corona ?

Ans. Factors affecting corona are (*i*) number of ions, size and charge per ion, mean free path depending on atmospheric conditions, (*ii*) line voltage, (*iii*) ratio of conductor spacing to conductor radius, (*iv*) contour of the conductor surface, and (*v*) state of the conductor surface.

Q. 3. What are the adverse effects of corona ?

Ans. Adverse effects of corona are (*i*) power loss but not very important except under abnormal weather conditions, (*ii*) corrosion due to production of ozone gas, and (*iii*) interference with neighbouring communication circuits.

Q. 4. What do you mean by dielectric strength of air ?

Ans. The value of potential gradient at which complete disruption of air occurs, is called the dielectric strength of air.

Q. 5. What is meant by disruptive critical voltage ?

Ans. The disruptive critical voltage is defined as the minimum phase to neutral voltage at which corona occurs.

Q. 6. What is visual critical voltage ?

Ans. Visual critical voltage is defined as the minimum phase to neutral voltage at which glow appears all along the line conductors.

Q. 7. What are the various factors which affect corona loss ?

Ans. The factors affecting corona loss are system frequency, system voltage, air conductivity, air density, conductor diameter, conductor surface condition, atmospheric conditions and load current.

Q. 8. How can the corona effect be minimised ?

Ans. Corona effect can be minimised in a transmission line by using large diameter conductors which may be accomplished by using hollow conductors with a hemp core, ACSR conductors or bundled conductors. Corona effect can also be reduced but to a limited extent, by increasing the spacing between the conductors.

PROBLEMS

1. A 3-phase line has conductors of radius 0.8 cm, spaced equilaterally 2.5 m apart. If the dielectric strength of air is 30 kV/cm, determine the disruptive critical voltage at which corona will occur. Take relative air density factor $\delta = 0.96$ and irregularity factor $m_0 = 0.94$.

[**Ans.** 152 kV (rms)]

2. Find the disruptive critical voltage for a transmission line having :

Conductor spacing = 1 *m*; conductor (stranded) radius = 1 cm;

barometric pressure = 76 *cm* of H*g*; temperature = 40 °C;

Air breakdown potential gradient (at 76 cm of Hg and at 25 °C) = 21.1 kV (rms)/cm.

[**Ans.** 136 kV (rms)]

3. Find the disruptive critical and visual corona voltage of a grid line operating at 132 kV.

Conductor diameter = 1.9 *cm*; conductor spacing = 3.81 *m*; temperature = 44° C; barometric pressure = 73.7 *cm*;

Conductor surface factor: fine weather = 0.8; rough weather = 0.66.

[**Ans.** 155.4 kV/phase, 157.6 kV/phase]

4. Find the critical disruptive voltage and the critical voltages for local and decided corona on a 3-phase line consisting of three stranded conductors spaced 2.5 *m* apart at the corners of an equilateral triangle. Air temperature and pressure are 21 °C and 73.6 cm of Hg respectively. The conductor diameter, irregularity factor, surface factor for local and decided corona are 10.4 mm, 0.85, 0.7 and 0.8 respectively.

[**Ans.** 97.89 kV (rms); 114.46 kV (rms); 130.82 kV (rms)]

5. Determine the corona loss for 3-phase line operating at 110 kV which has conductors of 1.25 cm dia arranged in a 3.05 m delta configuration. Assume polished wires and air density factor of 1.07.

[**Ans.** No corona loss]

6. A 3-phase, 220 kV, 50 Hz transmission line consists of 3.0 cm diameter conductors spaced 2 metres apart in equilateral triangle formation. If the temperature is 20 °C and atmospheric pressure 75 cm, determine the corona loss per km of the line. Take irregularity factor as 0.80.

[**Ans.** 0.425 kW/km]

7. A three phase equilaterally spaced transmission line has a total corona loss of 53 kW at 106 kV and a loss of 98 kW at 110.9 kV. What is the disruptive critical voltage between lines ? What is corona loss at 113 kV ?

[U.P. Technical Univ. Elements of Power System 2006-07]

[**Ans.** 92.5 kV; 121.3 kW]

❑❑❑

8

CHAPTER

ELECTROSTATIC AND ELECTROMAGNETIC INTERFERENCE WITH COMMUNICATION LINES

INSIDE THIS CHAPTER

8.1. Introduction **8.2.** Electromagnetic Effect **8.3.** Effects of Earth Fault Currents **8.4.** Electrostatic Effect **8.5.** Transposition of Lines.

8.1. INTRODUCTION

It is usual practice to run telephone lines along the same route as the power lines. The transmission lines transmit bulk power at relatively high voltages and, therefore, these lines give rise to electromagnetic and electrostatic fields of sufficient magnitude which induce currents and voltages in the neighbouring telephone lines. The currents so induced are superimposed on the true speech currents in the neighbouring telephone wires and set up distortion while the voltages so induced raise the potential of the communication circuit as a whole. In extreme cases the effect of these fields may make it impossible to transmit any message faithfully and may raise the potential of the telephone receiver above the ground to such an extent to render the handling of the telephone receiver extremely dangerous and in such cases elaborate precautions are required to be observed to avoid this danger.

8.2. ELECTROMAGNETIC EFFECT

8.2.1. Single Phase Single Circuit Line and Telephone Line

Consider two power conductors A and B of a single phase single circuit line and two telephone conductors C and D of a telephone line running on the same pole as the power conductors as shown in Fig. 8.1.

Let the distances between the power conductors and telephone conductors be d_{AC}, d_{AD}, d_{BC} and d_{BD} respectively.

First of all consider the loop formed by the conductors A and C. The self induction of this loop is given by one-half of the expression of the self induction of a pair of parallel conductors, because for this loop only conductor A is contributing to the emf induced in conductor C. Denoting this self-induction by L_{AC}, we have

$$L_{AC} = 0.2 \log_e \frac{d_{AC}}{r} \text{ mH/km}$$

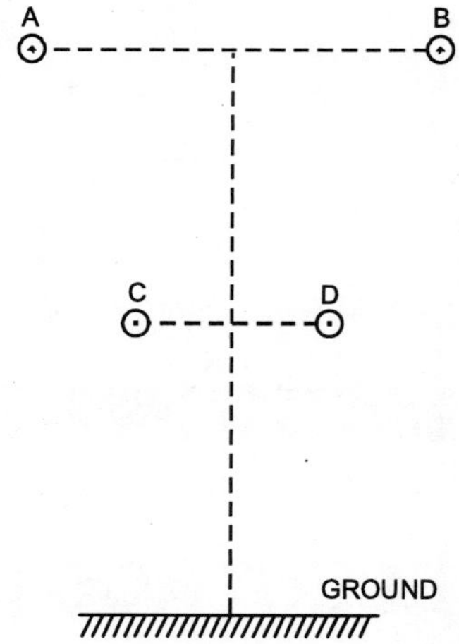

Fig. 8.1

Similarly the self-induction of the loop formed by conductors A and D,

$$L_{AD} = 0.2 \log_e \frac{d_{AD}}{r} \text{ mH/km}$$

Mutual inductance between conductor A and and the loop CD,

$$M_A = L_{AD} - L_{AC}$$

$$= \left(0.2 \log_e \frac{d_{AD}}{r} - 0.2 \log_e \frac{d_{AC}}{r}\right) \text{ mH/km}$$

$$= 0.2 \log_e \frac{d_{AD}}{d_{AC}} \text{ mH/km}$$

Similarly the mutual inductance between conductor B and the loop CD, M_B is given as

$$M_B = 0.2 \log_e \frac{d_{BD}}{d_{BC}} \text{ mH/km}$$

The net mutual inductance, which is phasor sum of two inductances is given as

$$\mathbf{M} = \mathbf{M_A} + \mathbf{M_B} = 0.2 \log_e \frac{d_{AD}}{d_{AC}} - 0.2 \log_e \frac{d_{BD}}{d_{BC}}$$

$\because$ M_A and M_B are in phase opposition

$$= 0.2 \log_e \frac{d_{AD} d_{BC}}{d_{AC} d_{BD}} \text{ mH/km}$$

$$= 0.4 \log_e \sqrt{\frac{d_{AD}\, d_{BC}}{d_{AC}\, d_{BD}}} \text{ mH/km}$$

The voltage induced in the loop CD is given by

$$V_m = 2\pi f \text{ MI volts per km}$$

where I is the current flowing through the power conductor and f is the supply frequency.

8.2.2. Three Phase Single Circuit Line and Telephone Line

Consider three power conductors A, B, and C of a 3-ϕ single circuit line and two telephone conductors P and Q of a telephone line running on the same transmission tower as the power conductors, as shown in Fig. 8.2.

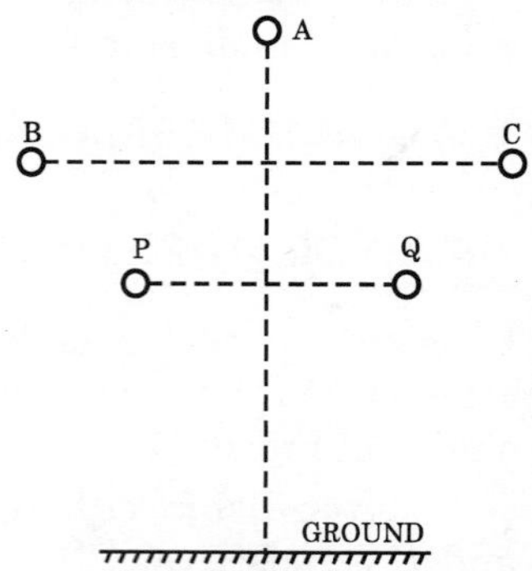

Fig. 8.2

Let the distances between the power conductors and telephone conductors be d_{AP}, d_{AQ}, d_{BP}, d_{BQ}, d_{CP} and d_{CQ} respectively.

First of all consider the loop formed by the conductors A and P. The self induction of this loop is given by one-half of the expression for the self-induction of a pair of parallel conductors, because for this loop only conductor A is contributing to the emf induced in conductor P. Denoting this self-induction by L_{AP} we have

$$L_{AP} = 0.2 \log_e \frac{d_{AP}}{r} \text{ m H/km}$$

Similarly the self-induction of the loop formed by the conductors A and Q.

$$L_{AQ} = 0.2 \log_e \frac{d_{AQ}}{r} \text{ m H/km}$$

∴ Mutual inductances between conductor A and the loop PQ,

$$M_A = L_{AQ} - L_{AP} = \left(0.2 \log_e \frac{d_{AQ}}{r} - 0.2 \log_e \frac{d_{AP}}{r}\right) \text{ m H/km} = 0.2 \log_e \frac{d_{AQ}}{d_{AP}} \text{ mH/km}$$

Similarly the mutual inductance between conductor B and the loop PQ, M_B and the mutual inductance between conductor C and loop PQ, M_C are given as

$$M_B = 0.2 \log_e \frac{d_{BQ}}{d_{BP}} \text{ m H/km}$$

$$\text{and } M_C = 0.2 \log_e \frac{d_{CQ}}{d_{CP}} \text{ m H/km}$$

These three mutual inductances are the result of fluxes which have a phase displacement of 120°. Therefore, the net effect of the magnetic field will be

$$\mathbf{M} = \mathbf{M_A} + \mathbf{M_B} + \mathbf{M_C}$$

where M is the net mutual inductance which is the phasor sum of the three inductances.

The voltage induced in the loop PQ is given by

$$V_m = 2\pi f \text{MI volts per km}$$

where I is the current flowing through the power conductor and f is the supply frequency.

It is to be noted that since the net mutual inductance M is the phasor sum of mutual inductances M_A, M_B, and M_C, therefore, there is a partial cancellation of the induced voltages due to fundamental of the power current and as the distance between the power conductors and the telephone conductors is increased, the mutual inductances, M_A, M_B and M_C progressively become equal in magnitude and thus M diminishes. But the presence of third harmonics and multiples of third harmonics will not cancel as they are in phase in all the power conductors and, therefore, are dangerous for communication circuits. These harmonics are troublesome in two ways: firstly because the induced voltage is proportional to the frequency, and secondly, because the higher frequencies come within the audible range.

Example 8.1. A single-phase, 50 Hz power line has horizontal configuration with 1.2 m spacing. A telephone line is run horizontally on the same support and is located symmetrically 0.85 m below the power line. Determine the voltage induced per km in the telephone line if the current in the power line is 120 A. Take the spacing between telephone conductors as 0.4 m.

[*U.P. Technical Univ. Elements of Power System 2003-04*]

Solution : Refer to Fig. 8.1

$$d_{AD} = \sqrt{0.85^2 + (0.6 + 0.2)^2} = 1.1673 \text{ m}$$

$$d_{AC} = \sqrt{0.85^2 + (0.6 - 0.2)^2} = 0.9394 \text{ m}$$

$$d_{BD} = d_{AC} = 0.9394 \text{ m}$$

$$d_{BC} = d_{AD} = 1.1673 \text{ m}$$

$$\text{Mutual inductance, M} = 0.4 \log_e \sqrt{\frac{d_{AD} d_{BC}}{d_{AC} d_{BD}}} \text{ mH/km}$$

$$= 0.4 \log_e \sqrt{\frac{1.1673 \times 1.1673}{0.9394 \times 0.9394}} \text{ mH/km} = 0.086883 \text{ mH/km}$$

The voltage induced per km in the telephone line,

$$V_m = 2\pi f \text{ MI} = 2\pi \times 50 \times 0.086883 \times 10^{-3} \times 120 = 3.275 \text{ V/km} \quad \textbf{Ans.}$$

8.3. EFFECTS OF EARTH FAULT CURRENTS

There may be a large voltage drop in the earth itself, due to high soil resistivity when an earth fault occurs from a line to earth and there is a large return fault current. In case the telephone-circuit earth is within the resistance area of the power system earth, the potential of the telephone-circuit earth may be raised well above true earth potential and so a serious hazard may be present.

8.4. ELECTROSTATIC EFFECT

We know that when a conductor, such as A (Fig. 8.3), runs parallel to an infinite plane (earth for example), the potential distribution between this conductor and the plane is exactly similar to that between its image and the plane. Hence if q is the charge per unit length of the power conductor A and h_A is the height of the conductor A from the ground, the potential of conductor A with respect to earth is given by

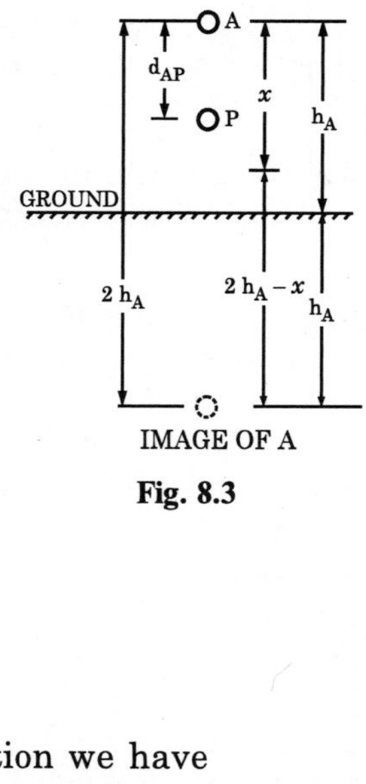

Fig. 8.3

$$V_A = \frac{1}{2\pi \epsilon_0} \int_r^{h_A} \left(\frac{q}{x} + \frac{q}{2h_A - x} \right) dx \text{ where } r \text{ is the radius of conductor A}$$

$$\text{or } V_A = \frac{q}{2\pi \epsilon_0} \log_e \frac{2h_A - r}{r} \qquad \ldots(8.1)$$

Similarly potential of conductor P due to the conductor A

$$V_{PA} = \frac{1}{2\pi \epsilon_0} \int_{d_{AP}}^{h_A} \left(\frac{q}{x} + \frac{q}{2h_A - x} \right) dx = \frac{q}{2\pi \epsilon_0} \log_e \frac{2h_A - d_{AP}}{d_{AP}}$$

Substituting $q = \dfrac{2\pi \epsilon_0 \cdot V_A}{\log_e \dfrac{2h_A - r}{r}}$ from Eq. (8.1) in the above equation we have

$$V_{PA} = \frac{2\pi \epsilon_0 \cdot V_A}{\log_e \dfrac{2h_A - r}{r}} \times \frac{1}{2\pi \epsilon_0} \log_e \frac{2h_A - d_{AP}}{d_{AP}} = V_A \frac{\log_e \dfrac{2h_A - d_{AP}}{d_{AP}}}{\log_e \dfrac{2h_A - r}{r}}$$

Similarly, we can determine the potential of telephone conductor P due to power conductors B and C and hence the potential of telephone conductor P due to power conductors A, B and C will be

$$V_P = V_{PA} + V_{PB} + V_{PC}$$

Similarly, the potential of telephone conductor Q due to power conductors A, B and C can be determined.

Example 8.2. A 3-phase, 50 Hz, 200 km long 132 kV transmission line delivers a total load of 28,000 kW at 0.85 power factor lagging. The conductor arrangement is shown in Fig. 8.4. The radius of each power conductor is 5 mm.

Determine the induced voltage at fundamental frequency in the telephone circuit due to electromagnetic effect. Also determine the potential of telephone conductor P above the earth due to electrostatic effect only.

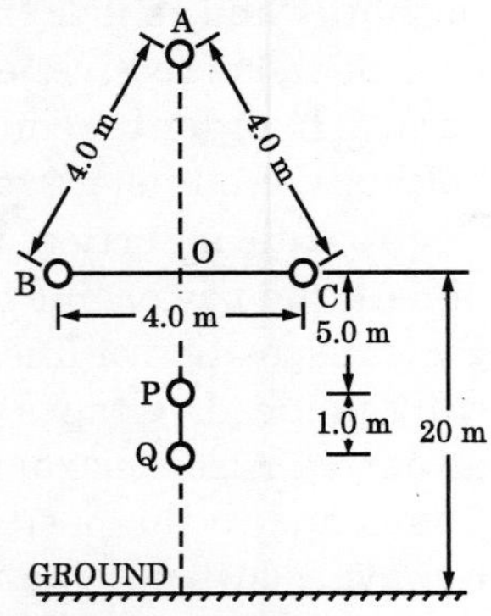

Fig. 8.4

Solution : From Fig. 8.4

$$AO = \frac{4\sqrt{3}}{2} = 3.46 \text{ m}$$

$$d_{AP} = 3.46 + 5 = 8.46 \text{ m}$$

$$d_{AQ} = 3.46 + 5 + 1.0 = 9.46 \text{ m}$$

$$d_{BP} = \sqrt{5^2 + 2^2} = 5.39 \text{ m}$$

$$d_{BQ} = \sqrt{6^2 + 2^2} = 6.32 \text{ m}$$

∴ Mutual inductance between conductor A and the loop PQ,

$$M_A = 0.2 \log_e \frac{d_{AQ}}{d_{AP}} \text{ mH/km} = 0.2 \log_e \frac{9.46}{8.46} = 0.022 \text{ m H/km}$$

$$M_B = M_C = 0.2 \log_e \frac{d_{BQ}}{d_{BP}} = 0.2 \log_e \frac{6.32}{5.39} = 0.032 \text{ m H/km}$$

Since, M_A, M_B and M_C are mutually displaced by 120°

$$\therefore M = (0.032 - 0.022) \text{ m H/km} = 0.01 \text{ mH/km}$$

Current flowing through the power conductors,

$$I = \frac{\text{Load in kW} \times 1{,}000}{\sqrt{3} \times \text{Line voltage} \times pf} = \frac{28{,}000 \times 1{,}000}{\sqrt{3} \times 1{,}32{,}000 \times 0.85} = 144 \text{ A}$$

Supply frequency, f = 50 Hz

∴ Induced voltage at fundamental frequency in the telephone circuit due to electromagnetic effect

$$V_m = 2\pi f M I = 2\pi \times 50 \times 0.01 \times 10^{-3} \times 144 = 0.452 \text{ V/km}$$

Thus for the whole route of 200 km the induced voltage at fundamental frequency in the telephone circuit due to electromagnetic effect

= 200 × 0.452 = 90.4 volts **Ans.**

For the electrostatic effect we have for conductor P,

$$V_{PA} = \frac{V_A \log_e \dfrac{2h_A - d_{AP}}{d_{AP}}}{\log_e \dfrac{2h_A - r}{r}} = \frac{132{,}000}{\sqrt{3}} \frac{\log_e \dfrac{2(20 + 3.46) - 8.46}{8.46}}{\log_e \dfrac{2(20 + 3.46) - 0.005}{0.005}} = 12{,}617 \text{ volts}$$

and
$$V_{PB} = V_{PC} = V_B \frac{\log_e \dfrac{2h_A - d_{BP}}{d_{BP}}}{\log_e \dfrac{2h_A - r}{r}} = \frac{132{,}000}{\sqrt{3}} \frac{\log_e \dfrac{46.92 - 5.39}{5.39}}{\log_e \dfrac{46.92 - 0.005}{0.005}} = 17{,}013 \text{ volts}$$

The potential of telephone conductor P with respect to earth, V_P being the phasor sum of above three components mutually displaced by 120°

$$V_P = 17{,}013 - 12{,}617 = 4{,}396 \text{ volts } \textbf{Ans.}$$

8.5. TRANSPOSITION OF LINES

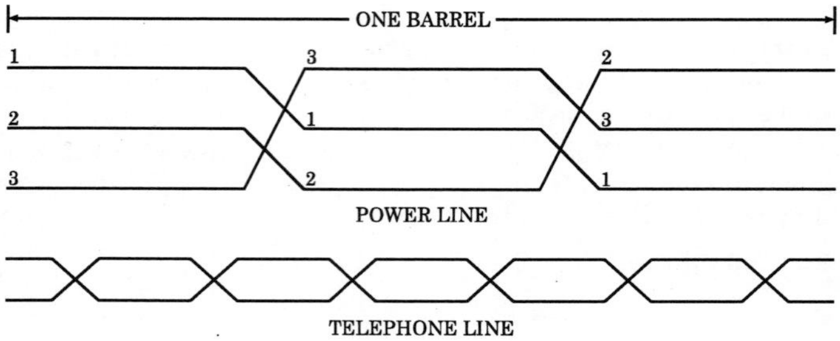

Fig. 8.5

As already said, in extreme cases the effects of the electromagnetic and electrostatic fields set up by power lines become so strong to make it impossible to transmit any message faithfully and render the handling of the telephone receiver extremely dangerous. In order to reduce the induced voltages due to electromagnetic and electrostatic induction, the conductors of both the power line and the telephone line are regularly transposed, as shown in Fig. 8.5. By transposition of lines the capacitances of the lines are balanced and, therefore electrostatically induced voltages balance out in the length of a complete set of transpositions, known as *barrel.* Also electromagnetically induced voltages on the conductors get diminished as the fluxes due to the positive and negative phase sequences cancel out along the barrel. This, however, does not apply to zero sequence currents and that is why the telephone line is also transposed along with the transposition of power line, as shown in Fig. 8.5. Each transposition of a telephone line consists of a complete cross-over of the two wires, while each transposition of a 3-ϕ line consists of a twist, in a plane at right angles to the run of the line, of one-third of a revolution. Thus three transpositions are then necessary to bring the phases back to their original positions.

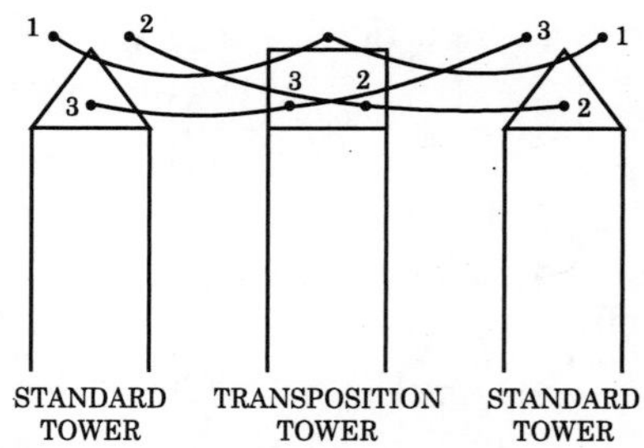

Fig. 8.6

Electrostatic charging of the telephone line also results in the flow of currents which tends to interfere with the clarity of the speech. This effect cannot be eliminated entirely by transposition of lines.

Although, transposition of conductors is helpful under normal operating conditions but does not provide complete solution of the problem under abnormal conditions caused due to such faults as a short-circuit to earth. In such a case high emf is induced due to electromagnetic induction since the fault results in large zero-sequence currents flowing along the wires in parallel and through the earth return.

With communication lines perfectly transposed and balanced no voltage between two conductors may exist and yet a considerable voltage will be induced between each conductor and earth. So it is necessary that the communication line insulation is so arranged that insulators can withstand such a voltage.

In extreme cases of electrostatic charging it may become necessary to isolate the telephone apparatus from the telephone line completely by means of highly insulated transformers, and also, to ensure the dissipation of the induced charges by means of such devices as earthed 'drainage' coils and lightning arrestors.

EXERCISES

1. Explain how does the power circuit interfere with the normal operation of the nearby communication circuit and what steps are taken to avoid this interference ?

2. Explain about electromagnetically induced emf in communication line induced by neighbouring power line currents. Mention how the effect can be reduced.

[U.P. Technical Univ. Elements of Power System 2003-04]

3. A single phase 50 Hz overhead power line (conductors *a* and *b*) is symmetrically supported on a horizontal cross-arm. A telephone line (conductors *c* and *d*) is also symmetrically supported on a horizontal cross-arm directly below the power line. Show that the mutual inductance per unit length between circuit *a-b* and *c-d* is given by

$$4 \times 10^{-7} \log_e \sqrt{\frac{D_{ad}\, D_{bc}}{D_{ac}\, D_{bd}}} \quad \text{H/m}$$

where for examples, D_{ad} denotes the distance in meters between conductors *a* and *d*.

[U.P. Technical Univ. Elements of Power System 2007-08]

4. Write short note on "inductive interference between power and communication lines."

PROBLEM

1. A 3-phase, 50 Hz, 100 km long 132 kV transmission line delivers a total load of 30 MW at 0.9 power factor lagging. The conductor arrangement is shown in the Fig. 1. The radius of each conductor is 6 mm.

Determine the induced voltage at fundamental frequency in the telephone circuit due to electromagnetic effect. Also determine the potential of telephone conductor P above the earth due to electrostatic effect.

[**Ans.** 28.67 V; 3,655 V]

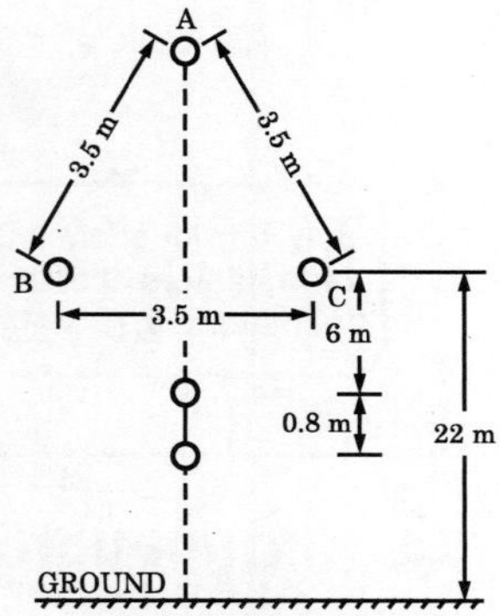

Fig. 1

□□□

9

CHAPTER

OVERHEAD LINE INSULATORS

INSIDE THIS CHAPTER

9.1. Introduction **9.2.** Insulator Materials **9.3.** Factors Involved in Insulator Design **9.4.** Types of Insulators **9.5.** Failure of Insulators **9.6.** Distribution of Potential Over a String of Suspension Insulators **9.7.** Methods of Improving String Efficiency **9.8.** Arcing Horn and Grading Ring.

9.1. INTRODUCTION

The overhead line conductors are bare and not covered with any insulating covering /coating. The line conductors are, therefore, secured to the supporting structures by means of insulating fixtures, called the *insulators*, in order that there is no current leakage to the earth through the supports. Insulators are mounted on the cross-arms and the line conductors are attached to the insulators so as to provide the conductors proper insulation and also provide necessary clearances between conductors and metal work. The insulators must provide proper insulation and necessary clearances against the highest voltage in worst atmospheric conditions to which the line is likely to be subjected. The insulators also prevent short-circuiting between the different phase conductors and provide necessary mechanical support for the line conductors. Thus insulator is undoubtedly one of the most important and vulnerable links in transmission and distribution practice and, therefore, proper selection is of utmost importance for the successful operation of overhead transmission and distribution system.

The important properties that an overhead line insulator must posses are:

1. *High mechanical strength* so as to bear the load due to the weight of line conductors, wind force and ice loading if any.
2. *High relative permittivity* so as to provide high dielectric strength.
3. *High insulation resistance* in order to prevent leakage of currents to earth.
4. *High ratio of rupture strength to flash-over voltage.*

5. *Ability to withstand large temperature variations i.e.*, it should not crack when subjected to high temperatures during summer and low temperatures during winter. The dielectric strength should remain unaffected under different conditions of temperature and pressure.

The material used should not be porous and should be imprevious to the fluids and gases in the atmosphere. Simultaneously it should be free from internal impurities and cracks etc as these lowers the value of dielectric strength to a great extent.

9.2. INSULATOR MATERIALS

The material most commonly used for overhead line insulators is porcelain but toughened glass, steatite and special composition materials are also used to a limited extent.

Porcelain is produced by firing at a controlled temperature a mixture of kaolin, feldspar and quartz. It is mechanically stronger than glass. It gives less trouble from leakage, and is less susceptible to temperature variations and its surface is not affected by dirt deposits. On the other hand, it is not so homogeneous as glass, owing to the fact that each component shell of a porcelain insulator is glazed during manufacturing process and its satisfactory performance in service depends to a considerable extent on the preservation of this glaze which is only of the order of 25 microns in thickness. Also fault cannot be detected easily as it is not transparent. In tension this material is usually weak and does not withstand tensile stresses exceeding 5 kg/mm^2. The dielectric strength and compressive strength of a mechanically sound porcelain insulator are about 6.5 kV/mm of its thickness and 700 kg/mm^2 respectively. Normally, it is difficult to manufacture homogeneous porcelain in the thickness required for some types of insulators and, therefore, for a particular operating voltage, a two or more piece construction is adopted in which each piece is fired and glazed separately and then they are cemented together.

If this insulating material is manufactured at lower temperature, its mechanical properties improve but the material remains porous and on putting it in service it may deteriorate. If this material is manufactured at high temperature, its porosity is reduced but the material becomes brittle. The porosity of this insulating material also reduces its dielectric strength, also any impurity or air bubble left within the material results in a lower dielectric strength. So a compromise is made between the mechanical strength and the porosity of the material and a suitable temperature of the kiln is designed.

Glass is cheaper than porcelain in the simpler shapes and if properly toughened and annealed gives high resistivity and dielectric strength (14 kV per mm of thickness of the material). Owing to high dielectric strength, the glass insulators have simpler design and even one piece design can be used. Glass is quite homogeneous material and can withstand higher compressive stresses as compared to porcelain. It has also a lower coefficient of thermal expansion which minimises the strains due to temperature changes and owing to its transparent nature flaws in the material can be readily detected by visual examination. The main disadvantage of glass is that moisture more readily condenses on its surface and facilitates the accumulation of dirt deposits, thus giving a high surface leakage. Also in large sizes the great mass of material combined with the irregular shape, may result in internal strains after cooling. Glass insulator however, can be used up to 25 kV under ordinary atmospheric conditions and well up to 50 kV in dry atmosphere.

Steatite is a naturally occurring magnesium silicate, usually found combined with oxides in varying proportions. It has a much higher tensile and bending stress than porcelain and can advantageously be used at tension towers or when the transmission line takes a sharp turn.

The **special artificial material** is used in insulators for low voltages and have an important advantage that these can be easily moulded into any shape without any internal stress. Metallic fittings can also be firmly embedded in the material without any difficulty. The disadvantage of insulators made from special artificial material is that they deteriorate rapidly in bad climatic conditions and on being subjected to flash-over their carbonised surface forms a conducting path.

9.3. FACTORS INVOLVED IN INSULATOR DESIGN

Insulators are required to withstand both electrical and mechanical stresses. In addition,the surface leakage path must have sufficiently high resistance so as to avoid any current leakage to earth. The design of the insulators must be such that the stress developed owing to contraction and expansion in any part of the insulator does not lead to any defect.

Electrical breakdown may occur either by flash-over or puncture. In a flash-over, an arc occurs between the line conductor and earth (*i.e.* supporting pin of the insulator) and the discharge jumps across the air-gaps in its path. In a puncture, the discharge occurs from conductor to pin through the body of the insulator. In case of electrical breakdown due to flash-over, the insulator continues to act in its proper capacity after the event unless fractured by the heat of the arc, but after a puncture, it is permanently damaged due to excessive heat. Thus it is very important to provide sufficient thickness of porcelain in the insulator to resist puncture by the combined effect of the line voltage and any probable transient voltage rise on the line. The ratio of puncture strength to flash-over voltage, termed as the *factor of safety*, must be high so as to provide a good margin for the protection of insulators from completefailure.

It is desirable that porcelain may not come in direct contact with a hard metal screw thread. Normally cement is used between metal and the porcelain. The cement so used must not cause any fracture by expansion or contraction.

9.4. TYPES OF INSULATORS

Various types of insulators used for overhead transmission and distribution lines are described below.

1. Pin Type Insulators. This type was amongst the earliest designs, and used for supporting line conductors. It provides the most economic, simple and efficient method of conductor and bus-bar support for voltages up to and including 33 kV. Modern pin type insulators are very reliable and inherent cracks in porcelain are very rare and never occur with toughened glass insulators. The life of modern porcelain insulators is relatively long (expected to be about 50 years). Pin type insulators are available for use up to 50 kV.

The pin type insulator is designed to be mounted on a pin which in turn is secured to the cross-arm of the pole. The insulator is screwed on the pin and the line conductor is placed in the groove at the top of the insulator and is tied down with soft copper or soft aluminium binding wire according to the conductor material. To avoid a direct contact between the porcelain and the metal pin, a soft metal (usually lead) thimble is employed. The insulator and its pin should be sufficiently mechanically stronger to withstand the resultant force due to combined effect of the weight of the conductor, wind pressure and ice loading, if any, per span length.

For lower voltages generally one piece type of insulators are used. For high-voltage transmission lines, stronger pin type insulators are used. The high-voltage pin type insulators differ in construction from low-voltage type in that they consist of two or three pieces of porcelain cemented together. These pieces form what we call *petticoats* or *rain sheds*. The multiple shells or petticoats are provided in order to have an adequate length of leakage path so that the flash-over voltage between the line conductor and the insulator pin is increased. The petticoats or rain sheds are so designed, that even when the outer surface is wet due to rain, sufficient leakage resistance is still provided by the inner dry surfaces. It is desirable that the horizontal distance between the tip of the lower most shell is less in comparison to the vertical distance between the same tip and the cross-arm, otherwise in case of an arc-over, the discharge will take place between the line conductor and cross-arm rather than line conductor and the pin of the insulator, thereby, the cross-arm will have to be replaced rather than the insulator.

Many engineers have a preference for one piece insulators up to and including 11 kV and use the multi-piece type for high voltages while others prefer multi-piece for all voltages. One-piece insulators are more prone to cracking, because of rain, and, therefore, their use may cause interruption of supply during the very first rain after the lines is commissioned. Multi-piece insulators have an advantage that a defect on one shed does not seriously affect the mechanical strength of the insulator, and it generally functions electrically at normal voltage until the defective unit is traced and replaced. For rural 11 kV lines, carrying less load and therefore, using light conductors, single piece insulators are employed for reasons of economy. For urban 11 kV feeders, specially feeding industrial loads, multi-piece insulators are preferred, being more reliable.

Two one-piece insulators are shown in Figs. 9.1 (*a*) and 9.1 (*b*). Leakage paths are shown by dotted lines in the Figs.

Two multi-part insulators for use on 33 kV and 66 kV are shown in Figs. 9.2 and 9.3 respectively. The flash-over distances are also shown when the insulators are wet and dry in Figs. 9.2 and 9.3 respectively.

Flash-over distances for dry insulators, as shown in Fig. 9.2,

$$= a + b + c + d$$

Flash-over distances for wet insulators, as shown in Fig. 9.3,

$$= a + b + c$$

Thus it is seen that flash-over distance in case of wet insulators is less as compared to that in case of dry insulators. The rain sheds are, therefore, made to keep the inner side of the insulator dry.

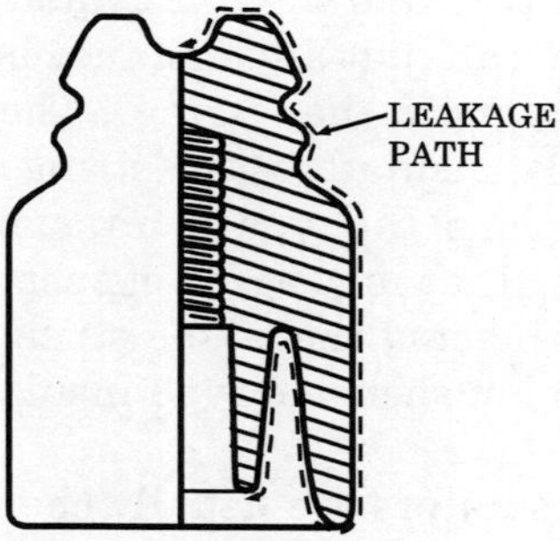

(*a*) *For 11,000 V*

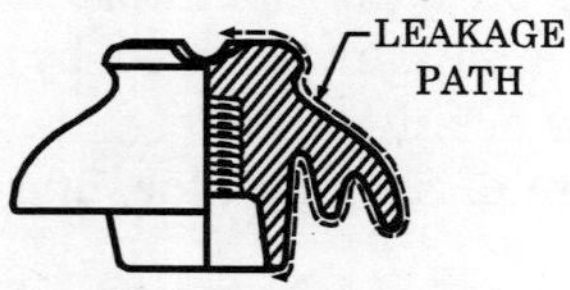

(*b*) *For 25,000 V*

Fig. 9.1. ***One-piece Pin Type Insulator***

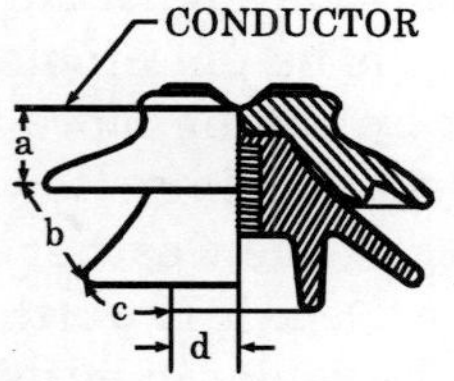

Fig. 9.2. ***33 kV Pin Type Insulator***

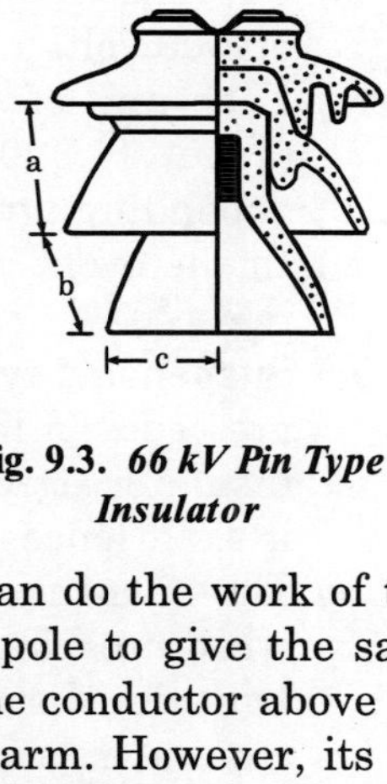

Fig. 9.3. ***66 kV Pin Type Insulator***

The main advantage of the pin type insulator is that it is cheaper. In many cases one pin insulator can do the work of two suspension insulators. Secondly the pin insulator requires a shorter pole to give the same conductor clearance above the ground since the pin insulator raises the conductor above the cross-arm while the suspension insulator suspends it below the cross-arm. However, its use beyond operating voltage of 80,000 volts become uneconomical. Pin type insulators become very bulky and cumbersome when designed for the higher voltages, and owing to the larger length of the supporting pin the bending moment near the point of attachment to the cross-arm tends to become excessive. The modern practice is not to use pin type insulators beyond 33 kV.

For these the ratio of the spark-over voltage to the working voltage (usually known as safety factor) must be high for low voltage than for high voltage. This is about 10, but these are so designed that spark-over takes place before these get punctured. Surface leakage current is due to the accumulation of dirt and in order to reduce its value the insulators are given a long leakage path by providing two or three petticoats or sheds. Insulators of this type are used on intermediate poles on straight run. These do not take tension.

2. Suspension Type Insulators. With the increase in operating voltage, the insulation required increases. Transmission lines use extremely high voltages, 400 kV, for example. At these voltages the pin type insulators become bulky, cumbersome and costly. Besides, the pin which must hold it would have to be inordinately long and large. In order to meet the problem of insulators for these high voltages, the suspension insulator was developed.

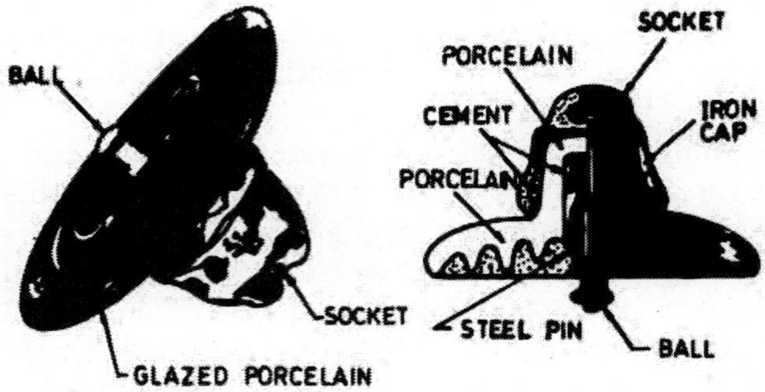

Fig. 9.4. Suspension Type Insulators

Suspension type insulators consist of a number of porcelain discs flexibly connected in series by metal links in the form of a string. The suspension insulator hangs from the cross-arm of the supporting structure and the line conductor is attached to its lower end. Because there is no pin problem, we can put any distance between the cross-arm and the conductor just by adding more insulators to the "string". The entire unit of suspension insulators in called a *string*. How many insulators the string consists of depends upon the working voltage, the weather conditions, the type of transmission construction, and the size of insulator used. It is worthnoting that in a string of suspension insulators one or more insulators can be replaced without replacing the whole string.

The main advantages of suspension type insulators over pin type ones are enumerated below :

(*i*) Suspension type insulators are usually cheaper in cost for operating voltages above 50,000 volts.

(*ii*) Each unit of suspension type insulators is designed for comparatively low voltage (about 11,000 volts) and can be used by connecting them in series, the number depending upon the working voltage.

(*iii*) In the event of failure of an insulator, one unit, instead of the whole string, can be replaced.

(*iv*) Suspension type insulators give more flexibility to the line and mechanical stresses are reduced in this arrangement. The connection at the cross-arm is such that the insulator string is free to swing in any direction, and thus takes up a position where it experiences only a pure tensile stress.

(*v*) The suspension type insulators, when used in conjunction with steel supporting structures, has the advantage of rendering the conductor less liable to be affected by lightning disturbances. At every point of support the wire is hung below the earthed cross-arm, thus enabling the tower to function as a lightning rod.

(*vi*) In case of rapid increase in load on transmission line, the increased demand can be met by raising the line voltage than to provide another set of conductors. With suspension type insulators, additional line insulation required can be obtained easily by adding one or more discs to the string.

(*vii*) In case of long spans (river or valley crossings) where heavy conductor load is to be sustained, two disc insulator strings can be yoked (Fig. 9.5). Such an arrangement is not possible with pin type insulators.

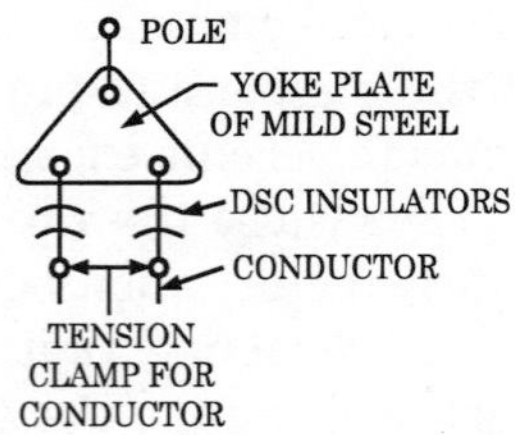

Fig. 9.5. ***Yoked Insulator Strings (Symbolic Figure)***

The disadvantage of suspension type insulators is that large spacings between conductors are required than with pin type insulators due to large amplitude of the swing of the conductors, but this is not a serious disadvantage.

There are three types of suspension insulators generally used namely, Hewlett or interlinking type, cemented-cap type and core and link type.

(*a*) Hewlett or Interlinking Type Suspension Insulators. The Hewlett suspension insulator is shown in Fig. 9.6 and is one of the earliest design. Each disc consists only of one piece of porcelain, the central bulbous portion of which is provided with two curved tunnels lying in planes at right angles to each other. The short steel strips forming the connection between individual discs are threaded through these tunnels and thus loop through each other, being separated by a layer of porcelain which is totally in compression.

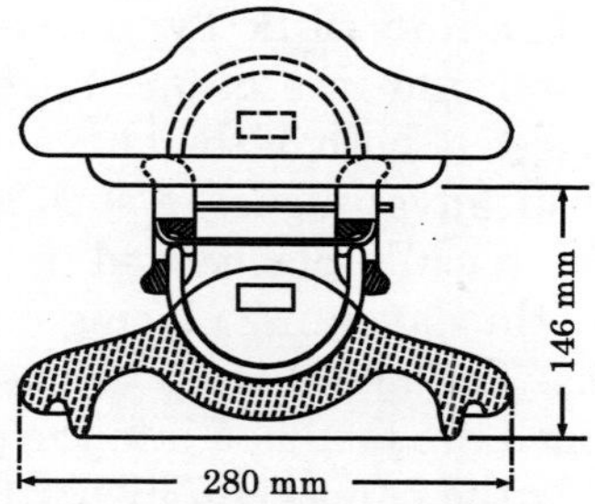

Fig. 9.6. ***Hewlett Type Suspension Insulator***

The main advantages of Hewlett type suspension insulators are :

(*i*) Simple in design.

(*ii*) High mechanical strength since the porcelain in between the two tunnels is under compression only.

(*iii*) No risk of breakage owing to the difference in expansion or contraction of the connecting links and the insulating material.

(*iv*) No risk of interruption to the service in case the porcelain between the links gets accidently broken, since the links keep the other units held together.

However, the Hewlett insulator appears to be rather more liable to puncture than other types of suspension insulators, owing to the high electrostatic stress in the material between the links.

(*b*) Cemented-Cap Type Suspension Insulators. Cemented-cap type suspension insulator shown in Fig. 9.7 is the most commonly used type and consists of a single disc-shaped piece of porcelain grooved on the under surface to increase the surface leakage path, and to a metal cap at the top, and to a metal pin underneath. The cap is recessed so as to take the pin of another unit and thus build up a string of any number of units. The cap is secured to the insulator by means of cement. A very uniform distribution of the electrostatic stress in the material between the connecting links is obtained in such insulators. The main drawback of this type of insulators has been that coefficients of cubical expansion of the three materials–porcelain, cement and steel are different and no provision is made for their expansion and the sudden temperature changes occurring in service are sufficient to set up internal stresses which ultimately crack the porcelain, leading to electrical failure. Further more, the cement itself, which is subject to volumetric changes depending on its moisture content, has often materially assisted in the process of failure of the insulator. The insulator manufacturers recognised these causes of ultimate failure of insulators and improved their designs from the point of view of reliability of service. One way of improving the design is by way of substituting the cementing of the

Fig. 9.7. ***Cemented Cap Type Suspension Insulator***

pin by purely mechanical fixing, such as the "spring-ring". In this method of fixing a spiral spring ring of steel wire carried on the stem of the pin is forced into the interior of the insulator head which is of bulbous shape. The ring immediately expands and is locked in position, the interior being then filled with a lead alloy to prevent any movement of the various parts and protect the fittings from the weather.

(c) Core and Link Type Suspension Insulators. Core and link type suspension insulator is illustrated in Fig. 9.8. Such an arrangement combines the advantages of both of the above two types of insulators and overcome their disadvantages. In such a construction, each insulator disc is symmetrically placed and it conforms to the electrostatic lines of force, thus avoiding placing materials of different permittivities in series. The metal work consists of pressed steel spiders, the legs of which are fastened into the porcelain by an alloy having approximately the same coefficient of cubical expansion as the porcelain. Thus high mechanical stresses on the porcelain, whether due to sudden temperature variations or to the employment of cement are completely eliminated. This is the recent type of construction and it allows discs to be formed out of quite thick porcelain thereby allowing the disc to be of one piece only. Added to the other advantages such an insulator has a high puncture strength.

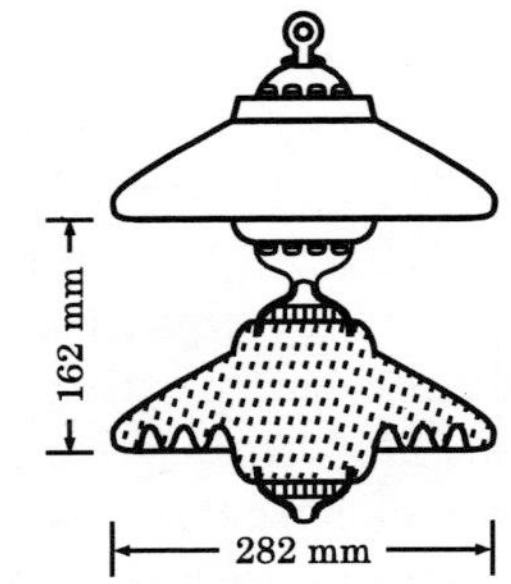

Fig. 9.8. *Core and Link Type Suspension Insulator*

3. Strain Insulators. Where there is a dead end of the line, or there is a corner or a sharp curve, or the line crosses river etc, the line is subjected to greater tension. Pin type insulators cannot be used in such situations because they cannot take conductor load in tension which often occurs in such situations. For low-voltage lines (say, up to 11,000 V), shackle insulators can be used, but for higher voltage transmission lines strain insulators consisting of an assembly of suspension type insulators are used, as illustrated in Figs. 9.9 and 9.10. Because of its peculiarly important job, a strain insulator must have considerable strength as well as necessary dielectric properties. Where the tension is exceedingly high, as at long river spans, two, three, or even four strings of insulators in parallel are used. The discs of strain insulators are employed in vertical plane where as the suspension insulators are used in horizontal plane.

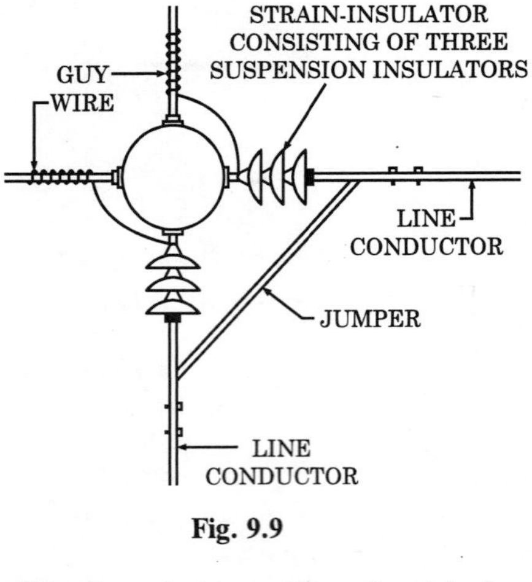

Fig. 9.9

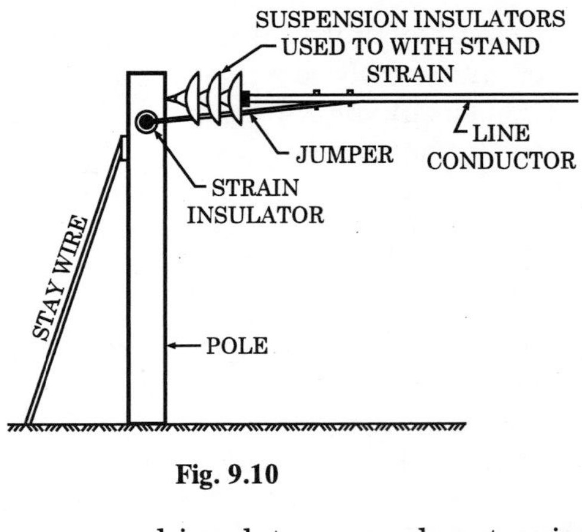

Fig. 9.10

4. Shackle Insulators. The shackle insulators or spool insulators are almost universally used on low voltage lines and provide a very neat, efficient and economical arrangement.

Fig. 9.11. *Shackle Insulator*

Its construction is shown is Fig. 9.11. Every insulator is coated with an extremely hard, smooth glaze, that reduces accumulation of surface deposits. The surface can be easily cleaned and it will not crack when subjected to temperature changes. The wet flash-over and dry flash-over voltages for shackle insulators are 10 kV and 25 kV respectively while the puncture voltage is about 35 kV. Its operating voltage is 1,000 V. Its weight, transverse mechanical load and total creepage distance are 0.5 kg, 1,150 kg and 63 mm respectively. The tapered hole in the shackle insulator distributes the load more evenly and reduces the possibility of breakage when heavily loaded.

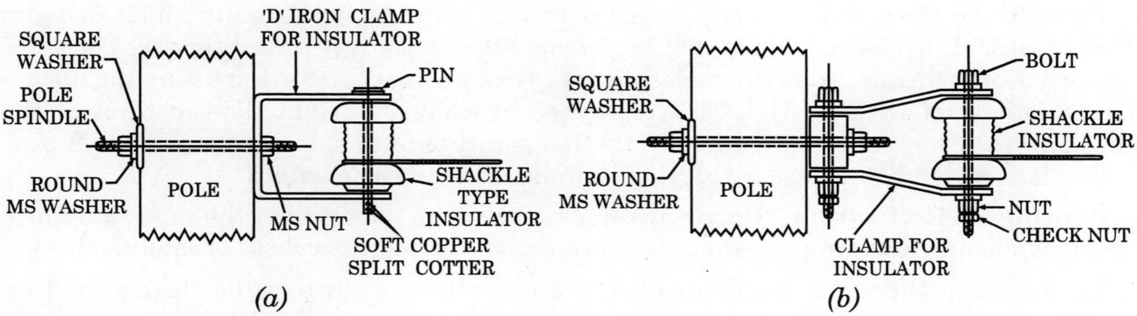

Fig. 9.12. *Shackle type Insulator Fitted To Poles*

Shackle insulators may either be mounted horizontally or vertically, and the conductors are fixed in the grooves by means of soft copper or aluminium binding wire according to the conductor material. They can be directly fixed to the pole with a bolt or to the cross-arm. The insulators are bell-mounted to prevent water being held in contact with the spindle. This type of insulator is used at all positions, either intermediate, terminal or angle. Where the angle exceeds 60° deviation they are generally used in conjunction with shackle straps.

5. Stay Insulators. Such insulators are of egg shape, also called *strain* or *guy* insulators, and are used in guy cables, where it is necessary to insulate the lower part of the guy cable from the pole for the safety of people and animals on the ground. This type of insulator consists of a porcelain piece pierced with two holes at right angles to each other through which two ends of the guy wires are looped. This is illustrated in Fig. 9.14. This keeps the porcelain between the two loops in compression and the guy wire remains in position even if the insulator breaks due to any reason. These insulators are provided at a height of about 3 m from the ground level. The size of insulator (small or large) used depends upon the tensile strength of stay wire.

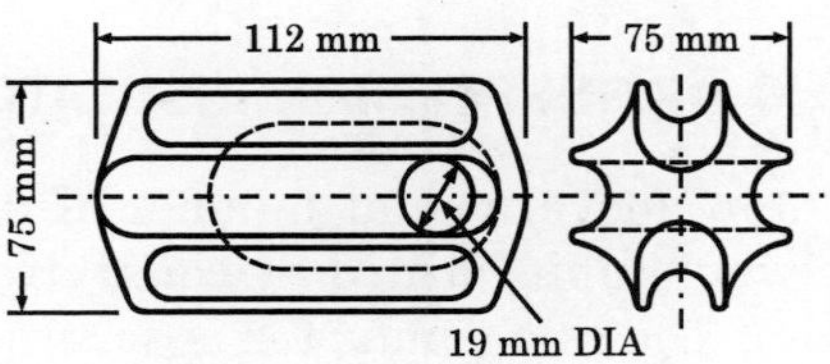

Fig. 9.13. *Stay Wire Insulator*

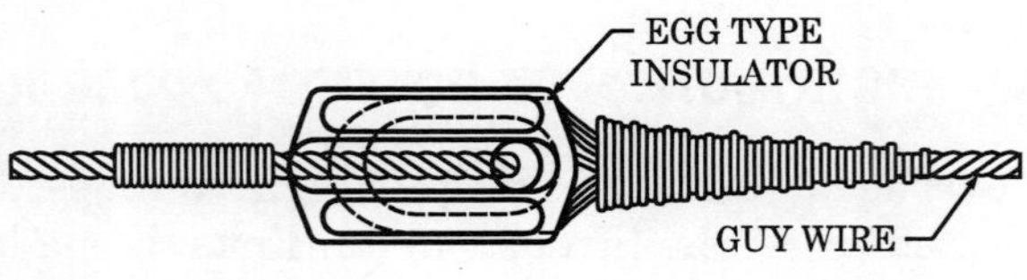

Fig. 9.14

9.5. FAILURE OF INSULATORS

Failure of insulators, resulting in a interruption in the continuity of power supply, is a matter of serious concern to the power engineers. It may occur due to cracking of porcelain, porosity, puncture, mechanical stresses, flash-over etc.

1. Cracking of Insulators. This failure is very common in case of pin type insulators and the cemented-cap type suspension insulators. This occurs due to unequal expansion of steel, porcelain and cement during the varying conditions of cold and heat and dryness and dampness. This develops high stresses in the porcelain near the joint and it results in tension failure. This can be avoided to some extent by using elastic cushions between the shells.

2. Porosity of Material. Porosity in the porcelain which may be due to under-firing or other causes, always leads to failure after a comparatively short period of service. The pores usually absorb moisture from the atmosphere or the cement, thereby decreasing the insulation resistivity of the material. This gives rise to leakage current flowing through the porcelain, resulting in a gradual rise in the temperature until the porcelain is punctured. Such a failure can be avoided by glazing the insulator, to some extent.

3. Improper Vitrification. This is another cause of the puncture of the material and it can be avoided by carrying out suitable routine tests during the course of manufacture.

4. Flash-Over. The most common cause of the insulator failure is the flash-over, that causes unequal expansion of the porcelain thereby shattering the insulator with big cracks and causing interruption of the supply. This can be avoided by providing arcing horns or rings which take up the arc and divert it away from the insulators.

5. Mechanical Stresses. Although the compressive strength of the porcelain is quite high, yet its tensile strength is not adequate and the insulator is always weak in tension and usually fails in that fashion. Such a failure is very rare because defective pieces are weeded out in the routine factory tests.

6. Short-Circuits. In the case of pin insulators, the birdage is very common. Birdage means the short-circuiting of the conductor to earth through the large birds or similar objects. Such a situation can be avoided by providing bird gaurds near the insulator on the cross-arm, by increasing the clearance of the conductor from earthed parts or using suspension insulators, instead of pin type, where the clearance between the earthed structure and the conductor is very large to warrant any birdage.

7. Deposition of Dust. If the insulator material is not properly glazed, the water will stick over it resulting into deposition of dust etc. over it which is partially conducting and reduces the flash-over distance. The deposits of dust and like matter *e.g.* salt, cement, dust etc. on the interior surfaces can actually cause much havoc under the condition of fog and mist. This can be easily avoided by cleaning the insulators periodically.

9.6. DISTRIBUTION OF POTENTIAL OVER A STRING OF SUSPENSION INSULATORS

For overhead lines operating at high voltages (33 kV and above) use of number of discs connected in series, through metal links, is made. The whole unit formed by connecting a number of discs in series is known as *string of insulators.* The line conductor is secured to the bottom disc of the string and the top disc is connected to the cross-arm of the pole or

---

* The insulator should flash-over at such voltages at which other equipment is likely to get damaged.

tower, as illustrated in Fig. 9.15. The number of discs connected in series in an insulator string depends upon the line operating voltage (higher the line operating voltage, the larger is the number of discs required for the insulator string, as given below in tabular form).

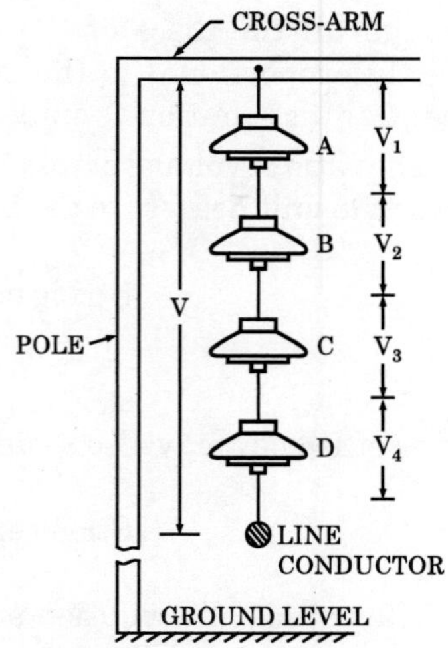

Fig. 9.15. *Arrangement of Insulator String*

TABLE 9.1. *Minimum Number of Insulator Discs Required For Transmission Lines*

Operating Voltage in kV	Number of Discs in	
	Suspension Assembly	**Tension or dead-end Assembly**
11	1	1
33	2	3
66	5	6
132	9	10
220	14	15
400	21	22

The number of discs indicated in above Table 9.1 is actually the usual number used. However, in the case of transmission lines operating at 66 kV or more, one disc less than the number indicated in Table 9.1 is used on about eight suspension structures near the substation. This is accomplished so that in the event of a lightning surge appearing on the line, the insulator string will flash-over and prevent the surge from travelling to the substation thus safe-guarding the equipment there.

It is found that the voltage impressed on a string of suspension insulators (the voltage applied between the line conductor and earth) does not distribute itself uniformly across the individual discs. The line unit (unit nearest the line conductor) has the maximum value across it, the figure progressively decreasing as the unit nearest the cross-arm is approached. The inequality of voltage distribution between individual units is all the more pronounced with a larger number of insulator units. This fact may be explained with the help of equivalent circuit of an insulator string (Fig. 9.16)

Each string insulator unit behaves like a capacitor having a dielectric medium between the two metallic parts (viz pin and cap), The capacitance due to two metal fittings on either side of an insulator is known as *mutual capacitance.* Further there is also a capacitance between metal fitting of each unit and the earthed pole or tower. The capacitance so formed is known as *shunt capacitance.* If a string of similar suspension insulators could be situated so far from neighbouring metal work that the capacitance between this metal work and the metal fitting of the insulators (*i.e.* shunt capacitance) would be negligibly small in comparison with the capacitance of each unit (*i.e.* mutual capacitance), then the charging current would have been the same through all of the discs, the discs being connected in series, and consequently the voltage across individual units would have been the same *ie.* applied voltage V divided by the number of units in the string. However, in practice this condition is not fulfilled because of nearness of the tower, the cross-arm, and the line. These shunt capacitances, sometimes called the *stray capacitances,* have an important effect on the voltage distribution between the units. Due to shunt capacitance, charging current is not the same through all the discs of the string (Fig. 9.16.). So voltage across individual units, being directly proportional to the current flowing through them, will be different. This unequal potential distribution is undesirable and is usually expressed in terms of string efficiency.

The ratio of voltage across the whole string and the product of the number of units and voltage across the unit nearest to the line conductor is known as the string efficiency *i.e.*

$$\text{String efficiency} = \frac{\text{Voltage across the string}}{n \times \text{voltage across the lower most unit}} \quad ...(9.1)$$

where n is the number of units in the string.

String efficiency may also be defined as

$$\text{String efficiency} = \frac{\text{Flash - over voltage of the string}}{n \times \text{flash - over voltage of one unit}}$$

The voltage distribution across different units of an insulator string and string efficiency can be mathematically determined with the help of an equivalent circuit of the insulator string (Fig. 9.16) as below. Fig 9.16 shows the equivalent circuit of a string of suspension insulators containing 4 units.

Let the mutual capacitance between the links be C and shunt capacitance between links and earth be C_1, voltage across the first unit (nearest the cross-arm) be V_1, voltage across the second unit be V_2, voltage across the third unit be V_3, voltage across the fourth unit (nearest the line conductor) be V_4 and voltage between conductor and earth be V volts

Let $\frac{C_1}{C} = K$ or $C_1 = KC$

Applying Kirchhoff's first law to node A, we get

$$I_2 = I_1 + i_1$$

or $\omega C V_2 = \omega C V_1 + \omega C_1 V_1$

or $\omega C V_2 = \omega C V_1 + \omega K C V_1$

or $V_2 = V_1 (1 + K)$ ...(9.2)

Applying Kirchhoff's first law to node B we get

$$I_3 = I_2 + i_2$$

or $\omega C V_3 = \omega C V_2 + \omega C_1 (V_1 + V_2)$

$\because$ Voltage across the second shunt capacitance C_1 from the top $= V_1 + V_2$

or $\omega C V_3 = \omega C V_2 + \omega K C (V_1 + V_2)$

or $V_3 = V_2 + K (V_1 + V_2) = K V_1 + V_2 (1 + K)$

or $V_3 = K V_1 + V_1 (1 + K)(1 + K)$

$\because$ From Eq. (9.2) $V_2 = V_1 (1 + K)$

or $V_3 = V_1 (1 + 3K + K^2)$ ...(9.3)

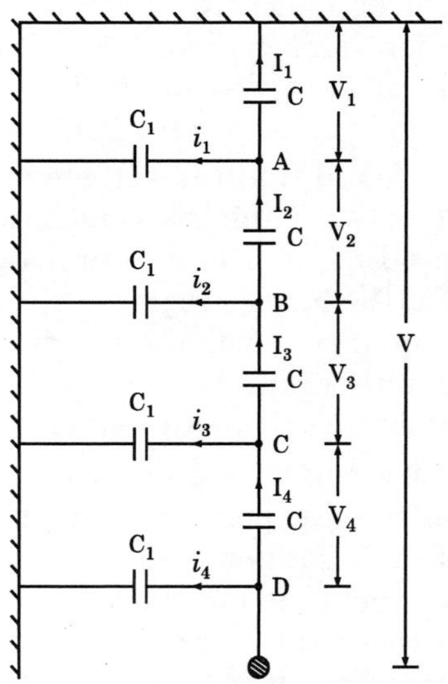

Fig. 9.16. *Equivalent Circuit For a 4-Unit String*

Applying Kirchhoff's first law to node C, we get

$$I_4 = I_3 + i_3$$

or $\omega C V_4 = \omega C V_3 + \omega C_1 (V_1 + V_2 + V_3)$

$\because$ Voltage across the third shunt capacitance C_1 from the top $= V_1 + V_2 + V_3$

or $\omega C V_4 = \omega C V_1 (1 + 3K + K^2) + \omega K C [V_1 + V_1 (1 + K) + V_1 (1 + 3K + K^2)]$

or $V_4 = V_1 (1 + 6K + 5K^2 + K^3)$ ...(9.4)

Finally voltage between line conductor and earth

$$\begin{aligned} V &= V_1 + V_2 + V_3 + V_4 \\ &= V_1 + V_1 (1 + K) + V_1 (1 + 3K + K^2) + V_1 (1 + 6K + 5K^2 + K^3) \\ &= V_1 (4 + 10K + 6K^2 + K^3) \end{aligned} \qquad ...(9.5)$$

$$\text{Thus} \quad \frac{V_1}{1} = \frac{V_2}{1+K} + \frac{V_3}{1+3K+K^2} = \frac{V_4}{1+6K+5K^2+K^3} = \frac{V}{4+10K+6K^2+K^3} \qquad ...(9.6)$$

The greatest voltage will be obviously V_4 which is given as

$$V_4 = \frac{1+6K+5K^2+K^3}{4+10K+6K^2+K^3} V \qquad ...(9.7)$$

$$\text{and Percentage string efficiency} = \frac{V}{n V_4} \times 100 = \frac{V}{4 \times \dfrac{1+6K+5K^2+K^3}{4+10K+6K^2+K^3}} \times 100$$

Since $n = 4$, and flash-over voltage of one unit = Greatest voltage across any unit *i.e.* V_4

$$\text{or Percentage string efficiency} = \frac{4+10K+6K^2+K^3}{4(1+6K+5K^2+K^3)} \times 100 \qquad ...(9.8)$$

Similarly derivation can be had for a string of insulators consisting of any number of units.

When the number of insulators in the string is large it becomes laborious to work out the voltage distribution across each unit, for such cases standard formula may be used.

In general case if there are n units in the string, V is the maximum voltage across the string, V_1, V_2, V_3 .....V_n denote the voltages across the insulator units starting from top, C is the capacitance between the links and KC be the shunt capacitance between the links and earth, the voltage distribution across the mth unit (counted from top) is given as

$$V_m = V \frac{2 \sinh\left(\frac{1}{2}\sqrt{K}\right) \cosh\left[\left(m - \frac{1}{2}\right)\sqrt{K}\right]}{\sinh\left(n\sqrt{K}\right)} \qquad ...(9.9)$$

and potential adjacent to the line conductor

$$V_n = V \frac{2 \sinh\left(\frac{1}{2}\sqrt{K}\right) \cosh\left[\left(n - \frac{1}{2}\right)\sqrt{K}\right]}{\sinh\left(n\sqrt{K}\right)} \qquad ...(9.10)$$

Graphical plot of how voltages are distributed across the units of an insulator string is shown in Fig. 9.17.

The following points may be noted.

1. The unit nearest to the line conductor is under maximum electrical stress and is likely to be punctured while the one nearest to the cross-arm is under minimum electrical stress.
2. The voltage distribution across various units depends upon the value of K and number of discs contained in the string. The greater the value of K, the more non-uniform is the voltage distribution across the discs and lesser is string efficiency. The inequality in voltage distribution increases with the number of discs in the string. Thus a shorter string has higher efficiency than the longer one.

When the insulators are wet the value of mutual capacitance C increases while C_1 remains constant (except for the unit nearest the cross-arm) so the value of K decreases, more uniform potential distribution is obtained and the string efficiency increases.

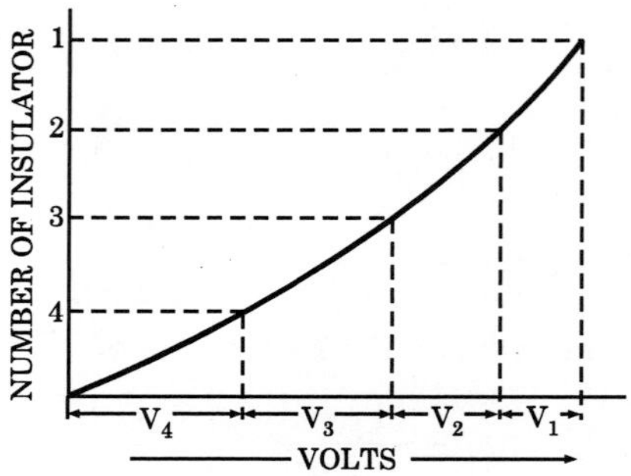

Fig. 9.17. ***Voltage Distribution Across Suspension Insulator String***

The value of K (the ratio of shunt capacitance C_1 to mutual capacitance C) varies and depends upon the length of the insulator string. The larger the number of insulator discs in a string, the longer will be the string. The longer the string, the greater must be the horizontal spacing between the insulator disc and the support (pole or tower) to make an allowance for conductor swing. The greater the horizontal spacing between the insulator string and the support, the lesser is the shunt capacitance C_1 and vice-versa. Thus the value of K is low for longer strings and high for shorter strings. In practice K varies from0.1 to 0.1667.

Example 9.1. In a string insulator having 5 units, the self capacitance of insulator units is 6 times to that between a unit and earth. Obtain the potential distribution across each unit in the string as percentage of potential of conductor with respect to earth. Also find the string efficiency.

[U.P. Technical Univ. Elements of Power System 2009-10]

Solution : The equivalent circuit of a 5-unit insulator string is shown in Fig. 9.18. Ratio of shunt capacitance to mutual capacitance, $K = \frac{C_1}{C} = \frac{1}{6}$

Let the voltage of conductor to earth be V volts

From Eqs. (9.2), (9.3), and (9.4) respectively, we have

$$V_2 = V_1 (1 + K) = V_1 \left(1 + \frac{1}{6}\right) = \frac{7}{6} V_1 = 1.16667 V_1 \quad ...(i)$$

$$V_3 = V_1 (1 + 3K + K^2) = V_1 \left(1 + \frac{3}{6} + \frac{1}{6^2}\right) = 1.5278 V_1 \quad ...(ii)$$

$$V_4 = V_1 (1 + 6K + 5K^2 + K^3) = 2.1435 V_1 \quad ...(iii)$$

If we apply Kirchhoff's first law to node D, we get

$$I_5 = I_4 + i_4$$

or $$\omega C V_5 = \omega C V_4 + \omega C_1 (V_1 + V_2 + V_3 + V_4)$$

$\because$ voltage across the fourth capacitance from the top $= V_1 + V_2 + V_3 + V_4$

or $$\omega C V_5 = \omega C V_1 (1 + 6K + 5K^2 + K^3) + \omega KC [V_1 + V_1 (1 + K) + V_1 (1 + 3K + K^2) + V_1 (1 + 6K + 5K^2 + K^3)]$$

or $$V_5 = V_1 (1 + 10K + 15K^2 + 7K^3 + K^4) = 3.1165 V_1 \quad ...(iv)$$

and voltage between line conductor and earth,

$$V = V_1 + V_2 + V_3 + V_4 + V_5$$
$$= V_1 + V_1 (1 + K) + V_1 (1 + 3K + K^2) + V_1 (1 + 6K + 5K^2 + K^3) + (1 + 10K + 15K^2 + 7K^3 + K^4]$$
$$= [5 + 20K + 21K^2 + 8K^3 + K^4] V_1 \quad ...(v)$$

So voltage across the first unit (nearest the cross-arm),

$$V_1 = \frac{V}{5 + 20\,K + 21\,K^2 + 8\,K^3 + K^4}$$

$$= \frac{V}{5 + 20 \times \frac{1}{6} + 21 \times \left(\frac{1}{6}\right)^2 + 8\left(\frac{1}{6}\right)^3 + \left(\frac{1}{6}\right)^4}$$

$$= \frac{V}{5 + 3.33 + 0.583 + 0.037 + 0.0008}$$

$$= \frac{V}{8.954} \text{ or } 11.168\% \text{ of V } \textbf{Ans.}$$

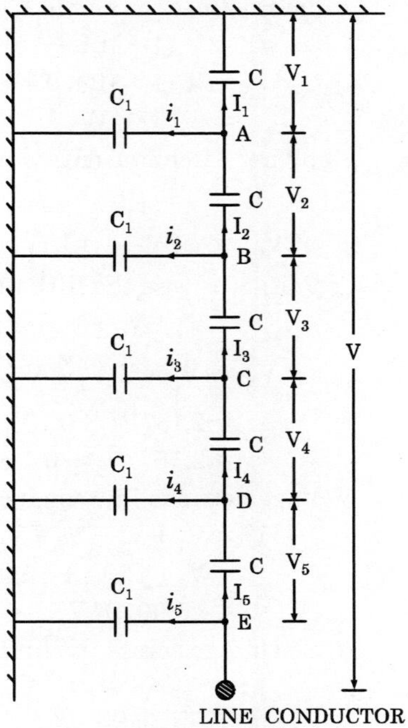

Fig. 9.18

Voltage across the second unit from top,

$$V_2 = 1.16667\ V_1 = 1.16667 \times 11.168\% \text{ of V} = 13.029\% \text{ of V } \textbf{Ans.}$$

Voltage across the third unit from top,

$$V_3 = 1.5278\ V_1 = 1.5278 \times 11.168\% \text{ of V} = 17.062\ \% \text{ of V } \textbf{Ans.}$$

Voltage across the fourth unit from top,

$$V_4 = 2.1435\ V_1 = 2.1435 \times 11.168\% \text{ of V} = 23.938\% \text{ of V } \textbf{Ans.}$$

Voltage across the bottom most unit,

$$V_5 = 3.1165\ V_1 = 3.1165 \times 11.168\% \text{ of V} = 34.804\% \text{ of V } \textbf{Ans.}$$

$$\text{String efficiency} = \frac{V}{nV_n} \times 100 = \frac{V}{\frac{5 \times 34.804\ V}{100}} \times 100 = 57.46\ \% \textbf{ Ans.}$$

String efficiency will be improved in rain due to decrease in value of K.

Example 9.2. Define string efficiency :

A string of 6 insulator units has a self-capacitance equal to 10 times the pin-to-earth capacitance. Find (*i*) the voltage distribution across various units as a percentage of total voltage across the string, (*ii*) the string efficiency. [*U.P. Technical Univ. Elements of Power System 2005-06*]

Solution : The equivalent circuit of a 6-unit insulator string is shown in Fig. 9.19.

Ratio of shunt capacitance to mutual capacitance,

$$K = \frac{C_1}{C} = \frac{1}{10} = 0.1$$

Let the voltage of conductor to earth be V volts.

From Eqs. (9.2), (9.3) and (9.4) respectively, we have

$$V_2 = V_1(1 + K) = V_1(1 + 0.1) = 1.1V_1 \qquad ...(i)$$

$$V_3 = V_1(1 + 3K + K^2) = V_1(1 + 3 \times 0.1 + 0.1^2) = 1.31\ V_1 \qquad ...(ii)$$

$$V_4 = V_1(1 + 6K + 5K^2 + K^3) = V_1(1 + 6 \times 0.1 + 5 \times 0.1^2 + 0.1^3) \qquad ...(iii)$$

$$= 1.651\ V_1$$

From Eq. (*iv*) of Example 9.1, we have

$$V_5 = V_1(1 + 10K + 15K^2 + 7K^3 + K^4)$$
$$= V_1(1 + 10 \times 0.1 + 15 \times 0.1^2 + 7 \times 0.1^3 + 0.1^4)$$
$$= 2.1571\ V_1 \quad ...(iv)$$

Applying Kirchhoff's first law to node E, we have

$$I_6 = I_5 + i_5$$

or $\omega CV_6 = \omega CV_5 + \omega C_1(V_1 + V_2 + V_3 + V_4 + V_5)$

$\because$ Voltage across the fifth capacitance from the top

$$= V_1 + V_2 + V_3 + V_4 + V_5$$

or $V_6 = V_5 + K(V_1 + V_2 + V_3 + V_4 + V_5)$

$$= 2.1571V_1 + 0.1(V_1 + 1.1V_1 + 1.31V_1 + 1.651V_1 + 2.1571V_1)$$
$$= 2.1571V_1 + 0.1 \times 7.2181V_1 = 2.87891V_1$$

Voltage between line conductor and earth

$$V = V_1 + V_2 + V_3 + V_4 + V_5 + V_6$$
$$= V_1 + 1.1V_1 + 1.31V_1 + 1.651V_1 + 2.1571V_1 + 2.87891V_1$$
$$= 10.09701V_1$$

(*i*) So voltage across the first unit (nearest the cross-arm)

$$V_1 = \frac{V}{10.09701} \text{ or } 9.904\% \text{ of V} \quad \textbf{Ans.}$$

Voltage across 2nd unit from top,

$$V_2 = \frac{1.1V}{10.09701} \text{ or } 10.894\% \text{ of V} \quad \textbf{Ans.}$$

Voltage across 3rd unit from top,

$$V_3 = \frac{1.31\ V}{10.09701} \text{ or } 12.974\% \text{ of V} \quad \textbf{Ans.}$$

Voltage across 4th unit from top,

$$V_4 = 1.651 \times 9.904\% = 16.351\% \text{ of V} \quad \textbf{Ans.}$$

Voltage across 5th unit from top

$$V_5 = 2.1571 \times 9.904\% = 21.364\% \text{ of V} \quad \textbf{Ans.}$$

Voltate across 6th unit from top

$$V_6 = 2.87891 \times 9.904\% = 28.513\% \text{ of V} \quad \textbf{Ans.}$$

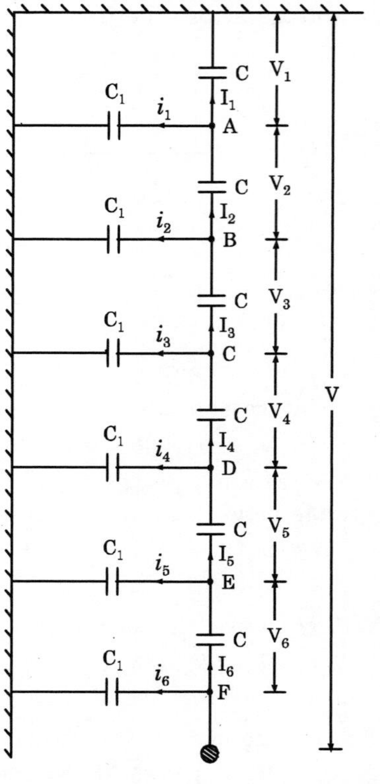

Fig. 9.19

(*ii*) String efficiency, $\eta = \dfrac{V}{nV_n} \times 100 = \dfrac{V}{6 \times 0.28513\ V} \times 100 = 58.45\%$ **Ans.**

Example 9.3. (*a*) Each conductor at 66 kV is suspended by a string of 5 similar insulators, the capacitance of each disc being 5 times the capacitor to earth. Determine the % string efficiency.

If in (*a*) the distance between the insulators and tower structure is increased so as to make capacitance of disc 6 times the earth capacitance, will the string efficiency improve and how much ?

Solution : Voltage between line conductor and earth, V = 66 kV

Number of units in insulator string, $n = 5$

(*a*) Shunt capacitance to mutual capacitance ratio, $K = \dfrac{1}{5} = 0.2$

Voltage across the top unit, $V_1 = \dfrac{V}{5 + 20\,K + 21\,K^2 + 8\,K^3 + K^4}$ refer to Eq. (*v*), example 9.1

$$= \frac{66\text{ kV}}{5 + 20 \times 0.2 + 21 \times (0.2)^2 + 8 \times (0.2)^3 + (0.2)^4} = \frac{66}{9.9056}\text{ kV} = 6.66\text{ kV}$$

Voltage across the bottom most or line unit,

$V_5 = V_1\,(1 + 10\,K + 15\,K^2 + 7\,K^3 + K^4)$ refer to Eq. (*iv*) example 9.1

$= V_1\,(1 + 10 \times 0.2 + 15 \times (0.2)^2 + 7\,(0.2)^3 + (0.2)^4]$

$= 3.6576\,V_1 = 3.6576 \times 6.66 = 24.37$ kV

Percentage string efficiency $= \dfrac{V}{n\,V_n} \times 100 = \dfrac{66}{5 \times 24.37} \times 100 = 54.16\%$ **Ans.**

(*b*) Shunt capacitance to mutual capacitance ratio, $K = \dfrac{1}{6}$

Voltage across the top unit, $V_1 = \dfrac{V}{8.954} = \dfrac{66}{8.954} = 7.37$ kV refer to example 9.1

Voltage across the bottom unit, $V_5 = 3.1165\,V_1 = 3.1165 \times 7.37 = 22.97$ kV refer to example 9.1

Percentage string efficiency $= \dfrac{V}{n\,V_n} \times 100 = \dfrac{66}{5 \times 22.97} \times 100 = 57.46\%$

i.e. string efficiency is improved and becomes 57.46% **Ans.**

Example 9.4. In a string of three identical suspension insulator units supporting a transmission line conductor, if the self capacitance of each unit is denoted as C farads, the capacitance of each connector pin to ground can be taken as 0.1 C farads. Determine the voltage distribution across the string if the maximum permissible voltage per unit is given as 20 kV. Also determine the string efficiency.

Solution : Number of units, $n = 3$

Ratio of shunt capacitance to mutual capacitance,

$$K = \frac{0.1\,C}{C} = 0.1$$

Voltage across bottom most unit, V_3 = Safe working voltage of the unit = 20 kV

So, voltage across top most unit, $V_1 = \dfrac{V_3}{1 + 3\,K + K^2} = \dfrac{20}{1 + 0.3 + 0.01} = 15.267$ kV **Ans.**

Voltage across middle unit, $V_2 = V_1\,(1 + K) = 15.267 \times 1.1 = 16.794$ kV **Ans.**

Maximum safe working voltage of the string,

$$V = V_1 + V_2 + V_3 = 15.267 + 16.794 + 20 = 52\text{ kV}$$

String efficiency $= \dfrac{V}{n\,V_n} \times 100 = \dfrac{52}{3 \times 20} \times 100 = 86.667\%$ **Ans.**

Example 9.5. Determine the maximum voltage that the string of suspension insulators in the figure can withstand if the maximum voltage per unit is 17.5 kV.

[*U.P. Technical Univ. Elements of Power System 2008-09*]

Solution : Number of units in insulator string, $n = 3$

Ratio of shunt capacitance to mutual capacitance, $K = \dfrac{C'}{C} = \dfrac{1}{8} = 0.125$

Voltage across bottom most unit, V_3 = Safe working voltage of the unit = 17.5 kV

$\because$ From Eq. (9.3), $V_3 = V_1\,(1 + 3K + K^2)$,

So, voltage across top most unit, $V_1 = \dfrac{V_3}{1 + 3K + K^2}$

$$= \frac{17.5}{1 + 3 \times 0.125 + 0.125^2} = 12.584 \text{ kV}$$

Voltage across mid unit, $V_2 = V_1(1 + K) = 12.584 \times 1.125$

$$= 14.157 \text{ kV}$$

Maximum safe working voltage, $V = V_1 + V_2 + V_3$

$$= 12.584 + 14.157 + 17.5$$

$$= 44.24 \text{ kV}$$ **Ans.**

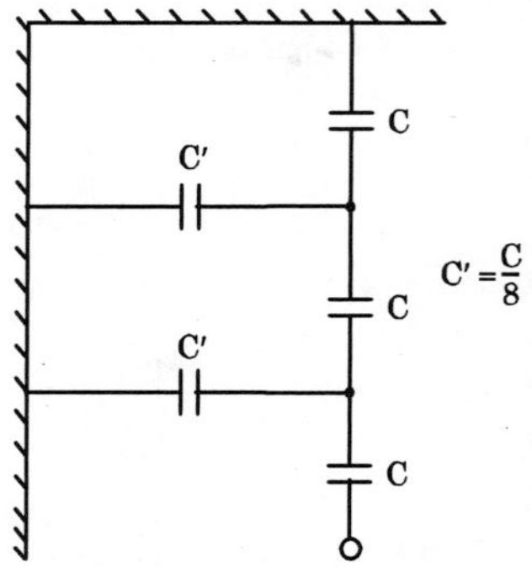

Fig. 9.20

Example 9.6. A string insulator has 4 units and each unit is rated for 12 kV. Find the maximum line voltage on which it can be operated safely. The mutual capacitance of unit is 10 times the capacitance between pin to earth. Find the string efficiency.

[U.P. Technical Univ. Elements of Power System 2007-08]

Solution : Number of units, $n = 4$

Ratio of shunt capacitance to mutual capacitance,

$$K = \frac{C_1}{10C_1} = 0.1$$

Voltage across bottom most unit, V_4 = Safe working voltage of the unit = 12 kV

$$\text{Voltage across top most unit, } V_1 = \frac{V_4}{1 + 6K + 5K^2 + K^3} = \frac{12}{1 + 0.6 + 0.05 + 0.001} = 7.27 \text{ kV}$$

Refer to Eq. (9.4)

Voltage across second unit from top, $V_2 = (1 + K)\, V_1 = 1.1 \times 7.27 = 7.99$ kV

Voltage across third unit from top, $V_3 = (1 + 3K + K^2)\, V_1 = (1 + 3 \times 0.1 + 0.01)V_1$

$$= 1.31 \times 7.27 = 9.52 \text{ kV}$$

Maximum safe working voltage for the string

$$V = V_1 + V_2 + V_3 + V_4 = 7.27 + 7.99 + 9.52 + 12$$

$$= 36.78 \text{ kV}$$ **Ans.**

$$\text{String efficiency, } \eta = \frac{V}{nV_n} \times 100 = \frac{36.78}{4 \times 12} \times 100 = 76.625\%$$ **Ans.**

Example 9.7. A string insulator has five units each rated for 11 kV. Find the maximum line voltage on which it can be operated safety. The mutual capacitance of unit is 10 times the capacitance between pin-to-earth. *[U.P. Technical Univ. Elements of Power System 2004-05]*

Solution : Number of units, $n = 5$

Ratio of shunt capacitance to mutual capacitance

$$K = \frac{C_1}{10C_1} = 0.1$$

Voltage across bottom most unit, V_5 = Safe working voltage of the unit = 11 kV

$$\text{Voltage across top most unit, } V_1 = \frac{V_5}{1 + 10K + 15K^2 + 7K^3 + K^4}$$ Refer to Eq. (*iv*) of Example 9.1

$$= \frac{11}{2.1571} = 5.1 \text{ kV}$$ Refer to Eq. (*iv*) of Example 9.2

Voltage across second unit from top, $V_2 = (1 + K)V_1 = 1.1 \times 5.1 = 5.6$ kV

Voltage across third unit from top, $V_3 = (1 + 3K + K^2)V_1 = 1.31 \times 5.1 = 6.7$ kV

Voltage across 4th unit from top, $V_4 = (1 + 6K + 5K^2 + K^3)V_1 = 1.651 \times 5.1 = 8.4$ kV

Maximum safe working voltage for the string

$$V = V_1 + V_2 + V_3 + V_4 + V_5$$
$$= 5.1 + 5.6 + 6.7 + 8.4 + 11 = 36.8 \text{ kV} \quad \textbf{Ans.}$$

Example 9.8. Each conductor of a 3-phase overhead transmission line is suspended from a cross-arm of a steel tower by a string of four suspension insulators. The voltage across the second unit is 15.0 kV and across the third 27.0 kV. Find the voltage between conductors and string efficiency.

Solution : Number of units, $n = 4$

Voltage across the second unit, $V_2 = 15.0$ kV

Voltage across the third unit, $V_3 = 21.0$ kV

From Eqs. (9.2) and (9.3)

$$\frac{V_3}{V_2} = \frac{(1 + 3K + K^2)\,V_1}{(1 + K)\,V_1}$$

$$\text{or } \frac{21.0}{15.0} = \frac{1 + 3K + K^2}{1 + K}$$

$$\text{or } 5K^2 + 8K - 2 = 0$$

$$\text{or } K = \frac{-8 \pm \sqrt{8^2 + 4 \times 2 \times 5}}{2 \times 5} = -0.8 \pm 1.0198$$

Taking positive value of K we have K = 0.2198

$$\text{Voltage across top most unit, } V_1 = \frac{V_2}{1 + K} = \frac{15}{1.2198} = 12.997 \text{ kV}$$

Voltage across the bottom most unit,

$$V_4 = (1 + 6K + 5K^2 + K^3)\,V_1 = 2.571 \times 12.297 = 31.616 \text{ kV}$$

Voltage between conductor and earth,

$$V = V_1 + V_2 + V_3 + V_4 = 12.297 + 15 + 21 + 31.616 = 79.9 \text{ kV}$$

Voltage between conductors, $V_L = \sqrt{3}\ V = \sqrt{3} \times 79.9 = 138.4$ kV **Ans.**

$$\text{String efficiency} = \frac{V}{n\,V_n} \times 100 = \frac{79.9}{4 \times 31.616} \times 100 = 63.18\% \quad \textbf{Ans.}$$

9.7. METHODS OF IMPROVING STRING EFFICIENCY

It has been seen that the voltage distribution across an insulator string is not uniform. The insulator unit nearest to the line conductor is highly stressed while that nearest to the tower is lightly stressed. The drawbacks of this non-uniform voltage distribution across the units of an insulator string are two-fold : if the voltage across the insulator unit nearest to the line conductor reaches beyond a prescribed safe value it may breakdown and the breakdown of other insulator units will take place in succession; at the same time if the voltage across the insulator unit farthest from the line conductor (or nearest to the cross-arm) is too low, that unit is not utilized to its optimum capacity resulting in poor efficiency of the insulator string. This necessitates the equalisation of voltage distribution across the various insulator units of the string and thereby improving the string efficiency.

The string efficiency indicates the extent of the wastage. Though string efficiency can never be made 100 per cent, an improvement in its value is necessary so as to reduce the wastage to the minimum. Some methods used for this purpose are given below.

1. By Using Insulators With Larger Discs or By Providing Each Insulator Unit With a Metal Cap. It is clear from the expression of string efficiency that the string efficiency increases with the decrease in value of K (*i.e.* the ratio of shunt capacitance to mutual capacitance). One method is to design the units such that the mutual capacitance (capacitance of each unit) is much greater than the shunt capacitance (capacitance to earth). This can be achieved by using insulators with larger discs or providing each insulator unit with a metal cap. The ratio K can be made $\frac{1}{6}$ to $\frac{1}{10}$ by this method

2. By Using Longer Cross-Arms. The ratio of shunt capacitance to mutual capacitance, K can alternatively be reduced by using longer cross-arms so that the horizontal distance from line support (pole or tower) is increased thereby decreasing the shunt capacitance. But the limitations of cost and mechanical strength of line supports do not allow the cross-arms to be too long and it has been found that in practice it is not possible to obtain the value of K less than 0.1.

3. By Capacitance Grading. It is seen that non-uniform distribution of voltage across an insulator string is due to leakage current from the insulator pin to the supporting structure. This current cannot be eliminated. However, it is possible that discs of different capacities are used such that the product of their capacitive reactance and the current flowing through the respective unit is same. This can be achieved by grading the mutual capacitance of the insulator units *i.e.*, by having lower units of more capacitance—maximum at the line unit and minimum at the top unit, nearest to the cross-arm, It can be shown that by this method complete equality of voltage across the units of an insulator string can be obtained but this method needs a large number of different-sized insulator units. This involves maintaining spares of all varieties of insulator discs which is contrary to the tendency of standardization. So this method is not used in practice below 200 kV.

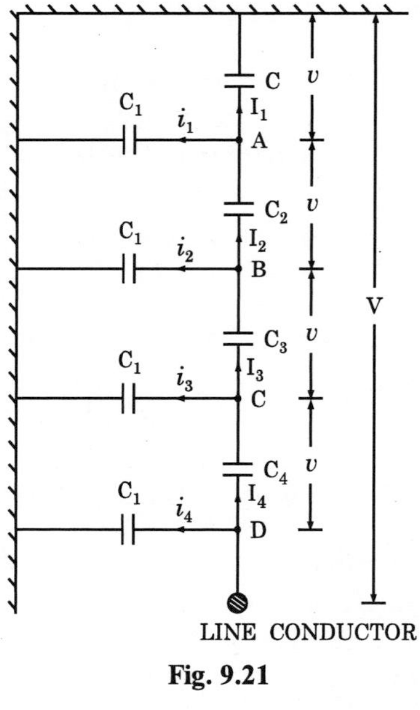

Fig. 9.21

Consider a 4-unit string. Let C be the capacitance of the top unit and let the capacitances of others units be C_2, C_3 and C_4, as shown in Fig. 9.21

Assume $C_1 = k\,C$

Applying Kirchhoff's first law to node A we get

$$I_2 = I_1 + i_1$$

$$\text{or } \omega C_2 v = \omega C v + \omega C_1 v$$

$$\text{or } C_2 = C + K C = C(1 + K) \quad ...(9.11)$$

Applying Kirchhoff's first law to node B we get

$$I_3 = I_2 + i_2$$

$$\text{or } \omega C_3 v = \omega C_2 v + \omega C_1 \times 2v$$

$$\text{or } C_3 = C_2 + 2KC = C(1 + K) + 2KC = C(1 + 3K) \quad ...(9.12)$$

Applying Kirchhoff's first law to node C we get

$$I_4 = I_3 + i_3$$

or $\omega C_4 v = \omega C_3 v + \omega C_1 \times 3 v$

or $C_4 = C_3 + 3 KC = C(1 + 3K) + 3 K C = C(1 + 6K)$ ...(9.13)

Thus it will be possible to equalise the potential across all the units, if their capacitances are in the ratio of 1 : (1 + K) : (1 + 3 K) : (1 + 6 K) and so on.

But in practice it is impossible to obtain such units which will have their capacitances in above ratio, although nearby results can be obtained by employing standard insulators for most of the units and employing larger units adjacent to the line.

4. By Static Shielding. In case of capacitance grading, insulator units of different capacitances are used so that the flow of different currents through the respective units produce equal voltage drop. In static shielding, pin to supporting structure charging currents are exactly cancelled so that the same current flows through the identical insulator units and produce equal voltage drops across each insulator unit. This arrangement is given is Fig. 9.22. In this method a *guard* or *grading ring,* which usually takes the form of a large metal ring surrounding the bottom unit and electrically connected to the metal work at the bottom of this unit, and therefore to the line conductor. The guard ring screens the lower units, reduce their earth capacitance C_1 and introduces a number of capacitances between the line conductor and the various insulator unit caps. These capacitances are greater for lower units and thus the voltages across them are reduced. With this method also it is impossible to obtain in practice an equal distribution of voltage but considerable improvements are possible.

Let the capacitances between the links and the shield be C_x, C_y and C_z respectively as shown in Fig. 9.22, and let v be the potential across each unit.

Since the capacitance of each unit is same, therefore, their charging currents I_1, I_2, I_3, and I_4 would be same, let it be I. If $C_1 = KC$

Applying Kirchhoff's first law to node A we get

$I + i_x = I + i_1$ or $i_x = i_1$ ...(9.14)

Similarly, $i_y = i_2$ ...(9.15)

and $i_z = i_3$ ...(9.16)

Also, $i_1 = \omega C_1 v = \omega K C v$ ...(9.17)

$i_2 = 2 \omega C_1 v = 2 \omega K C v$ ...(9.18)

$i_3 = 3 \omega C_1 v = 3 \omega K C v$ ...(9.19)

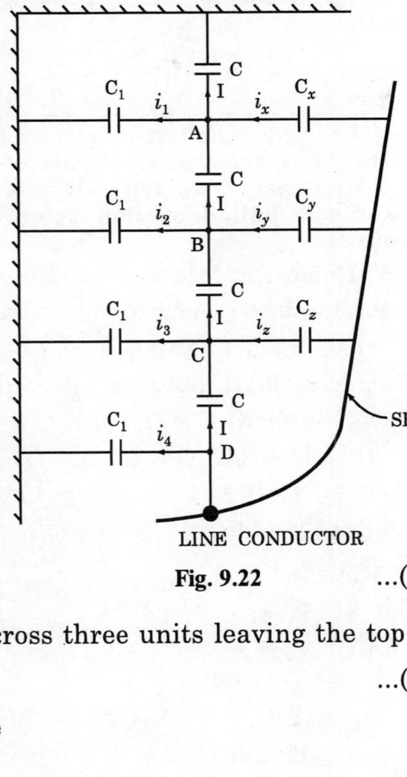

Fig. 9.22

The potential causing current i_x is $3v$ (voltage across three units leaving the top one)

So $i_x = \omega C_x \times 3 v = 3 \omega C_x v$ ...(9.20)

Comparing Eqs. (9.14), (9.17) and (9.20), we have

$$3 \omega C_x v = \omega K C v \text{ or } C_x = \frac{KC}{3} \quad ...(9.21)$$

The potential causing current y is $2v$ and therefore,

$i_y = 2 \omega C_y v$ ...(9.22)

Comparing Eqs. (9.15), (9.18) and (9.22) we have

$$2\,\omega\,C_y\,v = 2\,\omega K\,C\,v \quad \text{or} \quad C_y = K\,C \qquad ...(9.23)$$

The potential causing current i_z is v and therefore,

$$i_z = \omega\,C_z v \qquad ...(9.24)$$

Comparing Eqs. (9.16), (9.19) and (9.24), we have

$$\omega\,C_z v = 3\,\omega\,K\,Cv$$

$$\text{or } C_z = 3\,K\,C \qquad ...(9.25)$$

In general if there are n units

$$i_1 = \omega\,K\,C\,v \text{ and } i_x = (n-1)\,\omega\,C_x v$$

$$\text{or } C_x = \frac{K\,C}{n-1} \qquad ...(9.26)$$

$$\text{Similarly, } C_y = \frac{2\,K\,C}{n-2} \qquad ...(9.27)$$

$$\text{and } C_z = \frac{3\,K\,C}{n-3} \qquad ...(9.28)$$

or The capacitance of the pth metal link to the line is given as

$$C_p = \frac{p\,K\,C}{n-p} \qquad ...(9.29)$$

Example 9.9. A string of six insulators (suspension type) is to be graded to obtain uniform distribution of voltages across the string. If the capacitance of the top unit is 10 times the capacitance to ground of each unit, determine the capacitance of the remaining five units.

Solution : Number of insulator units, $n = 6$

Capacitance to ground to top unit capacitance ratio,

$$K = 0.1 \text{ or } C = 10\,C_1$$

From Eqs. (9.11), (9.12) and (9.13) we have

$C_2 = (1 + K)\,C = (1 + 0.1)\,C = 1.1 \times 10\,C_1 = 11\,C_1$ **Ans.**

$C_3 = (1 + 3\,K)\,C = 1.3\,C = 1.3 \times 10\,C_1 = 13\,C_1$ **Ans.**

$C_4 = (1 + 6\,K)\,C = 1.6\,C = 1.6 \times 10\,C_1 = 16\,C_1$ **Ans.**

Applying Kirchhoff's first law to node D we have

$$I_5 = I_4 + i_4$$

$$\text{or } \omega\,C_5\,v = \omega\,C_4\,v + \omega\,C_1 \times 4\,v$$

$$\text{or } C_5 = C_4 + 4\,K\,C = (1 + 6\,K)\,C + 4\,K\,C$$

$$= (1 + 10\,K)\,C$$

$= (1 + 10 \times 0.1)\,C = 2 \times 10\,C_1 = 20\,C_1$ **Ans.**

Applying Kirchhoff's first law to node E we have

$$I_6 = I_5 + i_5$$

$$\text{or } \omega\,C_6\,v = \omega\,C_5\,v + \omega\,C_1 \times 5\,v$$

$$\text{or } C_6 = C_5 + 5\,K\,C = (1 + 10\,K)\,C + 5\,K\,C$$

$= (1 + 15\,K)\,C = (1 + 15 \times 0.1) \times 10\,C_1 = 25\,C_1$ **Ans.**

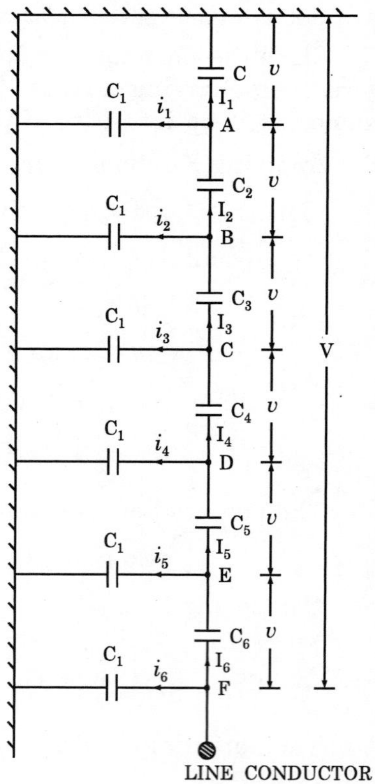

Fig. 9.23

Example 9.10. Explain the methods of equalizing the potential across the string insulator.

A string of eight suspension insulators is to be fitted with a grading ring. If pin to earth capacitance are equal to C, find the values of line to pin capacitances that would give a uniform voltage distribution over the string. [*U.P. Technical Univ. Elements of Power System 2006-07*]

Solution : Number of units, $n = 8$

Pin-to-either capacitance, $C_1 = C$

From Eq. (9.29) the capacitance of the pth metal link to the line is given as

$$C_p = \frac{p \times \text{pin-to-earth capacitance}}{n-p} = \frac{pC}{n-p}$$

Substituting $n = 8$, the values of the various pins to line are

$$C_1 = \frac{1 \times C}{8-1} = \frac{C}{7} \textbf{ Ans.}$$

$$C_2 = \frac{2 \times C}{8-2} = \frac{C}{3} \textbf{ Ans.}$$

$$C_3 = \frac{3 \times C}{8-3} = \frac{3}{5}C \textbf{ Ans.}$$

$$C_4 = \frac{4 \times C}{8-4} = 1C \textbf{ Ans.}$$

$$C_5 = \frac{5 \times C}{8-5} = \frac{5}{3}C \textbf{ Ans.}$$

$$C_6 = \frac{6 \times C}{8-6} = 3C \textbf{ Ans.}$$

$$\text{and } C_7 = \frac{7 \times C}{8-7} = 7C \textbf{ Ans.}$$

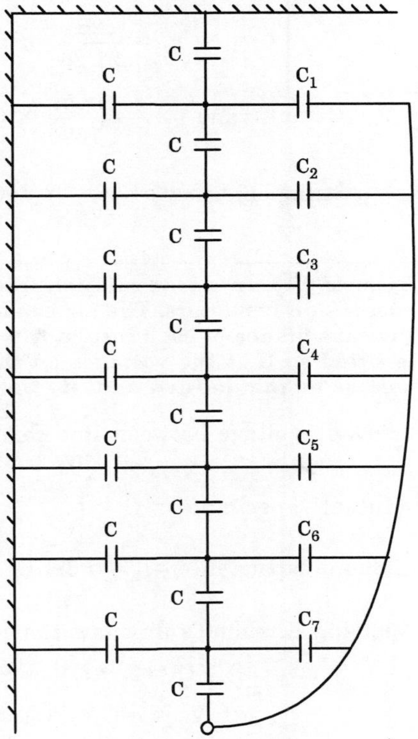

Fig. 9.24

Example 9.11. Define 'string efficiency' and calculate its value for a string of three insulator units if the capacitance of each unit to earth and line be 20% and 5% of the self capacitance of the unit. Derive any formula that might be used.

Solution : Equivalent circuit of a string of insulators is shown in Fig. 9.25.

Applying Kirchhoff's first law to node points A and B we have

$$I_1 + i_1 = I_2 + i_x$$

$$\text{and } I_2 + i_2 = I_3 + i_y$$

$$\text{or } \omega C v_1 + 0.2\, \omega C v_1 = \omega C v_2 + 0.05\, \omega C (v_2 + v_3)$$

$$\text{and } \omega C v_2 + 0.2\, \omega C (v_1 + v_2) = \omega C v_3 + 0.05\, \omega C v_3$$

$$\text{or } 1.2\, v_1 = 1.05\, v_2 + 0.05\, v_3 \quad ...(i)$$

$$\text{and } 0.2\, v_1 = -1.2\, v_2 + 1.05\, v_3 \quad ...(ii)$$

Solving equations (*i*) and (*ii*) in terms of v_1 we have

$$v_2 = \frac{25}{23.25} v_1 \text{ and } v_3 = \frac{1.65}{1.1625} v_1$$

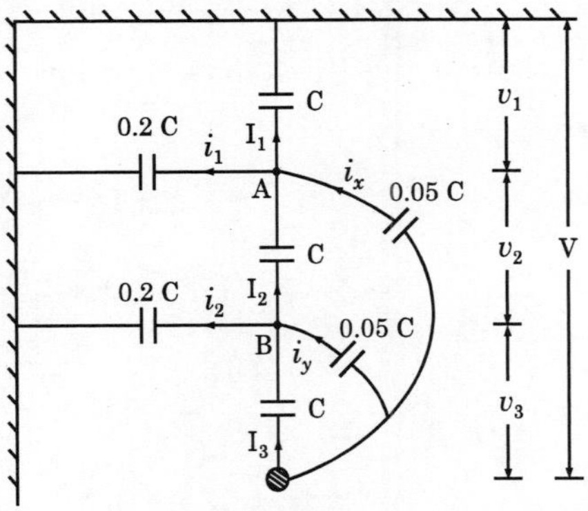

Fig. 9.25

Finally voltage between line conductor and earth

$$V = v_1 + v_2 + v_3 = v_1 + \frac{25}{23.25} v_1 + \frac{1.65}{1.1625} v_1 = 3.49462\ v_1$$

$$\text{So, } v_1 = \frac{V}{3.49462} = 0.286\ V$$

$$v_2 = \frac{25}{23.25} \times 0.286 = 0.308\ V$$

$$\text{and } v_3 = \frac{1.65}{1.1625} \times 0.286 = 0.406\ V$$

$$\text{String efficiency} = \frac{V}{3\, v_3} \times 100 = \frac{V}{3 \times 0.406\ V} \times 100 = 82.1\%\ \textbf{Ans.}$$

Example 9.12. In a transmission line each conductor is at 20 kV and is supported by a string of 3 suspension insulators. The air capacitance between each cap-pin junction and tower is one-fifth of the capacitance of each insulator unit. A guard ring, effective only over the line-end insulator unit is fitted so that the voltages on the two units nearest the line-end are equal. (*a*) Calculate the voltage on the line-end unit, (*b*) Calculate the value of capacitance C_x required.

Solution : Voltage between line conductor and earth,

$$V = 20\ \text{kV} \quad ...(i)$$

$$\text{Mutual capacitance} = C$$

$$\text{Shunt capacitance} = \frac{C}{5} = 0.2\ C$$

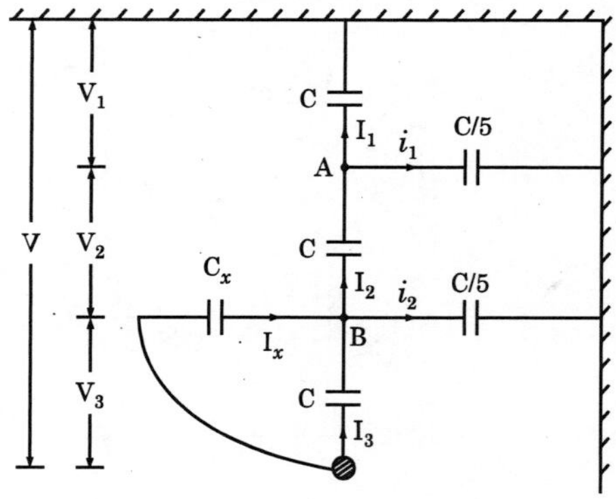

Fig. 9.26

Applying Kirchhoff's first law to node A

$$I_2 = I_1 + i_1$$

$$\omega C V_2 = \omega C V_1 + \omega \times 0.2\ C V_1$$

$$\text{or } V_2 = 1.2\ V_1 \quad ...(ii)$$

$$\text{Also } V_3 = V_2 \quad ...(iii)$$

$$\text{and } V = V_1 + V_2 + V_3$$

$$\text{or } 20 = V_1 + 1.2\ V_1 + 1.2\ V_1$$

$$\text{or } V_1 = \frac{20}{1 + 1.2 + 1.2} = 5.882\ \text{kV}$$

(*a*) Voltage on the line-end unit $= V_3 = 1.2\ V_1 = 1.2 \times 5.882 = 7.06$ kV **Ans.**

Applying Kirchhoff's first law to node B we have

$$I_3 + i_x = I_2 + i_2$$

$$\omega C V_3 + \omega C_x V_3 = \omega C V_2 + \omega \frac{C}{5}(V_1 + V_2)$$

$$\text{or } C \times 1.2\ V_1 + C_x \times 1.2\ V_1 = C \times 1.2\ V_1 + \frac{C}{5} \times 2.2\ V_1$$

$$\text{or } C_x = \frac{2.2}{5 \times 1.2} C = 0.3667\ C \text{ farads } \textbf{Ans.}$$

9.8. ARCING HORN AND GRADING RING

In the event of flash-over the insulator is cracked or broken up due to the heat of the arc. Grading ring, in addition to equalisation of voltage distribution across the insulator units, when used in conjunction with arcing horn fixed at the top end of the string serves the purpose of arcing shield and protects the insulator string from flash-over whenever over-voltage (under normal or abnormal condition) appears between the tower structure and the line conductor. They are designed to keep the arc away from the insulator string until it is interrupted by the device protecting the line. The arrangement of the arcing horns on a 7-unit string of suspension insulators is shown in Fig. 9.27. The combination of the arcing horn and grading ring provides path through the air medium and discharges the energy contained in the abnormal voltage and thus saves the insulator string.

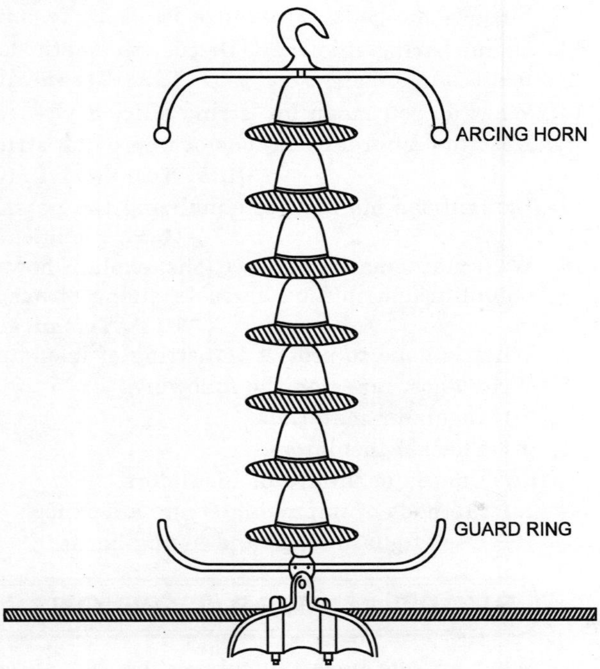

Fig. 9.27. ***Insulator String With Arcing Horn and Grading Ring***

EXERCISES

1. What electrical and mechanical characteristics are required for a good insulator for use in *hv* transmission lines ?
2. Name the various materials that are commonly used for overhead line insulators. Why glass insulators cannot be used above 50 kV ?
3. Describe different types of insulators used in transmission lines with their applications. **[U.P. Technical Univ. Elements of Power System 2004-05]**
4. Describe pin type, suspension type and strain type insulators. **[U.P. Technical Univ. Elements of Power System 2006-07]**
5. With neat diagrams, explain the construction and applications of suspension type insulators. **[U.P. Technical Univ. Elements of Power System 2002-03]**
6. Discuss the use and relative merits and demerits of pin type and suspension type insulators. **[U.P. Technical Univ. Elements of Power System 2003-04]**
7. What do you mean by string efficiency ?

   Describe the different types of insulators used in transmission system. **[U.P. Technical Univ. Elements of Power System 2004-05]**
8. Explain strain and shackle insulators with the help of neat sketches.
9. Mention appropriate locations where you would use (*i*) pin type (*ii*) suspension type (*iii*) strain (*iv*) shackle and (*v*) egg type insulators.
10. Describe about the following in brief :

    (*i*) String insulators. (*ii*) Pin insulators.

    (*iii*) Strain insulators. (*iv*) Shackle insulators.

    [U.P. Technical Univ. Elements of Power System 2009-10]

11. Name the various causes for the failure of overhead line insulators.
12. Why the voltage distribution across the insulators in an insulator string is not equal ? Suggest methods to equalize it. **[U.P. Technical Univ. Elements of Power System 2006-07]**
13. Define 'string efficiency'. Discuss the methods for improving string efficiency in a string of insulators. **[U.P. Technical Univ. Elements of Power System 2007-08]**
14. What do you mean by "string efficiency" ?
Describe different methods of improving string efficiency.
[U.P. Technical University Elements of Power System 2004-05]
15. Explain the methods of equalizing the potential across the string insulators.
[U.P. Technical Univ. Elements of Power System 2006-07]
16. With mathematical derivations, explain how the method of static shielding equalises the potential distribution across a string of suspensation insulators.
[U.P. Technical Univ. Elements of Power System 2002-03]
17. What is done to protect the string of insulators against over-voltages and lightning ?
18. Write short notes on the following:
(*i*) Insulator materials.
(*ii*) Shackle insulators.
(*iii*) Causes of failure of insulators.
(*iv*) Methods of improving string efficiency.
(*v*) Use of guard-rings and arcing-horns.

SHORT ANSWER TYPE QUESTIONS WITH ANSWERS

Q. 1. What are the important properties that an overhead line insulator must possess ?

Ans. The most important properties that an overhead line insulator must possess are : High mechanical strength, high relative permittivity, high insulation resistance, high ratio of rupturing strength to flash-over voltage and capability of withstanding temperature variations.

Q. 2. What types of insulators are used for overhead transmission and distribution lines ?

Ans. The insulators used for overhead line are pin type, suspension type, strain type and shackle insulators.

Q. 3. How do the insulators fail ?

Ans. Electric failure of insulators may occur either by puncture or flash-over. In the case of puncture the arc passes through the insulator body while flash-over is caused by an arc discharge between the line conductor and earth through air surrounding the insulator.

Q. 4. What is meant by birdage ?

Ans. Birdage means the short-circuiting of the line conductor to earth through the large birds or similar objects.

Q. 5. What do you mean by "string efficiency." **[U.P.T.U. 2004-05]**

or

Define string efficiency. **[U.P.T.U. 2005-06, 2007-08]**

Ans. The efficiency of an insulator string, called the *string efficiency*, is the ratio of voltage across the whole string and the product of the number of units and voltage across the unit nearest to the line conductor.

Q. 6. How are voltage distribution and the string efficiency affected by rain ?

Ans. More uniform voltage distribution is obtained and the string efficiency is improved because of reduction in value of K [when insulators are wet, mutual capacitance C increases while the shunt capacitance C_1 (except for the unit nearest the cross-arm) remains constant].

Q. 7. What are the methods, that are used for improving string efficiency ?

Ans. Various methods used for improving string efficiency are by using insulators with larger discs or by providing each insulator unit with a metal cap, by using longer cross-arms, by capacitance grading or by static shielding.

Q. 8. What are the two functions that are performed by grading ring ?

Ans. The grading ring performs two functions viz equalizes the voltage distribution across the insulator units and when used in conjunction with arcing horn fixed at the top end of the string serves the purpose of arcing shield and protects the insulator string from flash-over.

PROBLEMS

1. A 33 kV, 3-phase, OH line is supported on a string of 3 similar insulators, the mutual capacitance of which is 10 times the shunt capacitance between the unit and earthed frame work. Calculate the voltage across each insulator and string efficiency.

[**Ans.** V_1 = 5.587 kV, V_2 = 6.146 kV, V_3 = 7.319 kV, String efficiency = 86.77%]

2. For a string insulator with four discs, the capacitance of the disc is 10 times the capacitance between the pin and earth. Calculate the voltage across each disc when used on 66 kV line. Also calculate the string efficiency.

[**Ans.** V_1 = 7.529 kV, V_2 = 8.282 kV, V_3 = 9.863 kV, V_4 = 12.43 kV, string efficiency = 76.64%]

3. A string of six insulator units has self capacitances equal to 10 times the pin to earth capacitances. Find (*a*) the voltage distribution from top to bottom insulator as a percentage of the total voltage, (*b*) the string efficiency.

[**Ans.** 9.904%, 10.894%, 12.974%, 16.351%, 21.346%, 28.512% of V, 58.45%]

4. Each line of a 3-phase system is supported by a string of 3 similar insulators. If the voltage across the line unit is 17.5 kV, calculate the line-to-neutral voltage. Assume that the shunt capacitance between each insulator and earth is $\frac{1}{8}$th of the capacitance of the insulator itself. Also find the string efficiency.

[**Ans.** 44.24 kV, 84.26%]

5. An insulator string is made of three similar insulators. Find the maximum voltage that this string can withstand if the maximum voltage per unit is 17.5 kV. Assume $C' = C/8$.

[U.P. Technical Univ. Elements of Power System 2005-06] [**Ans.** 44.24 kV]

6. Discuss the use and relative merits and demerits of pin type and suspension-type insulators. An insulator string consists of three units, each having a safe working voltage of 15 kV. If the capacitance between each insulator pin and earth is 1/8 of self-capacitance of each insulator, find the maximum safe working voltage of the string.

[U.P. Technical Univ. Elements of Power System 2003-04] [**Ans.** 37.92 kV]

7. A string insulator has three units. The capacitance from each joint is 12.5% of the capacitance of each unit. The voltage across any unit should not exceed 11 kV. Find the maximum voltage for the string application.

[**Ans.** 27.8 kV]

8. Each conductor of a 3-phase high voltage transmission line is suspended from cross-arm of steel tower by a string of 4 suspension type disc insulators. If the voltage across the second unit is 13.2 kV and that across the third unit is 18 kV, calculate the voltage between the conductors.

[**Ans.** 70 kV]

9. Each of the three insulators forming a string has a self capacitance of C farads. The shunting capacitance of the connecting metal work of each insulator is 0.2 C to earth and 0.1 C to the line. Calculate the voltage across each insulator as a percentage of the line voltage to earth and also the string efficiency.

[**Ans.** 31%; 30.3% and 38.7% of V and 86.13%]

10. Each of the three insulators forming a string has a self capacitance of 'C' Farads. The shunting capacitance of the connecting metal work of each insulator is 0.2 C to earth and 0.1 C to the line. A guard ring increases the capacitance to the line of the metal work of the lowest insulator to 0.3 C. Calculate the string efficiency of this arrangement with the guard ring.

[**Ans.** 95.24%]

UNIT-IV

MECHANICAL DESIGN OF TRANSMISSION LINE AND INSULATED CABLES

CONTENTS IN THIS UNIT

MECHANICAL DESIGN OF TRANSMISSION LINE : Catenary Curve; Calculation of Sag and Tension; Effects of Wind and Ice Loading; Sag Template; Vibration Dampers.

INSULATED CABLES : Types of Cables and Their Construction, Dielectric Stress; Grading of Cables, Insulation Resistance; Capacitance of Single Phase and Three Phase Cables; Dielectric Loss; Heating of Cables.

CHAPTER 10. Mechanical Design of Transmission Lines.

CHAPTER 11. Insulated Cables.

10

CHAPTER

MECHANICAL DESIGN OF TRANSMISSION LINES

INSIDE THIS CHAPTER

10.1. Introduction **10.2.** Conductor Configurations, Spacings and Clearances **10.3.** Span Lengths **10.4.** Sag and Tension **10.5.** Stringing Chart **10.6.** Sag Template **10.7.** Equivalent Span **10.8.** Vibrations and Dampers.

10.1. INTRODUCTION

Transmission and distribution lines are vital links between generating stations and consumers as power from generating stations is transmitted at high voltage (such as 132, 220 or 400 kV) over long distances to the major load centres and then the power is distributed to various substations located at various places and localities through distribution lines. Because of tremendous industrial growth, requirement of power has increased manifold. Hence it becomes imperative that transmission and distribution of power from the generating stations to the various consumers is carried out with minimum possible loss and disturbance. This objective can be achieved only if the transmission and distribution system is so designed and constructed that it is an efficient, technically sound and reliable system. The line should have sufficient current carrying capacity so as to transmit the required power over a given distance without an excessive voltage drop and overheating. The line losses should be small and insulation of the line should be adequate to cope with the system voltage. The line should have sufficient mechanical strength to cope with the worst probable (not worst possible)* weather conditions and provide satisfactory service over a long period of time without the necessity of too much maintenance.

* The cost of the system which would be capable of withstanding a worst possible weather condition would be extremely large and from economical aspects it is better to bear the risk of failure under such extreme worst conditions.

10.2. CONDUCTOR CONFIGURATIONS, SPACINGS AND CLEARANCES

10.2.1. Conductor Configurations

Several conductor configurations are possible, but three configurations are the most common *i.e.* horizontal configuration (or horizontal disposition of conductors), vertical configuration and triangular configuration.

There is no special advantage in using the symmetrical *delta* or *triangular configuration* [Fig. 10.1(*a*)] and in most cases flat horizontal or vertical configurations are employed from mechanical considerations, particularly when suspension insulators are used. In *horizontal configuration,* all the conductors are mounted over one cross-arm, as shown in Fig. 10.1(*b*). Though such an arrangement of conductors needs supports of smaller height but needs a wider right of way. In certain congested areas where it is not possible to have horizontal arrangement of conductors, the conductors are placed in vertical formation (along the length of pole one below the other). The drawbacks of vertical formations are taller towers and more lightning hazards. There are places where both horizontal and vertical formations are applied.

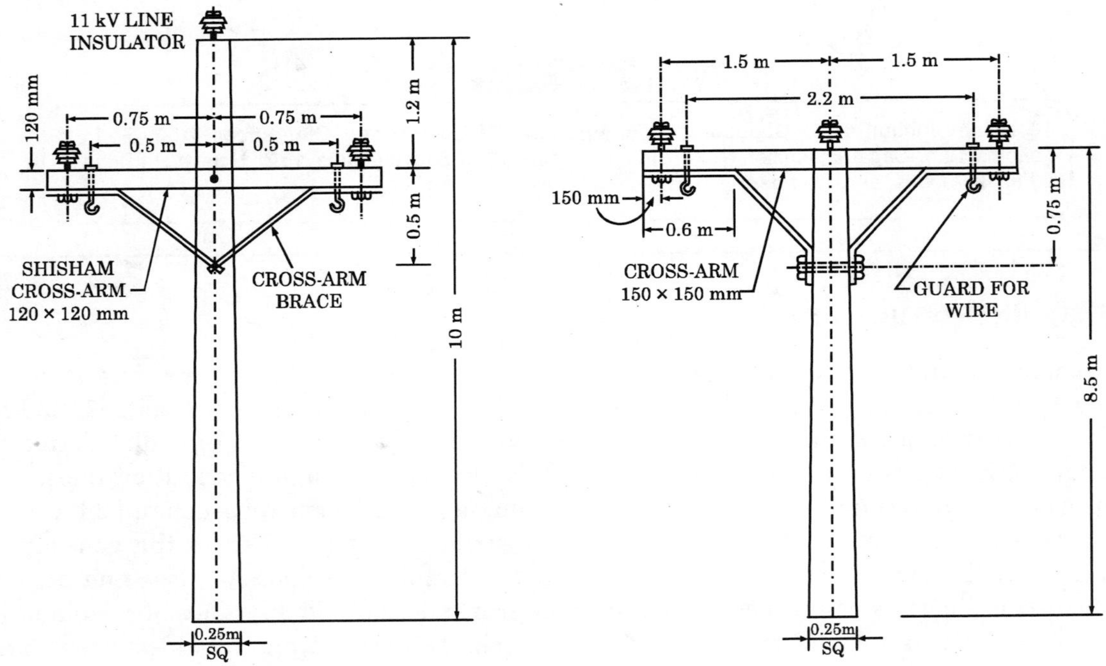

(*a*) *Triangular Formation* (*b*) *Horizontal Formation*

Fig. 10.1. ***Conductor Configurations***

In unsymmetrical arrangement of conductors, the conductors are usually transposed at regular intervals in order to balance the electrical characteristics of various phases, and prevent inductive interference with neighbouring communication circuits.

Experience shows that a vertical configuration is the most economical for double circuit lines and horizontal or L-type configuration for single circuit lines.

10.2.2. Conductor Spacings

The spacing of conductors is determined by considerations partly electrical and partly mechanical. Larger spacing causes increase in inductance of the line and hence the voltage drop, so that to keep the latter within a reasonable value the conductors should be as close together as is consistent with the prevention of corona.

The basic consideration regarding the minimum spacing between conductors is that the electrical clearances between conductors under the worst condition *i.e.* maximum temperature and wind pressure, shall not be less than the limits for safety, particularly at the mid spans. Owing to the action of the gusts of wind, conductor has got tendency to move about in an elliptical path, therefore, in case of suspension insulators, the minimum clearance to supporting structures should be calculated with a 45° swing of the suspension string towards the structure.

An empirical formula commonly employed for determination of spacing of conductors for an aluminium conductor line is given below.

$$\text{Spacing} = \sqrt{S} + \frac{V}{150} \text{ metres} \qquad ...(10.1)$$

where S is sag in metres and V is line voltage in kV.

Some typical spacing values are given below:

TABLE 10.1

Line voltage in kV	0.4	11	33	66	132	220	400	765
Spacing in metres	0.2	1.2	2.0	2.5	3.5	6.0	11.5	14

10.2.3. Conductor Clearances

The minimum vertical clearances between the ground and the conductor are recommended vide I E rule no. 77

These values are given below:

TABLE 10.2

Line Voltage in kV / **Clearance To Ground in metres**	0.4	11	33	66	132	220	400
Across Street	5.8	5.8	6.1	6.1	6.1	7.0	8.4
Along Street	5.5	5.5	5.8	6.1	6.1	7.0	8.4
Other Areas	4.6	4.6	5.2	5.5	6.1	7.0	8.4

10.3. SPAN LENGTHS

Neglecting the deciding influence upon the span length of such local conditions as the necessity for following the configurity of roads, canals or railways, it is interesting to note that there is one definite value for the span length which will give the minimum overall cost of the line. As the length of span increases, the number of insulators and supports decreases resulting in decrease in cost but at the same time the height of the support will go up to allow for more sag and also the length of the cross-arms will have to be increased to take up increased spacing, this will cause increase in cost. Moreover the length of the span depends upon the working voltage, higher the working voltage of the system, the greater will be the economical length of span owing to the higher relative cost of insulators to supports. Moreover the insulators constitute the weakest part of transmission line and reduction in number of towers per km with the use of longer span increases the reliability of the line. Thus it is not possible to give any hard and fast rule as to the best span length to be adopted, and the only way to determine it is to calculate the total cost per km for a number of different span lengths, and plot the results to get the most economical span length.

Many a time it happens that the conductor size determined from electrical calculations comes out rather small, it is possible to reduce the total cost of line by using a thicker and stronger conductor, and increasing the span length. Sometimes it is not feasible to determine the height of the line support and span length on the basis of line cost alone as lightning hazards increase greatly with the increase in height of conductors above the ground.

The usual spans are :

(*a*) With wooden poles : 40-50 m.

(*b*) With steel tubular poles : 50-80 m.

(*c*) With RCC poles : 80-200 m

(*d*) With steel towers : 200-400 m and above.

For river-crossings, etc. exceptionally long spans up to 800 m or so have been satisfactorily employed.

10.4. SAG AND TENSION

Overhead lines are supported on mechanical structures consisting of components like insulators, cross-arms, poles or towers, etc. The strength of these components must be such that there is no mechanical failure of line, even under the worst weather conditions.

The conductor is acted upon by the forces such as weight of the conductor itself, wind pressure and tension. A conductor stretched between two supports will have an ultimate strength at which it will fail and the ultimate strength of a conductor depends upon the type of conductor material used for overhead line.

While stringing overhead lines it is necessary to allow a reasonable factor of safety in respect of the tension to which the conductor is subjected. The tension in the conductor is normally expected to be less than 50% of its ultimate tensile strength even when there is 12.7 mm radial coating of ice and a wind pressure of the order of 380 N/m^2.

The tension in a conductor depends on the diameter of the conductor, length of the conductor between supports, material of conductor, sag in conductor, wind pressure and temperature. The relationship between tension and sag is dependent on the loading conditions and temperature variations. For instance, the tension increases when temperature decreases and there is a corresponding decrease in sag. Icing-up of the line and wind loading causes stretching of the conductor by an amount governed by the line tension.

10.4.1. Sag

A perfectly flexible wire of uniform cross-section, when strung between the two supports at the same level, will form a *catenary*. However, if the sag is very small compared to the span, its shape approximates a *parabola*.

The difference in level between the points of supports and the lowest point on the conductor is known as *sag*.

The factors affecting the sag in an overhead line are given below :

(i) *Weight of the Conductor*. This affects the sag directly. Heavier the conductor, greater will be the sag. In locations where ice formation takes place on the conductor, this will also cause increase in the sag.

(ii) *Length of the Span*. This also affects the sag. Sag is directly proportional to the square of the span length. Hence other conditions, such as type of conductor, working tension, temperature etc, remaining the same a section with longer span will have much greater sag.

(iii) *Working Tensile Strength*. The sag is inversely proportional to the working tensile strength of conductor if other conditions such as temperature, length of span etc. remain the same. Working tensile strength of the conductor is determined by multiplying the ultimate stress and area of x-section and dividing by a factor of safety.

(iv) *Temperature*. All metallic bodies expand with the rise in temperature and, therefore, the length of the conductor increases with the rise in temperature, and so does the sag.

The sag plays an important role in the design of overhead line. It is disadvantageous to provide either too high sag or too low sag. In case the sag is too high, more conductor material is required, more weight on the supports is to be supported, higher supports are necessary and there is a chance of greater swing-amplitude due to wind load. On the other hand in the case of too low sag, there is more tension in the conductor and thus the conductor is liable to break if any additional stress is to be taken, such as due to vibration of line or due to fall in temperature.

Normally two conditions should be investigated, when making sag-tension calculations :

(a) *At Minimum Temperature*. The lowest sag and maximum tension in conductor-section occurs when the temperature is minimum and wind maximum. Under these conditions, tension on the conductor should not exceed the breaking strength of the conductor divided by a factor of safety of 2.5, as prescribed by I E Rule No. 76.

(b) *At Maximum Temperature*. On the other hand maximum sag occurs when temperature is maximum and there is no wind pressure. Under these conditions, electrical clearances should be above the minimum values prescribed under I E Rule No. 77.

Approximate formulae for the calculation of sag is given below

$$S = \frac{wL^2}{T}$$

where w = weight of conductor in kg per metre length, L = length of the span in metres, and T = tension in the conductor in kg.

10.4.2. Calculation of Sag and Tension

As mentioned earlier, a conductor strung between two supports at the same level forms a catenary. For extra high voltage lines, when sag exceeds 10% of the span length, formulae based on catenary should be used,

Consider a portion OP of a curved length l of a wire hanging in still air with O as the lowest point on the wire [Fig. 10.2 (*a*)]. Let the weight of conductor per metre length be w kg.

If T_0 is the tension at point O (lowest point on the wire and where the curve is horizontal) and T the tension at point P (distant l from the lowest point O), the portion OP (length l) is in equilibrium under the action of three forces, namely T_0, T and the weight of the wire of length l acting vertically downward through centre of gravity, wl.

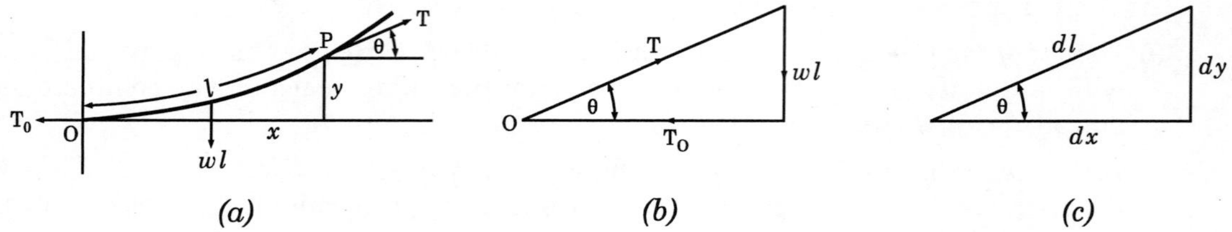

Fig. 10.2

The above three forces can be represented by a triangle shown in Fig. 10.2 (*b*) and from this triangle

$$\text{Tan } \theta = \frac{wl}{T_0} \quad \text{...(10.2)}$$

The horizontal and vertical distances of the length OP of the wire are x and y respectively, since the coordinates of the point P are (x, y).

Now from triangle shown in Fig. 10.2 (*c*),

$$\text{Tan } \theta = \frac{dy}{dx} \quad \text{...(10.3)}$$

$$\text{and also } dl^2 = dx^2 + dy^2 \quad \text{...(10.4)}$$

$$\text{or } \left(\frac{dl}{dx}\right)^2 = 1 + \left(\frac{dy}{dx}\right)^2 \quad \text{...(10.5)}$$

$$= 1 + \tan^2 \theta \qquad \because \text{ from Eq. (10.3) } \frac{dy}{dx} = \text{Tan } \theta$$

$$= 1 + \left(\frac{wl}{T_0}\right)^2 \qquad \because \text{ from Eq. (10.2) Tan } \theta = \frac{wl}{T_0}$$

$$\text{or } \frac{dl}{dx} = \sqrt{1 + \left(\frac{wl}{T_0}\right)^2}$$

$$\text{or } dx = \frac{dl}{\sqrt{1 + \frac{w^2 l^2}{T_0^2}}} \quad \text{...(10.6)}$$

Integrating both sides we get

$$x = \frac{T_0}{w} \sinh^{-1} \frac{wl}{T_0} + C \text{ where C is a constant of integration} \quad \text{...(10.7)}$$

From initial conditions, *i.e.* when $x = 0$, $l = 0$, we get C = 0 and Eq. (10.7) becomes

$$x = \frac{T_0}{w} \sinh^{-1} \frac{wl}{T_0} \quad \text{...(10.8)}$$

$$\text{or } \frac{wl}{T_0} = \text{Sinh} \frac{wx}{T_0} \quad \text{...(10.9)}$$

$$\text{or } l = \frac{T_0}{w} \sinh \frac{wx}{T_0} \quad \text{...(10.10)}$$

$$\text{Again } \frac{dy}{dx} = \text{Tan } \theta = \frac{wl}{T_0} = \frac{w}{T_0} \times \frac{T_0}{w} \sinh \frac{wx}{T_0} = \text{Sinh} \frac{wx}{T_0} \quad \text{...(10.11)}$$

Integrating both sides we get

$$y = \frac{T_0}{w} \cosh \frac{wx}{T_0} + D \text{ where D is another constant of integration ...(10.12)}$$

From initial conditions *i.e.* when $x = 0$, $y = 0$, we get $D = \frac{-T_0}{w}$ and Eq. (10.12) becomes

$$y = \frac{T_0}{w} \cosh \frac{wx}{T_0} - \frac{T_0}{w} = \frac{T_0}{w} \left(\cosh \frac{wx}{T_0} - 1 \right) \quad \text{...(10.13)}$$

Equation (10.13) is the equation of the curve, called the *catenary*.

The function cosh is the hyperbolic cosine and is such that

$$\text{Cosh } \frac{wx}{T_0} = 1 + \frac{\left(\frac{wx}{T_0}\right)^2}{\lfloor 2} + \frac{\left(\frac{wx}{T_0}\right)^4}{\lfloor 4} + \frac{\left(\frac{wx}{T_0}\right)^6}{\lfloor 6} + \ldots$$

$$\text{So } y = \frac{T_0}{w} \left[1 + \frac{\left(\frac{wx}{T_0}\right)^2}{\lfloor 2} + \frac{\left(\frac{wx}{T_0}\right)^4}{\lfloor 4} + \frac{\left(\frac{wx}{T_0}\right)^2}{\lfloor 6} - 1 \right]$$

If the fourth and higher order terms are neglected,

$$y \simeq \frac{T_0}{w} \cdot \left(\frac{wx}{T_0}\right)^2 \cdot \frac{1}{\lfloor 2} = \frac{wx^2}{2T_0} \quad \text{...(10.14)}$$

From Fig. 10.2(*b*) the tension T at point P is given by

$$T = \sqrt{(T_0)^2 + (wl)^2} = \sqrt{(T_0)^2 + \left(T_0 \sinh \frac{wx}{T_0}\right)^2} = T_0 \sqrt{1 + \left(\sinh \frac{wx}{T_0}\right)^2} = T_0 \cosh \frac{wx}{T_0} \quad \text{...(10.15)}$$

If the line is supported between two points A and B at the same level and the length of the span is L (Fig. 10.3), then at the supports $x = \pm \frac{L}{2}$ and

$$T = T_0 \cosh \frac{wL}{2T_0} \quad \text{...(10.16)}$$

The sag S is the value of y at A or B and is given by

$$S \simeq \frac{w}{2T_0} \left(\frac{L}{2}\right)^2 = \frac{wL^2}{8T_0} \quad \text{...(10.17) refer to Eq. (10.14)}$$

Length of line in a half span

$$= \frac{T_0}{w} \sinh \frac{wL}{2T_0} \quad \text{...(10.18) refer to Eq. (10.10)}$$

$$= \frac{T_0}{w} \left[\frac{wL}{2T_0} + \frac{w^3 L^3}{48 T_0^3} + \ldots \right] \quad \because \sinh \frac{wL}{2T_0} = \frac{wL}{2T_0} + \left(\frac{wL}{2T_0}\right)^3 \times \frac{1}{\lfloor 3} + \ldots$$

Neglecting terms of order exceeding cube we have

$$\text{Length of line in a half span} \simeq \frac{T_0}{w} \times \frac{wL}{2T_0} + \frac{T_0}{w} \times \frac{w^3L^3}{48\,T_0^3} \simeq \frac{L}{2} + \frac{w^2L^3}{48\,T_0^2} \quad ...(10.19)$$

$$\text{or Length of line in a full span length} \simeq L + \frac{w^2L^3}{24T_0^2} \quad ...(10.20)$$

$$\text{Tension, } T = T_0 \cosh \frac{wx}{T_0} \qquad \text{refer to Eq. (10.15)}$$

$$= T_0\left(1 + \frac{w^2x^2}{2T_0^2} + \frac{w^4x^4}{24T_0^4} + ...\right) \simeq T_0\left(1 + \frac{w^2x^2}{2T_0^2}\right)$$

Substituting $x = \frac{L}{2}$ in above equation we have

$$T = T_0\left(1 + \frac{w^2L^2}{8\,T_0^2}\right) = T_0 + \frac{w^2L^2}{8\,T_0} = T_0 + wS$$

$$\because \text{ from Eq. (10.17) } \frac{wL^2}{8T_0} = S$$

The maximum tension occurs when $x = \frac{L}{2}$ so maximum tension is given as

$$T_{max} = T_0 + wS \quad ...(10.21)$$

For very small sags the term wS may be neglected in comparison with T_0, and the tension is considered approximately uniform through out the conductor.

Approximate Formulae. Neglecting $\frac{w^2L^2}{8\,T_0}$, being very small compared with unity, T becomes nearly equal to T_0 and

$$\text{Sag, } S = \frac{wL^2}{8\,T} \quad ...(10.22)$$

$$\text{Total length of wire} = L + \frac{w^2L^3}{24\,T^2} \quad ...(10.23)$$

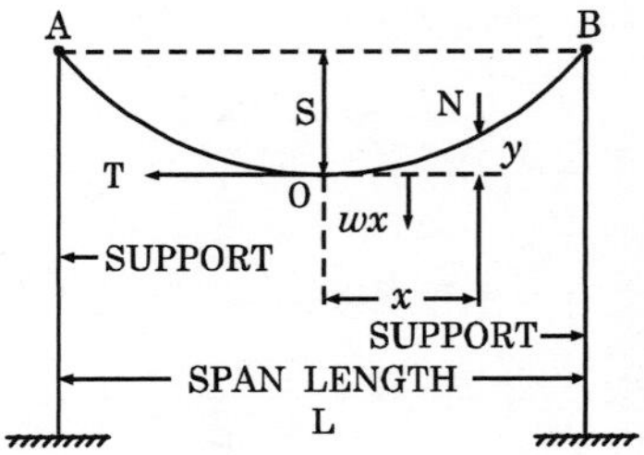

Fig. 10.3

The sag could also be very simply deduced as follows: Consider a conductor suspended between two, equal level supports A and B. The conductor is assumed to be flexible and sags below the level AB due to its own weight. Though the exact shape of the conductor is that of a catenary but except for lines with very long span and large sag, it is sufficiently accurate to assume that the shape of the hanging conductor is that of a parabola $y = ax^2$ where a is a constant for a given conductor and O is the origin. The curve at point O, being the lowest one, will be horizontal.

Let the length of span (*i.e.* horizontal distance between supports) be L metres, weight of conductor per metre length be w kg and tension in the conductor be T kg.

Assuming curvature very small, the length of conductor hanging between two supports may be taken equal to the length of span and maximum sag may be considered at half span length with equi-level supports.

Consider x metre length of conductor between mid-point O and point N. The two external forces acting on the portion ON of the conductor are

(*i*) the tension T acting at point O and

(*ii*) the weight of the conductor of x metres length *i.e.* wx acting at $\frac{x}{2}$ metres from point N.

Equating the moments of the above two forces about point N we get

$$\text{T}\,y = w\,x \times \frac{x}{2}$$

$$\text{or } y = \frac{wx^2}{2\,\text{T}} \qquad ...(10.24)$$

$$\text{At point B, } y = \text{S} \quad \text{and} \quad x = \frac{\text{L}}{2}$$

$$\text{or Sag, S} = \frac{w(\text{L}/2)^2}{2\,\text{T}} = \frac{w\text{L}^2}{8\,\text{T}} \qquad ...(10.25)$$

Supports at Different Levels. When transmission lines are run on steep inclines, as in case of hilly areas, the two supports A and B will be at different levels. The shape of the conductor strung between the supports may be assumed to be a part of the parabola. In this case, the lowest point of the conductor will not lie in the middle of the span.

Consider an overhead line conductor AOB, supported over the supports A and B, as illustrated in Fig. 10.4.

Let the difference in levels between the two supports be h and the lowest point O of the conductor be at a distance of x metres from the support at low level.

So, distance of O from higher level support

$$= (\text{L} - x) \text{ metres}$$

From Eq. (10.24)

$$\text{S}_1 = \frac{wx^2}{2\,\text{T}} \qquad ...(10.26)$$

$$\text{and } \text{S}_2 = \frac{w(\text{L} - x)^2}{2\,\text{T}} \qquad ...(10.27)$$

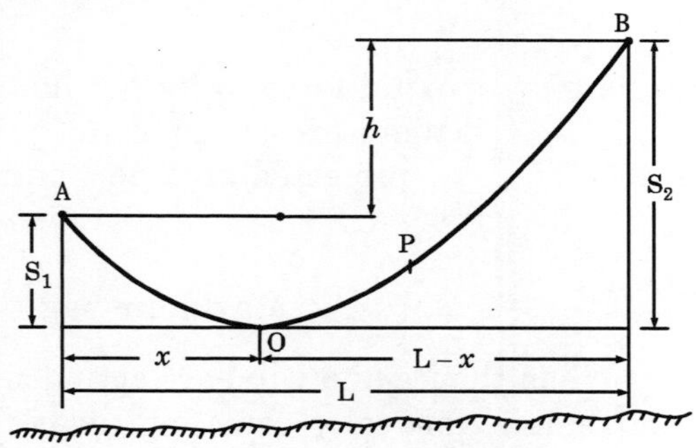

Fig. 10.4

Difference in two level of supports,

$$h = \text{S}_2 - \text{S}_1 = \frac{w(\text{L} - x)^2}{2\,\text{T}} - \frac{wx^2}{2\,\text{T}}$$

$$= \frac{w\text{L}^2 + wx^2 - 2w\text{L}x - wx^2}{2\,\text{T}} = \frac{w\text{L}^2 - 2w\text{L}x}{2\,\text{T}}$$

$$\text{or } 2\,\text{T}\,h = w\,\text{L}^2 - 2\,w\,\text{L}\,x$$

$$\text{or } 2\,w\,\text{L}\,x = w\,\text{L}^2 - 2\,\text{T}\,h$$

$$\text{or } x = \frac{\text{L}}{2} - \frac{\text{T}h}{w\text{L}} \qquad ...(10.28)$$

From the above Eqs. (10.26), (10.27) and (10.28) location of point O and thus the values of S_1 and S_2 can be determined.

Effect of Ice. In areas where it becomes too cold in winter, there is a possibility of formation of an ice coating on the line conductors. The formation of an ice coating on a line conductor has a two fold effect—increase in weight and effective diameter of the conductor.

In this condition the weight of conductor, together with weight of ice acts vertically downwards. Thus the total vertical weight acting on the conductor per metre length is $w_c + w_i$ where w_c is the weight of conductor in kg per metre length and w_i is the weight of ice coating per metre length, w_c is known and w_i is determined as follows:

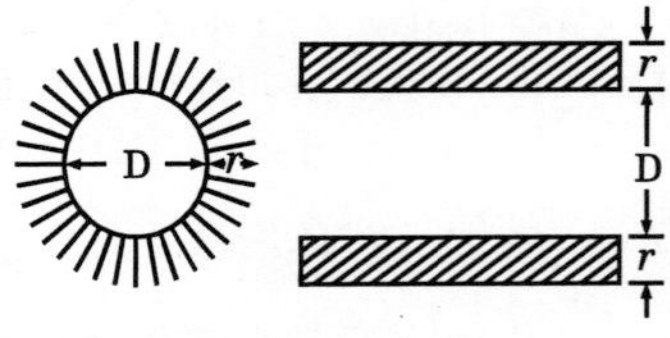

Fig. 10.5. *Ice Coating*

Let the diameter of conductor be D metres and radial thickness of ice coating be *r* metre, as illustrated in Fig. 10.5. The overall diameter of ice covered conductor, as obvious from Fig. 10.5, becomes equal to (D + 2*r*) metre.

$$\text{Volume of ice coating per metre length of conductor} = \frac{\pi}{4}\left[(D + 2r)^2 - D^2\right]$$

$$= \frac{\pi}{4}\left[4Dr + 4r^2\right] = \pi r (D + r) \text{ m}^3$$

The density of ice is approximately 920 kg per m^3, so

The weight of ice coating per metre length, $w_i = 920 \times \pi r (D + r)$ kg

$$= 2{,}890.3\, r\, (D + r) \text{ kg/m} \qquad ...(10.29)$$

Combined Effect of Wind and Ice. Due to weight of ice deposits on the line, and the wind pressure, the mechanical stress increases in the conductor and, therefore, the line must be designed to withstand these stresses and tensions. Under this condition, the weight of the conductor, together with weight of ice acts vertically downwards while the wind loading w_w acts horizontally, as shown in Fig. 10.6.

Resultant weight per metre length of conductor including ice coating and wind force,

$$w_r = \sqrt{(w_c + w_i)^2 + w_w^2} \qquad ...(10.30)$$

where w_w = Wind force in kg per metre length

= Wind pressure per m^2 of projected area × projected area per metre length

$= p \times (D + 2r)$ ...(10.31)

$$\therefore \text{ Maximum sag} = \frac{w_r L^2}{8T} \qquad ...(10.32)$$

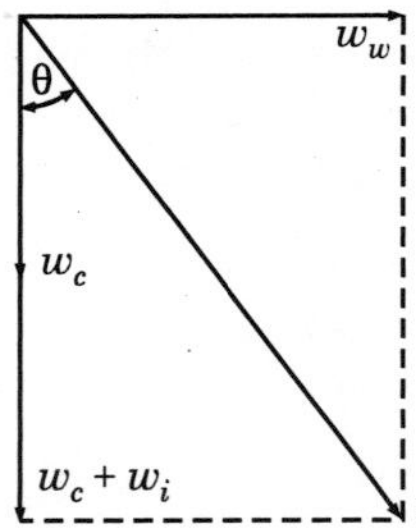

Fig. 10.6

When the ice and wind are acting simultaneously, the lowest point of the conductor does not remain vertically down but away from it at an angle θ given by the expression

$$\theta = \text{Cos}^{-1} \frac{w_c + w_i}{w_r} \qquad ...(10.33)$$

The sag calculated from Eq. (10.32) will not be vertical sag but will be *slant sag* and vertical sag will be obtained by multiplying the slant sag with cos θ

$$i.e. \text{ Vertical sag} = \frac{w_r L^2}{8T} \cos\theta \qquad ...(10.34)$$

Example 10.1. An overload line has a span of 200 m, the line conductor weights 0.7 kg/m. Calculate the maximum sag if the maximum allowable tension in the line is 1,400 kg. Prove formula used.

[*U.P. Technical Univ. Elements of Power System 2006-07*]

Solution :

Span length, L = 200 m

Weight of conductor per metre length, *w* = 0.7 kg

Allowable tension, T = 1,400 kg

$$\text{Maximum sag, S} = \frac{wL^2}{8T} = \frac{0.7 \times 200^2}{8 \times 1{,}400} = 2.5 \text{ m } \textbf{Ans.}$$

Example 10.2. A 132 kV transmission line has the following data :

Wt of conductor = 680 kg/km, Length of span = 260 m, Ultimate strength = 3,100 kg ; Safety factor = 2.

Calculate the height above ground at which the conductor should be supported. Ground clearance required is 10 m.

Solution : Working tension, $T = \frac{\text{Ultimate strength}}{\text{Safety factor}} = \frac{3,100}{2} = 1,550 \text{ kg}$

Weight of conductor per metre length, $w = \frac{680}{1,000} = 0.68 \text{ kg}$

$$\text{Maximum sag, } S = \frac{wL^2}{8\,T} = \frac{0.68 \times (260)^2}{8 \times 1,550} = 3.7 \text{ metres}$$

The height above ground at which the conductor should be supported

= Clearance + sag = 10 + 3.7 = 13.7 metres **Ans.**

Example 10.3. Calculate the horizontal component of tension and maximum sag for a span of 300 m if the maximum tension in the conductor be 3,500 kg and weight of conductor is 700 kg/km. Determine also the location of the points on the conductor at which the sag will be half of the above value.

Solution : Weight of conductor per metre length, $w = \frac{700}{1,000} = 0.7 \text{ kg}$

From Eqs. (10.17) and (10.21), we have

$$\text{Maximum sag, } S = \frac{wL^2}{8\,T_0} = \frac{0.7 \times (300)^2}{8T_0} = \frac{7,875}{T_0} \quad ...(i)$$

$$\text{and maximum tension, } T_{max} = T_0 + w\,S = T_0 + 0.7 \times \frac{7,875}{T_0} = T_0 + \frac{5,512.5}{T_0} \quad ...(ii)$$

where T_0 is the horizontal component of tension in kg

Substituting the value T_{max} = 3,500 kg in equation (*ii*) we have

$$3,500 = T_0 + \frac{5,512.5}{T_0}$$

or $T_0^2 - 3,500\,T_0 + 5,512.5 = 0$

$$\text{or } T_0 = \frac{3,500 \pm \sqrt{(3,500)^2 - 4 \times 5,512.5}}{2} = (1750 \pm 1,748) \text{ kg}$$

As $T_0 \simeq T_{max}$ hence + ve sign will be taken, so

T_0 = 1750 + 1,748.42 = 3,498.42 kg **Ans.**

$$\text{Maximum sag, } S = \frac{wL^2}{8\,T_0} = \frac{0.7 \times (300)^2}{8 \times 3,498.42} = 2.251 \text{ metres } \textbf{Ans.}$$

$$\text{Now, for sag, } y = \frac{S}{2} = \frac{2.251}{2} = 1.1255 \text{ m}$$

Since, from Eq. (10.14) $y = \frac{wx^2}{2T_0}$

$$\therefore x = \sqrt{\frac{2T_0 \times y}{w}} = \sqrt{\frac{2 \times 3,498.42 \times 1.1255}{0.7}} = 106 \text{ metres}$$

Hence, for half of maximum value of sag the points on the conductor will be located having value of x coordinate = 106 m on both sides of point O.

Example 10.4. A transmission line has a span of 150 m between level supports. Line conductor has a cross-sectional area of 1.25 cm^2 and it weighs 1.0 kg/m. If the breaking stress of conductor is 4,200 kg/cm^2, calculate the maximum sag for a safety factor of 4. Assume a maximum wind Apressure of 100 kg/m^2 of projected surface.

Solution : Span length, L = 150 m

Weight of conductor per metre length, w_c = 1.0 kg

$$\text{Working stress} = \frac{\text{Ultimate stress}}{\text{Factor of safety}} = \frac{4{,}200}{4} = 1{,}050 \text{ kg/cm}^2$$

$$\text{Working tension, T} = \text{Working stress} \times \text{cross-sectional area} = 1{,}050 \times 1.25 = 1{,}312.5 \text{ kg}$$

$$\text{Conductor diameter, } d = \sqrt{\frac{\text{Cross-sectional area}}{\pi/4}} = \sqrt{\frac{1.25 \times 4}{\pi}} = 1.2616 \text{ cm}$$

$$\text{Wind force per metre length of conductor, } w_w = \text{Wind pressure} \times \text{conductor diameter} = 100 \times 1.2616 \times 10^{-2} = 1.2616 \text{ kg}$$

$$\text{Resultant force per metre length, } w_r = \sqrt{w_c^2 + w_\omega^2} = \sqrt{1.0^2 + (1.2616)^2} = 1.61 \text{ kg}$$

$$\text{Sag, S} = \frac{w_r L^2}{8T} = \frac{1.61 \times 150^2}{8 \times 1{,}312.5} = 3.45 \text{ m} \textbf{ Ans.}$$

Example 10.5. An overhead line has the following data :

Span length = 160 m
Conductor diameter = 0.95 cm
Weight per unit length of the conductor = 0.65 kg/m
Ultimate stress = 4,250 kg/cm^2
Wind pressure = 40 kg/m^2 of the projected area
Factor of safety = 5

Calculate the sag. [*U.P. Technical Univ. Elements of Power System 2002-03, 2006-07*]

Solution : Span length = 160 m

Weight of conductor per metre length, w_C = 0.65 kg

$$\text{Working stress} = \frac{\text{Ultimate stress}}{\text{Factor of safety}} = \frac{4{,}250}{5} = 850 \text{ kg/cm}^2$$

$$\text{Working tension, T} = \text{Working stress} \times \text{cross-sectional area} = 850 \times \frac{\pi}{4}(0.95)^2 = 602.5 \text{ kg}$$

$$\text{Wind force per metre length of conductor, } w_w = \text{Wind pressure} \times \text{conductor diameter} = 40 \times 0.95 \times 10^{-2} = 0.38 \text{ kg/m}$$

$$\text{Resultant force per metre length, } w_r = \sqrt{w_C^2 + w_w^2} = \sqrt{0.65^2 + 0.38^2} = 0.7529 \text{ kg}$$

$$\text{Sag, S} = \frac{w_r L^2}{8T} = \frac{0.7529 \times 160^2}{8 \times 602.5} = 4 \text{ m} \textbf{ Ans.}$$

Example 10.6. The line supports of an overhead line are 180 m apart and are at the same level. The line conductor is of hard drawn copper having a diameter of 1.27 cm. Calculate the maximum sag (*a*) in still air; and (*b*) in a wind pressure of 33.7 kg per square metre of the projected area of the ice covered

line. The radial thickness of the ice is 1.25 cm. Assume that the maximum stress in the line is not to increase one-fifth of the ultimate strength of 4,220 kg per square cm. The weight of conductor per metre run is 1.13 kg.

Solution : Working stress, $f = \dfrac{\text{Ultimate strees}}{\text{Factor of safety}} = \dfrac{4,220}{5} = 844 \text{ kg/cm}^2$

Area of x-section of conductor, $a = \dfrac{\pi}{4} D^2 = \dfrac{\pi}{4} \times (1.27)^2 = 1.267 \text{ cm}^2$

Working tension, $T = f \times a = 844 \times 1.267 = 1,069.15$ kg

(*a*) Sag in still air $= \dfrac{w_c L^2}{8 T} = \dfrac{1.13 \times (180)^2}{8 \times 1,069.15} = 4.28$ m **Ans.**

(*b*) Weight of ice coating per metre length,

$w_1 = 2,890.3\, r\, (D + r)$ kg/m refer to Eq. (10.29)

$= 2,890.3 \times 1.25 \times 10^{-2} (1.27 + 1.25) \times 10^{-2} = 0.91$ kg

Wind force per metre length of conductor,

w_w = Wind pressure $p \times (D + 2\,r) = 33.7 \times (1.27 + 2 \times 1.25) \times 10^{-2} = 1.2705$ kg

Resultant force per metre length of conductor,

$$w_r = \sqrt{(w_c + w_i)^2 + w_w^2} = \sqrt{(1.13 + 0.91)^2 + (1.2705)^2} = 2.403 \text{ kg}$$

Maximum sag $= \dfrac{w_r L^2}{8 T} = \dfrac{2.403 \times (180)^2}{8 \times 1,069.15} = 9.104$ m **Ans.**

Example 10.7. A transmission line conductor has a diameter of 19.5 mm and weights 0.85 kg/m. The span is 275 meters. The wind pressure is 39 kg/m^2 of the projected area with ice coating of 13 mm. The ultimate strength of the conductor is 8,000 kg. Calculate the maximum sag if the factor of safety is 2 and ice weighs 910 kg/m^3. [*U.P. Technical Univ. Elements of Power System 2005-06*]

Solution : Working tension, $T = \dfrac{\text{Ultimate strength}}{\text{Factor of safety}} = \dfrac{8,000}{2} = 4,000$ kg

Weight of ice coating per metre length,

w_i = Density of ice $\times \pi r (D + r)$

$= 910 \times \pi \times 13 \times 10^{-3} \times (19.5 + 13) \times 10^{-3} = 1.208$ kg

Wind force per metre length of conductor,

$w_w = p \times (D + 2r) = 39(19.5 + 2 \times 13) \times 10^{-3} = 1.7745$ kg

Resultant force per metre length of conductor,

$$w_r = \sqrt{(w_c + w_i)^2 + w_w^2} = \sqrt{(0.85 + 1.208)^2 + (1.7745)^2} = 2.717 \text{ kg}$$

Maximum sag, $S = \dfrac{w_r L^2}{8T} = \dfrac{2.717 \times 275^2}{8 \times 4,000} = 6.422$ m **Ans.**

Example 10.8. A transmission line conductor is having a diameter of 20 mm and weighs 1.0 kg/m. The span is 280 m. The wind pressure is 40 kg/m^2 of projected area with ice coating of 10 mm. The ultimate strength of conductor is 10,000 kg. Calculate the maximum sag if the factor of safety is 2 and ice weighs 910 kg/m^3. [*U.P. Technical Univ. Elements of Power System 2004-05*]

Solution : Weight of conductor per metre length, $w_C = 1.0$ kg

Weight of ice per metre length, w_i = Density of ice $\times \pi r(D + r)$

$= 910 \times \pi \times 10 \times 10^{-3}(20 + 10) \times 10^{-3} = 0.85765$ kg

Wind force per metre length of conductor, $w_w = p \times (D + 2r) = 40 \times (20 + 2 \times 10) \times 10^{-3} = 1.6$ kg

Resultant force per metre length of conductor $= \sqrt{(w_C + w_i) + w_w^2} = \sqrt{(1 + 0.85765)^2 + (1.6)^2}$

$= 2.4517$ kg

$$\text{Maximum sag, } S = \frac{w_r L^2}{8T} = \frac{2.4517 \times 280^2}{8 \times \frac{10,000}{2}} \quad \because T = \frac{\text{Ultimate strength}}{\text{Factor of safety}}$$

$= 4.8$ metres **Ans.**

Example 10.9. An overhead line is erected across a span of 250 metres on level supports. The conductor has a diameter of 1.42 cm and has a dead weight of 1.09 kg/m. The line is subjected to a wind pressure of 37.8 kg per square metre of the projected area. The radial thickness of ice is 1.25 cm. The line is carried by insulator string 1.43 m long. Calculate (*i*) sag in an inclined direction (*ii*) sag in a vertical direction and (*iii*) the height of lowest cross-arm to give a minimum ground clearance of 7.62 m under bad weather conditions. Assume one cubic metre of ice weight 913.5 kg and a maximum stress of 1,050 kg/cm^2.

Solution : Working tension, $T = f \times \frac{\pi}{4} (D^2) = 1,050 \times \frac{\pi}{4} (1.42)^2 = 1,662.86$ kg

Weight of ice coating per metre length,

$$w_i = 913.5 \times \pi r (D + r)$$
$$= 913.5 \times \pi \times 1.25 \times 10^{-2} \times (1.42 + 1.25)\, 10^{-2} = 0.9578 \text{ kg}$$

Wind force per metre length of conductor,

$$w_w = p \times (D + 2r) = 37.8\,(1.42 + 2 \times 1.25) \times 10^{-2} = 1.48176 \text{ kg/m}$$

Resultant force per metre length of conductor,

$$w_r = \sqrt{(w_c + w_i)^2 + w_w^2} = \sqrt{(1.09 + 0.9578)^2 + (1.48176)^2} = 2.5277 \text{ kg}$$

(*i*) Sag in an inclined direction, $S = \frac{w_r L^2}{8T} = \frac{2.5277 \times 250^2}{8 \times 1,662.86} = 11.8756$ m **Ans.**

(*ii*) Sag in vertical direction, $S' = S \cos\theta = \frac{w_r L^2}{8T} \times \frac{w_c + w_i}{w_r}$

$$= \frac{(w_c + w_i)L^2}{8T} = \frac{(1.09 + 0.9578) \times 250^2}{8 \times 1662.86} = 9.62 \text{ m}$$ **Ans.**

(*iii*) Height of lowest cross-arm = Minimum ground clearance + vertical sag + insulator string length

$= 7.62 + 9.62 + 1.43 + 18.67$ m **Ans.**

Example 10.10. Derive the expression for the lowest point of the catenary curve if the supports are at different levels. Given that weight of conductor = 0.35 kg/m, Maximum allowable strength = 800 kg. Span length = 160 m and safety factor = 2, find the minimum point of catenary if supports are at 70 m and 65 m.

[*U.P. Technical Univ. Elements of Power System 2007-08*]

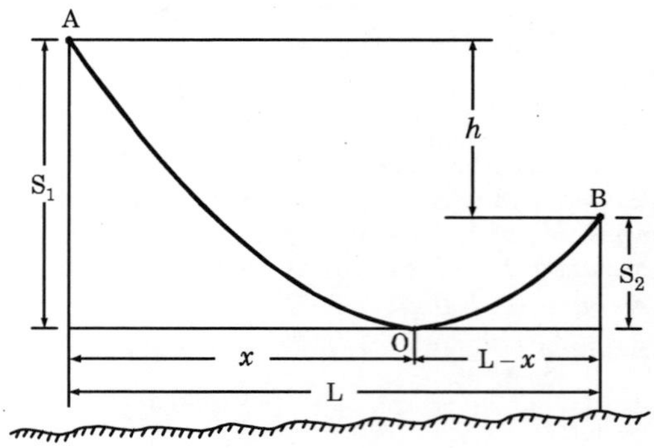

Fig. 10.7

Solution : Weight of conductor per metre length,

$w = 0.35$ kg/m

Maximum allowable stress = 800 kg

Safety factor = 2

$$\text{Allowable tension, T} = \frac{800}{2} = 400 \text{ kg}$$

$$\text{Difference in level of two supports, } h = 70 - 65 = 5 \text{ m}$$

$$\text{Span length, L} = 160 \text{ m}$$

Distance of lowest point of conductor O from the taller support,

$$x = \frac{L}{2} + \frac{Th}{wL} = \frac{160}{2} + \frac{400 \times 5}{0.35 \times 160} = 115.71 \text{ m} \quad \textbf{Ans.}$$

$$\text{Sag S}_1 = \frac{wx^2}{2T} = \frac{0.35 \times 115.71^2}{2 \times 400} = 5.8576 \text{ m}$$

The minimum point of catenary is at 70 – 5.8576, *i.e.* 64.1424 m above ground. **Ans.**

Example 10.11. An overhead line with stranded copper conductors is supported on two poles 200 m apart having a difference in level of 10 metres. The conductor diameter is 2 cm and weighs 2.3 kg per metre length. Calculate the sag at the lower support under the conditions if wind provides a pressure of 57.5 kg per square metre of the projected area and a factor of safety is 4. The maximum tensile strength of copper is 4,220 kg per square cm.

Solution : Horizontal wind force per metre length, w_w = Wind pressure × D = $57.5 \times 2 \times 10^{-2} = 1.15$ kg

$$\text{Resultant force per metre length, } w_r = \sqrt{w_c^2 + w_w^2} = \sqrt{2.3^2 + 1.15^2} = 2.5715 \text{ kg}$$

$$\text{Difference in level of two supports, } h = 10 \text{ metres}$$

$$\text{Working stress, } f = \frac{\text{Maximum tensile strength}}{\text{Factor of safety}} = \frac{4,220}{4} = 1,055 \text{ kg/cm}^2$$

$$\text{Allowable tension, T} = f \times \frac{\pi}{4}(D^2) = 1,055 \times \frac{\pi}{4}(2)^2 = 3,314.38 \text{ kg}$$

Distance of lowest point of conductor O (Fig. 10.4) from the support at low level,

$$x = \frac{L}{2} - \frac{Th}{w_r L} = \frac{200}{2} - \frac{3,314.38 \times 10}{2.5715 \times 200} = 35.556 \text{ m}$$

$$\text{Slant sag, S} = \frac{w_r x^2}{2\,T} = \frac{2.5715 \times (35.556)^2}{2 \times 3,314.38} = 0.4904 \text{ m} \quad \textbf{Ans.}$$

$$\text{Vertical sag, S}' = \frac{w_c x^2}{2\,T} = \frac{2.3 \times (35.556)^2}{2 \times 3,314.38} = 0.4386 \text{ m} \quad \textbf{Ans.}$$

Example 10.12. An overhead transmission line conductor having parabolic configuration weighs 1.925 kg per metre length, area of x-section 2.2 cm² and an ultimate strength of 8,000 kg/cm². When erected between supports 600 metre apart and having 15 metre difference in height, determine the vertical sag from the taller of the two supports which must be allowed so that the factor of safety shall be 5 with the wire loaded due to 1 kg of ice per metre and no wind pressure.

Solution : Effective weight of conductor per metre length,

$$w = 1.925 + 1 = 2.925 \text{ kg}$$

$$\text{Difference in level of two supports, } h = 15 \text{ metres}$$

$$\text{Allowable tension, T} = \frac{8,000 \times 2.2}{5} = 3,520 \text{ kg}$$

Distance of lowest point of the conductor, O (Fig. 10.4) from the support at low level

$$x = \frac{L}{2} - \frac{Th}{wL} = \frac{600}{2} - \frac{3,520 \times 15}{2.925 \times 600} = 270 \text{ metres}$$

Distance of lowest point of the conductor from the taller support

$$= L - x = 600 - 270 = 330 \text{ metres}$$

Vertical sag from the taller of the two supports

$$S_2 = \frac{w(L-x)^2}{2T} = \frac{2.925 \times (330)^2}{2 \times 3,520} = 45.25 \text{ m } \textbf{Ans.}$$

Example 10.13. A transmission line conductor at a river crossing is supported from two towers at heights of 50 m and 80 m above water level. The horizontal distance between the towers is 300 metres. If the tension in the conductor is 2,000 kg, find the clearance between the conductor and water level at a point midway between the towers. Weight of conductor per metre = 0.844 kg. Assume that the conductor takes the shape of parabolic curve.

Solution : Difference in level between the two supports, $h = 80 - 50 = 30$ m

Distance of lowest point of conductor, O from the support at low level

$$x = \frac{L}{2} - \frac{Th}{wL} = \frac{300}{2} - \frac{2,000 \times 30}{0.844 \times 300} = -86.967 \text{ m}$$

The negative value of x shows that support A is on the same side of O as support B (Fig. 10.8)

$$\text{Distance of midpoint P from O} = \frac{L}{2} - x = \frac{300}{2} - (-86.967) = 236.967 \text{ metres}$$

$$\text{Distance of point B from O} = L - x = 300 + 86.967 = 386.967 \text{ metres}$$

$$\text{Height of midpoint P above O, } S_{mid} = \frac{w\left(\frac{L}{2} - x\right)^2}{2T} = \frac{0.844 \times (236.967)^2}{2 \times 2,000} = 11.848 \text{ m}$$

$$\text{Height of point B from O, } S_2 = \frac{w \times (L-x)^2}{2T} = \frac{0.844 \times (386.967)^2}{2 \times 2,000} = 31.596 \text{ m}$$

Hence, mid point P is (31.596 – 11.848) *i.e.* 19.748 metres below point B or (80 – 19.748) *i.e.* 60.252 m above the water level **Ans.**

Example 10.14. An overhead transmission line at a river crossing is supported from two towers at heights of 25 m and 75 m above the water level. The horizontal distance between the towers is 250 m. If the required clearance between the conductor and the water mid way between the towers is 45 metres and if both the towers are on the same side of the point of maximum sag of the parabolic configuration, find the stringing tension in the conductor. The weight of the conductor is 0.7 kg/m.

Solution : Let T be the stringing tension and x be the distance between the lowest point O of the conductor and the support at low level, as shown in Fig. 10.8

$$\text{Since } h = S_2 - S_1$$

$$\text{or } h = \frac{w(L-x)^2}{2T} - \frac{wx^2}{2T}$$

$$= \frac{wL^2 - 2wLx}{2T}$$

$$\text{or } h = \frac{wL}{2T}(L - 2x) \quad \ldots(i)$$

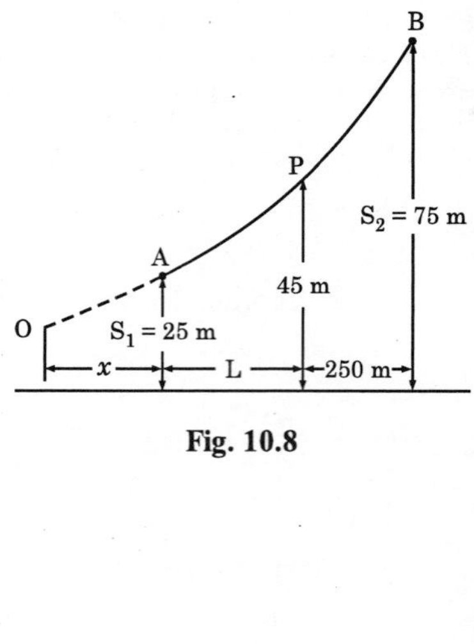

Fig. 10.8

For points A and B $\quad h = 75 - 25 = 50$ metres

$L = 250$ metres

and For points A and P (mid way point between the towers)

$h = 45 - 25 = 20$ metres

and $L = 125$ metres

Substituting the above values of h and L in Eq. (*i*) we get

$$50 = \frac{0.7 \times 250}{2T}[250 - 2x] \quad ...(ii)$$

$$20 = \frac{0.7 \times 125}{2T}[125 - 2x] \quad ...(iii)$$

Dividing Eq. (*ii*) by Eq. (*iii*), we get

$$2.5 = 2 \times \frac{250 - 2x}{125 - 2x}$$

$$\text{or } 2.5\ (125 - 2x) = 500 - 4x$$

$$\text{or } x = -\ 500 + 312.5 = -\ 187.5 \text{ metres}$$

Substituting, $x = -\ 187.5$ in Eq. (*ii*) we get

$$50 = \frac{0.7 \times 250}{2T}[250 - 2 \times (-\ 187.5)]$$

$$\text{or } T = \frac{0.7 \times 250}{2 \times 250}[250 + 375] = 1{,}093.75 \text{ kg } \textbf{Ans.}$$

Example 10.15. An overhead line with a span of 300 m and in a hilly area with uniform ground slope of 1 in 20 and having a conductor weight of 0.80 kg/m, is supported between two towers at equal height of 30 m above the ground level. Assuming the horizontal component of tension in the line to be 1,500 kg; find the clearance at the point where conductor has its lowest elevation : and (*ii*) the minimum clearance of the line from ground.

Solution : The lowest conductor AOB and the profile of the ground CE are shown in Fig. 10.9. Take origin O at the lowest point of conductor. Let OX and OY be the horizontal and vertical reference axis.

$$\text{Slope} = \tan \alpha = 1/20$$

$$\text{ED} = \text{CD} \tan \alpha = \text{L} \tan \alpha = 300 \times \frac{1}{20} = 15 \text{ m}$$

Height of the lowest conductor from the ground = AC = BE = 30 m

Height BD = BE + ED = 30 + 15 = 45 m

The equation of parabola AOB is given by

$$y = \frac{wx^2}{2T_0}$$

$$\text{Thus } S_1 = \frac{wx_1^2}{2T_0}$$

$$\text{and } S_2 = \frac{w(L - x_1)^2}{2T_0}$$

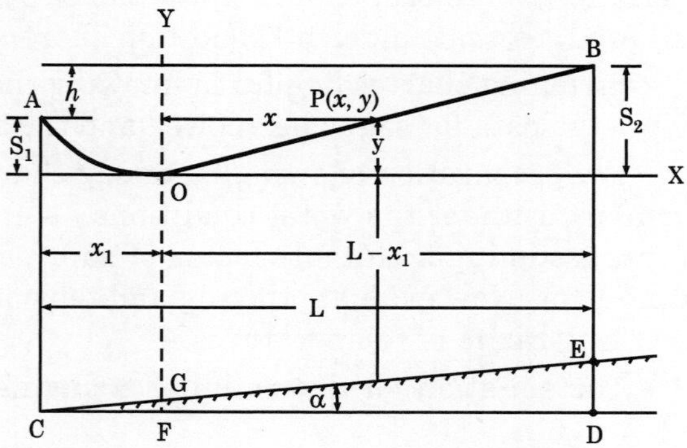

Fig. 10.9

Vertical distance between the supports

$$h = \text{ED} = 15 \text{ m}$$

$$\text{Also } h = S_2 - S_1 = \frac{w(L - x_1)^2}{2T_0} - \frac{wx_1^2}{2T_0} = \frac{w}{2T_0}(L^2 - 2Lx_1)$$

$$\text{or } x_1 = \frac{L}{2} - \frac{T_0 h}{wL} = \frac{300}{2} - \frac{1{,}500 \times 15}{0.8 \times 300} = 56.25 \text{ m}$$

$$S_1 = \frac{wx_1^2}{2T_0} = \frac{0.8 \times (56.25)^2}{2 \times 1{,}500} = 0.84375 \text{ m}$$

and $S_2 = \dfrac{w(L - x_1)^2}{2T_0} = \dfrac{0.8\,(300 - 56.25)^2}{2 \times 1,500} = 15.84375$ m

Clearance of the lowest point of the lowest conductor from the ground

$$= OG = AC - S_1 - GF = 30 - 0.84375 - CF \tan\alpha \text{ (or } x_1 \tan\alpha)$$

$$= 30 - 0.84375 - 56.25 \times \frac{1}{20} = 26.34375 \text{ m } \textbf{Ans.}$$

Equation of the slope of the ground is given as

$$y = mx + C = x \tan\alpha - OG = \frac{x}{20} - 26.34375$$

Clearance from ground at any point P (x,y)

$$C_1 = y - \left(\frac{x}{20} - 26.34375\right) = \frac{wx^2}{2T_0} - \frac{x}{20} + 26.34375$$

For minimum value of C_1, $\dfrac{dC_1}{dx} = 0$

or $\dfrac{2wx}{2T_0} - \dfrac{1}{20} = 0$ or $x = \dfrac{T_0}{20w} = \dfrac{1,500}{20 \times 0.8} = 93.75$ m

Hence minimum clearance of the line from ground will be for value of $x = 93.75$ m

and minimum clearance $= \dfrac{wx^2}{2T_0} - \dfrac{x}{20} + 26.34375 = \dfrac{0.8(93.75)^2}{2 \times 1,500} - \dfrac{93.75}{20} + 26.34375 = 24$ m **Ans.**

10.5. STRINGING CHART

Under statutory conditions, the sag is required to be determined for worst probable conditions and the minimum ground clearance is to be maintained for these conditions. At the time of erection the severe conditions do not prevail, the temperature is usually higher, the designer, therefore, should know the sag to be allowed and the tension in the line to be allowed, so that under no condition there should be any danger to the line.

Stringing chart is helpful in knowing the sag and tension at any temperature. This chart gives the data for sag to be allowed and the tension to be allowed at a particular temperature.

For preparation of stringing or sag chart first of all calculate the sag and tension on the conductor under the worst conditions *i.e.* maximum wind pressure and minimum temperature, assuming a suitable factor of safety in fixing the maximum working tension for the conductor. Now evaluate the sag and tension for a series of temperatures in steps within the working range of temperatures.

The equation for determining stringing chart of a line is derived as below:

$$\text{We know that } T = T_0;\; y = \frac{wx^2}{2T};\; S = \frac{wL^2}{8T} \text{ and span length, } l = L + \frac{w^2L^3}{24T^2}$$

Let w_1, f_1, l_1, S_1 and t_1 be the load per unit length, the stress, the span length, sag and temperature at the maximum load conditions (with the ice and wind and low temperature usually – 5.5°); w_2, f_2, l_2, S_2 and t_2 be the values under stringing conditions, a is the area of x-section of the conductor, α is the coefficient of linear expansion and E is the modulus of elasticity.

$$\text{Span length, } l = L + \frac{w^2L^3}{24T^2} = L + \frac{w^2L^3}{24a^2f^2} = L + \frac{8S^2}{3L} \qquad ...(10.35)$$

The span length at maximum load condition is

$$l_1 = L + \frac{w_1^2 L^3}{24 a^2 f_1^2} \quad ...(10.36)$$

The temperature rise from t_1 to t_2 causes an increase in the span length of $l_1 \alpha (t_2 - t_1)$ which is practically equal to $L \alpha (t_2 - t_1)$. The fall in stress from f_1 to f_2 causes a decrease in the length of $\left(\frac{f_2 - f_1}{E}\right) l_1 \simeq \left(\frac{f_1 - f_2}{E}\right) L$

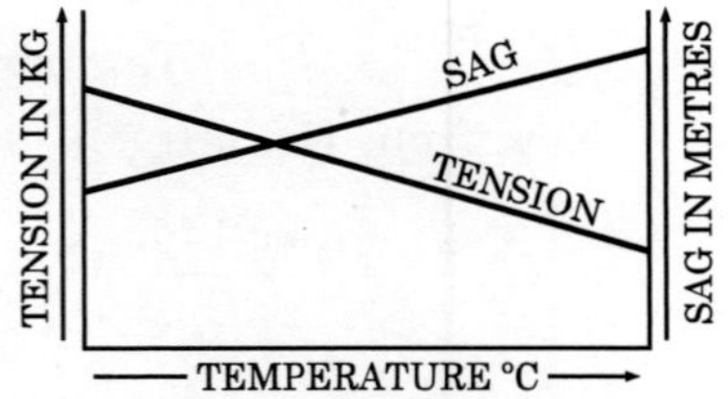

Fig. 10.10

The new length l_2 is thus given by

$$l_2 = l_1 - \frac{f_1 - f_2}{E} L + (t_2 - t_1) \alpha L \quad ...(10.37)$$

$$\text{But } l_2 = L + \frac{w_2^2 L^3}{24 a^2 f_2^2} \quad ...(10.38)$$

$$\therefore L + \frac{w_2^2 L^3}{24 a^2 f_2^2} = L + \frac{w_1^2 L^3}{24 a^2 f_1^2} - \frac{f_1 - f_2}{E} L + (t_2 - t_1) \alpha L$$

$$\text{or } f_2^2 \left[(f_2 - f_1) + \frac{w_1^2 L^2 E}{24 f_1^2 a^2} + (t_2 - t_1) \alpha E \right] = \frac{w_2^2 L^2 E}{24 a^2} \quad ...(10.39)$$

Equation (10.39) is a cubic equation and can be solved graphically or analytically. From this equation erection tension $T_1 = f_1 a$ can be determined such that tension $T_2 = f_2 a$ under worst probable conditions will not exceed the safe limit of tension. After determining f_2, the corresponding sag can be determined from the equation

$$S_2 = \frac{w_2 L^2}{8 f_2 a} \quad ...(10.40)$$

Various values of f_2 and S_2 are calculated using Eqs. (10.39) and (10.40) repeatedly for different temperatures. Now the graph of tension vs temperature and sag vs temperature can be plotted, as shown in Fig. 10.19. This graph is plotted for a fixed span and is called the stringing chart. The stringing chart is very useful while erecting the transmission line conductors for adjusting the sag and tension properly.

Example 10.16. An overhead line having a span of 250 m has ACSR conductors of 19.5 mm diameter; 2.25 cm^2 area and weighing 0.85 kg/m. This line is to be erected at a temperature of 35° in still air conditions. It is desired that a factor of safety 2 should be maintained under worst weather conditions (*i.e.* when the temperature is 5°C and the wind load is 38.5 kg/m^2 of projected area). Determine the sag and tension under erection conditions if breaking load is 8,000 kg, coefficient of linear expansion 18.44 × 10^{-6} per degree °C and Young's modulus 9,320 kg/mm^2.

Solution : *Under worst weather conditions*

Temperature, $t_2 = 5°$ C

$$\text{Stress, } f_1 = \frac{T_1}{a} = \frac{\text{Breaking load}}{\text{Factor of safety}} \times \frac{1}{\text{area of x-section}}$$

$$= \frac{8{,}000}{2 \times 2.25 \times 10^{-4}} = 1.778 \times 10^7 \text{ kg/m}^2$$

---

* Span length means length of conductor between two supports.

Wind force per unit length, w_w = Wind pressure × D = 38.5 × 1.95 × 10^{-2} = 0.75 kg

Effective load per unit length, $w_1 = \sqrt{w_c^2 + w_w^2} = \sqrt{0.85^2 + 0.75^2} = 1.1336$ kg

Under erection conditions

Temperature, t_2 = 35°C and w_2 = 0.85 kg/m

Now from Eq. (10.39)

$$f_2^2\left[(f_2 - f_1) + \frac{w_1^2 L^2 E}{24 f_1^2 a^2} + (t_2 - t_1)\,\alpha\, E\right] = \frac{w_2^2 L^2 E}{24\, a^2}$$

$$\text{or } f_2^2\left[(f_2 - 1.778\times10^7) + \frac{(1.1336)^2\times(250)^2\times 9.32\times10^9}{24\times(1.778\times10^7)^2\times(2.25\times10^{-4})^2} + (35-5)\times 18.44\times10^{-6}\times 9.32\times10^9\right]$$

$$= \frac{(0.85)^2\times(250)^2\times 9.32\times10^9}{24\times(2.25\times10^{-4})^2}$$

or $f_2^2\ [(f_2 - 1.778\times10^7 + 0.195\times10^7 + 0.5156\times10^7] = 3{,}463.84\times10^{17}$

or $f_2^2\ (f_2 - 1.0674\times10^7) = 3{,}463.84\times10^{17}$

or $f_2^3 - 1.0674\times10^7\times f_2^2 - 3{,}463.84\times10^{17} = 0$

By hit and trial, $f_2 = 1.279\times10^7$ kg/m²

$$\text{Sag at erection} = \frac{w_2 L^2}{8 f_2 a} = \frac{0.85\times(250)^2}{8\times1.279\times10^7\times2.25\times10^{-4}} = 2.3 \text{ metres } \textbf{Ans.}$$

10.6. SAG TEMPLATE

In the initial planning stages a survey of the proposed route enables an estimated line profile to be drawn, as illustrated in Fig. 10.11(*a*). Such a profile is constructed with the horizontal scale much more reduced (say to 1/10) in comparison to vertical scale. This profile should meet the minimum clearance requirements, and the location of the supports should be such that some horizontal adjustment is possible without departing from the standard or tangent tower lengths.

Use is made of *sag templates* drawn to the same scale as the line profile, to ensure the required clearance. The sag template is usually made on celluloid or tracing cloth. In sag template, shown in Fig. 10.11(*b*), the upper curve I represents the conductor line. The middle curve (*i.e.* curve II) is below the upper curve I by a uniform vertical distance equal to the desired minimum vertical clearance to ground. This clearance to ground is governed by the operating voltage and is given, according to IE rules, in Table 10.2. The lower curve (curve III) is below the upper curve by a uniform vertical distance equal to the height of a standard tower measured to the point of support of the conductor. If the location of the left tower has been decided , the location of the right hand tower can be determined by adjusting the sag template so

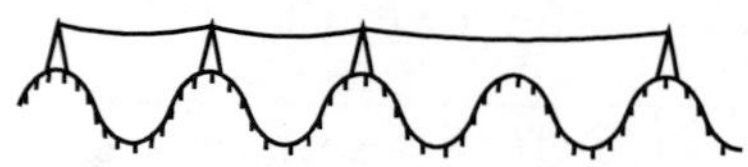

(*a*) Line Profile

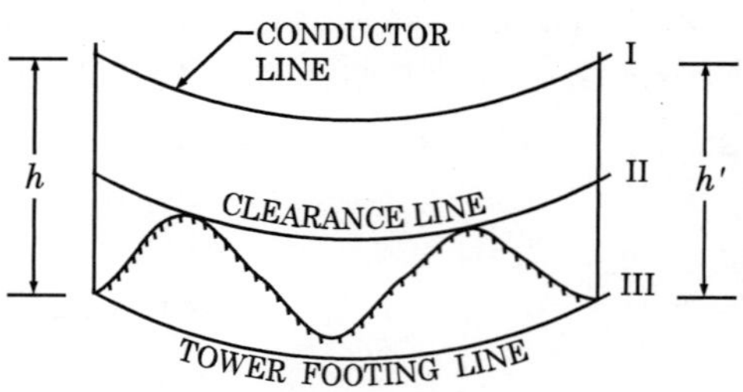

(*b*) Sag Template

Fig. 10.11

that the conductor line passes through the point of support on the left hand tower and the clearance line is tangent to ground at one or more points.

In the above particular case, illustrated the points of conductor support are upon the same horizontal level, but the same process applies when the route is a steeply sloping one. However, the shape and position of the conductor will always be represented by curve I, as shown in Fig. 10.11(*b*).

Sag template is very a convenient method for allocating the positions and height of the towers/supports correctly on the profile.

10.7. EQUIVALENT SPAN

It may not be possible to have a section of transmission line consisting of successive spans of equal lengths because the location of the towers depends upon the profile of land along which the transmission line is to be laid. Sometimes also, the towers are forced to be located to give spans of different lengths so that minimum interference is caused with the use of land. When the successive spans are of unequal lengths changes in tension in load or temperature will cause unequal changes in tension in the different spans.

It is very tedious to make calculations of sag and tension for each and every span individually and then to make adjustment while erecting the transmission line. In the erection of a transmission line the conductors are run out through snatch blocks attached to the support arms equally tensioned at each end of a section of five or six blocks. When the conductors are clamped to suspension insulator strings, the equal tension is maintained by insulator swing. When the conductors are bound to pin-type insulators, the flexibility of the supports ensures equal tension.

It is often convenient to make sag and tension calculations in terms of a *hypothetical equivalent span,* this tension being applied to each span within the section of overhead line between the tensioning points. If there are n spans of length L_1, L_2, L_3, etc. which are to be given an equivalent span L_e, then the strung length of the equivalent line must be the same as that of the individual spans. This may be expressed as

$$n\left(L_e + \frac{w^2 L_e^3}{24\, T_0^2}\right) = \Sigma\, L + \frac{w^2(\Sigma\, L^3)}{24\, T_0^2} \qquad \text{...(10.41)}$$

Since $\Sigma L \equiv nL_e$

$$\text{So } L_e^2 = \frac{\Sigma L^3}{nL_e} = \frac{\Sigma L^3}{\Sigma L} \text{ or } L_e = \sqrt{\frac{L_1^3 + L_2^3 + L_3^2 + \ldots}{L_1 + L_2 + L_3 + \ldots}} \qquad \text{...(10.42)}$$

If the line tension T_0 is determined for the equivalent span, the sag for the individual spans may be calculated from Eqs. (10.17) using the approximate value of span.

It is to be noted that the method of sag template for locating the towers should not be used for long spans as well as where the slope of the profile is very steep. In such cases, actual calculations for sag and tension should be made.

10.8. VIBRATIONS AND DAMPERS

The overhead transmission line experiences vibrations in the vertical plane and there are two types of such vibrations, in addition to normal swinging in wind, called the *aeoline vibrations* or *resonant vibrations* or *high frequency oscillations* and *galloping* or *dancing* or *low frequency vibrations.*

Simple swinging of conductors is harmless provided the clearance is sufficiently large so that the conductors do not approach within the sparking distance of each other.

Aeoline vibrations are high frequency (5–100 Hz) and low amplitude (20 mm to 50 mm) vibrations. They are caused by vortex phenomenon in light winds (5-20 km/hour). The line conductor vibrates in a number of loops. The length of the loop (half-wave length) depends on tension T and weight of conductor w per metre length and is given by $\frac{1}{2f}\sqrt{\frac{T}{w}}$. The loop length varies from 1 to 10 metres depending upon T, w and f. These vibrations are common to all conductors and are more or less always present. The harmful effects of such vibrations take place at clamps or supports where the conductor suffers fatigue and breaks eventually.

Low frequency vibrations (about one Hz) occur during sleet storms with a strong wind. The amplitude is very large, about 6 metres or more, and the conductors are said to 'dance'. Operation is almost impossible for the conductors touch one another, since the 'dancing' takes place horizontally and vertically. It is considered that the phenomenon takes place due to the fact that the line conductor receives an irregular coating of sleet. If the cross-section of the coating is considered to be an ellipse, the line will experience a drag; further more, if the major axis of an ellipse makes an angle of 45° to the wind, there will be a considerable aerodynamic lift (or downward thrust) such as is experienced by an aerofoil. The conductor will thus dance horizontally and vertically with a large amplitude and in an irregular manner owing to the irregular deposition of sleet. The stranding of conductors also contributes to these vibrations. The travel of conductor follows the path of an ellipse. There is no method for prevention of these (low-frequency) vibrations. However, danger due to such vibrations can be reduced if horizontal conductor configuration is used.

The conductors are protected by dampers which prevent the resonant vibrations from reaching the conductors at the clamps or supports.

The stock bridge damper, shown in Fig. 10.12, consists of two weights attached to a piece of stranded cable 0.3 or 0.5 m long, which is clamped to the line conductor. The energy of vibration is absorbed by the stranded cable, and the vibration is rapidly damped out.

Fig. 10.12. ***Stock Bridge Damper***

Another successful damper consists of a box containing a weight resting on a spring. In this case the spring absorbs the energy of vibration. The fatigue may be reduced by reinforcing the conductor for a few metres on either side of the clamp by binding metal rods or a length of the same conductor to the main conductor outside the clamp.

EXERCISES

1. Prove that a transmission line conductor between two supports at equal heights takes the form of a catenary.
2. Show that an overhead line conductor strung between two level supports takes up the shape of a parabola. Obtain expressions for the sag and maximum tension when the supports are at the same level.

3. Discuss in brief, the factors on which sag of the OH line depends. Write the expression of sag (approximate formula) and explain the terms with units.
4. Derive the expression for sag when the conductor is supported by supports at equal height. **[U.P. Technical Univ. Elements of Power System 2009-10]**
5. Derive expression for sag and tension in a power conductor strung between two supports at equal heights, taking into account the wind and ice loading also. **[U.P. Technical Univ. Elements of Power System 2003-04; 2008-09]**
6. Derive the expression of vertical sag in situations where the supports are at differents levels, consider the effect of wind and ice also. **[U.P. Technical Univ. Elements of Power System 2005-06]**
7. Explain the necessity of a stringing a chart for a transmission line and show how such a chart can be constructed. **[U.P. Technical Univ. Elements of Power system 2008-09]**
8. What is stringing chart. How is it plotted for different temperatures ? **[U.P. Technical Univ. Elements of Power System 2007-08]**
9. Discuss the usefulness of stringing chart. How it is plotted ? **[U.P. Technical Univ. Elements of Power System 2004-05]**
10. What is the use of a sag template ? Describe it in detail. **[U.P. Technical Univ. Elements of Power System 2009-10]**
11. Why do the vibrations get generated in conductors ? How are they damped. **[U.P. Technical Univ. Elements of Power System 2009-10]**
12. Explain various vibration dampers. **[U.P. Technical Univ. Elements of Power System 2004-05]**
13. Describe the vibration of power conductors and explain the methods used to damp out these vibratiuons. **[U.P. Technical Univ. Elements of Power System 2007-08]**
14. Write short notes on the following:
    (*i*) Sag template.
    (*ii*) Conductor vibrations.
    (*iii*) Vibration dampers.
    (*iv*) Method of damping vibration of power conductors.

SHORT ANSWER TYPE QUESTIONS WITH ANSWERS

Q. 1. What are the factors on which conductor spacings and ground clearance depend ?

Ans. Factor affecting conductor spacings is voltage while factors affecting ground clearance are voltage and area (living or non-living, urban or suburban, along the streets or across the streets, river crossings/hill crossings, railway track/bridge, etc).

Q. 2. What is sag ?

Ans. The difference in level between the points of line supports and the lowest point on the conductor is known as sag.

Q. 3. What are the factors affecting the sag in a transmission line ?

Ans. Factors affecting the sag in a transmission line are weight of conductor, span length (distance between the poles), working tensile strength, ice coating, wind pressure and temperature.

Q. 4. What is stringing chart ?

Ans. Stringing chart gives the data for sag and tension to be allowed at a particular temperature. It is very useful while erecting the transmission line conductors for adjusting the sag and tension properly.

PROBLEMS

1. Find the sag of a transmission line with following particulars :
   Span-500 m, Conductor weight per m = 1 kg, Tension 1,500 kg. [**Ans.** 20.833 m]
2. An OH line has a span of 200 metres between level supports. The conductor diameter is 1 cm and weighs 0.65 kg/metre length. The allowable tension is 550 kg. Calculate the maximum sag. [**Ans.** 5.91 m]
3. An overhead line has a span of 150 m between level supports. The conductor diameter is 0.94 cm and weighs 0.62 kg/m length. The allowable tension is 586 kg. Calculate the sag if the wind pressure is 39.2 kg/m^2 of a projected area. [**Ans.** 3.462 m]
4. A 132 kV OH line uses ACSR conductor 19.53 mm diameter, weight 844 kg per km, ultimate strength 7,950 kg. Calculate the height above ground at which the conductors with a span of 300 metres shall be supported, the factor of safety being 2, wind pressure 75 kg per square metre of projected area. Ground clearance required is 7 metres. [**Ans.** 4.784 m]
5. An overhead line is erected across a span of 250 m on level supports. The conductor has a diameter of 1.42 cm and has a dead weight of 1.09 kg per metre. The line is subjected to a wind pressure of 37.8 kg per square metre of the projected area. The radial thickness of ice is 1.25 cm. Calculate the sag (*a*) in an inclined direction, (*b*) in a vertical direction. Assume a maximum working stress 1,050 kg per square cm. One cubic metre of ice weighs 913.5 kg. [**Ans.** (*a*) 11.874 m; (*b*) 9.621 m]
6. A transmission line conductor having a diameter of 19 mm weighs 0.82 kg/m. The span is 250 m. The wind pressure is 39 kg/m^2 of projected area with ice coating of 12 mm. The ultimate strength of the conductor is 8,000 kg. Calculate the maximum sag if the factor of safety is 2.5 and ice weighs 900 kg/m^3. [**Ans.** 6.135 m]
7. An overhead line having a span of 300 m has ACSR conductors of 1.95 cm diameter. The allowable tension is 35,600 N. Find (*i*) sag in still air condition with no ice covering. (*ii*) vertical sag when there is an ice coating of 0.8 cm and a horizontal wind pressure of 400 N/m^2 of projected area. Ice density 8,900 N/m^3. (*iii*) What should be the height of the lowest cross-arm to give a minimum ground clearance of 7.8 m under bad weather conditions ?
   Line is carried by insulator string 1.6 m long. Weight of conductor = 0.85 kg/m
   [**Ans.** (*i*) 2.635 m (*ii*) 4.58 m (*iii*) 13.98 m]
8. A transmission line conductor consists of hard drawn copper 240 mm^2 in x-section and has a span of 160 metres, the supporting structures being level. The conductor has an ultimate strength of 42.2 kg/mm^2 and the desired factor of safety is 5. Find the vertical sag with a wind pressure of 1.35 kg/metre length and an ice coating of 1.25 cm. Take the density of hard drawn copper as 8.9 gm/cc and weight of ice as 915 kg/m^3. [**Ans.** 4.22 metres]
9. An overhead conductor consists of 7 strands of phosphor bronze having an ultimate tensile strength of 8,000 kg/cm^2 and an area of 2.2 cm^2 when erected between supports 600 metres apart having 15 metre difference in level. Find the vertical sag which must be allowed so that the factor of safety is 5. Assume the wire to be loaded with 1 kg/m of ice. The wire weighs 2.03 kg/m. [**Ans.** 16.2 metres from the lower support]
10. An overhead transmission line at a river crossing is supported, from two towers 600 metres apart and having a level difference of 60 metres. The conductors consist of 7 strands of silicon bronze wire having an ultimate strength of 8,000 kg/cm^2 and an area of 2.2 cm^2. Find

the vertical sag which must be allowed so that factor of safety shall be 4 with the wires loaded with 1.0 kg of ice per metre and a wind pressure of 1.75 kg per metre. Specific gravity of silicon bronze is 8.735. Assume the curve to be a parabola. [**Ans.** 69.65 metres]

11. A transmission line conductor at a river crossing is supported from two towers at heights of 40 m and 90 m above water level. The horizontal distance between towers is 275 m. If the tension in the conductors is 1,362 kg. Find the clearance between the conductor and water at a point mid-way between the towers. Weight of conductor per metre run is 0.8 kg.

    [**Ans.** 59.46 m]

12. An overhead transmission line spanning over a river is supported at two ends by towers at 45 m and 75 m. Find the clearance between the conductor and water at a point midway between the towers. The tension is to be limited to 2,500 kg. The weight of the conductor is 0.9 kg/m and the distance between the towers is 300 metres. [**Ans.** 55.95 m]

13. A transmission line conductor at a river crossing is supported from two towers at heights of 30 m and 90 m above water level. Horizontal distance between the towers is 300 m. If the tension in the conductor is 1,760 kg*f* and the conductor weight is 1 kg per metre, find the clearance between the conductor and the water at a point midway between towers. Assume parabolic configuration. [**Ans.** 68.61 m]

14. A hilly terrain has a gradient of 1 in 20 and two towers of 36 m height are considered with a horizontal distance 300 m. An overhead line is supported by these towers wherein the lowest conductor is fixed 6 m below the top of each tower. The weight of each line conductor is 1.5 kg*f*/m and the tension 2,200 kg*f*. Assume parabolic configuration of sag and deduce and develop a method to determine the clearance of the lowest point of the conductor from the ground and the minimum clearance. [**Ans.** 24.17 m, 22.337 m]

❑❑❑

11

CHAPTER

INSULATED CABLES

INSIDE THIS CHAPTER

11.1. Introduction **11.2.** General Construction of Cable **11.3.** Requirements of Cable **11.4.** Cable Conductors **11.5.** Insulating Materials For Cables **11.6.** Classification of Cables **11.7.** Insulation Resistance of a Single Core Cable **11.8.** Capacitance of a Single Core Cable **11.9.** Dielectric Stress in a Single Core Cable **11.10.** The Most Economical Diameter of Conductor **11.11.** Grading of Cables **11.12.** Capacitance of Three-core Belted Type Cables **11.13.** Heating of Cables **11.14.** Operating Problems Associated With Underground Cables.

11.1. INTRODUCTION

As already discussed, in chapter 2, the merits and demerits of transmission and distribution of electric power by underground cables in comparison with that by overhead system, the initial heavy cost* is the only factor which discouraged the use of underground cables for the purpose of transmission and distribution of electric power inspite of its numerous advantages. Underground cables are used for transmission and distribution of power where it becomes impracticable to make use of overhead construction. Such location may be congested urban area where right-of-way cost would be excessive or local ordinances prohibit overhead lines for reasons of safety, or around plants and substations; or crossings of wide bodies of water which for various reasons would not permit the overhead crossings. The type of cable used will depend upon the voltage and service requirements. Recent improvements in design and manufacture have led to the development of cables suitable for use at the highest voltages associated with modern transmission systems.

An electric cable may be defined as a single conductor insulated through its full length; or two or more conductors each provided with its own insulation and laid up together under one outer protective covering.

11.2. GENERAL CONSTRUCTION OF CABLE

The general construction of an underground cable is given below:

The underground cable employed for transmission of power at high voltage consists of one central core or a number of cores (two, three or four) of tinned stranded copper or aluminium

---

* The approximate cost ratios are as below :

System Operating Voltage in kV	0.4	11	33	66	132	220	400
Cost Ratio of UG/OH System	2	3	5	7	9	13	24

conductors insulated from each other by paper or varnished cambric or vulcanized bitumen or impregnated paper. A metallic sheath of lead or alloy or of aluminium is provided around the insulation to protect it against ingress of moisture, gases or other damaging liquids (acid or alkalies) in the soil and atmosphere. A metal is essential for sheath because no organic material is sufficiently impervious to moisture. The main advantages of lead sheaths are the comparative ease with which they are made in lead presses, their flexibility and high corrosion resistance while they have got drawbacks of large specific gravity, low mechanical strength, fluidity and small resistance to vibrations. The hardness, mechanical strength and resistance to vibrations of lead sheaths can be considerably improved by adding alloying mixtures (about 1 per cent of copper, tin, bismuth, etc). Aluminium sheaths being cheaper (costing 3.5 times less than the lead sheaths) and having much greater mechanical strength than lead sheaths, low weight and low fluidity are now being increasingly employed.

For the protection of metallic sheath against corrosion and from mechanical injury a layer of *bedding* consisting of paper tape compounded with a fibrous material (such as jute, cotton or hessian tape) is provided over the metallic sheath. The compound used should be of such a nature that it does not react with the material of the armouring and the lead sheath but simultaneously it is adhesive enough so as to stick both to the lead sheath and the armouring. Bedding is employed in paper insulated lead covered cables. It is not required in PVC cables.

Over the layer of bedding *armouring* consisting of one or two layers of galvanized steel wire or steel tape is provided to save the cable from mechanical injury and over the armouring a layer of fibrous material similar to that of bedding, known as *serving,* is provided in order to protect the armouring. Steel tape armour consists of two steel tapes coated with preservative compound, applied hellically in the same direction over a fibrous bedding, the outer layer covering the spaces between the turns of the inner layer. This provides sufficient protection against mechanical damage but steel wire armour is recommended where additional longitudinal stresses may occur during installation or in service. Double wire armours are employed for high degree of mechanical protection.

Single core cables for ac systems are not provided with armouring because eddy currents induced in the steel armour causes additional power losses. In such cables mechanical protection is provided by plastic wrap. Aluminium sheathed cables, owing to comparative hardness of sheath, need not to be armoured in the same way as the lead sheathed cables but for special conditions armouring is provided for these cables too.

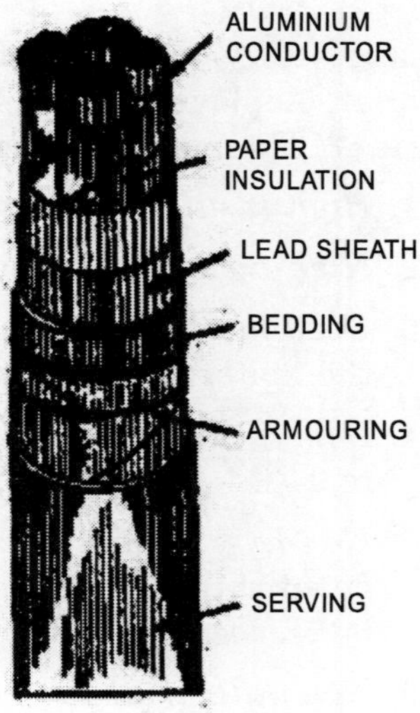

Fig. 11.1

Fig. 11.1 illustrates the various parts of an underground cable.

11.3. REQUIREMENTS OF CABLE

(*i*) The conductor (usually annealed copper with about 99% purity or aluminium) used in cables should be stranded one in order to provide flexibility to the cables and should be of such x-sectional area that it may carry the desired load current without overheating and causing excessive voltage drop.

(*ii*) The insulation provided should be of such thickness that it may give high degree of safety and reliability at the working voltage for which it is designed.

(*iii*) The cables should be provided with a mechanical protection so that it may withstand the rough usage in laying it.

(*iv*) Materials used in manufacture of cables should be such as to give complete chemical and physical stability throughout.

11.4. CABLE CONDUCTORS

For all cables the conductor was, at one time, almost universally stranded copper but because of scarcity of copper, only aluminium is allowed to be used in power cables in India. Copper conductors are only used for cables used in control circuits.

The stranding is necessary to provide flexibility to cables and the conductor numbers are 3, 7, 19, 37, and so on, all numbers except 3 having a centrally disposed conductor with all the others around it. Thus a 7-strand has a central conductor with 6 around it. The various conductors are spiralled round the central conductor, and when there is more than one layer alternate layers are spiralled in opposite directions. This is to prevent 'bird-caging' when the conductor is bent.

The spiralling of the conductors obviously causes the increase in resistance, since the length of each spiralled conductor is greater than the central strand, and the flow of current is along the various conductors. It is therefore necessary to be particular in the designation of the cross-section.

11.5. INSULATING MATERIALS FOR CABLES

The satisfactory operation of a cable depends to a great extent upon the characteristics of insulation employed. So the proper choice of insulating material for cables is considerably important. The main requirements of the insulating materials used for cables are :

1. *High insulation resistance* to avoid leakage current.
2. *High dielectric strength* to avoid electrical breakdown of the cable.
3. *Good mechanical properties i.e.* tenacity and elasticity. Good tenacity is required in the material to withstand the mechanical handling of cables.
4. *Immune to attacks by acids and alkalies,* over a range of temperature of about – 18°C to 94°C
5. *Non-hygroscopic* because the dielectric strength of any material goes very much down with moisture content. In case the insulating material is hygroscopic, it must be enclosed in a water tight covering like lead sheath etc.
6. *Non-inflammable.*
7. *Low coefficient of thermal expansion.*
8. *Low permittivity.*
9. *Capability of withstanding high rupturing voltages.*

The last point but not the least is that a dielectric should not be *expensive* and it should be possible to handle it from the point of view of the manufacturer.

No one insulating material possesses all the above mentioned qualities, so the type of insulating material used in a cable depends upon the service for which the cable is required. The various insulating materials used in manufacture of cables are rubber, VIR, paper, polyvinyl chloride, varnished cambric, polyethylene, vulcanized bitumen, gutta-percha, silk, cotton, enamel etc.

1. *Rubber.* Rubber may be natural and synthetic. Natural rubber is obtained from the milky sap of tropical trees. Synthetic rubber is produced from alcohol or oil products. Its relative permittivity is between 2 and 3 and dielectric strength is 30 kV/mm. Though it possesses high insulating qualities but it absorbs moisture readily, softens when heated to a temperature of 60 or 70° C, swells under the action of mineral oils and ages when exposed to light. Hence pure rubber cannot be used as insulating material.

2. *Vulcanized India Rubber (VIR).* It is prepared by mixing India rubber with mineral matter such as sulphur, zinc oxide, red lead etc. Though vulcanization process improves the non-hygroscopic and other properties like strength, durability, resilience, elasticity etc; but it slightly reduces its specific insulation resistance. It has dielectric strength of about 10-20 kV/mm and dielectric constant of about 2.5. The main drawback of VIR is that owing to sulphur content it attacks copper and therefore in cables using VIR insulation the copper conductor is tinned before providing this insulation. Sometimes a layer of pure rubber is also given on the conductor to protect it from sulphur.

As compared to impregnated paper, the performance of VIR is poor because VIR cables are expensive, less durable and have poor mechanical properties such as electrostatic capacity, elasticity and tenacity. VIR is used as insulating covering for low and medium voltage cables. Such type of cables are normally covered with cotton tape and jute braiding soaked in the compound in order to protect the cables against ingress of moisture.

3. *Impregnated Paper.* It is prepared from wood pulp rags or plant fibres by a suitable chemical process. It is quite cheap, has low capacitance, high dielectric strength (30 kV/mm), high insulation resistivity (of the order 10^7 Ω–m), high thermal conductivity, better thermal withstand capability and high durability. The only disadvantage is that it is hygroscopic, and even if it is impregnated with compound, it absorbs moisture, which lowers its insulation resistance. That is why paper insulated cables are always provided with some sort of protective covering and are never left unsealed. Its ends should be temporarily covered with wax or tar, even if it is required to be left unused on site while being laid. The maximum safe temperature of paper insulated cable is about 95° C. The compound usually employed for impregnation purposes is a heavy-grade mineral oil mixed with resin, the actual proportions of oil and resin depending on the particular manufacturer.

For providing insulation over the conductor, the paper tape is lapped on to the conductor until the required thickness* is obtained, it is then dried and impregnated with insulating compound. The insulating compounds are required to have, besides high electrical insulation properties, a high viscosity at operating temperatures (50–80° C), a low viscosity at impregnation temperatures of 50–120° C, a smooth change in viscosity with temperature and very low coefficient of thermal expansion. The impregnation of paper insulation increases its dielectric strength from about 3–5 to 10–80 kV/mm. The dielectric constant of impregnated paper insulation is about 3.5.

Paper insulated cables are used for conveying large blocks of power, in transmission and distribution, and particularly for distribution at low voltage in congested areas where the joints are to be provided only at the terminal apparatus or where the joints are rare, owing to cheapness and durability over VIR cables.

---

* The thickness of insulation depends on the mechanical and electrical requirements which it has to meet in the service.

4. ***Polyvinyl Chloride.*** This is a synthetic compound. For obtaining this material as a cable dielectric or sheathing it is processed with certain materials known as *plasticizer* and its type will depend upon the use of the finished product *i.e.* PVC. It is inert to oxygen and almost inert to oils and to many alkalies and acids and therefore its use is preferred over VIR in extreme environments such as in cement factory or chemical factory. PVC has dielectric strength of 17 kV/mm and dielectric constant of 5. Maximum continuous temperature rating is 75° C. The mechanical properties (*i.e.* elasticity and and recovery from stretching) of PVC are not so good as those of rubber. PVC insulated cables are usually employed for low and medium voltage domestic and industrial light and power installations.

5. ***Polythene.*** Polythene is a straight chain polymer derived from ethylene. Its electrical properties are very good—low permittivity (dielectric constant), high insulation resistivity and very low power factor at all frequencies. The thermal dissipation properties are better than those of impregnated paper and the impulse strength compares favourably with an impregnated paper-insulated cable. The maximum operating temperature under short-circuits is 100° C. It has limited use in power cables. However, a 3-core 11 kV self supporting aerial cable, which can be employed in rural distribution, has been developed.

6. ***Cross-Linked Polythene.*** Low density polythene, when vulcanized under controlled conditions, results in cross-linking of carbon atoms and the compound produced is a new material having extremely high melting point with light weight, small dimensions, low dielectric constant and good mechanical strength. By using chemical process it has been made technically possible to cross-link polythene in conventional equipment for the manufacture of rubber. This is why the product is said to be "vulcanized" 'or cross-linked' polythene.

Cross-linked polythene (XLPE) cables have high maximum continuous temperature rating of 90° with dielectric strength of about 20 kV/mm. Because of high temperature withstand capability and very low water absorption, such cables can be buried directly in soil. These cables have been found very suitable for all voltages up to 33 kV.

7. ***Varnished Cambric (or Empire Tape).*** This is a cotton cloth impregnated and coated with varnish. It has very smooth surface. The cambric is lapped in the form of a tap on to the conductor and its surfaces are coated with petroleum jelly compound to give easy sliding of the surfaces. Such cables require protective covering like lead sheath because this insulating material is hygroscopic. Its dielectric strength is 4 kV/mm and its dielectric constant is 2.5 to 3.8. Such cables do not require sealing.

8. ***Gutta-percha.*** It is similar to rubber but it becomes soft at about 65°C. It is non-hygroscopic but cannot withstand even medium voltages. It is mostly used for submarine cables for telephone and telegraph purposes.

9. ***Silk and Cotton.*** This is used in low voltage cables. The conductor may have a single layer or double layer covering depending upon the requirements of service. Silk or cotton covered wires are usually used for instrument and motor winding.

10. ***Enamel Insulation.*** *Enamelled* wires are also used for instruments and motor winding. The wires are cheaper than silk and cotton covered wires and therefore for low voltage machines and instruments enamelled wires are used. Enamel insulation is liable to crack.

11.6. CLASSIFICATION OF CABLES

The type of cable to be used at a particular location is determined by the mechanical considerations and the voltage at which it is required to operate. Usually the operating voltage determines the type of insulation and the cables are placed in various categories depending upon the voltage for which they are designed. According to voltage, the cables are classified as :

(i) Low voltage (LT) cables for operating voltage up to 1,000 V.
(ii) High voltage (HT) cables for operating voltage up to 11,000 V.
(iii) Super-tension (ST) cables for operating voltage up to 33,000 V.
(iv) Extra high tension (EHT) cables for operating voltage up to 66,000 V.
(v) Extra super-voltage cables for operating voltage beyond 132,000 V.

A cable may be a single core or multicore depending upon the type of service for which it is intended to be used. A multicore cable may be two-core, three-core or four core. For 3-phase, 3-wire systems either three single-core cables or three-core cables can be used depending upon the operating voltage and load demand. For 3-phase, 4-wire systems 4-core cables may be used.

1. Low Tension (or LT) Cables. These cables are meant for use up to 1,000 V. For voltages up to 6,600 V, the electrostatic stresses developed in cables are very small and thermal conductivity is also of not much importance so no special construction is required. The insulating materials used may be impregnated paper, varnished cambric, vulcanized rubber, or vulcanized bitumen.

The *lt* cables are of two types *viz* single core and multi-core cables. The former type has the advantages of simplicity of construction and availability of larger conductor section. The various sizes of *lt* cables are given below:

Single core cable with aluminium conductors	: 1.5 – 625 mm^2
Two-, three-, three-and-a-half and four core cables with aluminium conductors	: 1.5 – 625 mm^2
Control cables up to 61 cores with copper conductors	: 1.5 and 2.5 mm^2.

Single core *lt* cable consists of one circular core of tinned stranded copper (or aluminium) insulated by layers of impregnated paper or varnished cambric over it and a lead sheath over the insulation, as shown in Fig. 11.2. The lead sheath protects the cable against ingress of moisture and mechanical handling. If the cables are to be buried directly then an overall serving of compounded fibrous material or hessian tape is provided over the sheath in order to protect the metallic sheath against corrosion. Single core cables are usually not provided with armouring in order to avoid excessive loss in the armour.

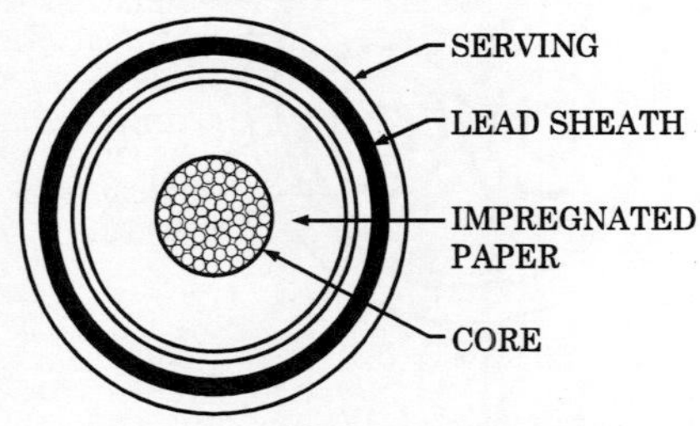

Fig. 11.2. *Single Core LT Cable*

The multicore cables for use up to 11,000 volts are of belted type. The belted type cable consists of either circular shaped or oval or sector shaped (oval or sector shaped core improve the copper space factor) cores of stranded copper (or aluminium) conductors wrapped around by impregnated paper and then wormed together. The interstices are filled with packing to make the cable of circular cross-section. Insultating belt of impregnated paper is provided surrounding the three cores. A lead sheath enclosing the whole is provided for its protection against the entry of moisture. The cable is provided with two layers of steel tape

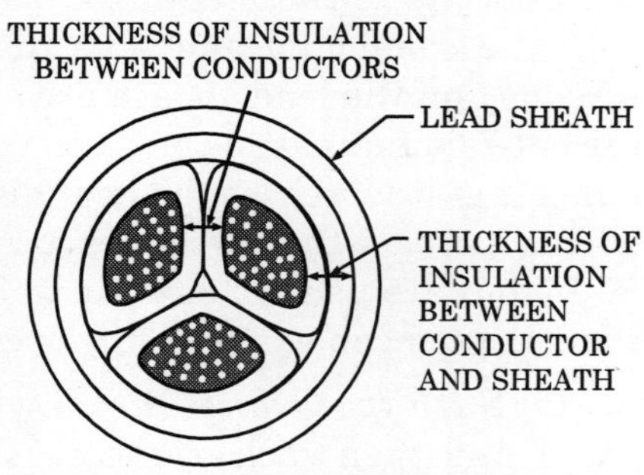

Fig. 11.3. *Three-Core Belted Cable (Up to Lead Sheath)*

pre-dipped in compound. The serving over the armouring consists of a compound and then one layer of impregnated hessian tape. In order to prevent adhesion, a coating of lime wash is applied to the outside of the cable.

The belted type construction is not suitable for the cables used for voltages above 22 kV because of development of both the radial and tangential stresses. The tangential stresses act along the layers of insulation. The electric resistance, therefore, dielectric strength of the impregnated paper is much higher across the layers than along the layers. The leakage current on account of tangential stresses along the impregnated paper insulation causes power loss at the centre filling and local heating resulting in breakdown at any moment. Further owing to non-homogenity of dielectric in belted construction, when cables are loaded and unloaded some portions of the dielectric are less stressed whereas some portions are overstressed resulting in formation of vacuous spaces and voids. These vacuous spaces are ionised when voltage is applied and ultimately deteriorate the cable insulation.

The above difficulties have been overcome in the screened cables where leakage currents are conducted to earth through metallic sheaths.

2. Super-Tension Cables. For cables above 11,000 V a special construction is adopted. For use up to 33 kV, the cables used are screened cables where leakage currents are conducted to earth through metallic sheaths. Screened cables are of two types *viz* H-type and S L type.

(a) H-Type Cables. In the "H" type cable, invented by Hochstadter, no belt insulation is used, but each of the core is insulated with paper to the desired thickness and over this is provided a layer of metallized paper perforated to facilitate the process of impregnation, the coefficient of expansion and contraction is same as that of dielectric. Additional layer of the cotton tape with fine wires of copper is wrapped round all the three cores. Over this tape,

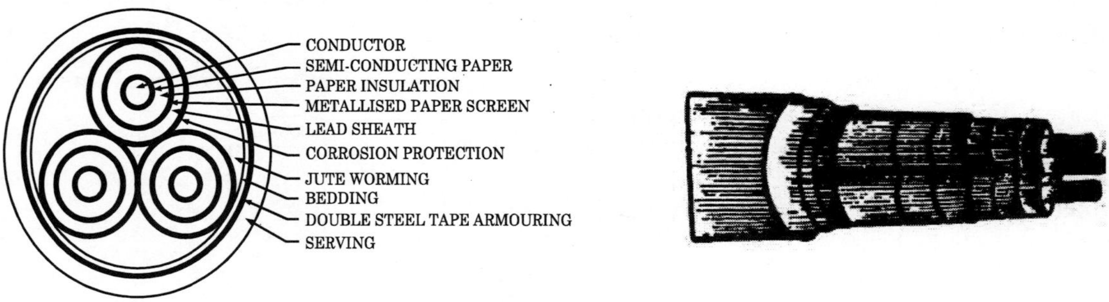

Fig. 11.4. ***H-Type Cable***

no belt insulation is provided, but the lead sheath and armouring are provided as usual. The perforations in the metallized paper sheaths assist in the complete impregnation of cable with the compound and thus the possibility of air-duct formation is eliminated. All four screens and the lead sheath are at earth potential, with the result that the electric stresses are entirely radial and the dielectric losses are therefore reduced. Another advantage of metal foil is that it increases the heat dissipating power and there are no sheath losses. All of these effects increase the current carrying capacity of the cable. These cables have been used up to 66 kV but are usually used up to 33 kV because pressurized cables have better performance than H type cables for voltages exceeding 33 kV.

(b) S L Type Cables. In S L type cables each core is first insulated with an impregnated paper and then each of them, is separately lead sheathed. Now the three cores are just

* It should not be confused with charging current.

equivalent to three separate cables, each having its own lead sheath. The three cables are laid up with fillers in the ordinary way, armoured and served overall with impregnated hessian tape as usual. No lead covering is provided surrounding all the three cores in addition to their individual lead sheaths. The advantages of S L type cables over H-type cables are :

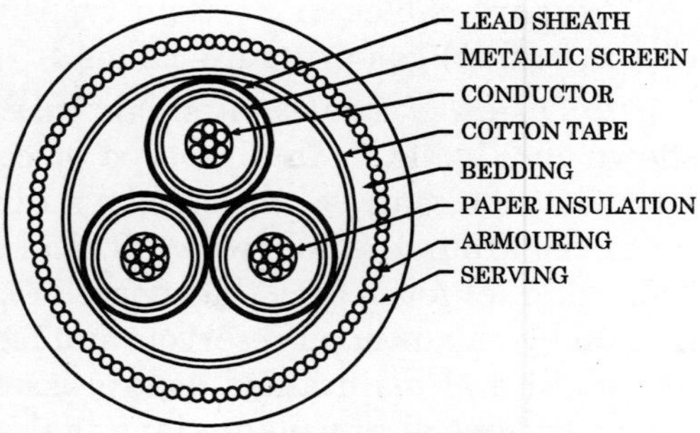

Fig. 11.5. ***SL Type Cable***

(*i*) Bending of cable becomes possible owing to no overall lead sheath.

(*ii*) Less tendency for oil drainage on hilly routes owing to elimination of filler spaces containing compound.

The disadvantage of SL type cable is that the manufacturing is difficult because of thinner lead sheaths.

SL type cables can be used up to 66 kV.

(*c*) *HSL Type Cables.* Such a cable is a combination of H-type and SL type cables. In HSL type cables each conductor is insulated, sheathed with metalized paper and in then provided with lead sheath. The three cores are then laid up and provided with filler, braided, armoured and finally served.

Advantages of Screened Type Cables Over Belted Types. (*i*) The metal sheathed core cable has a greater core to core thickness for a given overall diameter than a plain cable so the possibility of core to core faults are reduced to a large extent.

(*ii*) Electric stresses are uniformly radial in all sections of the dielectric.

(*iii*) As there is no worming or packing so the dielectric subjected to electric stress is only paper which is quite homogeneous and therefore possibility of formation of voids within the electric field is not there.

(*iv*) The metal sheaths help in dissipation of heat, so the current carrying capacity of the cables is increased.

3. Extra High Tension Cables. The cables considered up till now are also called *solid type* and in such cables it was assumed that the dielectric is homogeneous and there are no voids in the layers. With this assumption it is necessary to stick to maximum safe dielectric stress of about 4 or 5 kV per mm and also to maximum operating temperature of 50° or 60°C in order to ensure safety from breakdown because the dielectric is far from homogeneous and with the normal manufacturing methods, it is almost impossible to avoid voids in the layers of dielectric. The formation of these voids or vacuous spaces causes unequal voltage stresses (which may exceed the safe limits) and also temperature rise due to leakage current. Again, even if a new cable could be constructed free from voids owing to development of modern techniques, it would be impossible to avoid their formation under the influence of the movements of the compound and expansion and contraction of the paper during normal working. When these voids are subjected to the electrostatic stresses, ionization takes place, which gives rise to certain chemical action, resulting in the deterioration of the chemical compound. Thus, these voids are sometimes the primary cause of a breakdown and therefore are required to be avoided. In order to meet the increased voltage demand the extra high tension and extra super-voltage power cables, useful for 132 kV and above, have been developed. In all such cables, the voids have been eliminated by increasing the pressure of the compound and that is why such cables are also called *pressure cables.*

Pressure cables are of two types viz., (*a*) oil-filled cables, and (*b*) gas-pressure cables.

(*a*) *Oil-filled Cables.* A single core oil-filled cable is shown in Fig. 11.6. In this type of cable a channel is formed at the centre of the core by stranding the conductor wires around a hollow cylindrical steel spiral tape. This channel formed at the centre of core is filled with thin oil by means of oil reservoir and feeding tanks along its length and maintained at a pressure, not below atmospheric one at any point along the cable. This oil is the same light mineral oil of very low viscosity as used for initial impregnation. The system is so designed that when the oil gets expanded on account of increase in temperature of cable, the extra oil collects in the external reservoir, which sends it back during contraction on account of fall in temperature during light load conditions. Thus formation of voids in the dielectric is rendered impossible. Such cables are known as *single core conductor channel cables.* Its disadvantage is due to the fact that the channel is at the middle of the cable and is at full voltage with respect to earth, so requiring a complicated system of jointing. It is advantageous from the point of view of potential gradient due to larger diameter of conductor and its hollow construction.

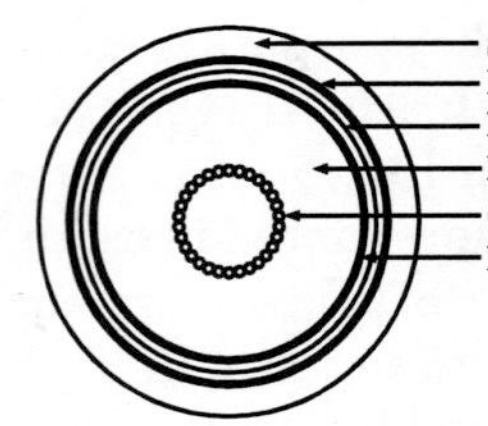

Fig. 11.6. ***Single Core Conductor Channel Oil-Filled Cable***

The other type of single core oil-filled cable is *sheath channel cable* (Fig. 11.7). In this type of cable, the conductor is made solid similar to that of solid cables and is paper insulated. In this type of cables the oil channels are produced either by grooving the sheath or by arranging spacers between dielectric and lead sheath. The latter arrangement is more sound mechanically in comparison to former one but due to high resistance to oil flow (6 to 8 times that of former type) requires large number of feeding points. In this case since the channels are at earth potential so system of joints and installation are simpler.

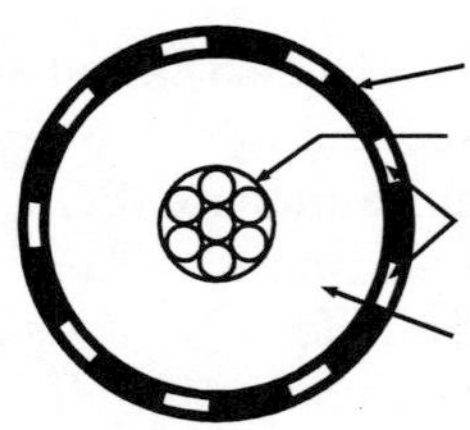

Fig. 11.7. ***Single Core Sheath Channel Oil-Filled Cable***

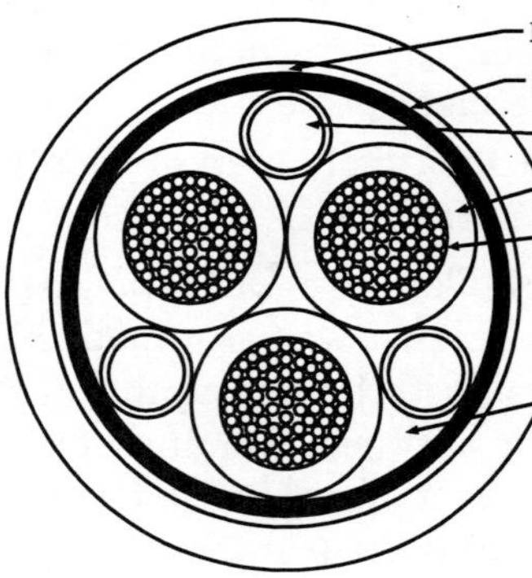

Fig. 11.8. ***Three Core Filler Space Channel Oil-Filled Cable***

In 3-core oil-filled cables (Fig. 11.8) the oil ducts are accommodated in the hollow filler spaces. These channels are made of perforated metal-ribbon tubing and are at earth potential. While jointing such a cable great care is required to be taken.

Another design of 3-core oil-filled cable is flat type, as illustrated in Fig. 11.9. The flat sides are reinforced with metallic tapes and binding wires so that during increase in oil pressure, due to heating, the flat side is deformed and the cable section becomes slightly elliptical.

In another construction of 3-core oil filled cables use of 3-core paper insulated cable without a lead sheath is made. The cable is pulled into a steel pipe which is then filled with oil. Pumps are then used to maintain a specified oil pressure and permit it to expand and contract with the loading cycle.

Leakage of oil in oil-filled cables is a very serious problem. Automatic signalling is provided to indicate the decrease in oil pressure in any of the phases. The pressure within the cable must not go below atmospheric and a lower limit of about 21,700 N/m^2 gauge is set. The transient pressure, developed due to sudden increase of load, should not increase beyond about 8.6 × 10^6 N/m^2 gauge. The joints in the cable are to be properly made so that there is no restriction to the oil flow.

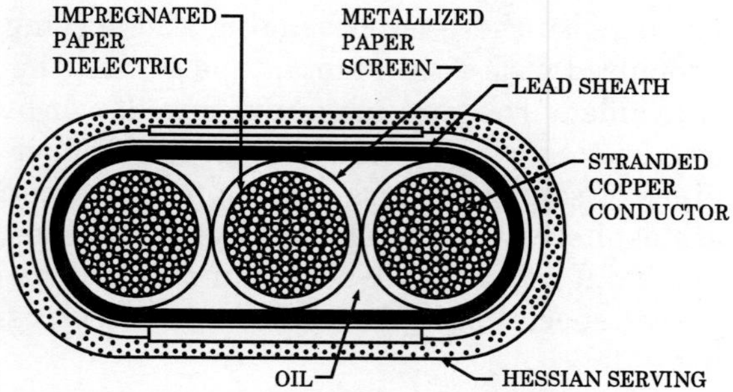

Fig. 11.9. ***Flat Oil-Filled Pressure Cable***

The breakdown inception stress for oil-filled cables is 30–40 kV/mm which is much higher than that for a solid type cable. So the normal operating stresses are raised considerably. Maximum stresses up to 13 kV/mm are acceptable for 275 kV and 400 kV oil-filled cables. The current carrying capacity of oil-filled cables is determined by allowing a normal maximum conductor temperature of 85° C.

The advantages and disadvantages of oil-filled cables are given below:

Advantages. (*i*) Smaller overall size and smaller weight for given voltage and kVA rating due to reduction in the thickness of dielectric required.

(*ii*) No ionisation, oxidation and formation of voids.

(*iii*) More perfect impregnation.

(*iv*) Possibility of increased temperature range in service.

(*v*) Smaller thermal resistance due to decrease in dielectric thickness, so higher current rating.

(*vi*) Impregnation possible after sheathing.

(*vii*) More maximum permissible stresses.

(*viii*) Detection of fault easy, as it will come to notice immediately as oil will start leaking.

Disadvantages. (*i*) Greater cost.

(*ii*) Complicated laying of cables and maintenance.,

(*b*) *Gas Pressure Cables.* Gas pressure cables are of two types (*i*) external pressure cables and (*ii*) gas-filled cables.

(*i*) *External Pressure Cables.* The external pressure cable, is a serious rival to the oil-filled type, and is being developed for the highest voltages. In such a cable the pressure, is applied externally and raised to such an extent that no ionisation can take place. At the same time the radial compression due to this increased pressure tend to close any voids. Also the working power factor of such a cable is improved.

Fig. 11.10. ***66 kV, Three-core Pressure Cable***

The external pressure cable originally designed by Hochstadter, Vogel and Bowden is illustrated is Fig. 11.10. This cable is similar in construction to that of ordinary solid type except that it is triangular instead of circular in section and thickness of lead sheath is 75% of that of solid type cable. The triangular section reduces the weight, gives a low thermal resistance and at the same time lead sheath acts as a pressure membrane. Also there is no bedding and serving so thermal resistance is reduced. The cable is armoured with a thin metal tape so that the formation of any abnormal ties over its surface is avoided. The core may have circular or oval section but modern practice is to employ an oval section, which gives an improved copper space factor. The cable is laid in gas-tight metal pipes of somewhat larger section. The pipe is filled with nitrogen at a pressure of 12 to 15 atmospheres which continually compresses the cable radially from outside so that radial breathing of the cable occurs and any voids etc. are closed. The steel pipes used are coated with a special paint in order to avoid corrosion and it is further protected with an impregnated felt.

In comparison with a normal cable, such cables can carry 1.5 times the load current, double the operating voltage and thus transmit 3 times power. The maximum potential gradient is 10 kV/mm and the dielectric power factor at 15°C is 0.6 per cent. The steel pipes provide mechanical protection to the cables. The nitrogen in steel pipes help in quenching any flame. Moreover, maintenance cost of such cables is small. The disadvantage of such cables is that cost is very high.

The above cable is pipeline type. The second type is self contained type, in which an additional reinforced lead sheath is employed, the principle being the same as that of former type. Dried nitrogen is fed into the small annular space between dielectric and sheath from gas cylinders, and finding its way along the cable maintains the whole of the dielectric at a pressure of about 14 atmospheres.

(*ii*) *Internal Pressure Cables.* Such cables are of 3-types :

1. High-pressure gas-filled cables.
2. Gas-cushion cables.
3. Impregnated pressure cables.

In high pressure gas-filled cables spaces are provided in the dielectric itself for the gas which is inert gas like nitrogen at pressure of about 12 atmospheres for super-voltage cables and about 6 atmospheres for extra high tension cables. Pressure is retained by means of a lead sheath which in the case of single core cables, has a diametral clearance of about 0.63 mm. This facilitates the axial flow of gas, which also passes along the unimpregnated strand. In the case of multicore cables this clearance is not essential; the filler spaces and strands providing a sufficiently low resistance path for the flow of gas.

In gas-cushion cable, screened space is provided all along the length of cable, in between the lead sheath and the dielectric. The screened space is sub-divided by means of barriers and thus inert gas is stored at various points along the run of a cable. This facilitates the jointing of cable without losing gas from the entire cable, because while cutting the cable, loss of gas will be local only. The special feature of the cable is thus that no arrangement is required for the transmission of pressure to the cables from outside, so that the cable is a complete unit with its own armouring, requiring no external pipe protection whatever.

Impregnated pressure cable is similar to solid cable except that such a cable consists of mass-impregnated paper dielectric and this is maintained under a pressure of 14 atmospheres by means of nitrogen. Special reinforcement is provided to cater for the large hoop and longitudinal stresses set up. This reinforcement consists of longitudinal and circumferential metallic tapes.

The advantages of the internal pressure cables are :

(*i*) External accessories are eliminated.

(*ii*) With suitable designs the cable can be used for vertical run without any fear of drainage.

(*iii*) There is marked improvement in the power factor of the cable dielectric with the increased pressure.

Gas SF_6, because of its good thermal characteristics and high dielectric strength, is also used for insulating the cables. Cables insulated with SF_6 gas can be matched to overhead lines and can be operated corresponding to their surge impedance loading. Such cables can be employed for transmission of thousands of MVA even at ultra high voltage while the use of conventional cables is limited to 100 MVA and 500 kV.

11.7. INSULATION RESISTANCE OF A SINGLE CORE CABLE

The cable conductor is provided with an insulation of suitable thickness so as to avoid leakage of current. The path for leakage current is radial, as shown in Fig. 11.11, through the insulating material. The opposition offered by the insulation to the leakage current is called the *insulation resistance* of the cable.

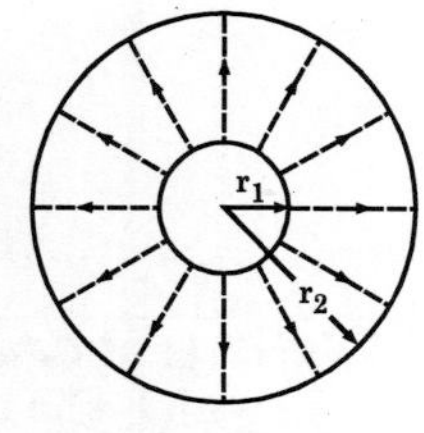

Fig. 11.11

Consider a single core cable of conductor radius r_1, internal sheath radius r_2, length l and insulation material resistivity ρ.

Let an elementary cylindrical section of the insulation of radius r and thickness dr be considered. Now the length through which the leaking current will flow is dr and area of x-section is $2\pi rl$.

The insulation resistance offered to the leakage current by elementary cylindrical section of the insulation under consideration is $\dfrac{\rho dr}{2\pi rl}$

Insulation resistance of the cable,

$$R_{INS} = \int_{r_1}^{r_2} \frac{\rho\, dr}{2\pi r l} = \frac{\rho}{2\pi l}\left[\log_e r\right]_{r_1}^{r_2} = \frac{\rho}{2\pi l}\log_e \frac{r_2}{r_1} \qquad ...(11.1)$$

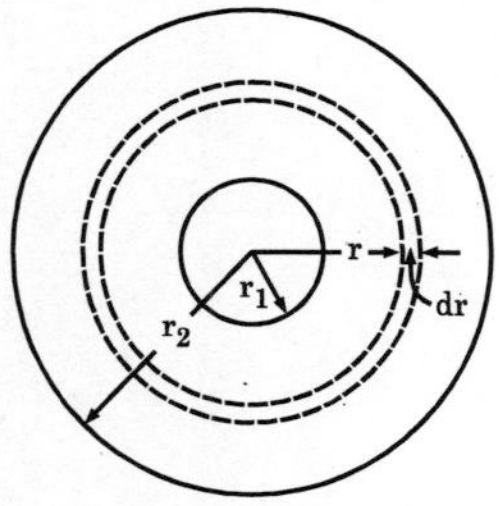

Fig. 11.12

i.e. the insulation resistance of the cable varies inversely as the length of the cable $\left(R_{INS}\ \alpha\ \dfrac{1}{l}\right)$.

If ρ is in Ω-*m*, r_2, r_1, and l are in metres then insulation resistance of the cable is in ohms.

Average value of ρ for impregnated paper is about $5 \times 10^{12} - 8 \times 10^{12}$ Ω-*m* at 15°C and decreases exponentially with temperature so that

$$\rho_t = \rho_0 e^{-\alpha t} \qquad ...(11.2)$$ α is about 0.04 or 0.05.

Example 11.1. Derive the formula for insulation resistance of a cable. Calculate insulation resistance of 5 km length of single core cable whose insulation resistance is 5×10^{14} Ω-cm, insulation thickness is 1 cm and radius of conductor is 1.25 cm.

[*U.P. Technical Univ. Elements of Power System 2009-10*]

Solution : Specific resistance of insulation, $\rho = 5 \times 10^{14}\ \Omega\text{-cm} = 5 \times 10^{12}\ \Omega\text{-m}$

Length of cable, $l = 5$ km $= 5{,}000$ m

Core radius, $r_1 = 1.25$ cm

Internal sheath radius, $r_2 = r_1 +$ insulation thickness $= 1.25 + 1 = 2.25$ cm

$$\log_e \frac{r_2}{r_1} = \log_e \frac{2.25}{1.25} = 0.5878$$

$$\text{Insulation resistance of cable} = \frac{\rho}{2\pi l}\log_e \frac{r_2}{r_1} = \frac{5 \times 10^{12}}{2\pi \times 5{,}000} \times 0.5878 = 93.55\ \text{M}\Omega \ \textbf{Ans.}$$

Example 11.2. Derive the expression for insulation resistance of a cable. Calculate the insulation resistance for a 5 km length of a 1-core cable. Specific resistance of insualtion (impregnated paper) is 5 × 10^14 Ω-cm, insulation thickness is 1 cm and radius of conductor is 2.5 cm.

[*U.P. Technical Univ. Elements of Power System 2004-05*]

Solution : Specific resistance of insulation, $\rho = 5 \times 10^{14}\ \Omega\text{-cm} = 5 \times 10^{12}\ \Omega\text{-m}$

Length of cable, $l = 5 \text{ km} = 5{,}000 \text{ m}$

Core radius, $r_1 = 2.5 \text{ cm}$

Internal sheath radius, $r_2 = 2.5 + 1 = 3.5 \text{ cm}$

$$\log_e \frac{r_2}{r_1} = \log_e \frac{3.5}{2.5} = 0.3365$$

$$\text{Insulation resistance of cable} = \frac{\rho}{2\pi l} \log_e \frac{r_2}{r_1} = \frac{5 \times 10^{12}}{2\pi \times 5{,}000} \times 0.3365 = 53.55\ \text{M}\Omega \textbf{ Ans.}$$

Example 11.3. Calculate the resistivity of the dielectric in a cable, the core diameter of which is 1.5 cm and sheath diameter is 5 cm. For a length of 3,000 metres of cable, the insulation resistance is 1,820 MΩ.

Solution : Core radius, $r_1 = \frac{1.5}{2} = 0.75$ cm

Sheath radius, $r_2 = \frac{5}{2} = 2.5$ cm

$$\log_e \frac{r_2}{r_1} = \log_e \frac{2.5}{0.75} = 1.2$$

Installation resistance, $R_{INS} = 1{,}820\ \text{M}\Omega = 1{,}820 \times 10^6\ \Omega$

$$\text{Resistivity, } \rho = \frac{R_{INS} \times 2\pi l}{\log_e \frac{r_2}{r_1}} = \frac{1{,}820 \times 10^6 \times 2\pi \times 3{,}000}{1.2} = 28.6 \times 10^{12}\ \Omega\text{-m} \textbf{ Ans.}$$

11.8. CAPACITANCE OF A SINGLE CORE CABLE

The cable can be considered to be two coaxial cylinders of inner diameter *d* and outer diameter D. In actual cable, *d* represents the diameter of core and D represents the inner diameter of lead sheath which is at earth potential. Let the relative permittivity of dielectric in between the core and sheath be ϵ_r.

Let the charge per metre length of the cable on the outer surface of the core be +Q coulombs and on the inner surface of the lead sheath be –Q coulombs. For all practical purposes, the charge of +Q coulombs/m on the surface of the core can be assumed to be located along its axis. The metal sheath is earthed.

Surface area of the coaxial cylinder of radius *x* metres and length one metre is $2\pi x$ metres^2.

Therefore, electric field intensity at a point *x* metres from the centre of the inner cylinder,

$$E_x = \frac{Q}{2\pi \epsilon_0 \epsilon_r x} = \frac{Q}{2\pi \epsilon_0 \epsilon_r} \cdot \frac{1}{x} \text{ V/m}$$

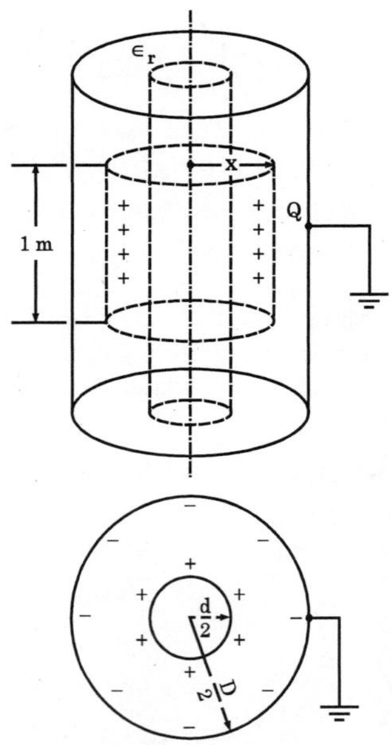

Fig. 11.13

Potential difference between the capacitor plates, (between core and sheath),

$$V = \int_{d/2}^{D/2} E_x \, dx = \int_{d/2}^{D/2} \frac{Q}{2\pi \epsilon_0 \epsilon_r} \cdot \frac{1}{x} dx = \frac{Q}{2\pi \epsilon_0 \epsilon_r} \log_e \frac{D}{d} \text{ volts}$$

Capacitance of cable,

$$C = \frac{Q}{V} = \frac{Q}{\frac{Q}{2\pi \epsilon_0 \epsilon_r} \log_e \frac{D}{d}} \text{ F/m} = \frac{2\pi \epsilon_0 \epsilon_r}{\log_e \frac{D}{d}} = \frac{2\pi \times 8.854 \times 10^{-12} \epsilon_r}{2.303 \log_{10} \frac{D}{d}}$$

$$= \frac{\epsilon_r 10^{-9}}{41.4 \log_{10} \frac{D}{d}} \text{ F/m or } \frac{0.024 \varepsilon_r}{\log_{10} \frac{D}{d}} \mu\text{F/km} \quad ...(11.3)$$

11.9. DIELECTRIC STRESS IN A SINGLE CORE CABLE

Under operating conditions, the insulation of a cable is subjected to electrostatic stress, called the *dielectric stress.* Potential gradient at any point is defined as the rate of increase of potential at that point and is the same as the dielectric stress at that point.

Since cable is a form of cylindrical condenser, therefore, electric intensity at a distance x from the centre O of the cable is given by the expression

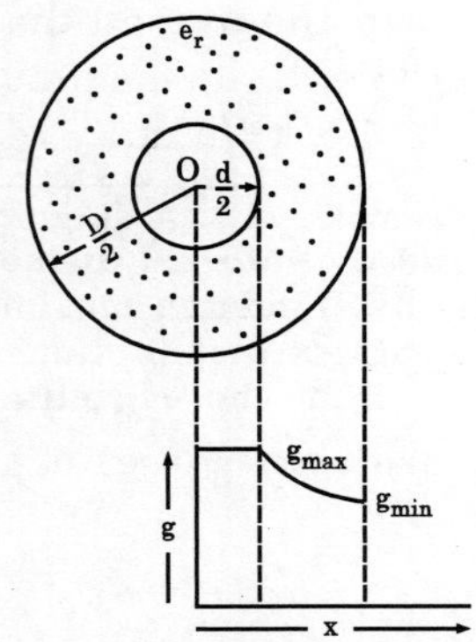

Fig. 11.14

$$E_x = \frac{Q}{2\pi \epsilon_0 \epsilon_r} \cdot \frac{1}{x} \text{ V/m}$$

Since potential gradient = Electric intensity

$$\therefore g = \frac{Q}{2\pi \epsilon_0 \epsilon_r} \cdot \frac{1}{x} \text{ V/m} \quad ...(11.4)$$

From foregoing Art. 11.8

$$V = \frac{Q}{2\pi \epsilon_0 \epsilon_r} \log_e \frac{D}{d}$$

Substituting the value of Q from above expression in Eq. (11.4) we have

$$g = \frac{2\pi \epsilon_0 \epsilon_r V}{\log_e \frac{D}{d}} \cdot \frac{1}{2\pi \epsilon_0 \epsilon_r} \cdot \frac{1}{x} = \frac{V}{x \log_e \frac{D}{d}} \text{ volts/metre} \quad ...(11.5)$$

Since potential gradient g varies inversely as x (as obvious from the above expression), therefore, potential gradient will be maximum when x is minimum *i.e.*, $x = \frac{d}{2}$ and potential gradient will be minimum when x is maximum *i.e.* $x = \frac{D}{2}$

Maximum and minimum values of potential gradient are given by

$$g_{max} = \frac{2V}{d \log_e \frac{D}{d}} \text{ volts/metre} \quad ...(11.6)$$

$$\text{and } g_{max} = \frac{2V}{D \log_e \frac{D}{d}} \text{ volts/metre} \quad ...(11.7)$$

Hence the ratio between the maximum potential gradient and the minimum potential gradient *i.e.*

$$\frac{g_{max}}{g_{min}} = \frac{2\,V / d \log_e \frac{D}{d}}{2\,V / D \log_e \frac{D}{d}} = \frac{D}{d} \qquad ...(11.8)$$

It is to be noted here that the Eq. (11.6) for the dielectric stress at the conductor surface is obtained on the assumption of a smooth cylindrical conductor, and thus only provides a nominal value for this stress. With the ordinary stranded conductors in common use, the dielectric stresses near the conductor are increased by about 20 per cent due to the greater curvature of the surface of the individual wires. This, however, can be avoided in practice by covering the conductor with a thin smooth metal sheath.

In high voltage cables the value of g is taken to the highest possible limit in order to keep down the overall diameter, but since the dielectric losses increase rapidly with potential gradient, the maximum permissible value of g is only about one-fifth of the breakdown value viz 4-5 kV/mm.

Example. 11.4. A single core cable for use on 11 kV, 50 Hz system has conductor area of 0.645 cm^2 and the internal diameter of sheath is 2.18 cm. The permittivity of the dielectric used in the cable is 3.5. Find the maximum and minimum electrostatic stresses in the cable. Show also the curve of stress distribution in the cable v/s distance from the centre of the conductor.

Find the capacitance of the above cable per km length. Find the charging current.

Solution : Area of x-section of conductor, $a = 0.645$ cm^2

$$\text{Diameter of conductor, } d = \sqrt{\frac{4a}{\pi}} = \sqrt{\frac{4 \times 0.645}{\pi}} = 0.906 \text{ cm or } 0.00906 \text{ m}$$

Internal diameter of sheath, D = 2.18 cm or 0.0218 m

$$\text{System peak voltage per phase, } V = \frac{11,000}{\sqrt{3}} \times \sqrt{2} = 8,981 \text{ V assuming 3-phase system}$$

$$\text{Maximum electrostatic stress, } g_{max} = \frac{2\,V}{d \log_e \frac{D}{d}}$$

$$= \frac{2 \times 8,981}{0.00906 \log_e \frac{0.0218}{0.00906}} = 2,257,937 \text{ V/m or } 22.58 \text{ kV/cm } \textbf{Ans.}$$

$$\text{Minimum electrostatic stress, } g_{min} = \frac{2\,V}{D \log_e \frac{D}{d}}$$

$$= \frac{2 \times 8,981}{0.0218 \log_e \frac{0.0218}{0.00906}} = 938,390 \text{ V/m or } 9.384 \text{ kV/cm } \textbf{Ans.}$$

Stress distribution curve is shown in Fig. 11.14

Capacitance of the cable per km length,

$$C = \frac{0.024\, \epsilon_r}{\log_{10} \frac{D}{d}}\ \mu F = \frac{0.024 \times 3.5}{\log_{10} \frac{0.0218}{0.00906}} \times 10^{-6} \text{ F} = 0.22 \times 10^{-6} \text{ F } \textbf{Ans.}$$

Charging current per phase per km length of cable,

$$I_C = 2\,\pi f\, CV_P = 2\,\pi \times 50 \times 0.22 \times 10^{-6} \times \frac{11,000}{\sqrt{3}} = 0.44 \text{ A } \textbf{Ans.}$$

Example 11.5. A 33 kV, 3-phase, 50 Hz underground line, 3.4 km long, uses three single core cables. Each cable has a core diameter of 2.5 cm and the radial thickness of insulation is 0.6 cm. The relative permittivity of the dielectric is 3.1. Find (*i*) maximum stress and (*ii*) total charging kVAR.

Solution : System peak voltage per phase,

$$V = \frac{33,000}{\sqrt{3}} \times \sqrt{2} = 26,944 \text{ V}$$

Inner diameter of sheath, D = d + 2 × radial thickness of insulation = 2.5 + 2 × 0.6

= 3.7 cm or 0.037 m

Maximum electrostatic stress, $g_{max} = \dfrac{2\text{ V}}{d \log_e \dfrac{D}{d}} = \dfrac{2 \times 26,944}{0.025 \log_e \dfrac{0.037}{0.025}} = 5.5 \times 10^6$ V/m **Ans.**

Capacitance of cable, $C = \dfrac{0.024\, \epsilon_r}{\log_{10} \dfrac{D}{d}} \times l \times 10^{-6} = \dfrac{0.024 \times 3.1}{\log_{10} \dfrac{0.037}{0.025}} \times 3.4 \times 10^{-6} \text{ F} = 1.4857 \times 10^{-6}$ F

Charging current of cable, $I_C = 2\pi f C V_P = 2\pi \times 50 \times 1.4857 \times 10^{-6} \times 19,052 = 8.893$ A

Total charging kVA = $\sqrt{3}\, V_L I_C \times 10^{-3} = \sqrt{3} \times 33 \times 10^3 \times 8.893 \times 10^{-3} = 508.23$ kVAR **Ans.**

11.10. THE MOST ECONOMICAL DIAMETER OF CONDUCTOR

For a given voltage and overall diameter of a single core cable there is a certain value of the conductor radius (or core) diameter that gives a minimum potential gradient at the surface. A small conductor radius allows a greater insulation thickness, but on the other hand, the smaller radius of curvature tends to increase the stress; while the effect of too large a diameter of core is to cause an increase of the stress through reduction of the total thickness of the insulation.

For fixed values of the voltage V and overall diameter D, the g_{max} will be minimum when $d \log_e \dfrac{D}{d}$ is maximum and $d \log_e \dfrac{D}{d}$ will be maximum if

$$\frac{d}{dd}\left[d \log_e \frac{D}{d}\right] = 0 \text{ or } \log_e \frac{D}{d} = 1 \text{ or when } \frac{D}{d} = e = 2.71828 \text{ or when } d = \frac{D}{2.71828} \quad ...(11.9)$$

$$\text{and } g_{max} = \frac{2\text{ V}}{d} \text{ volts/metre} \quad ...(11.10)$$

For the low and medium voltage cables, the diameter of the conductor determined from the above consideration, *i.e.* $d = \dfrac{2V}{g_{max}}$ is somewhat less than that determined from the consideration of the safe current density, therefore, the main criterion for determining conductor diameter for such cables is the current carrying capacity. On the other hand for high voltage cables the value of core diameter d determined from the above expression would, in general, give a conductor of cross-section much larger than required from the consideration of current carrying capacity. The desired overall diameter of the conductor without increasing the cross-sectional area can be had in two ways, by making the conductor hollow or by building it with a lead sheath. The latter method has the advantage of eliminating at the same time the increase of stress due to stranding which may be as high as 20%.

Example 11.6. Show that for a given voltage V and maximum stress E_{max} in the dielectric of a single core cable, the sheath diameter D is a minimum when D: *d* = *e* where *d* is conductor diameter and *e* is the base of natural logarithms. Determine D and *d* for V = 10 kV and E_{max} = 23 kV/cm.

Solution : Maximum stress, $g_{max} = E_{max} = 23$ kV/cm $= 23 \times 10^5$ V/m (assume rms)

Applied voltage, V = 10,000 V

$$\therefore \text{ Diameter of conductor, } d = \frac{2\,V}{g_{max}} = \frac{2 \times 10{,}000}{23 \times 10^5} = 0.0087 \text{ m or } 8.7 \text{ mm } \textbf{Ans.}$$

Internal diameter of sheath, D $= e \times d = 2.71828 \times 8.7 = 23.64$ mm **Ans.**

Example 11.7. Find the most economical value of the diameter of a single core cable to be used on a 132 kV, 3-phase system. Find also the overall diameter of the insulation if the peak permissible stress is not to exceed 60 kV/cm.

Solution : Maximum stress, $g_{max} = 60$ kV/cm (peak) $= \frac{60}{\sqrt{2}}$ kV/cm (*rms*) or 42.426×10^5 V/m

$$\text{RMS value of phase voltage, } V = \frac{132}{\sqrt{3}} \times 1{,}000 = 76{,}210 \text{ V}$$

$$\text{Diameter of conductor, } d = \frac{2\,V}{g_{max}} = \frac{2 \times 76{,}210}{42.426 \times 10^5} = 0.0359 \text{ m or } 35.9 \text{ mm } \textbf{Ans.}$$

Internal diameter of sheath, D $= e\,d = 2.71828 \times 35.9 = 97.65$ mm **Ans.**

11.11. GRADING OF CABLES

It has been observed that the electrostatic stress in a single core cable is maximum at the conductor surface and decreases towards the sheath. In a cable having homogeneous insulation, the maximum voltage which can be safely applied to a cable is limited by the electrostatic stress at the conductor surface. If a dielectric of high strength is used in a cable, then no doubt it is useful just over the conductor where the stress is maximum, but as we go away from the conductor surface, the value of dielectric stress decreases, so the dielectric will be unnecessarily strong and expensive. If, however, by some means the stress can be so distributed that its value in the outer layers of the dielectric is increased without increasing the stress near the conductor, insulation of less thickness will be required for any given working voltage, and an important economy will be a affected. *The process of achieving uniformity in dielectric stress is known as grading of cables.*

There are two methods of grading *viz.* (*i*) capacitance grading and (*ii*) intersheath grading.

(*i*) ***Capacitance Grading.*** In this method of cable grading, the uniformity in dielectric stress is achieved by using various layers of different dielectrics in such a manner that the permittivity, ϵ_r of any layer is inversely proportional to its radius of distance from the centre, *i.e.*

$$\epsilon_r \propto \frac{1}{x} \text{ or } \epsilon_r\, x = \text{constant or } g \propto \frac{Q}{2\pi\, \epsilon_0} \quad \text{which in constant}$$

Thus we see that if such a condition is attained, the value of dielectric stress at any point is constant and is independent of distance from the centre, and the grading will be ideal one. But it is not possible to use infinite number of dielectrics for a single cable and in practice two or three dielectrics are used in such a manner that the permittivity of the dielectric near the core has got highest value and then it decreases and finally it has got minimum value in the outer most layer.

Now consider a cable in which 3-layers of dielectrics of outer diameters d_1, d_2 and D and of relative permittivity ϵ_{r1}, ϵ_{r2} and ϵ_{r3} have been used. If the permittivities are such that $\epsilon_{r1} > \epsilon_{r2} > \epsilon_{r3}$ and if the maximum stress to which each dielectric is subjected is constant then

$$\epsilon_{r1}\, d = \epsilon_{r2}\, d_1 = \epsilon_{r3}\, d_2$$

For the *pd* across the inner layer we have

$$V_1 = \int_{d/2}^{d_1/2} g\, dx = \int_{d/2}^{d_1/2} \frac{Q}{2\pi \epsilon_0 \epsilon_r}\, dx$$

$$= \frac{Q}{2\pi \epsilon_0 \epsilon_r} \log_e \frac{d_1}{d} = g_{max} \frac{d}{2} \log_e \frac{d_1}{d}$$

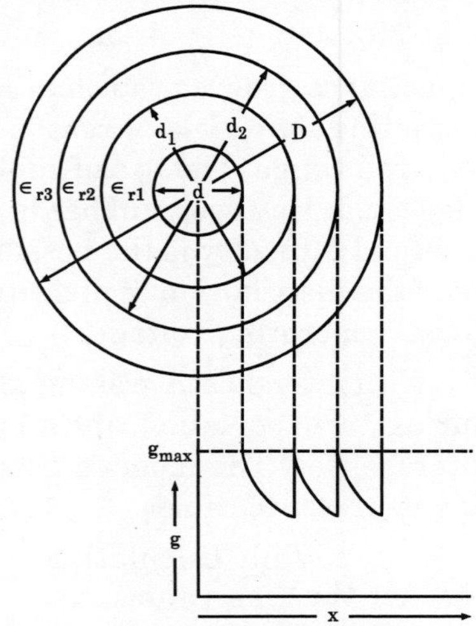

Fig. 11.15. *Capacitance Grading of Cables*

Similarly $V_2 = g_{max} \dfrac{d_1}{2} \log_e \dfrac{d_2}{d_1}$

and $V_3 = g_{max} \dfrac{d_2}{2} \log_e \dfrac{D}{d_2}$

and the *pd* between core and earthed sheath,

$$V = V_1 + V_2 + V_3$$

$$= \frac{g_{max}}{2}\left(d \log_e \frac{d_1}{d} + d_1 \log_e \frac{d_2}{d_1} + d_2 \log_e \frac{D}{d_2} \right) \quad ...(11.11)$$

The advantage of grading is that if the overall diameter is same for a non-graded and graded cable, then the permissible safe potential of the cable is increased *i.e. for* the same permissible safe potential, the size of the graded cable will be smaller than that of the non-graded cable.

(ii) ***Intersheath Grading.*** In this method of cable grading a homogeneous dielectric is used, which is divided into various layers, by suitably placing the metallic intersheaths. These metallic intersheaths are held at certain potentials which are in between the core potential and the earth potential. There is thus a definite potential difference between the inner and outer layers of each sheath, so that each sheath can be treated like a homogeneous single core cable and we have

$$g_{1max} = \frac{V_1}{\frac{d}{2} \log_e \frac{d_1}{d}}, \quad g_{2max} = \frac{V_2}{\frac{d_1}{2} \log_e \frac{d_2}{d_1}}, \quad g_{3max} \frac{V_3}{\frac{d_2}{2} \log_e \frac{D}{d_2}} \quad ...(11.12)$$

But for a homogeneous dielectric we have

$$g_{1max} = g_{2max} = g_{3max}$$

$$\therefore \frac{V_1}{\frac{d}{2} \log_e \frac{d_1}{d}} = \frac{V_2}{\frac{d_1}{2} \log_e \frac{d_2}{d_1}} = \frac{V_3}{\frac{d_2}{2} \log_e \frac{D}{d_2}} = g_{max} \text{ (say)} \quad ...(11.13)$$

Again, since all these potential differences are in phase, since the cable is, in effect, a series of capacitors in series, we also have

$$V = V_1 + V_2 + V_3 \quad ...(11.14)$$

Grading in Practice. Grading is only useful for very high voltage cables for which the ratio D/*d* is large. If $\frac{D}{d}$ is 2, grading will increase the permissible safe working voltage by only 30% but if $\frac{D}{d}$ = 5, grading will increase the safe working voltage by 100%. With capacitance grading the chief handicap is due to the fact that there are such few high grade insulating materials of reasonable cost, whose permittivities can be made to vary over the required range. The use of rubber ($\in_r = 4 - 6$) and impregnated paper ($\in_r = 3 - 4$) have been suggested. However, rubber is very expensive and the cost of dielectric becomes prohibitive. A second difficulty is the possibility of change of permittivity with time that may completely alter the distribution of dielectric stress and lead to the breakdown of insulation even at the normal operating voltage.

With intersheath grading, the main objection is the possibility of damage to the intersheath during transportation and installation. Due to charging currents there are losses in the intersheaths. The modern trend is to avoid grading as far as possible and employ oil-filled or gas-pressure cables.

Note : While calculating the safe potential, the peak values of voltages are to be taken and not the rms values.

Example 11.8. A conductor of 1 cm diameter passes centrally through a porcelain cylinder of internal diameter 2 cm and external diameter 7 cm. The cylinder is surrounded by a tightly fitting metal sheath. The permittivity of porcelain is 5 and the peak voltage gradient in air must not exceed 34 kV/cm. Determine the maximum safe working voltage.

[*U.P. Technical Univ. Elements of Power System 2007-08*]

Solution : The configuration is shown below in Fig. 11.16.

Let *q* be charge per unit length of the conductor.

Maximum value of potential gradient (at the surface of conductor),

$$g_{1\max} = \frac{q}{2\pi\varepsilon_0 r}$$

Maximum value of potential gradient (at the surface of porcelain cylinder)

$$g_{2\max} = \frac{q}{2\pi\varepsilon_0\varepsilon_r r_1}$$

Thus, $\quad r g_{1\max} = \varepsilon_r r_1 g_{2\max}$

or $$g_{2\max} = \frac{r}{r_1\varepsilon_r} \times g_{1\max} = \frac{0.5}{1 \times 5} \times 34 = 3.4 \text{ kV/cm}$$

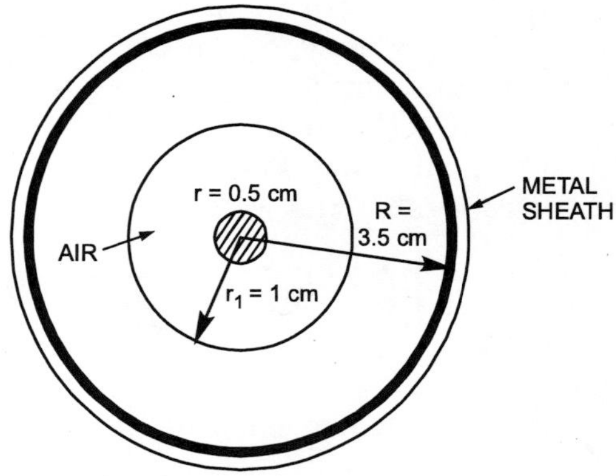

Fig. 11.16

Maximum safe working voltage (peak value),

$$V_{peak} = r\, g_{1\max} \log_e \frac{r_1}{r} + r_1\, g_{2\max} \log_e \frac{R}{r_1}$$

$$= 0.5 \times 34 \log_e \frac{1}{0.5} + 1 \times 0.34 \log_e \frac{3.5}{1}$$

$$= 11.7835 + 4.2594 = 16.043 \text{ kV}$$

RMS value of maximum safe working voltage,

$$V_{rms} = \frac{16.043}{\sqrt{2}} = 11.44 \text{ kV} \ \textbf{Ans.}$$

Example. 11.9. A single-core lead sheathed cable has a conductor of 1 cm diameter and a sheath of 5 cm inside diameter. Two dielectric materials having permittivities of 4 and 2.5 and permissible potential gradients of 60 kV/cm and 50 kV/cm respectively are used. Determine the thickness of the dielectric materials and the maximum working voltage.

Solution : Maximum permissible stress, $g_{1max} = 60$ kV/cm

Maximum permissible stress, $g_{2\,max} = 50$ kV/cm

Let the diameter over the insulation of inner dielectric be d_1 cm

$$\text{Now } \frac{g_{1max}}{g_{2max}} = \frac{Q/2\pi \in_0 \in_{r1} d}{Q/2\pi \in_0 \in_{r2} d_1} = \frac{\in_{r2} d_1}{\in_{r1} d} = \frac{2.5 \times d_1}{4 \times 1} = 0.625\, d_1$$

$$\text{or } \frac{60}{50} = 0.625\, d_1$$

$$\text{or } d_1 = \frac{1.2}{0.625} = 1.92 \text{ cm}$$

$$\text{Radial thickness of inner dielectric} = \frac{1.92 - 1}{2} = 0.46 \text{ cm or 4.6 mm } \textbf{Ans}$$

$$\text{Radial thickness of outer dielectric} = \frac{5 - 1.92}{2} = 1.54 \text{ cm or 15.4 mm } \textbf{Ans}$$

$$\text{Peak voltage of cable} = \frac{g_{1max}}{2} d \log_e \frac{d_1}{d} + \frac{g_{2max}}{2} d_1 \log_e \frac{D}{d_1}$$

$$= \frac{60}{2} \times 1 \times \log_e \frac{1.92}{1} + \frac{50}{2} \times 1.92 \log_e \frac{5}{1.92}$$

$$= 19.57 + 45.94 = 65.51 \text{ kV}$$

$$\text{Maximum working voltage } (rms) = \frac{65.51}{\sqrt{2}} = 46.32 \text{ kV } \textbf{Ans.}$$

Example 11.10. A single core cable employing three layers of insulation with dielectric constants $\epsilon_{r1} = 5$, $\epsilon_{r2} = 4$ and $\epsilon_{r3} = 3$ respectively has a conductor of radius 1 cm. Assuming that all the three insulating materials are worked at the same maximum potential gradient, workout the potential difference in kV between the core and earthed sheath. The inner radius of the sheath is 2.5 cm and the maximum potential gradient is 40 kV/cm.

Solution : Diameter of conductor, $d = 2 \times 1 = 2.0$ cm

The diameter over the insulation, $D = 2 \times 2.5 = 5.0$ cm

Maximum potential gradient, $g_{max} = 40$ kV/cm

Let the diameters over the insulations of relative permittivities 5 and 4 be d_1 and d_2 respectively.

As the maximum stress in the three dielectrics is the same

$$\therefore \in_{r1} d = \in_{r2} d_1 = \in_{r3} d_2$$

$$\text{or } d_1 = \frac{\in_{r1}}{\in_{r2}} d = \frac{5}{4} \times 2.0 = 2.5 \text{ cm}$$

$$\text{and } d_2 = \frac{\in_{r1}}{\in_{r3}} \times d = \frac{5}{3} \times 2.0 = 3.33 \text{ cm}$$

$$\text{Permissible peak voltage for the cable} = \frac{g_{max}}{2}\left(d \log_e \frac{d_1}{d} + d_1 \log_e \frac{d_2}{d_1} + d_2 \log_e \frac{D}{d_2}\right)$$

$$= \frac{40}{2}\left(2 \log_e \frac{2.5}{2} + 2.5 \log_e \frac{3.33}{2.5} + 3.33 \log_e \frac{5}{3.33}\right)$$

$$= 20\,(0.4463 + 0.7192 + 1.3516) = 50.34 \text{ kV}$$

$$\text{So working voltage } (rms) \text{ for the cable} = \frac{50.34}{\sqrt{2}} = 35.6 \text{ kV } \textbf{Ans.}$$

Example 11.11. A single core lead sheathed cable has a conductor of 1 cm diameter and two layers of different insulating materials each 1 cm thick. The relative permittivities are 3.0 (inner) and 2.5 (outer). Calculate the potential gradient at the surface of the conductor and at other points. System voltage is 66 kV.

Solution : Diameter over the insulation of relative permittivity 3, $d_1 = 1.0 + 2 \times 1.0 = 3.0$ cm

Diameter over the insulation of relative permittivity 2.5, $D = 3.0 + 2 \times 1.0 = 5.0$ cm

Let the potential gradient at the surface of the conductor be g_{1max} kV/cm

Then the maximum stress in outer dielectric is given by

$$g_{2\,max} = g_{1\,max} \times \frac{d \in_{r1}}{d_1 \in_{r2}} = g_{1\,max} \times \frac{1 \times 3}{3 \times 2.5} = \frac{g_{1\,max}}{2.5}$$

System peak voltage per phase, $V_{max} = \frac{66}{\sqrt{3}} \times \sqrt{2} = 53.89$ kV

Since system peak voltage, V_{max} is given by the expression

$$V_{max} = g_{1\,max} \times \frac{d}{2} \times \log_e \frac{d_1}{d} + g_{2\,max} \times \frac{d_1}{2} \log_e \frac{D}{d_1}$$

$$\therefore\ 53.89 = g_{1\,max} \times \frac{1.0}{2} \times \log_e \frac{3.0}{1.0} + \frac{g_{1\,max}}{2.5} \times \frac{3.0}{2} \log_e \frac{5.0}{3.0}$$

$$= g_{1\,max}\ (0.5 \log_e 3.0 + 0.6 \log_e 1.6667)$$

$$\text{or } g_{1\,max} = \frac{53.89}{0.5 \log_e 3.0 + 0.6 \log_e 1.6667} = 63 \text{ kV/cm } \textbf{Ans.}$$

The maximum stress in the outer dielectric, $g_{2\,max} = \frac{g_{1max}}{2.5} = \frac{63}{2.5} = 25.2$ kV/cm **Ans.**

The stress at the surface of the outer dielectric $= \frac{g_{2\,max} \times d_1}{D} = \frac{25.2 \times 3}{5} = 15.12$ kV/cm **Ans.**

Example 11.12. What do you mean by Grading of Cables ? What are the different methods ?

A 66 kV single core lead sheathed cable is graded by using two dielectrics of relative permittivity 5 and 3, the thickness of each being 1 cm. The core diameter is 2 cm. Determine the maximum stress in the two dielectrics. [*U.P. Technical Univ. Elements of Power System 2003-04*]

Solution : Conductor diameter, $d = 2$ cm

Diameter over the insulation of relative permittivity 5, $d_1 = 2 + 2 \times 1 = 4$ cm

Diameter over the insulation of relative permittivity 3, $D = 4 + 2 \times 1 = 6$ cm

Let the potential gradient at the conductor surface be $g_{1\,max}$ kV/cm

Then the maximum stress in outer dielectric is given by

$$g_{2max} = g_{1max} \times \frac{d\varepsilon_{r_1}}{d_1\varepsilon_{r_2}} = g_{1max} \times \frac{2 \times 5}{4 \times 3} = \frac{5}{6} g_{1max}$$

System peak voltage per phase, $V_{max} = \frac{66}{\sqrt{3}} \times \sqrt{2} = 53.89$ kV

Since, system peak voltage, V_{max} is given by the expression

$$V_{max} = g_{1max} \times \frac{d}{2} \times \log_e \frac{d_1}{d} + g_{2max} \times \frac{d_1}{2} \log_e \frac{D}{d_1}$$

$$\text{so } 53.89 = g_{1max} \times \frac{2}{2} \times \log_e \frac{4}{2} + \frac{5}{6} g_{1max} \times \frac{4}{2} \log_e \frac{6}{4}$$

$$= g_{1max}\,(0.6931 + 0.6758) = 1.36888\, g_{1max}$$

$$\text{or } g_{1max} = \frac{53.89}{1.36888} \text{ kV/cm} = 39.37 \text{ kV/cm } \textbf{Ans.}$$

Maximum stress in the outer dielectric,

$$g_{2max} = \frac{5}{6} g_{1max} = \frac{5}{6} \times 39.37 = 32.8 \text{ kV/cm } \textbf{Ans.}$$

Example 11.13. A single core lead covered cable is to be designed for 66 kV to earth. Its conductor radius is 0.5 cm and its three insulating materials A, B and C have relative permittivities of 4, 2.5 and 4 with maximum permissible stresses of 50, 30 and 40 kV/cm respectively. Determine the minimum internal diameter of lead sheath.

Solution : Peak value of cable voltage, $V_{max} = 66\sqrt{2}$ kV

In order to have minimum internal diameter of the lead sheath and consistency in the dielectric stresses, it is necessary that the material having maximum of the product of dielectric stress and the permittivity (*i.e.* in this case the material A having this product of 50 × 4 = 200 as the maximum) is placed near the surface of the core. Material C having the product of dielectric stress and the permittivity of 40 × 4 = 160 (second maximum) be placed next and then the material B

Thus $g_{1\,max} = 50$ kV/cm; $g_{2\,max} = 40$ kV/cm and $g_{3\,max} = 30$ kV/cm

$\epsilon_{r1} = 4$; $\epsilon_{r2} = 4$ and $\epsilon_{r3} = 2.5$.

For a given cable we have

$$g_{1\,max} = \frac{Q}{2\pi\,\epsilon_0\,\epsilon_{r1}\,r};\ g_{2\,max} = \frac{Q}{2\pi\,\epsilon_0\,\epsilon_{r2}\,r_1} \text{ and } g_{3\,max} = \frac{Q}{2\pi\,\epsilon_0\,\epsilon_{r3}\,r_2}$$

From the above equations we have

$$Q = 2\pi\,\epsilon_0\,\epsilon_{r1}\,r\,g_{1\,max} = 2\pi\,\epsilon_0\,\epsilon_{r2}\,r_1\,g_{2\,max} = 2\pi\,\epsilon_0\,\epsilon_{r3}\,r_2\,g_{3\,max}$$

$$\text{or } \epsilon_{r1}\,r\,g_{1\,max} = \epsilon_{r2}\,r_1\,g_{2\,max} = \epsilon_{r3}\,r_2\,g_{3\,max}$$

$$\text{or } r_1 = \frac{\epsilon_{r1}\,r\,g_{1\,max}}{\epsilon_{r2}\,g_{2\,max}} = \frac{4 \times 0.5 \times 50}{4 \times 40} = 0.625 \text{ cm}$$

$$\text{and } r_2 = \frac{\epsilon_{r2}\,r_1\,g_{2\,max}}{\epsilon_{r3}\,g_{3\,max}} = \frac{4 \times 0.625 \times 40}{2.5 \times 30} = 1.333 \text{ cm}$$

$$\text{Since, } V_{max} = g_{1\,max} \times r \log_e \frac{r_1}{r} + g_{2\,max} \times r_1 \log_e \frac{r_2}{r_1} + g_{3\,max}\,r_2 \log_e \frac{R}{r_2}$$

$$\text{or } 66\sqrt{2} = 50 \times 0.5 \log_e \frac{0.625}{0.5} + 40 \times 0.625 \log_e \frac{1.333}{0.625} + 30 \times 1.333 \log_e \frac{R}{r_2}$$

$$93.338 = 5.5786 + 18.942 + 40 \log_e \frac{R}{r_2}$$

$$\text{or } \log_e \frac{R}{r_2} = \frac{1}{40}[93.338 - 5.5786 - 18.942] = 1.7204$$

$$\text{or } R = r_2 \times 5.58676 = 1.333 \times 5.58676 = 7.45 \text{ cm}$$

and inner diameter of lead sheath, D = 2 R = 2 × 7.45 = 14.9 cm **Ans.**

Example 11.14. A 66 kV, single core metal sheathed cable is to be graded by means of a metallic intersheath. Calculate the diameter of the intersheath and the voltage at which it must be maintained inorder to obtain minimum overall cable diameter. The maximum voltage gradient at which the insulating material can be worked is 60 kV/cm.

Solution : RMS value of cable voltage, $V_{max} = 66$ kV

Peak value of cable voltage, $V_{max} = 66\sqrt{2}$ kV

Maximum permissible potential gradient of dielectric, $g_{max} = 60$ kV/cm

Let d be the conductor diameter, d_1 be the outer diameter of the inner insulating layer and D be the outer diameter of the outer insulating layer.

If V_1 = potential difference between core and sheath

V_2 = potential difference between intersheath and the outer sheath

$$\text{then, } g_{max} = \frac{V_1}{\frac{d}{2}\log_e \frac{d_1}{d}} = \frac{V_2}{\frac{d_1}{2}\log_e \frac{D}{d_1}}$$

For minimum overall diameter of cable,

$$\text{Conductor diameter, } d = \frac{2\,V_{max}}{e \times g_{max}} = \frac{2 \times 66\sqrt{2}}{2.71828 \times 60} = 1.144 \text{ cm}$$

Diameter of intersheath, $d_1 = e \times d = 2.71828 \times 1.144 = 3.1$ cm **Ans.**

$$\text{Voltage between conductor and intersheath, } V_1 = \frac{V}{e} = \frac{66}{2.71828} = 24.28 \text{ kV}$$

Voltage between the intersheath and sheath overall = $V - V_1 = 66.0 - 24.28 = 41.72$ kV **Ans.**

Example 11.15. A single-core cable has a conductor diameter of 2.5 cm and a sheath of inside diameter 6 cm. Calculate the maximum stress. It is desired to reduce the maximum stress by using two intersheaths. Determine their best positions, the maximum stress and the voltage on each. System voltage is 66 kV, 3-phase.

Solution : System peak voltage per phase, $V_{max} = \frac{66}{\sqrt{3}} \times \sqrt{2} = 53.8888$ kV

$$\text{Without intersheath the maximum stress, } g_{max} = \frac{2\,V}{d \log_e \frac{D}{d}} = \frac{2 \times 53.8888}{2.5 \log_e \frac{6}{2.5}} = 49.24 \text{ kV/cm } \textbf{Ans.}$$

For same maximum and minimum stresses,

$$\frac{d_1}{d} = \frac{d_2}{d_1} = \frac{D}{d_2} = \alpha \text{ (say)}$$

$$\text{So } \frac{d_1}{d} \times \frac{d_2}{d_1} \times \frac{D}{d_2} = \frac{D}{d} \text{ or } \alpha^3 = \frac{6}{2.5} \text{ or } \alpha = 1.339$$

The diameter of intersheaths, $d_1 = \alpha\, d = 1.339 \times 2.5 = 3.347$ cm

and $d_2 = 1.339 \times 3.347 = 4.481$ cm

From Eqs. (11.14) and (11.13), we have

$$V_1 + V_2 + V_3 = V$$

$$g_{max} \times \left[\frac{d}{2}\log_e \frac{d_1}{d} + \frac{d_1}{2}\log_e \frac{d_2}{d_1} + \frac{d_2}{2}\log_e \frac{D}{d_2}\right] = V$$

$$\text{or} \quad g_{max} \times \left[\frac{2.5}{2}\log_e \alpha + \frac{3.347}{2}\log_e \alpha + \frac{4.481}{2}\log_e \alpha\right] = 53.8888 \text{ kV}$$

$$\text{or} \quad g_{max} \times \frac{53.8888}{5.164 \log_e 1.339} = 35.75 \text{ kV/cm}$$

$$\text{Peak voltage } V_{1max} = g_{max}\frac{d}{2}\log_e \frac{d_1}{d} = 35.75 \times \frac{2.5}{2}\log_e 1.339 = 13 \text{ kV}$$

$$\text{Peak voltage, } V_{2max} = g_{max}\frac{d_1}{2}\log_e \frac{d_2}{d_1} = 3.5.75 \times \frac{3.347}{2} \times \log_e 1.339 = 17.46 \text{ kV}$$

Peak voltage on conductor surface, $V_{max} = 53.8888$ kV

Peak voltage on first intersheath = $V_{max} - V_{1max} = 53.8888 - 13 = 40.8888$ kV **Ans.**

Peak voltage on second intersheath = Peak voltage on first intersheath – V_{2max}

$= 40.8888 - 17.46 = 23.4288$ kV **Ans.**

11.12. CAPACITANCE OF THREE-CORE BELTED TYPE CABLES

The capacitance of a cable system is of much greater importance than that of an overhead line of the same length firstly due to the narrow spacing between the conductors themselves and between the conductors and earthed sheath and secondly due to their separation by a dielectric medium of higher permittivity as compared to air.

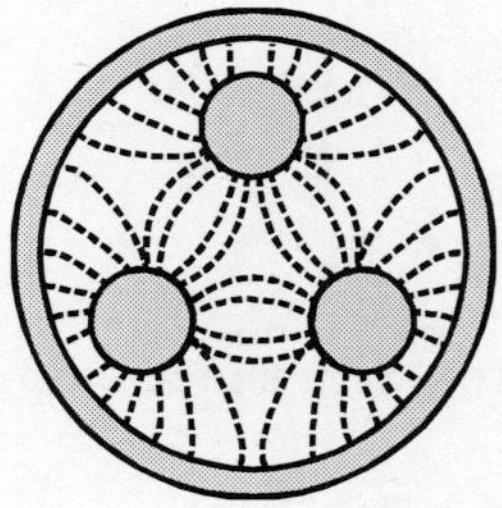

Fig. 11.17. ***Electrostatic Field of a 3-core Belted Cable***

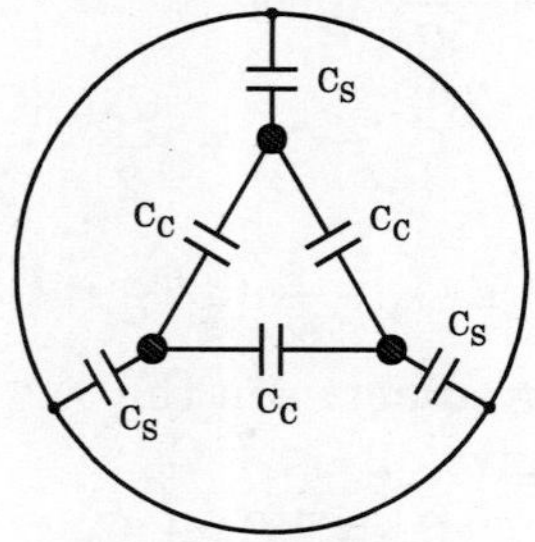

Fig. 11.18.

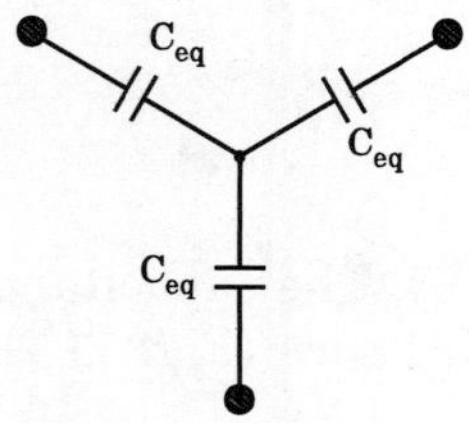

Fig. 11.19.

Since there is a potential difference between pairs of conductors and between each conductor and sheath there will be a system of electrostatic fields in the cross-section of the cable somewhat as shown in Fig. 11.17. Consequently the three-core cable has capacitance between the cores and each core has capacitance with sheath, as shown in Fig. 11.18.

Let the capacitance between cores be C_C and capacitance between cores and sheath be C_S.

The delta-connected capacitances can be converted into equivalent star-connected capacitances, as shown in Fig. 11.18.

$$\text{The } C_{eq} = 3\, C_C \qquad \text{...(11.15)}$$

The star point may be assumed to be at zero potential and if the sheath is also at zero potential, the capacitance of each core to neutral

$$C_N = C_{eq} + C_S = 3\, C_C + C_S \qquad \text{...(11.16)}$$

If V_P is the phase voltage, then the charging current is given

$$I_C = 2\,\pi f V_P C_N \text{ amperes} \qquad \text{...(11.17)}$$

The values of capacitances C_C and C_S may be determined by measurement as follows:

1. Capacitance is measured between the three cores bunched together and the earthed sheath. (Fig. 11.20 *a*). This gives $3C_S$ because three capacitances C_C are eliminated leaving the three capacitances C_S in parallel

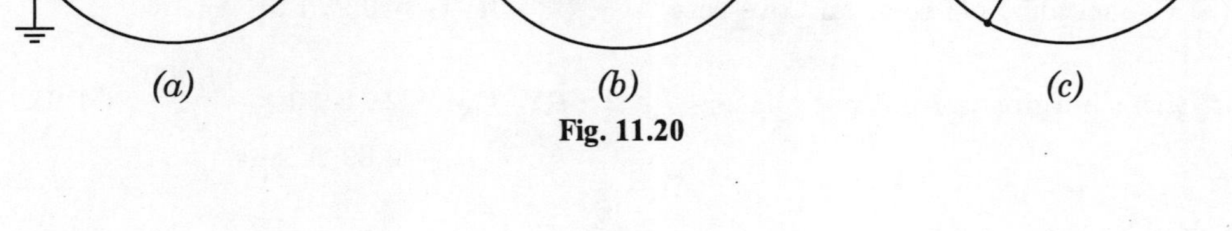

(*a*) (*b*) (*c*)

Fig. 11.20

$$i.e.\ C_1 = 3\,C_S$$

$$\text{or } C_S = \frac{C_1}{3} \qquad \ldots(11.18)$$

2. Capacitance between the two cores or lines is measured with the third core being either insulated or connected to sheath (Fig. 11.20 *b*). This eliminates one of the capacitors C_S and measured value C_2 gives $\frac{C_N}{2}$ directly.

$$C_2 = C_C + \frac{C_C}{2} + \frac{C_S}{2}$$

$$\text{or } C_2 = \frac{1}{2}(3\,C_C + C_S) = \frac{1}{2}\,C_N \qquad \ldots(11.19)$$

3. The capacitance is measured between the two cores shorted with the sheath and the third core (Fig. 11.20 *c*). This gives $2\,C_C + C_S$

$$i.e.\ C_3 = 2\,C_C + C_S \qquad \ldots(11.20)$$

From the above equations values of C_C and C_S can be determined.

It is very difficult to determine the capacitance C_N of the cable from the geometry of the figure. However the empirical formulae given below may be used for determining the value of C_N.

$$C_N = \frac{0.03\,\epsilon_r}{\log_e\left[1 + \frac{T+t}{d}\left(3.84 - 1.7\frac{t}{T} + 0.52\frac{t^2}{T^2}\right)\right]}\ \mu\text{ F/km} \qquad \ldots(11.21)$$

where d = diameter of conductor, t = thickness of belt insulation and T = thickness of conductor insulation.

The value of C_N is usually determined by experiments, as described above, not by the emperical formula.

Example 11.16. An 11 kV, 2.5 km, 3-phase, 3-wire belted cable gave the following results in 'capacitance' testing.

(1) Capacitance between two conductors joined to sheath and the third conductor = 1.5 μF.

(2) Capacitance between all three conductors joined and sheath = 1.8 μF. Find (*i*) capacitance of each core to neutral (*ii*) capacitance between any two core and (*iii*) charging current.

Solution : Capacitance between all three conductors joined and sheath [Fig. 11.20 (*a*)] gives $3C_S$

$$\text{So, } 3\,C_S = 1.8\ \mu\text{F or } C_S = \frac{1.8}{3} = 0.6\ \mu\text{F}$$

Capacitance between two conductors joined to sheath and the third conductor [Fig. 11.20 (*c*)] gives $2C_C + C_S$

$$\text{So, } 2C_C + C_S = 1.5\ \mu\text{F}$$

$$\text{or } C_C = \frac{1.5 - C_S}{2} = \frac{1.5 - 0.6}{2} = 0.45\ \mu\text{F}$$

(*i*) Capacitance of core to neutral, $C_N = 3C_C + C_S = 3 \times 0.45 + 0.6 = 1.95\ \mu\text{F}$ **Ans.**

(*ii*) Capacitance between any two core $= \frac{1}{2}\,C_N = \frac{1.95}{2} = 0.975\ \mu\text{F}$ **Ans.**

(*iii*) Charging current per phase, $I_C = 2\,\pi\,f\,V_P\,C_N = 2\pi \times 50 \times \frac{11{,}000}{\sqrt{3}} \times 1.95 \times 10^{-6}$

$= 3.89$ A **Ans.**

Example 11.17. Calculate the kVA taken by a 10 km long, 3-phase, 3-core, cable, if the capacitance measured between any two cores is 0.3 μF/km when it is connected to 10 kV, 50 Hz bus-bar.

[U.P. Technical Univ. Elements of Power System 2006-07]

Solution : Capacitance between any two core for 10 km length,

$$C_2 = \frac{1}{2} C_N = 0.3 \times 10 = 3 \ \mu F$$

$$\text{or } C_N = 2 \times 3 \times 10^{-6} = 6 \times 10^{-6} \text{ F}$$

$$\text{kVA taken by the cable} = \frac{3 V_P I_C}{1,000} = \frac{3 V_P \times 2\pi f V_P C_N}{1,000}$$

$$= \frac{\frac{3 \times 10,000}{\sqrt{3}} \times 2\pi \times 50 \times \frac{10,000}{\sqrt{3}} \times 6 \times 10^{-6}}{1,000} = 188.5 \text{ kVAR } \textbf{Ans.}$$

Example 11.18. A 3-core, 3-phase metal-sheathed cable had (*i*) capacitance of 1 μF between shorted conductors and sheath and (*ii*) capacitance between two conductors shorted with the sheath and the third conductor 0.6 μF. Find the capacitance (*i*) between any two conductors and (*ii*) between any two shorted conductors and the third conductor.

Solution : (*i*) Capacitance between shorted conductor and sheath [Fig. 11.20 (*a*)] gives $3C_S$

So $3 C_S = 1 \ \mu F$

$$\text{or } C_S = \frac{1}{3} = 0.333 \ \mu F$$

(*ii*) Capacitance between two conductors shorted with sheath and the third conductor [Fig. 11.20 (*c*)] gives $2 C_C + C_S$

So $2C_C + C_S = 0.6 \ \mu F$

$$\text{or } C_C = \frac{0.6 - C_S}{2} = \frac{0.6 - 0.333}{2} = 0.1333 \ \mu F$$

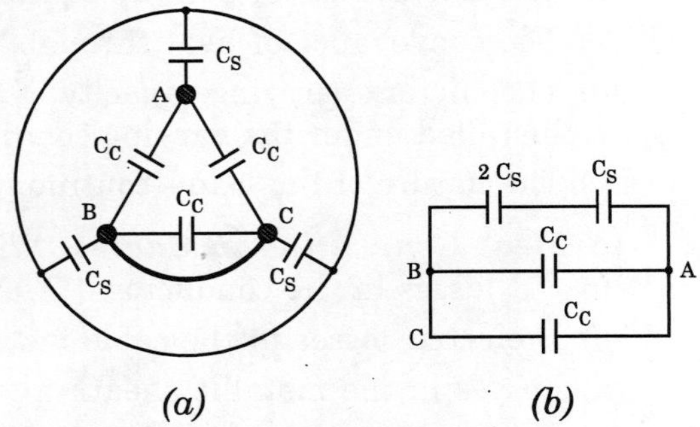

Fig. 11.21

$$\text{Capacitance between any two cores} = \frac{1}{2}\left[3C_C + C_S\right] = \frac{1}{2} [3 \times 0.1333 + 0.333] = 0.3666 \ \mu\text{F } \textbf{Ans.}$$

Capacitance between any two shorted conductors and the third conductor (Fig. 11.21) gives

$$2C_C + \frac{2}{3} C_S = 2 \times 0.1333 + \frac{2}{3} \times 0.333 = 0.488 \ \mu \text{F } \textbf{Ans.}$$

Example 11.19. A 33 kV, 3-phase, 50 Hz underground line, 3.4 km long, uses three single core cables. Each cable has a core diameter of 2.5 cm and the radial thickness of insulation is 0.6 cm. The relative permittivity of the dielectric is 3.1. Find (*i*) maximum stress and (*ii*) total charging kVAR.

[U.P. Technical Univ. Elements of Power System 2005-06]

Solution : Diameter of conductor, d = 2.5 cm

Internal diameter of sheath, D = 2.5 + 2 × 0.6 = 3.7 cm

$$\text{Capacitance of cable, C} = \frac{0.024\varepsilon_r}{\log_{10} \frac{D}{d}} \times \text{length of cable } \mu F$$

$$= \frac{0.024 \times 3.1}{\log_{10} \frac{3.7}{2.5}} \times 3.4 \times 1,000 = 1,486 \times 10^{-6} \text{ F/phase}$$

Charging current, $I_C = 2\pi f CV = 2\pi \times 50 \times 1,486 \times 10^{-6} \times \frac{33 \times 10^3}{\sqrt{3}}$ A/phase = 8.895 A/phase

Total charging kVAR $= 3 \times V_P \times I_C \times 10^{-3} = 3 \times \frac{33,000}{\sqrt{3}} \times 8.895 \times 10^{-3}$ = 508.4 kVAR **Ans.**

Maximum stress in the cable,

$$E_{max} = \frac{V}{\frac{d}{2}\log_e \frac{D}{d}} = \frac{33,000/\sqrt{3}}{\frac{2.5}{2}\log \frac{3.7}{2.5}} = 38.88 \text{ kV/cm (rms)}$$ **Ans.**

11.13. HEATING OF CABLES

The temperature rise of a body depends upon the rate of generation and dissipation of heat by the body, The temperature goes on rising until the rate of heat generation becomes equal to that of heat dissipation.

The temperature rise of a cable under operating conditions depends upon the following factors :

(*a*) Total heat produced within the cable *i.e.* up to its periphery.

(*b*) The conveyance of heat up to the periphery of the cable.

(*c*) The conveyance of heat through the medium, and thus away from the cable.

(*d*) The current carrying capacity of the cable, or cables, singly and collectively, where installed under the varying conditions encountered in practice.

(*e*) The nature of the load—continuous, intermittent or possibility of short-circuits etc.

(a) Heat Generation in Cables. Within the cables the sources of heat generation are

(*i*) I^2R losses in the conductor.

(*ii*) Dielectric losses in the cable insulation,

(*iii*) Losses in the metallic sheath and armourings.

(*i*) *Copper losses in the conductors.* Resistance of conductor at an operating temperture of 65°C (assumed) is determined from the resistance given in standard tables (usually at 15.6°C) from the following relation.

$$R_h = R_a\,[1 + \alpha_a\,(65 - 15.6)]$$

where R_h, R_a and α_a are the hot resistance (resistance at 65° C), resistance at ambient temperature of 15.6° C and the temperature coefficient of resistance at ambient temperature of 15.6° C.

To allow for stranding, the resistance of a single-core cable is multiplied by the factor 1.02. In case of multicore cables the resistance is further multiplied by the factor 1.02 in order to allow for the lay of the whole conductor.

(*ii*) *Dielectric Losses.* The energy losses occuring in the dielectric of cables are due to leakage and so called *dielectric hysteresis.* The former loss, is due to passing of current by conduction through the resistance of dielectric, and is independent of supply frequency and, therefore, it occurs both with d c and a c. The leakage current is proportional to the applied voltage and, therefore, the loss is proportional to the square of the applied voltage.

With alternating voltages there is a further loss of energy, usually known as *hysteresis loss*, and this loss under usually operating conditions is very much higher than the leakage loss. Apparently some energy is consumed in reversing the stresses in solid dielectrics, and this appears as heat, causing increase in temperature of the dielectric. At normal operating stresses the dielectric hysteresis loss is proportional to the square of the electrostatic stresses, that is to the square of the applied voltage.

In order to take into account the above losses, the charging current of a cable I is assumed to have two components : One being true capacitance current which is equal to ω C V and leads the applied voltage by 90° and the other being the energy component which is in phase with the applied voltage and represents the dielectric loss component of current.

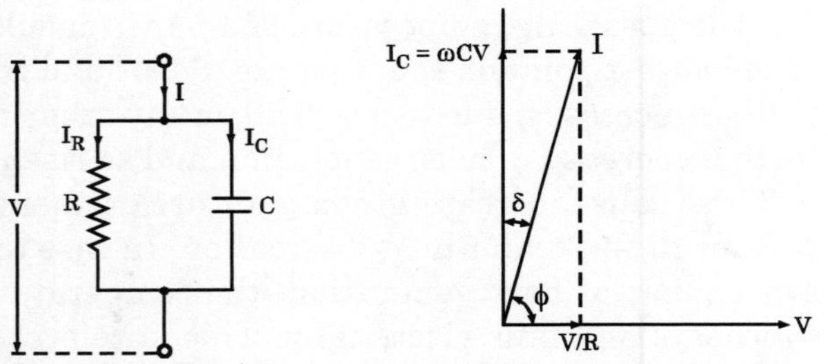

(a) Equivalent Circuit of a Cable *(b) Phasor Diagram of (a)*

Fig. 11.22

The equivalent circuit for this system is represented by a parallel combination of leakage resistance R, representing dielectric power loss, and a capacitance C, as shown in Fig.11.22 (*a*). The phasor diagram is shown in Fig. 11.22 (*b*).

If V is the applied voltage C is the capacitance of cable, φ is the phase angle between voltage and current called the *power factor angle of the cable* and δ is the loss angle of the dielectric,

$$\text{Charging current, } I_C = \frac{V}{X_C} = \frac{V}{1/\omega C} = \omega CV$$

$$\text{Dielectric loss per phase} = V\,I\cos\phi = VI\sin\delta \qquad \because \phi = 90^\circ - \delta$$

$$= V\,\frac{I_C}{\cos\delta}\sin\delta = V\omega CV\tan\delta = \omega CV^2\tan\delta$$

Sine, δ is normally very small, so tan δ can be taken equal to δ (in radians) and dielectric loss per phase is given by the expression

$$P = V^2\,\omega\,C\,\delta \text{ watts} \qquad \ldots(11.22)$$

This shows that the dielectric losses are proportional to the square of the voltage, with the result that, although these are of little importance in low-voltage cables, these are of very great importance in high voltage (say over 22,000 V) cables.

It has been found that, over the working temperature range, the dielectric losses in most cables fall to a minimum at about 40°C and then rise again. The variation when plotted gives rise to what is known as the 'V' curve of the dielectric losses, and the minimum temperature point corresponds very closely to the temperature of liquefaction of the resinous impregnating compound. A typical curve for the paper-insulated cable is shown in Fig. 11.23.

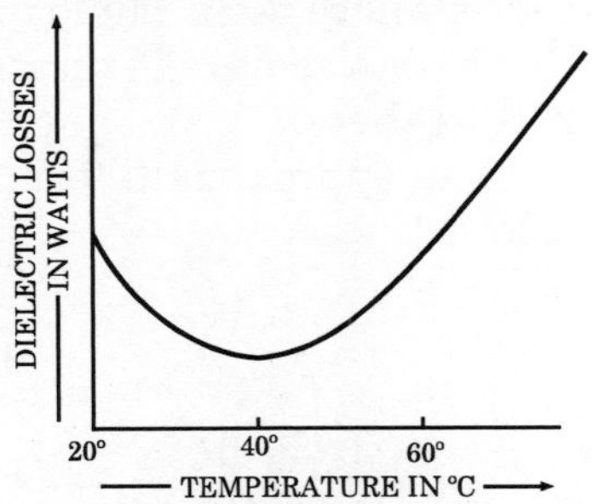

Fig. 11.23

Since the variation of capacitance current with temperature is small, the dielectric power factor–temperature curve is of the same general shape as the dielectric loss–temperature curve. In this connection, it seems preferable to exhibit dielectric – loss data in terms of dielectric power factor rather than in power loss per km of cable in order to facilitate comparison between different types and sizes of cables.

The actual shape of 'V' curve depends on several factors and may be deliberately changed by using different papers, impregnating compounds, and methods of manufacture.

The operating temperature of a paper–insulated cable is about 65°C which is to the right of the lowest point on the 'V' curve. So around the operating temperature, if the temperature further increases, due to overload or any other reason, dielectric losses will increase causing further increase in heat generation and so rise in temperature. The increase in temperature will also cause increase in temperature gradient between the cable and the atmosphere that will result in greater heat dissipation. In case the rate of heat dissipation is less in comparison to that of heat generation, the temperature will go on increasing until the dielectric overheats and fails electrically. This is referred to as *thermal instability*.

Fortunately, action taken to reduce dielectric losses, makes the dielectric loss-temperature curve flat and reduces the tendency towards thermal instability.

(*iii*) *Sheath and Armouring Losses.* In case of power transmission by single-core cables, the presence of lead sheath around each conductor influences the electrical characteristics of the circuit. In 3-core cables the effect is negligible but for single core cables the effect is of great importance. The electromagnetic fields produced by the currents flowing through the conductors induce emfs in the sheaths, and under certain conditions heavy currents are set up therein. The actual current flowing along the sheath depends, amongst other things, on the magnitude and frequency of the current in the conductor, the arrangement and spacing of the cables, the sheath resistance and upon the conditions that the sheath is bounded (two different cables having sheath electrically connected at both ends) or unbounded (cables having one end or no end electrically short-circuited).

The induced sheath currents are of two types:

(*i*) The currents, which have both outward and inward directions, called the *sheath eddies*.

(*ii*) The currents, which have outward and inward current paths in separate sheaths, as shown in Fig. 11.24, called the *sheath-circuit eddies*.

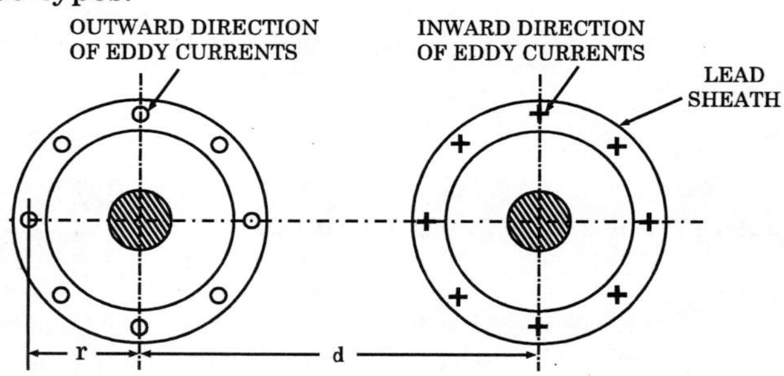

Fig. 11.24

Thus for unbounded cables sheath eddies will be there, but for bounded cables both sheath eddies and sheath circuit eddies will be there.

The approximate formulae for eddy loss for unbounded cables given by Arnold is as follows:

$$\text{Sheath loss} = I^2 \left[\frac{78\,\omega^2}{R_s} \left(\frac{r}{d} \right)^2 \times 10^{-9} \right] \text{ watts/phase} \quad \text{...(11.23)}$$

where, I is the current per conductor, r is the mean radius of the sheath, d is the inter-axial spacing of conductors, and R_s is the sheath resistance in ohms.

If $f = 50$ Hz, then $78\,\omega^2 \times 10^{-9} = 78\,(2\pi \times 50)^2 \times 10^{-9} = 0.0077$

$$\text{and sheath loss} = \frac{0.0077\, I^2}{R_s} \left(\frac{r}{d} \right)^2$$

$$\text{Conductor loss} = I^2 R \qquad \text{where R is the resistance of conductor}$$

$$\therefore \quad \frac{\text{Sheath loss}}{\text{Conductor loss}} = \frac{78\,\omega^2}{R\,R_s} \left(\frac{r}{d} \right)^2 \times 10^{-9}$$

The sheath loss in open-circuited sheath is about 2% of the total loss and can be neglected.

The voltage induced in the sheath = I × ω M = I X_m
where M is the mutual inductance between conductor and sheath.

The value of M between the conductor and sheath can be determined in a similar manner to that of self inductance of the two conductors. For two sheaths with respect to the core conductors (Fig. 11.24) the mutual inductance is given as

$$M = 4 \log_e \frac{d}{r} \times 10^{-7} \text{ H/m}$$

Large voltages are induced in the sheaths if they are open-circuited, and it is very probable that arcing will occur between them. It is, therefore, standard practice to bond the sheaths at both ends so that high voltages are avoided. The impedance of the sheath current path is due to the sheath resistance R_s and the sheath-self inductance, which is equal to M. Thus if the sheaths are bonded the sheath current is given by the expression

$$I_{sh} = \frac{E_{sh}}{\sqrt{(R_s^2 + X_m^2)}} = I \frac{X_m}{\sqrt{(R_s^2 + X_m^2)}}$$

The magnitude of the sheath current is independent of the distance between the bonds, since X_m and R_s, both, are proportional to the length.

$$\text{The sheath losses/phase} = \frac{I^2 X_m^2 R_s}{R_s^2 + X_m^2}$$

Therefore the effective resistance per phase is the conductor resistance plus $\frac{X_m^2 R_s}{R_s^2 + X_m^2}$

$$\text{and ratio of sheath losses to copper losses} = \frac{\dfrac{I^2 X_m^2 R_s}{R_s^2 + X_m^2}}{I^2 R} = \frac{X_m^2 R_s}{R\left(R_s^2 + X_m^2\right)}$$

where R is the resistance of the conductors.

With large conductors and close spacing X_m is very small in comparison with R_s and therefore, the ratio of sheath losses to copper losses

$$\simeq \frac{X_m^2 R_s}{R R_s^2} \simeq \frac{X_m^2}{R R_s} \quad ...(11.24)$$

Example 11.20. A single core cable 1.0 km long has a core diameter of 0.5 cm and diameter undersheath 2.0 cm. The relative permittivity of the insulating material is 3.5. The power factor on open circuit is 0.05 and the supply voltage is 11 kV, 50 Hz. Determine (*i*) the capacitance of the cable (*ii*) the charging current per conductor (*iii*) the dielectric loss and (*iv*) the equivalent insulation resistance.

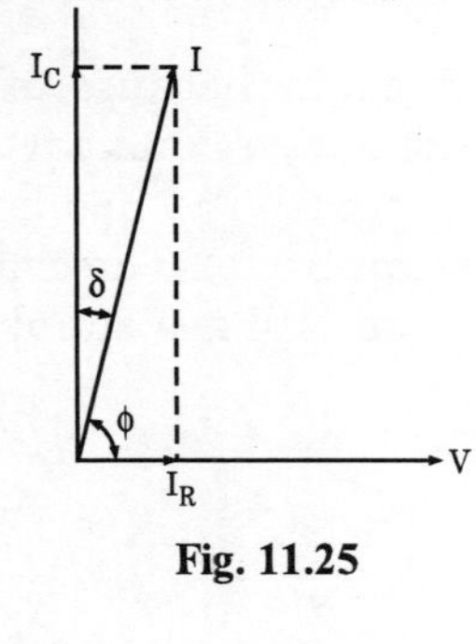

Fig. 11.25

Solution : (*i*) Capacitance of the cable, $C = \frac{0.024 \in_r}{\log_{10} \frac{D}{d}} \times 10^{-6}$ F/km

$$= \frac{0.024 \times 3.5}{\log_{10} \frac{0.02}{0.005}} \times 10^{-6} = 0.1395 \ \mu\text{F} \textbf{ Ans.}$$

(*ii*) Charging current, $I_C = \omega C V = 2\pi \times 50 \times 0.1395 \times 10^{-6} \times \frac{11,000}{\sqrt{3}} = 0.278$ A **Ans.**

$$\text{Power factor on open circuit, } \cos\phi = 0.05$$

$$\therefore \phi = \text{Cos}^{-1}\, 0.05 = 87.134$$

$$\text{Dielectric loss angle, } \delta = 90° - 87.134 = 2.866°$$

(*iii*) Dielectric loss $= \omega C V^2 \tan\delta = 2\pi \times 50 \times 0.1395 \times 10^{-6} \times \left(\frac{11,000}{\sqrt{3}}\right)^2 \tan 2.866° = 88.5$ watts **Ans.**

(*iv*) The equivalent insulation resistance of the cable $= \frac{V^2}{\text{Dielectric loss}} = \frac{(11,000/\sqrt{3})^2}{88.5} = 0.456\ \text{M}\Omega$ **Ans.**

Example 11.21. For a certain length of a 3-core lead sheathed belted type cable following test results were obtained.

(*i*) With one core shorted to sheath capacitance between the other two core 2.5 μF and

(*ii*) With one core shorted to sheath resistance between the other two core : 810 k Ω.

Estimate the loss angle of the cable and the no-load current drawn by it when the cable is connected to a 3-phase, 11 kV 50-Hz supply.

Solution : The capacitance between two cores with one core shorted to sheath gives $\frac{1}{2} C_N$

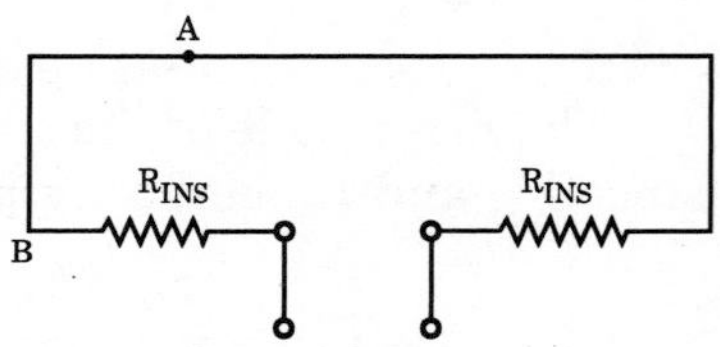

Fig. 11.26

$$\text{So } \frac{1}{2} C_N = 2.5\ \mu\text{F or } C_N = 5\ \mu\text{F}$$

$$\text{Charging current, } I_C = 2\pi V f V_P C_N$$

$$= 2\pi \times 50 \times \frac{11 \times 1,000}{\sqrt{3}} \times 5 \times 10^{-6} = 9.976 \text{ A}$$

The equivalent resistance circuit for the cable with one core shorted to sheath is shown in Fig. 11.26. Measured resistance between two cores gives twice the insulation resistance of each core *i.e.*

$$2 R_{INS} = 810\ \text{k}\Omega \text{ or } R_{INS} = 405\ \text{k}\Omega$$

$$\text{Loss angle } \delta = \text{Tan}^{-1} \frac{V/R_{INS}}{\omega CV} \text{ or } \text{Tan}^{-1} \frac{1}{R_{INS}\, \omega C}$$

$$i.e.\ \delta = \text{Tan}^{-1} \frac{1}{405 \times 10^3 \times 2\pi \times 50 \times 5 \times 10^{-6}} = \text{Tan}^{-1}\, 0.00157 = 0.09° \textbf{ Ans}$$

$$\text{Energy component of current, } I_e = \frac{V_p}{R_{INS}} = \frac{11 \times 1,000}{\sqrt{3} \times 405 \times 10^3} = 0.0157 \text{ A}$$

$$\text{No-load current drawn by cable, } I = \sqrt{I_C^2 + I_e^2} = \sqrt{(9.976)^2 + (0.0157)^2} = 9.976 \text{ A } \textbf{Ans.}$$

11.14. OPERATING PROBLEMS ASSOCIATED WITH UNDERGROUND CABLES

A cable, because of narrow spacing between the conductors themselves and between the conductors and earthed sheath and due to their separation by a dielectric medium of higher permittivity, has large capacitance resulting in high charging current and reactive power (capacitive). The charging current and reactive power drawn by the cable of capacity C farads from 3-phase supply of V volts and frequency f Hz are given by the following expressions

$$I_C = 2\pi f C V_P = \frac{2\pi f C V_L}{\sqrt{3}} \text{ amperes} \qquad \text{...(11.25)}$$

$$\text{Reactive power} = \sqrt{3}\, V_L I_C = \sqrt{3}\, V_L \times \frac{2\pi f C V_L}{\sqrt{3}} = 2\pi f C V_L^2 \text{ VARs (capacitive)} \quad ...(11.26)$$

Flow of charging current causes heating of cable. This reduces the load current capability of the cable. Sometimes current carrying capability of the cable is also reduced due to dielectric losses resulting in increase in temperature rise. The use of cable, in practice, is reduced to a length of less than 50 km because of these factors.

Because of capacitive current/capacitive reactive power drawn from the system, an increase in receiving-end voltage may occur under no-load conditions (Ferranti effect). Thus the receiving-end voltage may vary considerably from full-load to no-load conditions.

Generation of capacitive reactive power may hamper the stability conditions of large generators as large size generators have high synchronous reactance (about 1.5 pu) and their leading MVAR capacity is low.

The interruption of cable capacitive currents may give rise to over-voltage.

EXERCISES

1. Describe the general construction of an underground cable with a neat sketch.
2. With a neat labelled diagram, show the various parts of a high voltage single core cable.
3. Discuss the qualities the dielectrics used for cable insulation must have. Also name three dielectrics with their properties used in the manufacture of underground cables.
4. Describe, briefly, the various materials used for insulation in the high voltage cables. **[U.P. Technical Univ. Elements of Power System 2002-03]**
5. Discuss the usefulness of various materials used in insulating cables. **[U.P. Technical Univ. Elements of Power Systems 2006-07]**
6. Discuss in brief about the types of cables. **[U.P. Technical Univ. Elements of Power System 2009-10]**
7. Mention criterion of the cable classification alongwith justification. Draw the extra high tension cable schematic diagram and explain the various terminology. **[U.P. Technical Univ. Elements of Power System 2005-06]**
8. Write about the commonly used insulating materials for underground cables.
   Describe with a neat sketch, the construction of a 3-core belted type cable. **[U.P. Technical Univ. Elements of Power System 2003-04]**
9. Write about the commonly used insulating materials for underground cables. Describe, with a neat sketch, the construction of a 3-core belted-type cable. **[U.P. Technical Univ. Elements of Power System 2005-06]**
10. Describe with a neat sketch the construction of a 3-core belted type cable. Discuss the limitations of such a cable. **[U.P. Technical Univ. Elements of Power System 2007-08]**
11. Derive an expression for the insulation resistance of a single core metal sheathed cable.
12. Show that the insulation resistance of a cable is inversely proportional to the length of the cable. **[U.P. Technical Univ. Elements of Power System 2004-05]**
13. Find expression for the capacitance of a cable per km length.
14. Find expression for the electrostatic stress in a single core cable, Where does maximum stress occur and where is it minimum and why ?
15. For a single-core lead sheathed cable derive the expressions for (*i*) maximum dielectric stress and (*ii*) capacitance.
16. Show that for a given voltage V and maximum stress E_{max} in the dielectric of a single core cable, the sheath diameter D is a minimum when D; $d = e$ where d is the conductor diameter and e is the base of natural logarithms.

17. Show that the most economical size of conductor in a single core cable is obtained when radius of cable sheath (R) equals er where e is the base of natural log and *r* is the radius of conductor. Explain dielectric loss and heating of a cable.

[U.P. Technical Univ. Elements of Power System 2006-07]

18. Derive relation between conductor radius '*r*' and inner radius of sheath R for stable and satisfactory working of a cable. Discuss the effect of *r*/R on the stability of the cable.

[U.P. Technical Univ. Elements of Power System 2006-07]

19. What do you mean by grading of cables ? Discuss different methods of cable grading.

[U.P. Technical Univ. Elements of Power System 2004-05]

20. Why is capacitance grading required in cables ? How is it done ?

[U.P. Technical Univ. Elements of Power System 2009-10]

21. Discuss the capacitance grading when insulating materials have some safety factor.

[U.P. Technical Univ. Elements of Power System 2004-05]

22. With appropriate derivations compare the performance of cables using (*i*) capacitance grading (*ii*) intersheath grading. **[U.P. Technical Univ. Elements of Power System 2005-06]**

23. Prove that the ratio of gradient with and without intersheath will be $\frac{2}{1+\alpha}$, when there is only one larger, where $\alpha = \frac{r_1}{r} = \frac{R_1}{r_1}$. The core radius is r and outer sheath radius is R. Intersheath is at r_1 distance from the centre. **[U.P. Technical Univ. Elements of Power System 2007-08]**

24. What is the purpose of using intersheaths in a cable ? Show that in a cable using two intersheaths the maximum stress in the (homogeneous) dielectric is reduced to $\frac{3}{1+\alpha+\alpha^2}$ of maximum stress without intersheaths where $\alpha = \frac{d_1}{d} = \frac{d_2}{d_1} = \frac{D}{d_2}$ (D and *d* are the external sheath diameter and core diameter and d_1 and d_2 are the internal diameters of intersheath 1 and 2, respectively.

[U.P. Technical Univ. Elements of Power System 2008-09]

25. Explain, how you can find the capacitance of a three core cable.

[U.P. Technical Univ. Elements of Power System 2002-03]

26. Describe an experiment to determine the capacitance of a belted cable.

27. Explain the dielectric power factor.

[U.P. Technical Univ Elements of Power System 2004-05]

28. Write short notes on the following :

(*i*) General construction of underground cable.

(*ii*) Insulating materials used in hv underground cables.

(*iii*) Classification of underground cables.

(*iv*) Dielectric loss.

(*v*) Heating of cables.

SHORT ANSWER TYPE QUESTIONS WITH ANSWERS

Q. 1. For what purpose metallic sheath is provided in underground cables ?

Ans. The metallic sheath is provided around the insulation to protect it against ingress of moisture, gas or other damaging liquids (acid or alkalies) from the soil and atmosphere.

Q. 2. Why is a layer of bedding provided over the metallic sheath in underground cables ?

Ans. A layer of bedding is provided over the metallic sheath for the protection of metallic sheath against corrosion and from mechanical injury.

Q. 3. Why are copper conductors not used in power cables ?

Ans. Copper conductors are not used in power cables because of scarcity of copper.

Q. 4. What are the main requirements of insulating materials used for underground cables ?

Ans. The main requirements of the insulating materials used for underground cables are high insulation resistivity, high dielectric strength, good mechanical properties, non-hygroscopic, non-inflammable, immune to attacks by acids and alkalies.

Q. 5. Why single core cables are usually not provided with armouring ?

Ans. Single core cables are usually not provided with armouring in order to avoid excessive loss in the armour.

Q. 6. On what factors does the insulation resistance of a cable depend ?

Ans. The insulation resistance of a cable depends upon core radius, insulation thickness, length of cable and resistivity of insulating material.

Q. 7. How does the insulation resistance of a cable vary with its length ?

Ans. Insulation resistance of a cable varies inversely as the length of the cable.

Q. 8. On what factors does the capacitance of a single core depend ?

Ans. The capacitance of a single core depends upon its length, relative permittivity of its dielectric and the ratio of core diameter to sheath inner diameter.

Q. 9. How does electric stress vary in the dielectric of a cable ?

Ans. The electric stress in the dielectric of a cable varies inversely as the distance from the centre of the cable and hence minimum at the sheath and maximum at the conductor surface.

Q. 10. For what value of R/r the minimum value of stress is obtained in a cable ?

Ans. For value of $\frac{R}{r} = e$ *i.e.* 2.71828 the minimum value of stress is obtained in a cable.

Q. 11. What is the modern practice adopted to avoid grading of cables ?

Ans. The modern practice adopted to avoid grading of cables is to employ oil-filled or gas-pressure cables.

Q. 12. Why loss angle of a cable should be very small ?

Ans. As the dielectric losses occurring in a cable are very small, the value of loss angle δ is small.

Q. 13. How does dielectric loss vary with the change in voltage, frequency of supply and capacitance of cable ?

Ans. The dielectric loss in a cable varies in proportion with the capacitance of a cable and frequency of supply and varies as the square of the voltage.

Q. 14. What are the factors, which are to be considered while selecting a cable for a particular service ?

Ans. The factors to be considered in selection of a cable are continuous current rating, voltage grade, permissible voltage drop and short-circuit current rating (in case of *ht* cables only) and operating cost.

PROBLEMS

1. A single core cable has the core diameter of 2.5 cm, thickness of insulation as 1.25 cm. Calculate the insulation resistance per km assuming the resistivity of insulation as 1.5×10^{14} Ω-cm. Derive any formula used. [**Ans.** 495 MΩ]
2. Three single core cables are used for a 132 kV three-phase underground line. The core diameter of the cable is 2.5 cm and the insulation thickness is 4 cm. Calculate the maximum and minimum electrical stress in the dielectric. Derive formula used if any. [**Ans.** 4.248×10^6 V/m; 1.01×10^6 V/m]
3. A single core lead-sheathed cable has core diameter of 2.5 cm and the sheath diameter is 5.3 cm. The oil impregnated paper having a relative permittivity of 3.2 has been used as

insulation between the core and sheath. Determine the maximum and minimum voltage gradients in the dielectric when the voltage applied between the conductor and the sheath is 75 kV (rms). **[Ans.** 83.3 kV/cm; 39.2 kV/cm]

4. Calculate the charging current taken by an oil-filled cable of 7 km length having a core diameter 3.5 cm and inside sheath diameter of 7.8 cm. The cable is used as one phase of a 3-phase, 66 kV, 50 Hz system. Permitivity of the insulating material is 2.8. **[Ans.** 17.4 A]
5. Calculate the maximum field strength of a concentric single core cable, operating at 33 kV, having an insulation of 1.0 cm and a conductor radius of 0.4 cm. For the same maximum potential gradient, find the most economical size of the conductor.
   [Ans. 66 kV; 1 cm diameter]
6. Find the overall diameter of single core cable and its most economical core diameter when working on 220 kV, 3-phase system. The maximum permissible stress in the dielectric is not to exceed 250 kV cm. **[Ans.** 3.9 cm 1.44 cm]
7. Determine the 'economic' overall diameter of a-1-core cable metal sheathed for an operating voltage of 85 kV, assuming the dielectric strength of insulation to be 65 kV/cm. Derive any formula used. **[Ans.** 7.109 cm]
8. A 110 kV single core metal sheathed cable is to be graded by means of a metallic inter-sheath. Find the diameter of the inter-sheath and the voltage at which it must be maintained in order to obtain the minimum overall diameter D. The insulating material can be worked up to 55 kV/cm. **[Ans.** 5.66 cm; 69.5 kV]
9. A single core lead covered cable is designed for 66 kV to earth. Its conductor radius is 1.0 cm and its 3-insulating materials A, B and C have permittivities of 5, 4 and 3 respectively with corresponding maximum safe working stress of 38 kV/cm (rms), 26 kV/cm and 20 kV/cm respectively. Find the minimum diameter of the lead sheathed cable. **[Ans.** 8.3 cm]
10. Estimate the charging current drawn by a cable with 3-cores and protected by a metal sheath when switched on to 11 kV, 3-phase, 50 Hz supply. The capacitance measured between two cores with the third connected to the sheath is 3.7 μF. **[Ans.** 14.78 A]
11. A kilometre of a 3-core, 3-phase, metal-sheathed cable gave the following results on a test for capacitance (*i*) Capacitance between bunched conductors and sheath is 1.0 μF, (*ii*) Capacitance between two conductors bunched with the sheath and third conductor 0.6 μF. With the sheath insulated find the capacitance (*a*) between any two cores (*b*) between any two bunched conductors and the third conductor. **[Ans.** (*a*) 0.367 μF (*b*) 0.489 μF]
12. The capacitance of a 3-core belted type cable are measured and found to be as follows:
    (*i*) Between cores bunched together and the sheath = 8 μF
    (*ii*) Between conductor and the other two connected together to the sheath = 5 μF
    Calculate the capacitance to neutral and the total charging kVA when the cable is connected to an 11 kV, 50 Hz, 3-phase system. **[Ans.** $\frac{37}{6}$ μF; 234.4 kVAR]

❑❑❑

UNIT-V

NEUTRAL GROUNDING AND EHVAC AND HVDC TRANSMISSION

Contents in This Unit

NEUTRAL GROUNDING : Necessity of Neutral Grounding; Various Methods of Neutral Grounding; Earthing Transformer; Grounding Practices.

ELECTRICAL DESIGN OF TRANSMISSION LINE : Design Consideration of EHV Transmission Lines, Choice of Voltage, Number of Circuits, Conductor Configuration, Insulation Design, Selection of Ground Wires.

EHVAC AND HVDC TRANSMISSION : Introduction to EHVAC and HVDC Transmission and Their Comparison; Use of Bundle Conductors; Kinds of DC Links and Incorporation of HVDC into AC System.

CHAPTER 12. Neutral Grounding.

CHAPTER 13. Extra High Voltage (EHV) AC Transmission.

CHAPTER 14. High Voltage Direct Current (HVDC) Transmission.

12

CHAPTER

NEUTRAL GROUNDING

INSIDE THIS CHAPTER

12.1. Power System Earthing **12.2.** Definitions **12.3.** Neutral Earthing **12.4.** Isolated Neutral or Ungrounded Systems **12.5.** Ground Neutral Systems **12.6.** Effectively Grounded System **12.7.** Solid or Effective Grounding **12.8.** Resistance Grounding **12.9.** Reactance Grounding **12.10.** Arc Suppression Coil Grounding (or Resonant Grounding) **12.11.** Voltage Transformer Grounding **12.12.** Grounding Transformer **12.13.** Harmonic Suppressors **12.14.** Neutral Grounding Practice.

12.1. POWER SYSTEM EARTHING

The term '*earthing*' means connecting of the non-current carrying parts of the electrical equipment or the neutral point of the supply system to the general mass of earth in such a manner that at all times an immediate discharge of electrical energy takes place without danger. Earthing is achieved by electrically connecting the respective parts in the installations to some system of electrical conductors or electrodes placed in intimate contact with the soil some distance below the ground level. This contacting assembly is called the *earthing*. The metallic conductors connecting the parts of the installations with the earthing are called *electrical connections*. When speaking of the earthing and the earthing connections as a whole, it is customary to use the term the *earthing system*.

The earthing is provided with the following objectives :

1. For the safety of the personnel from the electric shock–insuring that non-current carrying parts, such as equipment frames are always safely at ground potential even though insulation fails.
2. For the safety of the equipment and personnel against lightning and voltage surges providing the discharge path for surge arresters, gaps and other similar devices.
3. For providing the ground connections for grounded neutral systems.
4. For providing a means of positively discharging and de-energizing feeders or equipment before proceeding with maintenance on them.

Power system earthing is highly important. A *substantial and adequate ground* that will not burn off or permit dangerous rise in voltage under abnormal conditions is essential. Extensive damage and dangerous conditions have arisen when inadequate grounds have been provided. Multiple grounds and multiple connections to them are usually desirable to ensure ground protection, even though one ground or connection opens owing to burn-off or other condition such as high resistance. The station earthing system should have low resistance.

The earthing can be divided into *neutral* and *equipment* earthing. Neutral earthing (or grounding) deals with earthing of the system neutral to ensure system security and protection. Neutral earthing is also called the *system earthing*. Equipment earthing or safety earthing deals with earthing of non-current carrying parts of the equipment to ensure safety of personnel and protection against lightning.

Equipment earthing also helps in the earth fault protection. The earth fault current from the equipment flows through the earthing system to the earth and is sensed by protection system and circuit breakers are opened. The faulty equipment is then repaired and recommissioned. The earthed parts remain at approximately earth potential even during flow of fault current. In *neutral earthing* (or *neutral grounding*), the neutral points (or star points) of star-connected 3-phase windings of power transformers, generators, motors, earthing transformers are connected to low resistance earth. Neutral earthing has numerous advantages and will be discussed later on.

System earthing and equipment earthing have common earth mat* from which flat iron risers are taken out to connect all the non-current carrying metal parts of the equipment. At the time of fault the earth mat conductors rise to a voltage, which is equal to the resistance of the earth mat multiplied by ground fault current. This difference of potential results in voltage gradients**. The earth mat of the substation is so designed that the *step potential* (potential difference between two feet of a person at the time of fault) and *touch potential* (potential difference to which body is subjected to when a person comes in contact with an earthed body) are with in safe limits. The average resistance of human body is 1,000 Ω and a current of 9 to 25 mA may be painful for the body for long duration. As fault is cleared very fast, *i.e.*, within 2 to 4 cycles (40 to 80 milliseconds) therefore 9 mA current is considered safe.

In electrical installations the following components must be earthed :

(*i*) The frames, tanks and enclosures of electric machines, transformers, and apparatus, lighting fittings, and other items of equipment;

(*ii*) the operating mechanism of the switchgear;

(*iii*) the frame work of the switchboards, control boards, individual panel boards, cubicles;

(*iv*) the structural steel work of indoor and outdoor substations, metal cable jointing boxes, the metal sheaths of the cable, the rigid metal conduit runs, and similar metal work.

12.2. DEFINITIONS

1. Earth Electrode. Any wire, rod, pipe, plate or an array of conductors, embedded in ground horizontally or vertically is known as the earth electrode. In distribution systems the earth electrode may consist of a rod, about 1 metre in length and driven vertically into the ground. In generating stations and substations, grounding mats are used rather than individual rods.

2. Resistance-To-Earth or Earth Resistance. The resistance offered by the earth electrode to the flow of current into the ground is known as resistance-to-earth. The term resistance-to-earth (or earth resistance) does not imply the contact resistance between the earthing electrode

* A number of rods when joined together through copper conductors constitute an earthing mat or earthing grid.

** The resistance of an earthing electrode is not concentrated at one point but is distributed over the soil around the electrode.

and the soil (which is insignificant), but mainly the resistance of soil between the electrode and the point of zero potential. Numerically it is equal to the ratio of the potential of earth electrode with respect to remote point, to the current dissipated by it.

Mathematically, earth resistance is given as $\frac{V}{I}$ where V is the measured voltage between the electrode and the voltage spike and I is the injected current during the earth resistance measurement through the electrode.

As a general rule the lower the value of earth resistance better it is but even then the following values of earth resistance (maximum permissible) values will give satisfactory results.

Large power stations — 0.5 Ω

Major power stations — 1.0 Ω

Small substations — 2.0 Ω

In all other cases — 8.0 Ω

3. Earthing Lead. The wire which connects earth wire or any other apparatus to be earthed to the earth electrode is known as the earthing lead.

4. Earth Current. The current dissipated by the earth electrode into the ground is called the earthing current.

12.3. NEUTRAL EARTHING

There are two ways in which three-phase system can be operated and they are :

(*i*) With isolated or free or ungrounded neutral and (*ii*) With grounded neutral

The 3-phase 50 Hz ac power systems with *neutral earthed (or grounded) at every voltage level* are used for generation, transmission, distribution and utilization of electrical power. The neutral or star points of star-connected 3-phase winding of power transformers, generators, motors, grounding transformers are connected to low resistance ground (earth electrode/earth mat). Such a connection is called *neutral earthing* or *neutral grounding*.

Before 1950 s the power systems were often without neutral grounding. Such systems are called *ungrounded systems* or *insulated systems*. The chief argument in favour of such a system is the possibility of maintaining continuity of supply in case one phase gets grounded until it is convenient to disconnect and repair the faulty line. This argument is particularly true for an overhead transmission system when the ground fault on one line is not likely to develop into a fault between two or more lines. But such systems experience repeated arcing grounds. The earth fault protection of ungrounded systems is difficult. Insulation failure may occur in several equipment and machines over entire voltage level during a single ground fault at remote location. The ungrounded systems must necessarily have equipment insulation withstand level corresponding to next higher system voltage so that cascade insulation failures can be avoided. Such a system needs a costlier insulation system of next higher voltage level (*e.g.*, 11 kV insulation for 6.6 kV busbars and motors, transformers, CTs, PTs etc.).

Ungrounded systems have advantage of negligible ground fault current but drawback of arcing grounds. Modern power systems are with grounded neutrals except some continuous process systems and essential protection/auxiliary supply systems where single phase to ground faults should not trip entire bus supply.

12.4. ISOLATED NEUTRAL OR UNGROUNDED SYSTEMS

The main feature of an ungrounded system is its ability, in some cases, to clear earth faults without interruption. The self clearing feature disappears when the length of the line becomes excessive.

A simple three phase system with isolated neutral is shown in Fig. 12.1 (*a*). The line conductors have capacitances between one another and to earth, the former being delta-connected, while the later are star-connected, as shown in the Fig. The effect of line capacitances on the grounding characteristic of the system is little and, therefore, can be neglected.

First of all consider a 3-phase line (perfectly transposed) having same capacitances to ground. In such a line the charging currents for each line to earth capacitor lead the phase voltage by 90° and are equal. The magnitude of these currents is given by $\frac{V_P}{X_C}$ where V_P is the phase voltage and X_C is the reactance due to the capacitance of the line to ground.

The charging currents I_{CR}, I_{CY} and I_{CB} are balanced and their resultant is zero and no current flows to the earth.

Under balanced conditions, the potential of neutral is held at ground due to the presence of the shunt capacitances of the system, shown in Fig. 12.1 (*a*). Under balanced condition the phasor diagram is shown in Fig. 12.1 (*b*).

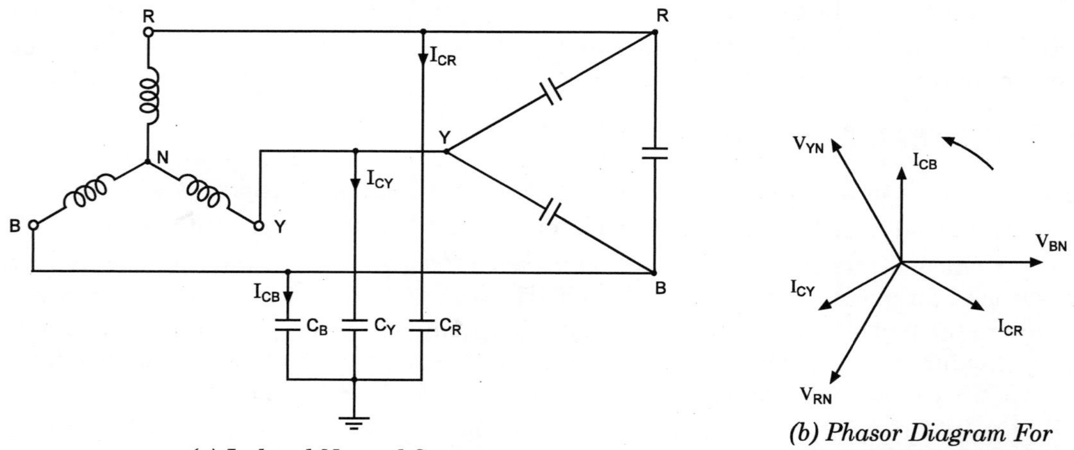

(a) Isolated Neutral System

(b) Phasor Diagram For Isolated Neutral System

Fig. 12.1

Now consider a phase to earth fault in line Y say at point F. Under these circumstances, the faulty line takes up the earth potential while the potentials of the remaining two healthy lines R and B rise from phase values to line value. The capacitance currents become unbalanced and fault current I_F flows through the faulty line, into the fault and returns to the system via earth and through the earth capacitances C_R and C_B. Thus fault current I_F has two components I_{CR} and I_{CB} which flow through capacitances C_R and C_B respectively under the potential differences of V_{RY} and V_{BY} respectively. These currents lead their respective voltages by 90° and their phasor sum is equal to fault current I_F.

$$I_{CR} = \frac{V_{RY}}{X_{CR}} = \frac{\sqrt{3}\,V_P}{X_C} \quad \text{...(12.1)}$$

$$\text{Similarly } I_{CB} = \frac{V_{BY}}{X_{CB}} = \frac{\sqrt{3}\,V_P}{X_C} \quad \text{...(12.2)}$$

$\because$ phase voltages are equal, say, equal to V_P and capacitances from line to ground are also equal, say, equal to X_C.

Since the fault current, I_F is equal to phasor sum of I_{CR} and I_{CB}, therefore, fault current,

$$I_F = \sqrt{3}\, I_{CR} = \sqrt{3} \times \frac{\sqrt{3}\, V_P}{X_C} = \frac{3\, V_P}{X_C} \qquad ...(12.3)$$

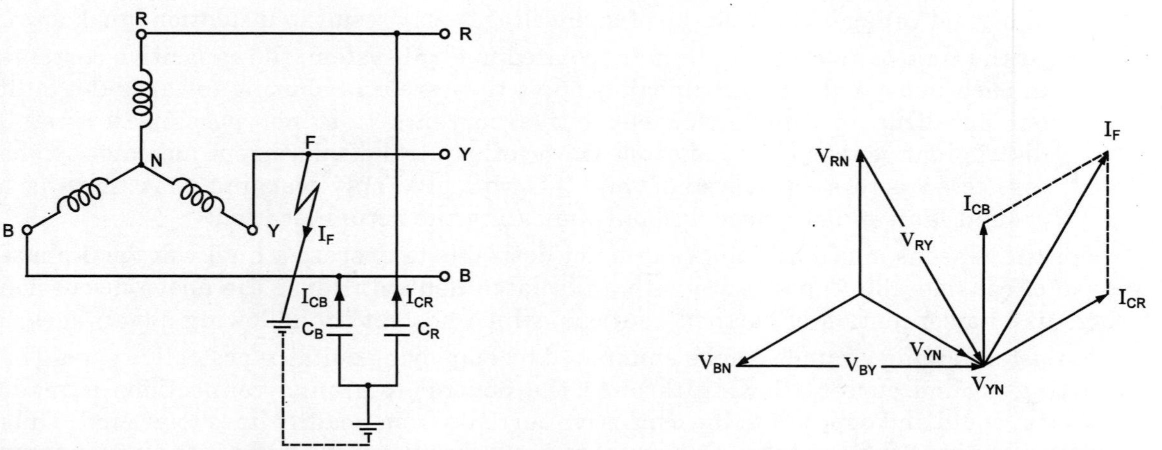

(a) Isolated Neutral System With Fault on One Phase *(b) Phasor Diagram For Fault on Phase Y*

Fig. 12.2

Phasor diagram for fault on phase Y is shown in Fig. 12.2 (*b*).

The following conclusions can be drawn from the above.

(*i*) There is no zero sequence current.

(*ii*) There will be little interference with communication lines because of the absence of zero sequence currents.

(*iii*) In case of one phase becoming earthed the voltages of the remaining two healthy phases to earth rise from their normal phase-to-neural voltage to full line value ($\sqrt{3}$ times their normal value). This causes stress on the insulation of all the machines and equipment connected to the system. The voltage rise of phase above earth is sustained and thereby insulation failure is likely to occur on connected machines, though fault current in arcing ground may be negligible.

(*iv*) The capacitance currents in the two healthy phases increase to $\sqrt{3}$ times their normal values.

(*v*) The capacitance current in the faulty phase becomes 3 times the normal value.

(*vi*) For operating the protective devices, it is essential that the magnitude of the current supplied should be adequate to operate them. But in case of an earth fault on an isolated neutral system, the fault current may be too small to actuate the protective devices. Thus in ungrounded systems the earth fault relaying becomes complicated.

(*vii*) The overvoltages due to induced static charges are not discharged to ground in isolated neutral systems. The voltage due to lightning surges do not find path to earth.

(*viii*) The danger to equipment on the occurrence of line-to-line ground fault is appreciable and danger to life in the proximity of fault is often prolonged.

(*ix*) A capacitive fault current I_F flows into the earth. Such a current if exceeds 4–5 amperes, is sufficient to maintain an arc in the ionized path of the fault, even though the medium causing the fault has cleared itself. The persistency of the arc due to the flow of capacitance current gives rise to a condition known as "*arcing ground*" in which cyclic charging and discharging of the system capacity through the fault results in high frequency oscillations being superimposed on the whole system and a build-up of very high voltages can occur. This results in phase voltage to rise to 5 to 6 times of normal voltage. The build-up of high voltages may result in insulation breakdown.

(*x*) At the time of an earth fault in an isolated neutral system, the capacitive currents in each phase will become unbalanced as they are contributing towards the fault current. Due to unbalanced capacitive currents it is not possible to install discriminative type fault indicator. However, neutral displacement indicator can be connected across such system, but this will give only warning that there is a ground fault in one phase without indicating the actual location.

For the reasons mentioned above it is not desirable to operate a high voltage 3-phase system of considerable capacitance with an isolated neutral. Hence the above discussion brings us to a conception of earthed neutrals, which has got the following advantages :

1. Persistent arcing grounds can be eliminated by employing suitable protective gear. The arcing ground current flowing through the neutral to ground connections is made almost equal and opposite to the capacitive currents from healthy lines to ground. Thus sum of current flowing through neutral to ground connections and capacitive currents is zero and arcing grounds are eliminated. The system is thus not subjected to overvoltage surge due to arcing grounds.
2. In this system the neutral point is not shifted (*i.e.*, stable neutral point).
3. The voltages of healthy phases with respect to ground remain at normal value. They do not increase to $\sqrt{3}$ time the normal value as in the case of isolated neutral system.
4. Earth faults can be utilized to operate protective relays to isolate the fault in case of grounded neutral system.
5. The induced static charges do not cause any disturbance as they are conducted to ground immediately.
6. There is a possibility of installing discriminative protective gear on such systems.
7. By employing resistance or reactance in ground-connections, the ground fault current can be controlled.
8. Improved service reliability due to limitation of arcing grounds and prevention of unnecessary tripping of circuit breakers.
9. Such system provides greater safety to personnel and equipment. This is because of operation of fuses or relays on ground fault and limitation of voltages.
10. The life of insulation is long as voltage surges caused by arcing grounds are eliminated. Thereby maintenance, repairs and breakdowns are reduced and continuity of supply is improved.
11. Life of equipments, machines, installation is improved due to limitation of voltage. Thus overall economy is achieved.

12.5. GROUNDED NEUTRAL SYSTEMS

In grounded neutral systems, the neutral point is grounded directly or through resistance, reactance etc., depending on particular requirement.

Neutral grounding can broadly be classified in two categories *viz.* effective grounding (or solid grounding) and non-effective grounding. For non-effective grounding of neutral any of the following four methods can be used.

1. Resistance grounding.
2. Reactance grounding.
3. Peterson coil or arc suppression coil grounding.
4. Voltage transformer grounding.

12.6. EFFECTIVELY GROUNDED SYSTEM

The term 'effectively grounded' is now used in place of old term 'solidly grounded' for the reason of definition.

According to IEEE definition, a system or a portion of a system can be said to be effectively grounded when for all points on the system or specified portion thereof, the ratio of zero sequence reactance to positive sequence reactance is not greater than three (*i.e.*, $\frac{X_0}{X_1} \not> 3$) and the ratio of zero sequence resistance to positive sequence resistance is not greater than one (*i.e.*, $\frac{R_0}{R_1} \not> 1$) for any condition of operation and for any amount of generator capacity.

The effective grounded systems are less expensive than any other type of grounding for any operating voltage because for such a system in the event of a single line-to-ground fault, the maximum voltage of healthy phase does not exceed 80% of line-to-line voltage while for all other grounded systems the voltage of healthy phases rises to about 100% line-to-line voltage.

12.7. SOLID OR EFFECTIVE GROUNDING

In solid grounding, also called the *effective grounding*, a direct metallic connection is made from the system neutral to one or more earth electrodes [rods, pipes or plates buried or driven into the ground, as shown in Fig. 12.3 (*a*)].

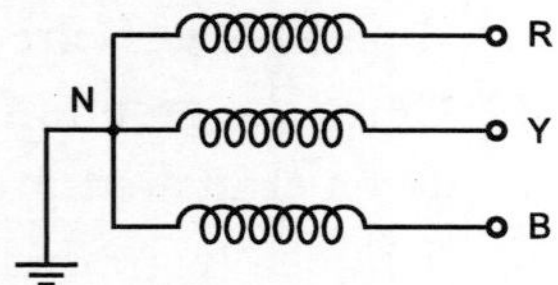

(a) Solid or Effective Grounding

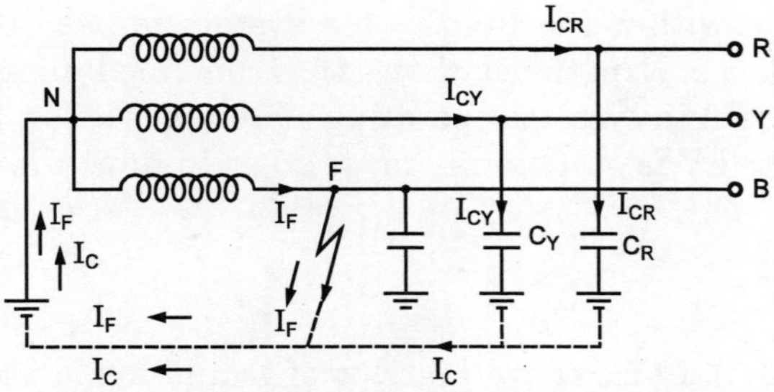

(b) Solidly Grounded System With Ground Fault on Line B

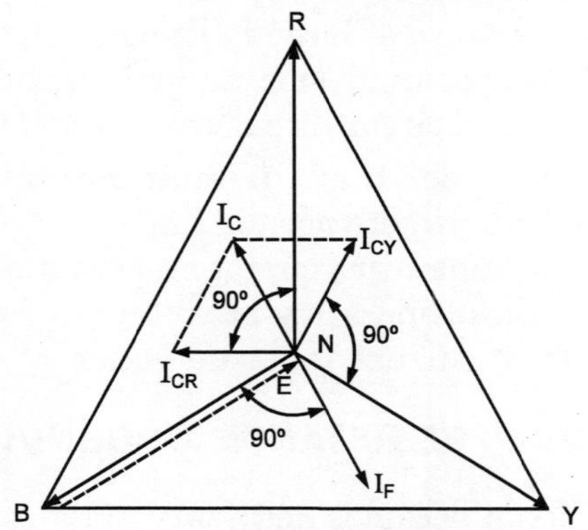

(c) Phasor Diagram For (b)

Fig. 12.3

When there is a ground fault over any phase, the phase to earth voltage of the grounded phase will become zero, but the voltage to earth of the remaining two healthy phases will be the normal phase voltages as in this case neutral point will not shift. Under an L-G fault on phase B, as shown in Fig. 12.3 (*b*), the neutral and terminal B are at earth potential. The phasor diagram for such a condition is shown in Fig. 12.3 (*c*). The reversed phasor is shown at V_B. Capacitive current I_{CR} leads V_{NR} by 90° and I_{CY} leads V_{NY} by 90°. The resultant capacitive current I_C will be phasor sum of I_{CR} and I_{CY}. It should be noted that in this system, in addition to capacitive currents the supply source also supplies the fault current I_F. This fault current will go to the fault point F through the faulty phase and then return back to supply source through the earth and neutral connection. The fault current I_F lags behind the faulty phase voltage by approximately 90° since the circuit is predominantly inductive (due to transformers, machines and line inductances). As illustrated in Fig. 12.3 (*c*), the fault current I_F will be in phase opposition to capacitive current I_C. Due to this effect, the capacitive current I_C will be fully neutralized by the large fault current.

The main features of effectively grounded neutral systems are given below :

1. Since fault current eliminates the effect of the capacitive currents, chances of occurrence of arcing grounds and overvoltages are eliminated up to a greater extent.
2. The flow of heavy fault current permits the use of discriminative protective gear.
3. Ground fault relaying is simple and satisfactory.
4. Since in this system the voltage of the healthy phases in case of a line-to-ground fault does not exceed 80% of the line-to-line voltage and is much less as compared to other types of systems, the equipments for all voltage classes are less expensive. An 84% lightning arrester instead of 105% can be used. On system operating at 115 kV and above additional savings are possible because of the transformers with the insulation graded towards the neutral are less costly.
5. The ground fault current is large. Its maximum values sometimes exceeds even the 3-phase short-circuit current.
6. Since the ground fault current is large, danger to personnel in the vicinity of fault is high.
7. Even transient ground faults may get converted into short-circuit.
8. The heavy ground fault current may cause considerable damage.
9. Because of large ground fault current, the interference due to electromagnetic induction with neighbouring communication circuits may be high.
10. Because of large fault current, the current handling in the circuit breakers is a difficult proposition. This problem is, however, overcome by employing high rupturing and high speed circuit breakers and fast operating protective relays.

In order that the fault current remains within the limits, this system is used on the networks where normal impedance is quite large. Experience shows that combined impedance of the apparatus, circuit and ground return path in systems operating at voltages below 3.3 kV and those operating at voltages exceeding 33 kV is sufficiently large so as to limit the value of fault current to a safe value.

12.8. RESISTANCE GROUNDING

When it becomes necessary to limit the earth fault current, this type of grounding is used. In this method of neutral grounding a resistor is inserted between the neutral and earth to limit the fault current to a safe value, as illustrated in Fig. 12.4 (*a*). The resistor used for neutral grounding is usually a metallic resistance mounted on insulators in a metallic frame. For voltages below 6.6 kV liquid resistors are also used. Metallic resistors do not alter with time and need little or no maintenance. They are, however, slightly inductive and this is a drawback with overhead lines exposed to lightning, because travelling waves or impulses are subjected to positive reflection and this may cause undue stress to the insulation of the equipment and result in breakdown. Liquid resistors are free from this drawback and are normally of simple robust construction. They may be mounted outdoor. The value of resistance used for neutral grounding should be such that the ground fault current does not exceed 3-phase short-circuit current. This is necessary so that the power loss in the grounding resistance is not excessive. Neutral grounding resistors are normally designed to carry their rated current for a short period, usually 30 seconds.

When there is a ground fault over any phase, neutral is displaced and the maximum voltage across the healthy phases may become equal to line-to-line value. Fig. 12.4 (*b*) shows a ground fault on phase B of a resistance grounded system. The phasor diagram for such a condition is illustrated in Fig. 12.4 (*c*). Capacitive currents I_{CR} and I_{CY} leads V_{BR} and V_{BY} respectively by 90°. Fault current I_F lags behind the phase voltage of the faulted phase by an angle ϕ, which depends upon the grounding resistance R and impedance of the system up to the fault point. The fault current I_F can be resolved into two components; one in phase with the faulty phase voltage and the other lagging the faulty phase voltage by 90° ($I_F \sin \phi$). The lagging component of fault current ($I_F \sin \phi$) is in phase opposition to the capacitive current I_C (phasor sum of I_{CR} and I_{CY}). By adjusting the value of grounding resistance R to a sufficiently low value it is possible to neutralize the effect of capacitive current I_C so as to avoid the occurrence of transient oscillations due to arcing ground. In case the grounding resistance R is made sufficiently large so that the lagging component of fault current becomes less than capacitive current I_C, then the system conditions approach that of the isolated neutral system with the risk of high transient voltage occurrence.

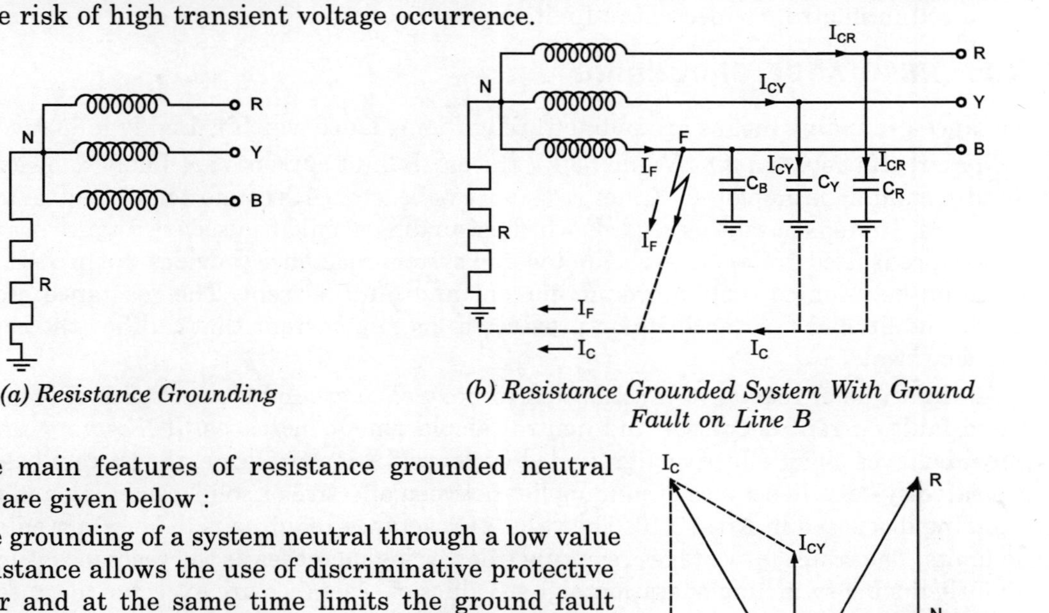

(a) Resistance Grounding

(b) Resistance Grounded System With Ground Fault on Line B

The main features of resistance grounded neutral system are given below :

1. The grounding of a system neutral through a low value resistance allows the use of discriminative protective gear and at the same time limits the ground fault current to a value that will cause no serious disturbance to the system and minimize the extent of damage at the fault location.
2. In general, a resistance grounded system will have ground fault current materially lower than that in effectively grounded system and therefore, the inductive interference to the neighbouring communication circuits is also lesser than that in effectively grounded system.

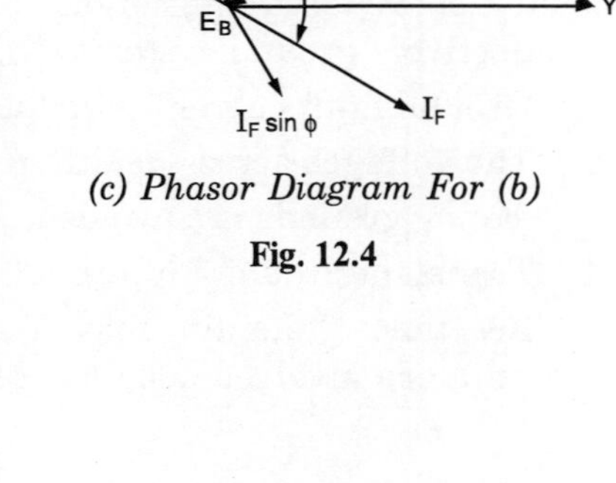

(c) Phasor Diagram For (b)

Fig. 12.4

3. The power dissipation in the grounding resistance may improve system stability because the power dissipation reduces the accelerating power.
4. Transient ground faults are converted into controlled current faults.
5. There is an enormous energy loss in the neutral grounding resistance.
6. The system neutral will almost invariably be displaced in the case of an earth fault, thereby requiring the use of lightning arresters at an increased cost and sacrifice in protective performance.
7. The cost of equipment, particularly of transformers, increases as graded insulation towards neutral cannot be used.
8. The provision of ground resistor involves extra investment in the system.

Hence resistance grounding is usually employed for the systems operating on voltage exceeding 3.3 kV but not exceeding 33 kV. For circuits below 3.3 kV (*i.e.*, say 400 V distribution networks), the external resistance in the neutral circuit is unnecessary because the voltage available between phase and ground is only 230 V. The earth resistance of earthing electrode, earth connections etc. is of the order of 1.5 Ω. The earth current is limited to $\frac{230}{1.5}$, *i.e.*, 153 A even if the grounding resistance is not used. For networks above 33 kV the capacitive currents are large enough to neutralize the reactive fault currents, and therefore no external resistance is required in neutral connection to limit the fault current.

12.9. REACTANCE GROUNDING

Reactance grounding means grounding through impedance which is highly inductive.

For circuits between 3.3 kV and 33 kV the earth fault currents are likely to be excessive, if solid grounding is employed. Either resistance or reactance is connected in neutral to ground connection. There is no rule as regards which grounding should be used–resistance or reactance. If resistance is used fault current is limited and system reactance provides the necessary phase opposition between capacitive ground current and fault current. The reactance grounding provides additional reactance which provides a lagging current that nulifies the capacitive ground current.

As the value of reactance X connected in neutral to ground connection is increased, the ground fault current decreases and neutral displacement increases. If X is very small, the system behaves as an effectively grounded system. If X is very large, the system behaves as an isolated system. Reactance grounding lies between effective or solid grounding and resonant grounding discussed in Art. 12.10. The value of reactance required is to keep currents within safe limits. The transient voltages resulting from arcing increases as the reactance is increased. Similarly during switching operation higher values of reactance are expected to cause higher values of surge voltages.

A reactance grounded system ensures satisfactory relaying, partial grading of apparatus insulation, reduced interference to neighbouring communication circuits as compared with that in effectively ground systems. Other features of reactance grounding are :

1. Ground fault current is reduced but is much larger than capacitive ground fault current.
2. The voltages across healthy phases are between 80 to 100 per cent of line-to-line voltage.
3. Arcing grounds are avoided.
4. Transient ground faults are converted into controlled current faults.

Reactance grounding may be used for grounding the neutral of circuits where high charging currents are involved such as transmission lines, underground cables, synchronous motors, synchronous capacitors, etc. For networks where capacitance is relatively low, resistance grounding is preferred.

12.10. ARC SUPPRESSION COIL GROUNDING (OR RESONANT GROUNDING)

In the arc suppression coil method of grounding, the arcing-ground danger has been eliminated and the system is approximated to the isolated neutral system, in which one or two healthy phases continue to supply power and complete shutdown on the system is avoided till the fault was located and isolated.

It operates on the principle that if an inductance of appropriate value is connected in parallel with the capacitance, the fault current can be reduced considerably or even it can be neutralized. Thus the magnitude of inductance of the coil depends upon the capacitive currents flowing into the ground capacitances.

An arc *suppression coil*, also known as *Peterson coil* or *Ground fault neutralizer*, is an iron core tapped reactor connected in neutral to ground connection. The reactor is provided with tappings so that it can be tuned to the system capacitance. The function of arc suppression coil is to make the arcing ground faults self extinguishing and, in case of sustained faults, to reduce ground fault current to a comparatively low value so that the system can be kept in operation with one line grounded.

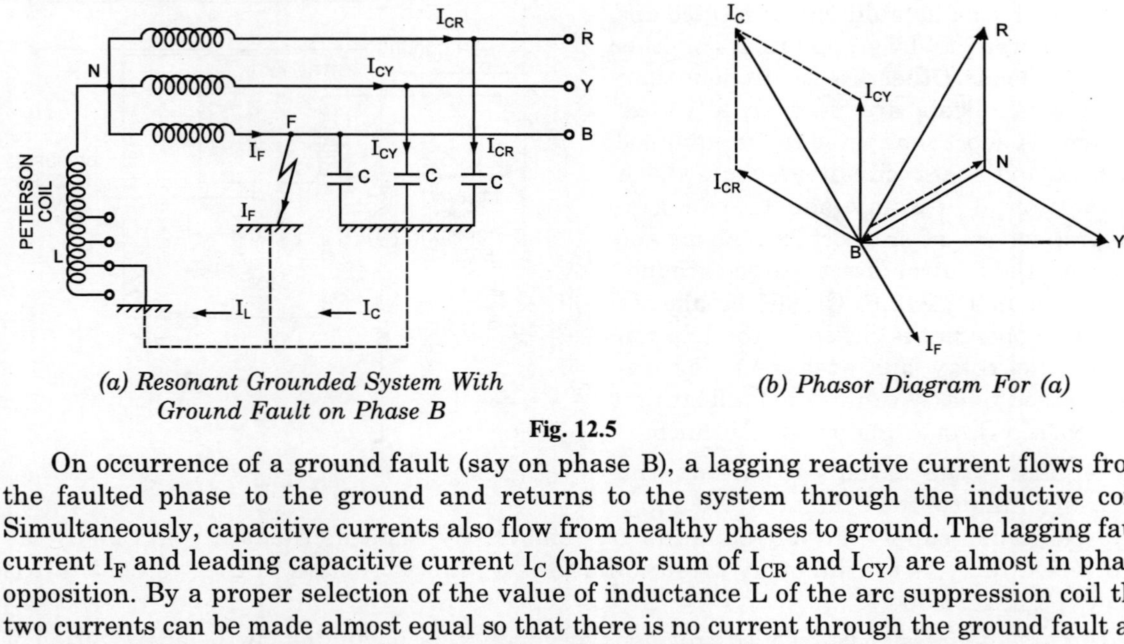

(a) Resonant Grounded System With Ground Fault on Phase B *(b) Phasor Diagram For (a)*

Fig. 12.5

On occurrence of a ground fault (say on phase B), a lagging reactive current flows from the faulted phase to the ground and returns to the system through the inductive coil. Simultaneously, capacitive currents also flow from healthy phases to ground. The lagging fault current I_F and leading capacitive current I_C (phasor sum of I_{CR} and I_{CY}) are almost in phase opposition. By a proper selection of the value of inductance L of the arc suppression coil the two currents can be made almost equal so that there is no current through the ground fault and so there will be no arc. The combination of neutral reactance L and line capacitance C acts as a parallel resonant circuit. The circuit is shown in Fig. 12.5(*a*). If V_P is the line-to-neutral voltage

$$I_{CR} = I_{CY} = \sqrt{3}\, V_P\, \omega\, C \qquad ...(12.4)$$

$$\text{Capacitive current, } \mathbf{I_C} = \mathbf{I_{CR}} + \mathbf{I_{CY}} = \sqrt{3} \times \sqrt{3}\; V_P\, \omega\, C = 3V_P\, \omega\, C \qquad ...(12.5)$$

$$\text{For balance condition, } I_L = I_C$$

$$\text{or } \frac{V_P}{\omega L} = 3\, V_P\, \omega\, C$$

$$\text{or } L = \frac{1}{3\omega^2 C} \qquad ...(12.6)$$

Phasor diagram is shown in Fig. 12.5 (*b*).

This method of neutral grounding is usually confined to medium voltage overhead transmission lines which are connected to the system generating source through power transformers. The reasons for this limitation are : (*i*) The higher insulation level is required on apparatus associated with arc suppression coil grounded system and it is comparatively difficult to provide on generators as compared on transformers. In addition, adequate winding protection cannot so readily be achieved for generators as for power transformers under arc suppression coil operating conditions. (*ii*) Overhead transmission lines are normally subject to transient ground faults due to lightning, birdage etc., of short duration.

It has been found that arc suppression coil grounding reduces the line outages from ground faults to 20 or 30 per cent of those obtainable with other types of grounding. As such this method of grounding is advantageously employed on radial lines as this avoids the construction of duplicate circuits for maintaining service continuity and thus there is overall economy.

Ground fault neutralizers should not be used where (*i*) fully graded insulation transformers are employed as the neutrals of such transformers are not sufficiently well insulated (*ii*) auto-transformers having a ratio greater than 0.95 to 1 are used.

The coils of the ground fault neutralizers are ten-minute time-rated on system where permanent ground faults can be located and removed promptly by ground relays or other suitable means. Otherwise, continuous time-rated neutralizers are used on all other systems. A short-time rated coil is equipped with an automatic circuit breaker, which by action of a relay shorts the arc-suppression coil after a certain time lag and connects the neutral directly to the ground, as shown in Fig. 12.6. Circuit breaker is normally open but is closed by the trip coil when the relay operates after a pre-determined time. With this the fault current is bypassed through the resistor branch.

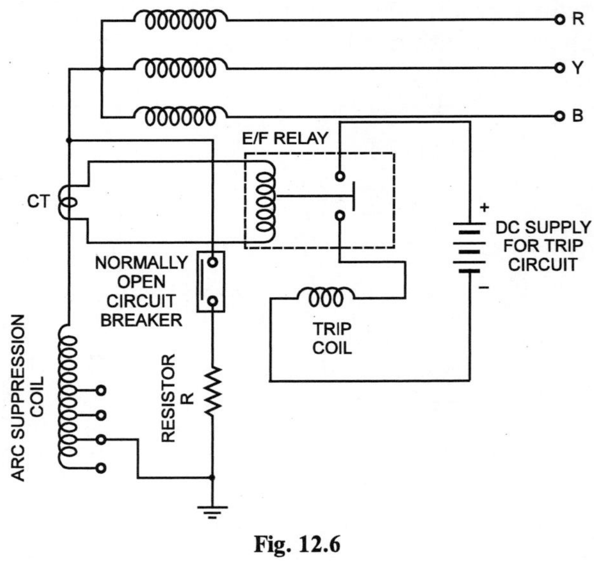

Fig. 12.6

A short-time rated coil allows the clearing of intermittent faults, without any interruption of supply. Prolonged faults are removed by isolating the faulty section by the protective relay. A continuously rated coil allows a fault to remain in the system till it is located and removed.

An arc-suppression coil is also provided with an auxiliary winding for the energization of a relay to operate the short-circuiting device of a short-time rated coil.

Special features of arc-suppression coil grounding are given below :

1. During a single line-to-ground fault, the voltages across healthy phases rise to line-to-line value.
2. Continuity of supply can be maintained for long periods with one line grounded till the faulty section is isolated.
3. During sustained ground fault, the current is a small fraction of capacitive ground fault current. Burning and heating effects are reduced to minimum and damage to the equipment is limited.

4. Transient ground faults are suppressed. More than two-third of such faults are cleared without necessitating the operation of the circuit breaker.
5. Voltage gradient on ground surface in the vicinity of fault is greatly reduced and so the danger to the life in the proximity of fault.
6. Inductive interference to neighbouring communication circuits is lesser in magnitude but duration may be long.
7. Relaying needs special provision and is usually restricted to indication.
8. During normal operation, the arc suppression coil has little effect on balance of system voltages to ground and losses occuring in it are negligible.

Example 12.1. Calculate the reactance of a coil suitable for a 33 kV, 3-phase transmission system of which the capacitance to earth of each conductor is 4.5 μF.

Solution : Supply frequency = 50 Hz (assume)

Capacitance of each conductor to earth, $C = 4.5 \times 10^{-6}$ F

Reactance of the suitable coil, $\omega L = \dfrac{1}{3\omega C} = \dfrac{1}{3 \times (2\pi \times 50) \times 4.5 \times 10^{-6}} = \mathbf{235.8\ \Omega}$ **Ans.**

Example 12.2. A 132 kV, 3-phase, 50 Hz transmission line 192 km long consists of three conductors of effective diameter 20 mm arranged in a vertical plane with 4 m spacing and regularly transposed. Find the inductance and kVA rating of the arc suppressor coil.

[U.P. Technical Univ. Elements of Power System 2006-07]

Solution : Conductor radius, r = 10 mm = 0.01 m

Spacings, $d_1 = 4$ m; $d_2 = 4$ m; $d_3 = 4 + 4 = 8$ m

Supply frequency, f = 50 Hz

Capacitance per phase is given by

$$C = \frac{2\pi\varepsilon_0}{\log_e \dfrac{\sqrt[3]{d_1 d_2 d_3}}{r}} \times \text{length of transmission line}$$

$$= \frac{2\pi \times 8.854 \times 10^{-12}}{\log_e \dfrac{\sqrt[3]{4 \times 4 \times 8}}{0.01}} \times 192 \times 10^3 = 1.717\ \mu\text{F}$$

Necessary inductance of Peterson coil

$$L = \frac{1}{3\omega^2 C} \qquad \text{Refer to Eq. (12.6)}$$

$$= \frac{1}{3 \times (2\pi \times 50)^2 \times 1.717 \times 10^{-6}} = 1.967\ \text{H} \quad \textbf{Ans.}$$

For ground fault, the current in the neutral is given by

$$I_L = \frac{V_P}{\omega L} = \frac{132{,}000/\sqrt{3}}{2\pi \times 50 \times 1.967} = 123.33\ \text{A}$$

$$\text{Rating of suppressor coil} = \frac{V_P I_L}{1{,}000}\ \text{kVA} = \frac{132{,}000}{\sqrt{3}} \times 123.33 \times \frac{1}{1{,}000}$$

= 9,400 kVA or 9.4 MVA per coil **Ans.**

12.11. VOLTAGE TRANSFORMER GROUNDING

In this system of neutral grounding the primary of a voltage transformer is connected between the neutral and the ground, as shown in Fig. 12.7. A low resistor and a relay combination is connected across the secondary of the voltage transformer. An earth fault on the primary side will produce a voltage across the relay that will cause the protective device to operate.

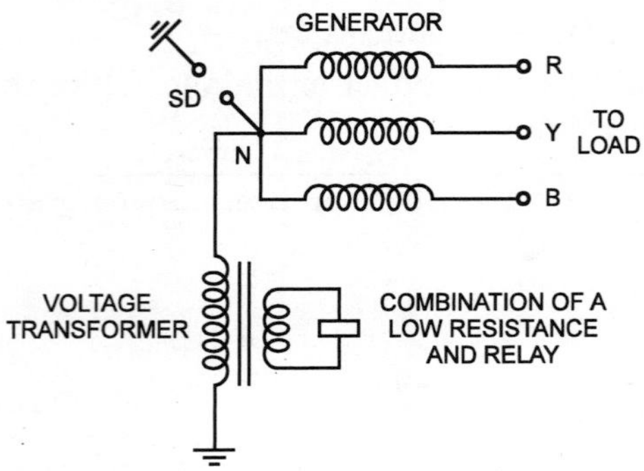

(a) Voltage Transformer Grounding

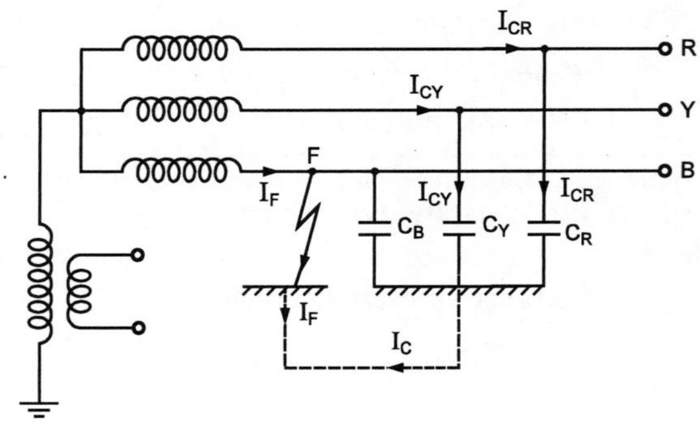

(b) Voltage Transformer Grounded System With Ground Fault on Phase B

This system of neutral earthing has got the following advantages :

1. The transient overvoltages owing to switching and arcing grounds are reduced because voltage transformer provides extremely high reactance to the earth path.
2. Arcing grounds are eliminated.
3. In this system of grounding, the neutral is grounded through a single phase voltage transformer and so the system operates virtually like an isolated neutral system.

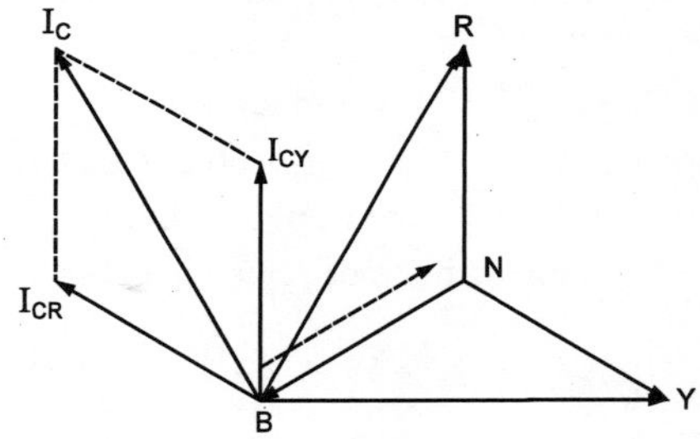

(c) Phasor Diagram For (b)

Fig. 12.7

The major drawback of this system is that the grounded neutral acts as a reflection point for the travelling waves passing through the machine winding and in order to avoid high voltage build-up connection of a surge diverter between the machine neutral point and ground, as shown in Fig. 12.7 (*a*), becomes imperative.

The voltage transformer acts as a fault indicator, the extent of the fault can be determined by measuring the voltage on its secondary.

The use of this system of neutral grounding is normally confined to the generator equipments which are directly connected to step-up transformers. The generating networks of such equipments are physically isolated from the main distribution network. This is because interconnecting cables between the generator and transformer windings are usually of very short length, the electrostatic capacity of the circuit will be negligible.

12.12. GROUNDING TRANSFORMER

The neutral or star point is usually available at every voltage level from generator or transformer neutral. However, if no such point is available due to delta connections or if neutral point is desired on bus-bars, a zig-zag transformer is most commonly used. Such a transformer has no secondary. It is a core type transformer having three limbs built up in the same fashion, as illustrated in Fig. 12.8. The connections are also shown in Fig. 12.9. It is seen that the currents in the two halves of the winding on each limb are in opposite directions. As a result there will not be any undesirable harmonic prevailing in the circuit and so stresses on the insulation of the transformer are considerably reduced.

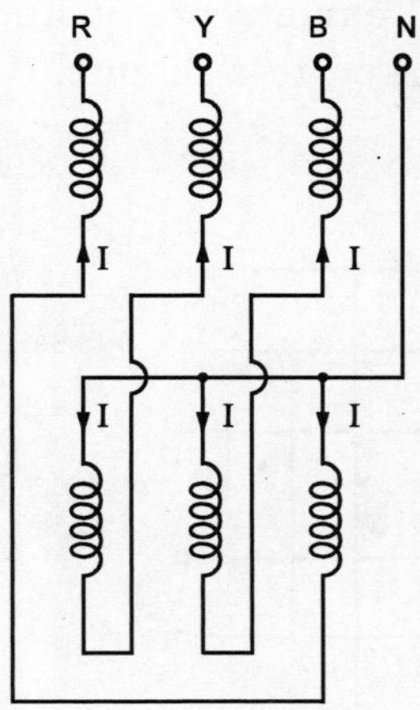

Fig. 12.8. *Zig-Zag Transformer*

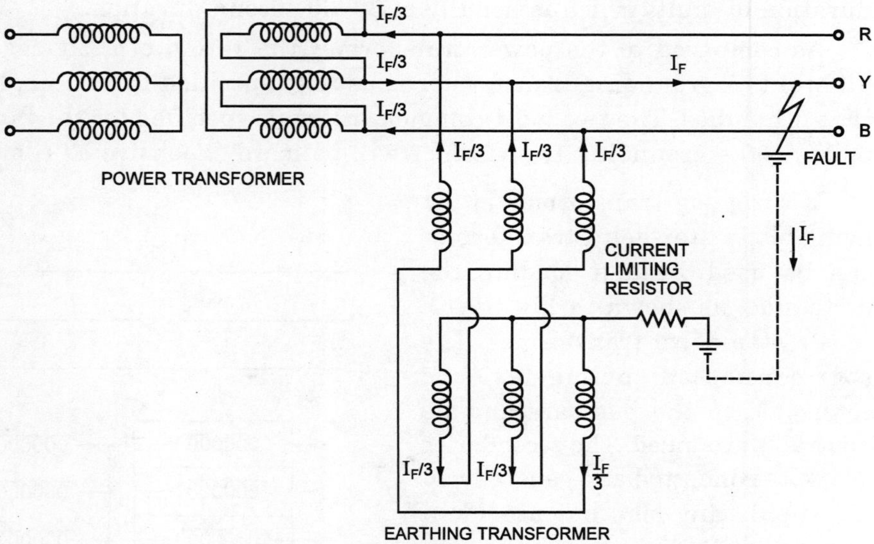

Fig. 12.9.

The fault current is quite high as the impedance of earthing transformer is quite low. The resistance or resistances are inserted either in the neutral circuit or in the windings of earthing transformer, as illustrated in Figs. 12.9 and 12.10 respectively in order to limit the magnitude of the fault current.

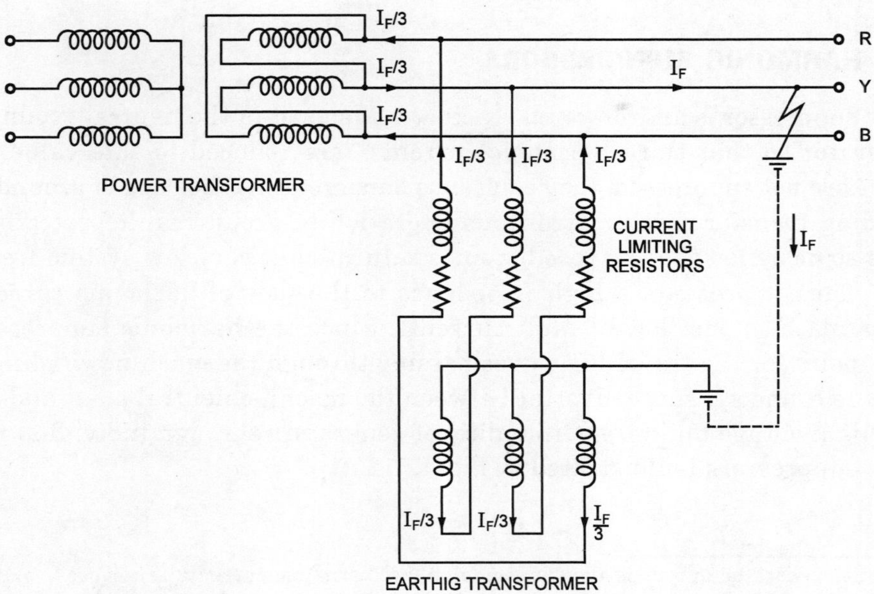

Fig. 12.10

Under normal condition, only iron losses will be continuously occurring but at the time of fault copper losses will also occur in addition, because of heavy fault current, in the earthing transformer. The duration of occurrence of copper losses is quite short, being the duration of fault, which is usually of 30–60 second duration.

As compared to the power transformer the rating of a zig-zag transformer is quite different. A grounding transformer is usually specified by the single phase fault current that it handles. The two most common intervals specified for the fault current are 30 and 60 s. A 60 s grounding transformer will be more expensive as compared to a 30 one.

If a zig-zag transformer is not available, a star-delta transformer can be used without loading the delta side, as shown in Fig. 12.11. It is a step-down transformer. The star-connected primaries are connected to the bus-bars and its neutral is grounded. The secondaries are delta-connected and generally do not supply any load but provide a closed path for triple harmonic currents to circulate in them. Under normal operation the current in the grounding transformer is only its own exciting current. However, in the event of a single line-to-ground fault condition, large current may flow. So it should be of sufficient rating to handle the fault current.

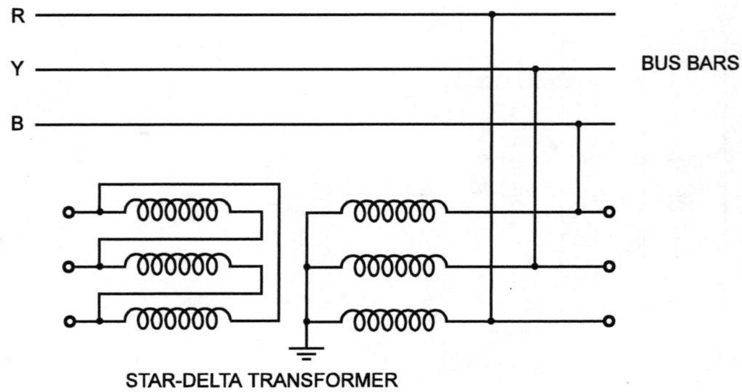

Fig. 12.11. ***Star-Delta Transformer Grounding***

12.13. HARMONIC SUPPRESSORS

Harmonic suppressor is an iron-cored reactor and is used in the neutral ground connection of a generator so that third harmonic currents* are reduced to safe value. If the third harmonics are not suppressed, they can cause dangerous overheating of grounding resistors of grounding transformers or erroneous operation of ground fault protective gear. The reactor is so designed that its core becomes saturated at very low system frequency fault currents. Thus it presents a high impedance to the flow of harmonic currents and low impedance path to the flow of fault currents. Since the harmonic suppressor acts as a reflection point for the travelling waves passing through the machine winding, it becomes necessary to connect a surge diverter between the machine neutral point and ground so as to avoid high voltage build-up. Grounding of generators through individual resistors and harmonic suppressors is illustrated in Fig. 12.12(*a*).

* In case the generator output is non-sinusoidal, third harmonic currents are caused, by third harmonic voltage, to flow from the generator line terminals through the system and return via comparatively low zero sequence reactance path of the system capacitance or the windings of the transformer or other generators to its neutral terminal.

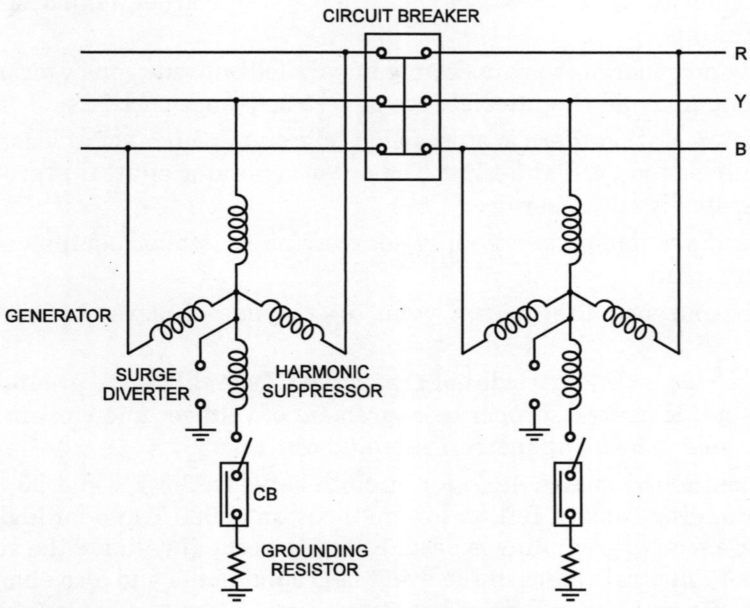

Fig. 12.12 (*a*) *Grounding of Generators Through Individual Resistors and Harmonic Suppressors*

12.14. NEUTRAL GROUNDING PRACTICE

1. Generally one-neutral ground is provided at each voltage level. Between generation and distribution there are several voltage levels. One ground is provided at each voltage level [Fig. 12.12 (*b*)]
2. The grounding is provided at source and not at load end.
3. Each major bus section is grounded.
4. (*a*) The neutrals of generators are grounded through a resistance to limit the stator fault current. The value of resistor determines the percentage of generator winding left unprotected. Synchronous motors and synchronous capacitors are provided with reactance grounding.

   (*b*) Where several generators are connected to a common neutral bus, the bus is connected to ground through a single

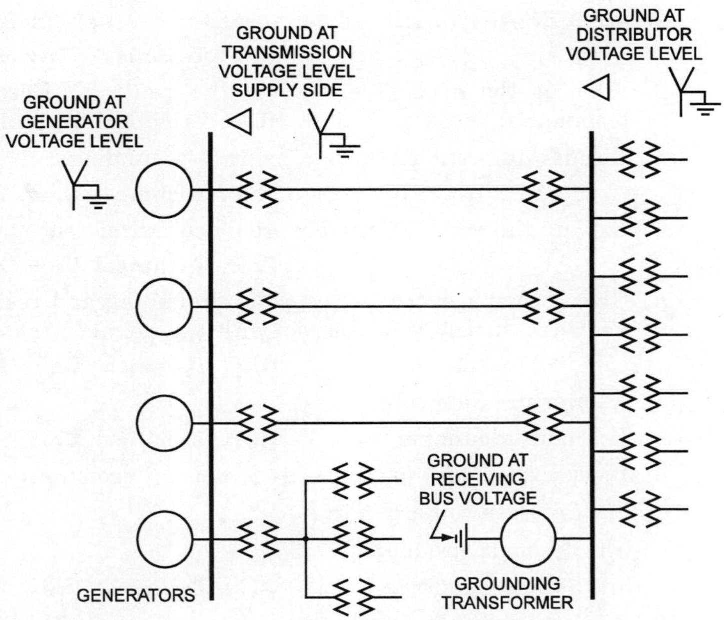

Fig. 12.12 (*b*) *Neutral Ground at Every Voltage Level at Source END*

grounding device. Disconnect switches can be used to ground the desired generators to the neutral bus.

(*c*) Where several generators are operating in parallel, only one generator neutral is earthed. This is done to avoid the interference of zero sequence currents.

(*d*) In generating station there is a provision to ground neutral of at least two generators, though only one is used at a time. The other generator neutral is grounded when the first generator is out of service.

(*e*) When there are one or two supply sources, no switching equipment is used in the grounding circuit.

5. For protection purpose, the neutral point of star side of power transformer is usually grounded.
6. The star-connected secondary sides of protective CTs and PTs are grounded at one point. This ensures stable neutral, proper measurement of voltages and currents, kWh and kVA on secondary side measuring instruments and controls.
7. As already mentioned in Art. 12.7, for circuits between 3.3 kV and 33 kV resistance or reactance grounding is used. But for low voltages up to 600 V and for high voltages above 33 kV solid or effective grounding is used. Effective grounding limits the voltages of healthy phases to line-to-neutral values in the event of ground faults and also eliminates the arcing grounds. The effective neutral grounding causes ground fault currents of high magnitudes but the modern protection systems are very sensitive and fast operating so that the faults are cleared in very short times.

EXERCISES

1. Discuss the advantages of (*i*) grounding the neutral (*ii*) keeping the neutral isolated of a power system. **[U.P. Technical Univ. Elements of Power System 2005-06]**

2. What are the various advantages of neutral grounding ?

[U.P. Technical Univ. Elements of Power System 2007-08]

3. What is the need of grounding the neutral ? Describe briefly the various grounding techniques. **[U.P. Technical Univ. Elements of Power System 2009-10]**

4. Discuss different methods of neutral grounding.

[U.P. Technical Univ. Elements of Power System 2007-08]

5. Explain the term "Effectively grounded" in reference to neutral grounding in power system.

[U.P. Technical Univ. Elements of Power System 2002-03]

6. Give a comparison between solid grounding and resistance grounding of power systems neutrals. Justify your answer with the help of neat phasor diagrams.

[U.P. Technical Univ. Elements of Power System 2006-07]

7. Answer the following :

Resonant grounding. **[U.P. Technical Univ. Elements of Power System 2009-10]**

8. Discuss the following methods of neutral grounding :

(*i*) Peterson coil grounding

(*ii*) Resonant grounding

(*iii*) Solid grounding. **[U.P. Technical Univ. Elements of Power System 2008-09]**

9. With the help of a neat phasor diagram and mathematical equations explain the method and advantages of resonant grounding.

[U.P. Technical Univ. Elements of Power System 2005-06]

10. Explain the advantages of grounding power system neutrals. With the help of circuit and phasor-diagrams, explain the function of a Peterson coil in a 3-phase system.
[U.P. Technical Univ. Elements of Power System 2003-04]

11. Identify the criterion of grounding at 400 kV substation. If the total safety is to be ensured what will be the layout of grounding ? Explain alongwith diagram.
[U.P. Technical Univ. Elements of Power System 2005-06]

12. What is the necessity of neutral grounding ?
Write a note on 'earthing transformer'.
[U.P. Technical Univ. Elements of Power System 2004-05]

13. What is an earthing transformer ? Explain the same in brief.
[U.P. Technical Univ. Elements of Power System 2006-07]

14. Explain the role of earthing transformer in neutral grounding.
[U.P. Technical Univ. Elements of Power System 2006-07]

15. What is the difference between reactance and resonant grounding systems ? What are grounding practices. **[U.P. Technical Univ. Elements of Power System 2004-05]**

16. Write short notes on the following :
    (*i*) Methods of neutral grounding in a power system and their relative merits and demerits.
    (*ii*) Peterson coil.
    (*iii*) Earthing transformer.
    (*iv*) Harmonic suppressors.

SHORT ANSWER TYPE QUESTIONS WITH ANSWERS

Q. 1. What is difference between system earthing and equipment earthing ?

Ans. System earthing deals with earthing of the system neutral to ensure system security and protection while equipment earthing deals with earthing of non-current carrying parts of the equipment to ensure safety of personnel and protection against lightning.

Q. 2. What is meant by an earth electrode ?

Ans. Any wire, rod, pipe, plate and an array of conductors embedded in ground horizontally or vertically is called the earth electrode.

Q. 3. What is meant by system grounding ?

Ans. The process of connecting some part of the electrical power system (e.g. neutral point of a star-connected system, one conductor of the secondary of a transformer etc.) to ground (*i.e.* soil) is known as the system grounding.

Q. 4. What is neutral grounding ?

Ans. The process of connecting neutral point of a 3-phase system to ground (i.e. soil) either directly or through some circuit element (e.g. resistance, reactance etc.) is called the neutral grounding.

Q. 5. What is the chief argument that favours use of isolated neutral system ?

Ans. The chief argument that goes in favour of isolated neutral system is the possibility of maintaining continuity of supply in case one phase gets grounded until it is convenient to disconnect and repair the faulty line.

Q. 6. What are the various methods of neutral grounding ? **[U.P.T.U. 2006-07]**

Ans. The methods commonly used for for grounding the neutral point of a 3-phase system are
  (*i*) Solid or effective grounding
  (*ii*) Resistance grounding.
  (*iii*) Reactance grounding

(*iv*) Peterson coil or arc suppression coil grounding

(*v*) Voltage transformer grounding.

Q. 7. What is effectively grounded system ?

Ans. When the neutral point of a 3-phase system (e.g. 3-phase generator, 3-phase transformer etc.) is directly connected to ground (*i.e.* soil) through a wire of negligible resistance and reactance, it is called the solid grounding or effective grounding.

According to IEEE definition, a system or a portion of a system can be said to be effectively grounded when for all points on the system or specified portion thereof, the ratio of zero sequence reactance to positive sequence reactance is not greater than three (*i.e.*, $\frac{X_0}{X_1} \not> 3$) and the ratio of zero sequence resistance to positive sequence resistance is not greater than one (*i.e.*, $\frac{R_0}{R_1} \not> 1$) for any condition of operation and for any amount of generator capacity.

Q. 8. What is arc suppression coil grounding ?

Ans. In the arc suppression coil method of grounding, the arcing-ground danger has been eliminated and the system is approximated to the isolated neutral system, in which one or two healthy phases continue to supply power and complete shutdown on the system is avoided till the fault was located and isolated.

An arc *suppression coil*, also known as *Peterson coil* or *Ground fault neutralizer*, is an iron core tapped reactor connected in neutral to ground connection. The reactor is provided with tappings so that it can be tuned to the system capacitance. The function of arc suppression coil is to make the arcing ground faults self extinguishing and, in case of sustained faults, to reduce ground fault current to a comparatively low value so that the system can be kept in operation with one line grounded.

This method of neutral grounding is usually confined to medium voltage overhead transmission lines which are connected to the system generating source through power transformer.

It has been found that arc suppression coil grounding reduces the line outages from ground faults to 20 or 30 per cent of those obtainable with other types of grounding. As such this method of grounding is advantageously employed on radial lines as this avoids the construction of duplicate circuits for maintaining service continuity and thus there is overall economy.

Q.9. What is an earthing transformer ? **[U.P.T.U. 2006-07]**

or

What is the use of earthing transformer ? **[U.P.T.U. 2007-08]**

or

What is the role of earthing transformer in neutral grounding ?

Ans. In cases where the neutral points of a three-phase system are not accessible or where the transformers or generators are delta connected, an artificial neutral point is required to be created. This is accomplished by means of a zig-zag transformer, called the earthing transformer.

As compared to the power transformer the rating of a zig-zag transformer is quite different. A grounding transformer is usually specified by the single phase fault current that it handles. The two most common intervals specified for the fault current are 30 and 60 s. A 60 s grounding transformer will be more expensive as compared to a 30 one.

Q. 10. What is harmonic suppressor ?

Ans. Harmonic suppressor is an iron-cored reactor and is used in the neutral ground connections of a generator so that third harmonic currents are reduced to safe value.

PROBLEMS

1. A 33 kV, 50 Hz network has a capacitance to neutral of 1.0 μF per phase. Calculate the reactance of an arc suppression coil suitable for the system to avoid adverse effect of arcing ground. [**Ans.** 1,061 Ω]
2. Calculate the reactance of a coil suitable for a 33 kV, 3-phase transmission system of which the capacitance to earth of each conductor is 4.5 μF. [**Ans.** 3.75 H; 308 kVA]
3. A 230 kV, 3-phase, 50 Hz, 200 km transmission line has a capacitance to earth of 0.01 μF/km per phase. Calculate the inductance and kVA rating of the Peterson coil used for earthing the above system. [**Ans.** 1.7 H; 33,238 kVA]
4. Derive an expression for the reactance of the Peterson coil in terms of the capacitance of the protected line. Calculate the reactance of a coil suitable for a 33 kV, 3-phase transmission system of which the capacitance to earth of each conductor is 5.0 μF. [**Ans.** 212.2 Ω]

❑❑❑

13

CHAPTER

EXTRA HIGH VOLTAGE (EHV) AC TRANSMISSION

INSIDE THIS CHAPTER

13.1. Introduction **13.2.** Necessity of EHV Transmission **13.3.** Problems Associated With EHV Transmission **13.4.** Limitations and Design Aspects of EHV-AC Power Transmission Systems **13.5.** Flexible AC Transmission System (FACTS)

13.1. INTRODUCTION

The basic function of a transmission system is to transfer electrical power from one place to another place or from one network to another network. A transmission system includes terminal substations, transmission lines and intermediate substations.

Transmission systems are required for (*i*) transfer of bulk power from large group of generating stations up to the main transmission network, (*ii*) the main transmission network (*iii*) system interconnection and (*iv*) transfer of power from the main transmission network to the secondary substations.

A transmission system is employed either for transfer of power from sending end to the receiving end or for system interconnection for exchange of electrical power between independently controlled networks.

The increasing need of transmitting greater and greater amounts of power over long distances has led to a continuous increase in the transmission voltages. As a rule, larger the amount of power to be transferred, higher is the requirement of transmission voltage. Similarly longer the lines, higher is the transmission voltage.

Up to 1970's, the choice was exclusively in favour of ac high voltage (up to 220 kV) and extra high voltage (above 220 kV and up to 750 kV). The present trend is to use transmission voltage from high voltages (up to 300 kV), extra high voltages (300-765 kV) to ultra high voltages (above 765 kV), depending upon the technical and economical considerations, for the transmission of large amounts of power over long distances.

In India, the transmission voltages in use are 33, 66, 110, 132, 220 and 400 kV. The next higher voltage of 765/800 kV has been selected for transmission of huge blocks of power from generating stations to the load centres.

13.2. NECESSITY OF EHV TRANSMISSION

As already mentioned above in Art. 13.1, modern trend is to use extra high voltage (EHV) and ultra high voltage (UHV) for transmission of huge blocks of power over long distances. The reasons for adopting of EHV/UHV range for transmission purposes are given below :

1. ***Reduction of electrical losses, increase in transmission efficiency, improvement of voltage regulation and reduction in conductor material requirement***—For transmission of given amount of power over a given distance through the conductors of a given material and at a given power factor as the transmission voltage increases,

(*a*) line losses are reduced since line losses are inversely proportional to the transmission voltage,

(*b*) transmission efficiency increases because of reduction in line losses,

(*c*) voltage regulation is improved because of reduction of percentage line drop, and

(*d*) lesser conductor material is required being inversely proportional to the square of transmission voltage. (Refer to Art. 2.7).

2. Economic considerations have led to the construction of power stations of large capacity and so need of transfer of bulk power over long distances arose. Transmission of bulk power from generating stations to the load centres is technically and economically feasible only at voltages in the EHV/UHV range.

3. Generating stations (Steam-, hydro- and nuclear-power stations) are located in remote areas (far away from load centres) because of the reasons of economy, feasibility and from the point of view of safety and environmental conditions. EHV transmission is, therefore, inevitable for transmission of huge blocks of power over long distances from these power plants to load centres.

4. ***Flexibility for future system growth***—There is flexibility of future system growth.

5. ***Increase in transmission capacity of the line***—Power transferred is expressed as

$$P = \frac{V_S \cdot V_R}{X} \sin \delta \qquad ...(13.1)$$

where V_S and V_R are the two terminal voltages, δ is the load angle and X is the line reactance.

Thus the power transmission capacity of a transmission line increases with the increase in transmission voltage. No doubt the cost of transmission line and terminal equipment also increases with the increase in the transmission voltage but in general these costs are proportional to the transmission voltage rather than the square of the transmission voltage. Moreover there is also a saving in cost due to reduction in energy losses occuring in transmission lines. As a consequence the total cost of transmission decreases with the increase in transmission voltage, as depicted in Fig. 13.1.

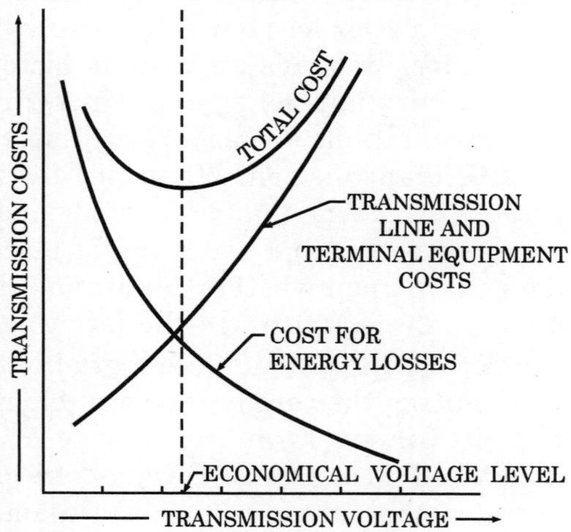

Fig. 13.1

6. ***Possibility of interconnections of power systems***—It is practically not possible to have interconnections of two or more power systems, which is necessary to achieve sharing of installed reserves and for development of integrated systems and grids, without EHV transmission.

7. ***Increase of surge impedance loading***—Load carrying capability of a line is usually expressed in terms of "*surge impedance loading*" (SIL). Surge impedance loading

(SIL) is the power that a line carries when each phase is terminated by a load equal to the surge impedance of the line. For a transmission line, the surge impedance is given as $Z_C = \sqrt{\frac{L}{C}}$ where L and C are respectively the series inductance and shunt capacitance per unit length. The surge impedance loading (SIL), for a transmission line is given as $\frac{3\,V^2}{Z_C}$ where V is the line-to-neutral voltage.

It is evident that SIL varies as the square of the operating voltage, and, therefore, with the increase in voltage level, SIL itself increases. Thus power transfer capability of the line increases with the increase in voltage level.

The surge impedance of a line can be determined from its conductor configuration. The approximate values of surge impedances for lines with single, double, triple and quadruple conductors are 400, 300, 280 and 260 ohms respectively.

8. ***Reduction in right-of-way***—In some countries 'rights-of-way' are paid for at a rate proportional to the total width of the transmission lines. Even in countries where right-of- way is not directly paid, there are usually strong pressures from the public towards fewer and fewer transmission lines. With the passage of time right-of-way becomes either more costlier or difficult to obtain and therefore it is becoming necessary to have fewer transmission lines operating at EHV/UHV. The worthnoting point here is that with the increase in operating voltage, number of circuits and requirement of land is reduced considerably.

13.3. PROBLEMS ASSOCIATED WITH EHV TRANSMISSION

The major problems associated with EHV transmission are discussed below :

1. Corona Loss and Radio Interference. As we know that corona is not only a source of power loss but it is also a source of interference with radio and television. The corona loss depends on various factors such as system frequency, system voltage, air conductivity, air density, conductor radius, conductor surface, load conditions, atmospheric conditions etc. as already explained in Art. 7.9. The problem is more acute in case of EHV transmission. When the electric field at the surface of an energized conductor exceeds 2-3 kV/mm, audible and sometimes visible corona discharge takes place, causing a loss of power and radiation of electrical noise. Corona loss varies through the year depending upon weather conditions. The corona loss under bad weather conditions may be as high as 100 times the fair weather condition loss.

Corona inception voltage gradient is an important parameter for conductor design. For limiting the corona loss, audible noise and radio interference it is necessary to limit the electric stress at the surface of the conductor to 1.8 kV/mm (rms), preferably to 1.5 kV/mm (rms). The corona loss can be reduced either by increasing the spacing between the conductors or the diameters of conductors but the spacing between the conductors cannot be increased to a large extent as already explained in Art. 7.10. Large diameter conductors (hollow conductors or ACSR conductors) have been used to bring down the corona loss and radio interference but the cost of manufacture of such conductors is high and their handling is both difficult and expensive. Another difficulty with the use of such conductors is large wind and snow loadings. However, use of ACSR conductors is quite economical for line voltage up to 400 kV. Corona and radio-interference are also kept within permissible limits for such lines with this type of conductors. Bundled conductors, explained in Art. 4.5, are invariably employed for EHV (400 kV or higher) lines. The use of bundled conductors reduces the voltage gradient in the vicinity of line and thus, reduces the possibilities of corona discharge. Although the bundled conductors are used on EHV transmission lines primarily to reduce corona loss and radio-interference, but they have several other advantages, as explained in Art. 4.5. *The choice of configuration and cross-section of ac conductors with reference to corona considerations is generally above the economic cross-section based on thermal considerations or transient stability considerations i.e. the design of ac conductors based on corona limitations gives a cross-section much larger than that with respect to economical power transfer limit imposed by stability limit.*

The corona loss is quite large in EHV transmission lines. For example, in a twin-Moose bundled conductor, 2 × 235 mm^2 (copper equivalent area) at 1.618 kV/mm of the 400 kV line, designed by UPSEB (Now U.P. P.C.L.) the annual corona loss per km of line is about 3,700 kWh. If the bundled-conductors were not have been employed in EHV lines, then corona loss would have been comparatively very high. It is quite difficult to estimate corona loss with reasonable accuracy, as such loss depends on weather conditions. Attempts are made towards designing the transmission lines such that the fair weather corona loss is kept low and bad weather conditions do not seriously affect the line performance.

When bundled conductors are employed, spacers have to be installed at intervals of 80-100 m in each span to maintain the spacing between sub-conductors. This is necessary to prevent the conductors from getting stuck-up due to wind forces or electromagnetic induction introduced by heavy currents.

The increase in SIL with the increase in the spacing between the sub-conductors is small and, therefore, the intra bundle spacings are not very critical and 0.45 m is the most commonly used value.

Transmission lines with bundled conductors need somewhat stronger towers. It is due to the fact that bundled conductors are subjected to large wind and snow loadings.

2. Heavy Supporting Structures and Erection Difficulties. EHV transmission lines have large mechanical loading on towers because of use of bundled conductors, as already mentioned above, large air and ground clearances, considerable dynamic forces due to broken conductors etc. Transmission line towers with fabricated steel members are usually employed in EHV transmission. A typical standard suspension tower for a 500 kV line is shown in Fig. 13.2.

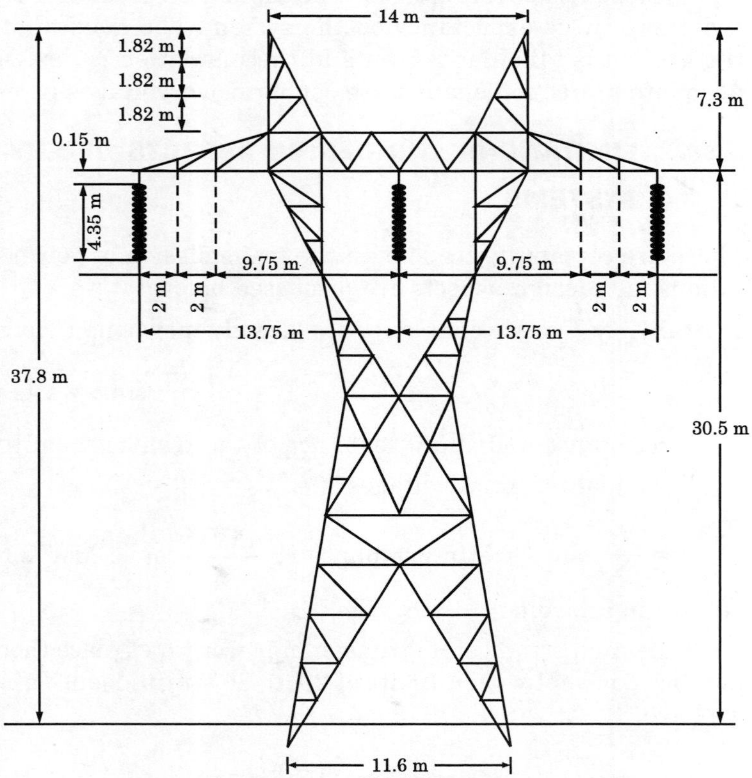

Fig. 13.2. *Standard Suspension Tower For 500 kV Line Showing Various Arrangements For Locations of Conductors and Ground Wires*

The transmission lines are made more wind-resistant as they are to bear out the wind pressures during storms and cyclones. Recent designs for line supports include any of the following concepts :

(*i*) Composite design employing standard structural steel for lightly loaded members and high tensil steel for heavily loaded members.

(*ii*) H-frame structures supported by steel guy cables.

(*iii*) Guyed aluminium towers in place of steel towers.

(*iv*) Suspension towers using reinforced concrete tubes.

Problems of transportation and erection arise as the supporting structures are to be transported over long distances and high standard workmanship is required for erection of EHV transmission lines.

3. Insulation Requirements. The level of insulation required depends upon the magnitude of likely voltage surges due to internal causes (switching operations) or due to external causes (lightning etc.) The lines are usually protected against lightning etc. by use of ground wire and rapid auto-reclosing circuit breakers. The ground wire, the line insulation and the tower footing are properly coordinated for adequate lightning protection. Switching surges are, however more dangerous as they may cause overvoltages of 2-3 times the normal operating voltage. With the developments in the design of relay-breaker systems, however it is possible to control and minimise switching overvoltages.

Line insulation is designed to take care of switching overvoltages, temporary overvoltages and atmospheric overvoltages. The insulation level of a transmission line is based on the switching surge expectancy on the system. The maximum switching surge overvoltage to the ground is taken as 2.5 pu and the insulation is designed for this voltage. In addition adequate protection against atmospheric overvoltages is provided.

13.4. LIMITATIONS AND DESIGN ASPECTS OF EHV-AC POWER TRANSMISSION SYSTEMS

There are certain limitations in the transmission of power by EHV ac system. These limitations and design aspects are discussed below :

1. Stability Considerations. As already mentioned, power transferred is expressed as

$$P = \frac{V_S V_R}{X} \sin \delta \text{ watts per phase}$$

if resistance and shunt leakance of the transmission line are neglected.

For equal terminal voltages (*i.e.* $V_S = V_R = V$)

$P = \frac{V^2}{X} \sin \delta$ watts per phase or $\frac{(kV)^2}{X} \sin \delta$ MW for 3 phases where kV is the line voltage in kilovolts.

Maximum steady state transfer of power takes place theoretically for $\delta = 90°$ but in actual practice δ is kept within limits of 20° to 30° at full load. Taking $X = 0.3475$ Ω/km and $\delta = 30°$, we have

$P = \frac{kV^2}{0.3475\, l} \sin 30° = 1.43 \frac{(kV)^2}{l}$ MW where l is the length of line in km.

If the line operates at natural load given by the expression

$$P = VI = V \times \frac{V}{Z_C} \text{ watts or } \frac{(kV)^2}{Z_C} \text{ MW}$$

where Z_C is surge impedance and is equal to $\sqrt{\frac{L}{C}}$ ohms.

then $\frac{1.43 \times (kV)^2}{l} = \frac{(kV)^2}{Z_c}$ or $l = 1.43\ Z_C$ kilometres

Thus we see that the theoretical length of a transmission line in km, that can be operated on its natural load and load angle of 30° without loss of stability only is 1.43 times surge impedance (which comes out to be about 500 km). The transmission distance in EHV ac systems is, therefore, limited. This limitation is overcome by use of special equipment such as series capacitors or shunt reactors.

2. Current Carrying Capacity. The loading of overhead line conductors does not depend on the thermal considerations. However, for overhead transmission lines operating at voltage up to 220 kV, the conductor size is determined on the basis of its continuous and short-term current carrying capacity (*i.e.* on the basis of thermal considerations). In case of underground cables, however, both thermal current limit and charging current are very important and that is why the underground cable is never operated beyond its natural load. Also the cost of cable is very high in comparison to that of an overhead line, therefore use of underground cables is limited in high voltage systems.

3. Ferranti Effect. We know that with the capacitive load on the line the receiving-end voltage is higher than sending-end voltage. This increase is of the order of 1.5 per cent for 160 km, 13 per cent for 500 km and 100 per cent for 960 km. Also there is a rise in sending-end voltage whenever the load on the generator is thrown off suddenly. Care is also to be taken from these aspects in case of ac transmission system.

Voltage rise, due to Ferranti effect, is controlled by using shunt reactors at the load end.

4. Surge Impedance Loading. Surge impedance loading (SIL) of a transmission line is defined as the load at the receiving end which is equivalent to $\sqrt{\frac{L}{C}}$. The surge impedance loading for a transmission line is given as $\frac{V^2}{Z_0}$ watts per phase. SIL is also called the natural load. When the line carries natural load, the voltage along the entire length of line is the same.

However, in practice a line is not always operated at its natural load. In such a case there will be a voltage drop along the line for inductive load and a voltage rise for capacitive load. Although the reactive voltage drop/rise and the natural load do not impose any restriction on the distance over which power may be transmitted but fix the voltage that may be employed for the transmission of a certain amount of power. Typical values of surge impedance and natural load for overhead transmission line are given below in tabular form.

Voltage (Line) in kV	132	220	275	380	400	500	765
Surge impedance, Z_0 in ohms	400	375	370	335	320	312	345
Natural Load in MW	44	130	205	430	500	800	1,700
Current in Amperes	190	340	430	655	720	925	1,280

5. Mechanical Vibrations and Oscillations. With the increase in number of sub-conductors in bundled conductor transmission lines, there may be a considerable effect on vibrations of conductors. Hence the mechanical design of the system (tower dimensions, phase spacing, sub-conductor spacing, conductor height etc.) needs modifications so as to counter the problem of conductor vibrations. These modifications affect the electrical design such as voltage gradient.

Other problems include Aeolian vibrations and galloping. The problem of galloping can, however, be controlled by applying weights at different locations of the line span.

6. Environment and Biological Aspects. The biological effect of electric field of EHV lines and EHV substations has been studied extensively during 1970s. The electrostatic and electromagnetic fields produced by EHV/UHV transmission lines induce currents and voltages in animals, birds and human beings. EHV/UHV lines are designed such that maximum electrostatic field gradient does not exceed 9 kV/m at mid span under the line near the ground level.

It has been found that very low frequency magnetic fields even of weak intensities can affect certain cellular processes. Induced current densities below 10 mA/m^2 have no significant biological effect. Currents 10-100 mA/m^2 cause minor biological effects. Current densities exceeding 100 mA/m^2 cause health hazards.

Safe line to ground clearance of 20 m at mid span is recommended for 400 kV lines and 24 m for 1,100 kV lines. This permits movement of vehicles safely.

7. Audible Noise. EHV transmission lines and substations also produce audible noise. The sources of noise generation are corona, humming of transformers, cooling systems and mechanical and electrical auxiliaries.

The audible noise (AN) generated by a transmission line is a function of the following :

(*i*) Voltage gradient on the surface of a conductor.

(*ii*) Number of sub-conductors in a bundle.

(*iii*) Conductor diameter.

(*iv*) Atmospheric conditions.

(*v*) Lateral distance between the line and the point of measurement of noise.

The audible noise is caused by vibrations produced in air due to change in the air pressure.

Below 500 kV, audible noise (AN) levels do not exceed the permissible noise but the lines of 500 kV and above must be designed for a satisfactory audible noise level.

The width of right-of-way for the line corridor has a reference to the decision about AN. Line geometry is based upon 50 dB at the edge of ROW.

8. Radio Interference and Television Interference. Operation of EHV transmission lines and substations can cause radio interference. Since radio noise is associated with corona, it mainly depends on the potential gradients at the conductors. It is also influenced by all those factors which influence corona, such as air density, humidity, wind contaminants, imperfections and precipitation. The radio noise in bad weather is about 10-25 dB higher than the radio noise during fair weather. The other causes for radio interference are partial discharges on insulators and spark across gaps.

The noise due to corona in EHV lines does not cause a serious problem for frequency modulated (FM) radio-receivers or TV receivers. However, the TV reception in the immediate vicinity of a UHV line may get affected especially during bad weather.

Radio interference and television interference is caused by electromagnetic waves in the frequency range of broad cast frequencies.

Radio interference : 0.5–1.6 MHz

Television interference : 54–216 MHz

The line is designed such that radio noise within the width of the line corridor should be below permissible limits, say 40 dB at 1 MHz. Radio interference is more important factor in the design of line and in deciding right of way. The radio interference can be eliminated and/or minimised by appropriate design of line conductor and hardware. By using bundled conductors, surface voltage stress, corona and radio interference can be reduced.

9. Route Clearance. EHV/UHV transmission lines are generally constructed over different terrain including dense forests, hilly areas and agricultural land, and clearance of minimum ROW is very important. Generally the trees are not taller than 10 m. Only dry trees need to be cleared from the ROW of transmission lines because a live tree is found to be well grounded through its sap, has small power loss and little chance of burning. It has also been found that the dry tree burning can occur if an induced voltage exceeding 100 kV occurs and potential of 100 kV can exist within a radius of 6 m from 765 kV line. So a minimum clearance of 6 m from 765 kV line conductors is essential.

10. Conductor Material. ACSR conductors are universally employed for overhead transmission lines. Necessary data about ACSR bundle conductors used for ac transmission lines is given below in tabular form.

Normal voltage in kV	Highest Voltage in kV	Phase Spacing in Metres	Sub-conductor Diameter in mm	Number of Sub-conductors Per Phase
230	245	6	25	1
345	365	8.3		2
500	550	11		3
765	800	15.3		4
1,100	1,200	20		8

11. Span, Number of Circuits, Conductor Configuration. The considerations governing the choice of span has already been discussed in Art. 10.3. For a high voltage line the economical value of span is 200-300 m, whereas for 400 kV line it is 350-400 m.

If two three-phase lines are on the same tower and are identical in construction and electrically in parallel then they have the same reactance X. The inductive reactance of the single equivalent circuit becomes $\frac{X}{2}$ and hence the power transfer capability is doubled. Double-circuit lines are in common use. They have additional advantage that continuity of supply is maintained over one line, though with reduced capability, while the other line is out of service for maintenance or repair.

Each circuit of a double circuit line is usually designed for 75% of the line capacity. In India, both single and double circuit lines exist in the high voltage and extra high voltage class (66, 132, 220 and 400 kV). The number of circuits is determined from the SIL.

Vertical configuration has been found the most economical for double circuit lines. For single circuit lines, horizontal or L type configurations are most suitable.

12. Insulation Coordination. Transmission lines and substations are subjected to overvoltages due to switching, lightning, faults, resonance and other causes. The surge arrestors are provided at strategic locations to protect the line insulation and substation equipment from transient and temporary overvoltages.

In general there are three types of overvoltages viz. switching overvoltages, temporary overvoltages and overvoltages due to lightning. Switching overvoltages are caused due to energization of lines, reclosing, fault interruption, load throw, out of phase switching and sudden switch-off of a line. The temporary overvoltages at power frequency are due to sudden loss of load, disconnection of inductive loads, connections of capacitive loads, etc.

For EHV and UHV transmission lines the switching overvoltages become the limiting factor in the design of insulation levels and arrester characteristics. Once the line is designed adequately against switching overvoltages, it is capable of encountering temporary overvoltages. It has also been found that overvoltages due to lightning are not very important for EHV/UHV transmission lines.

The waveshapes of switching surges are practically infinite. Since the waveshape has a definite influence on the flash-over strength, a standard waveshape has been selected for analysing the line performance. This standard switching surge has a *crest time* (*i.e.* the time to attain the peak value) of 175 μs, and a *tail time* (*i.e.* the time to attain half the peak value on the tail) of 3,200 μs. It has been found desirable and possible to reduce the switching surges for EHV systems to about 2.5 pu and insulation of EHV system is usually designed for this strength. The temporary overvoltages can be limited to 1.5 pu if shunt reactors are employed.

The insulation characteristics of various equipment is defined in terms of rated voltage, standard switching impulse, standard lightning impulse, power frequency withstand voltage etc. These are co-related with characteristic of surge arresters.

The insulation levels of various substation apparatus and the transmission line should be coordinated with various surge arresters such that the overvoltages are discharged to earth without causing damage to the equipment insulation.

13. Series Compensation. Series compensation is an important method of improving the performance of EHV transmission lines. It consists of capacitors connected in series with the line at suitable locations and thus opposes directly the effect of series inductive reactance of the line. It increases the power handling capacity and reduces the voltage regulation as explained below :

The power transfer capability of a transmission line is given as

$$P = \frac{V_S V_R}{X_L} \sin \delta \qquad \text{...Refer to Eq. (13.1)}$$

With the insertion of capacitors having capacitive reactance X_C in the line, the net reactance of the line becomes $(X_L - X_C)$ and the power transfer capability of the line is given as

$$P' = \frac{V_S V_R}{X_L - X_C} \sin \delta \qquad \text{...(13.2)}$$

From above Eqs. (13.1) and (13.2) it is obvious that for the same magnitudes of V_S, V_R and δ, P' is much higher than P and the increase in power transfer capability is given as

$$\frac{P'}{P} = \frac{X_L}{X_L - X_C} = \frac{X_L}{X_L\left(1 - \frac{X_C}{X_L}\right)} = \frac{1}{1 - K} \qquad \text{...(13.3)}$$

where K is the *degree of compensation* and equals $\frac{X_C}{X_L}$.

For example if the line is compensated to an extent of 50% *i.e.* K = 0.5, the power transfer capability of the line becomes $\frac{1}{1-0.5}$ *i.e.* double of that without series compensation.

Economic Considerations. The growing cost of ROW, aesthetic considerations calling fewer transmission lines, increase in the lengths of transmission lines and in transmission voltages, decrease in series capacitor costs, have all contributed to an increase in the rate of use of series capacitors in India and other countries.

In view of the line cost, the effect of lines operating in parallel and problems of system protection, the economic degree of compensation is found to be in the range of 0.4 to 0.7.

Improvement of System Stability. From the power flow equation it is evident that for the transfer of same amount of power and the same values of sending-end and receiving- end voltages V_S and V_R, the power angle δ is lower for compensated line as compared to that in case of an uncompensated line. Lower values of power angle means better system stability.

Various measures have been adopted for improvement of power system stability. These include reducing the reactance of synchronous generators and transformers, increasing the number of parallel lines, use of bundled conductors and series compensation. Series capacitors usually offer the most economic solution with respect to both the steady state and transient stability.

Location of Series Capacitors. Series capacitors can be located either along the lines, in intermediate substations or switching stations. Line location has many advantages such as better voltage profile along the line, reduced short-circuit current through the capacitor during a fault. The small short-circuit current contributes to a much simpler protection of the capacitor. The capacitor stations are generally unattended type.

Installation of series capacitors in intermediate attended type substations or switching stations may appear advantageous, since staff for maintenance and service, auxiliary power etc. are available. However, these advantages must be weighted against the higher cost for protection and control and sometimes also for the space required as well as the increase of transmission losses.

Capacitor Bank. A capacitor bank is built up of a number of capacitor units connected in series and in parallel. The units for series capacitors are designed, manufactured and tested with due regard for the specific service conditions, such as high overvoltages and capacitor discharge currents.

A capacitor unit consists of a number of capacitor elements in a common container. The units are equipped with fuses, which may be either external ones for each unit or internal individual element fuses. The capacitor units are mounted in simple frames, called the racks, placed on supporting insulators and stacked on top of each other.

Problems Associated With Series Compensation. There are some major problems associated with series compensation. These problems are summarized below.

(i) Subsynchronous Resonance. The series capacitor introduces a subsynchronous frequency, proportional to the square root of the compensation in the system. In some cases this frequency may interact with turbo-generator shafts and develop high torsional stresses. The risk of subsynchronous resonance in hydro-generators is quite small as the torsional frequencies are about 10 Hz or even less.

(ii) Ferroresonance. When a lightly loaded transformer is energized through a series compensated line, ferroresonance may occur. This can be suppressed by using shunt resistors across the capacitors or by short-circuiting the capacitor temporarily through an isolator or bypass breaker.

(iii) Maloperation of Line Protection. Series compensation may lead to maloperation of the protection system leading to unnecessary trippings.

(iv) High Recovery Voltage. High recovery voltage may develop across the circuit breaker contacts due to series compensation.

14. Shunt Compensation. In EHV transmission systems, shunt compensation is invariably essential. Shunt compensation with capacitive VARs is used to inject reactive power and control the receiving-end voltage whereas shunt reactor compensation is used to neutralize the Ferranti effect. For shunt compensation, static compensation employing capacitors and reactors or synchronous compensation using synchronous phase modifier may be employed. Static compensation is being preferred to the synchronous compensation because of its inherent advantages of higher speed of response, absence of fault infeed to the system, lower maintenance, low cost, greater reliability and simpler erection. In addition to shunt capacitors, shunt reactors and phase modifiers, a thyristor controlled static shunt compensator to meet reactive power generation and absorption demand has appeared in recent years.

One primary reason for using shunt reactors or reactive control devices on EHV lines is to control steady-state overvoltages when energizing the long EHV lines or when operating under light-load conditions. If the shunt reactors are not employed, the reactive power generated by the capacitance can cause high voltages at the receiving end of the line (Refer to Ferranti effect Art. 5.6). To restrict insulation stresses caused by overvoltages following sudden load rejection a substantial part of the shunt reactive compensation is usually left permanently connected.

It should be noted that with shunt reactors, the power transfer capability of the line is reduced owing to increase of transfer reactance X. Thus at fault load when stability is most important, the shunt compensation reduces the maximum power limit of the long line. In case line shunt reactors are switched-out under heavy load conditions, the maximum power transfer capability can be considerably increased, but voltage variations due to sudden load thrown off are likely to be unacceptably high. In actual practice some of the shunt reactors are kept connected permanently so as to avoid voltage increase due to sudden fall in load from heavy load conditions.

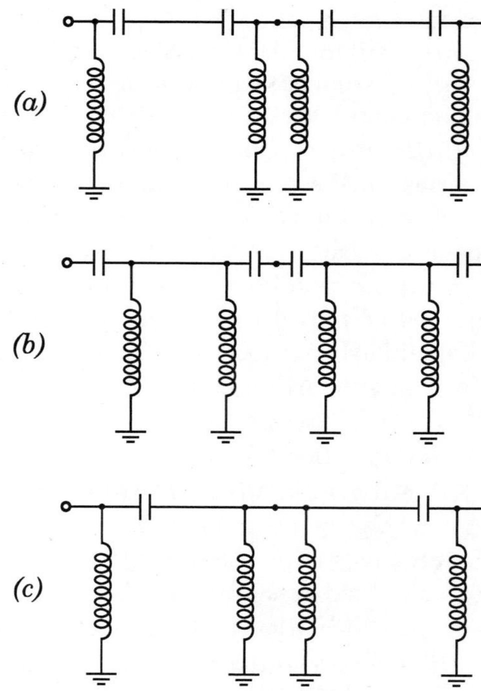

Fig. 13.3. *Series and Shunt Compensation Schemes*

15. Series and Shunt Compensation. In EHV long transmission lines both series capacitors and shunt reactors are provided. The purpose of providing series capacitors is to artificially reduce series reactance of the line, so as to improve stability and efficiency of transmission. The shunt reactors are provided to artificially reduce the susceptance of the line so as to improve voltage regulation under light load conditions.

Practical and economical reasons lead to concentration of the compensating elements at a few points along the line. For a given total MVARs of series and shunt compensation of a transmission system, maximum power transfer capability, voltage control conditions and efficiency of power transmission depend on the number, location and circuit schematic of the series capacitor and the shunt reactor stations. Series and shunt compensation schemes are illustrated in Figs. 13.3. There are also other possible schemes.

13.5. FLEXIBLE AC TRANSMISSION SYSTEM (FACTS)

With ever increasing demand of electric power, the existing transmission networks are found to be weak resulting in poor quality of unreliable supply. Also, it is realized that for expanding or enhancing the power transfer capability of the present transmission networks, huge sum of finances are required and sometimes even difficulties are encountered in finding right-of way for the new transmission lines. Lot of research has been done for developing new technologies over the past few years to gain increased efficiency and reliability from the existing transmission network. As a result, new transmission system, known as *Flexible AC Transmission System* (abbreviated as FACTS) has been developed. The new technologies make use of high speed thyristors for switching in or out transmission line components such as capacitors, reactors or phase shifting transformer for some desirable performance of the sysetm. The main objective of this system is to replace the existing slow acting mechanical controls needed to react to the changing system conditions by rather fast acting electronic controls.

These recently developed systems are likely to be employed for ac interconnecting lines and long ac lines. A long flexible ac transmission system (FACTS) is illustrated in Fig. 13.4. It has intermediate substations at an interval of 250–350 km. 1, 1 are controllable series capacitors and 1′, 1′ are controllable shunt compensation (SVS).

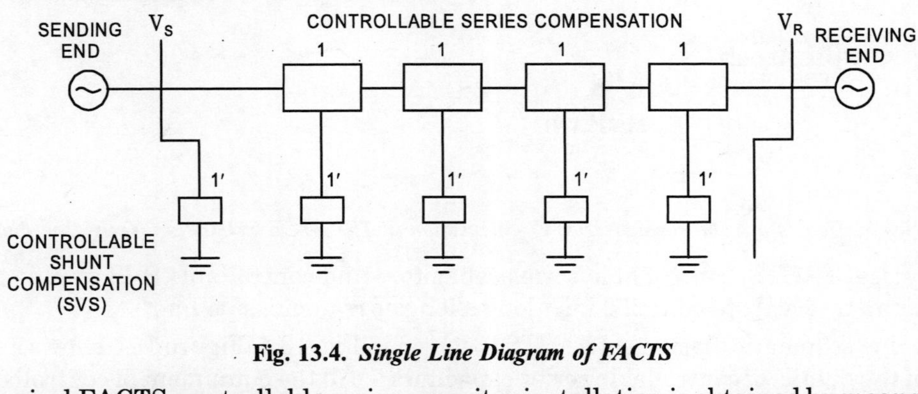

Fig. 13.4. ***Single Line Diagram of FACTS***

In a typical FACTS, controllable series capacitor installation is obtained by means of bypass thyristor switch (Fig. 13.5). The amount of series compensation is controlled by controlling the bypass current I_B through the thyristor controlled switch (TCS) and current I_C through series capacitor C_{se}

Power flow is given by expression

$$P = \frac{V_S V_R}{X_L - X_{CF}} \sin \delta$$

where X_L is series reactance of the transmission line and X_{CF} is the controllable series reactance of series capacitor of FACTS.

So, by controlling X_{CF} by means of bypass thyristor switch, the power flow can be controlled.

The control over power flow P and power angle δ is provided by the phase control of bypass thyristor switch. The basic limitation of ac transmission line and series compensation had been the lack of controllability, which has been overcome by controllable series capacitors of FACTS. In addition to controllable series capacitor; controllable shunt compensation is obtained by means of SVS in FACTS. Thus, voltage at both ends of transmission line is controlled dynamically. This provides voltage stability and improves dynamic stability of the transmission link.

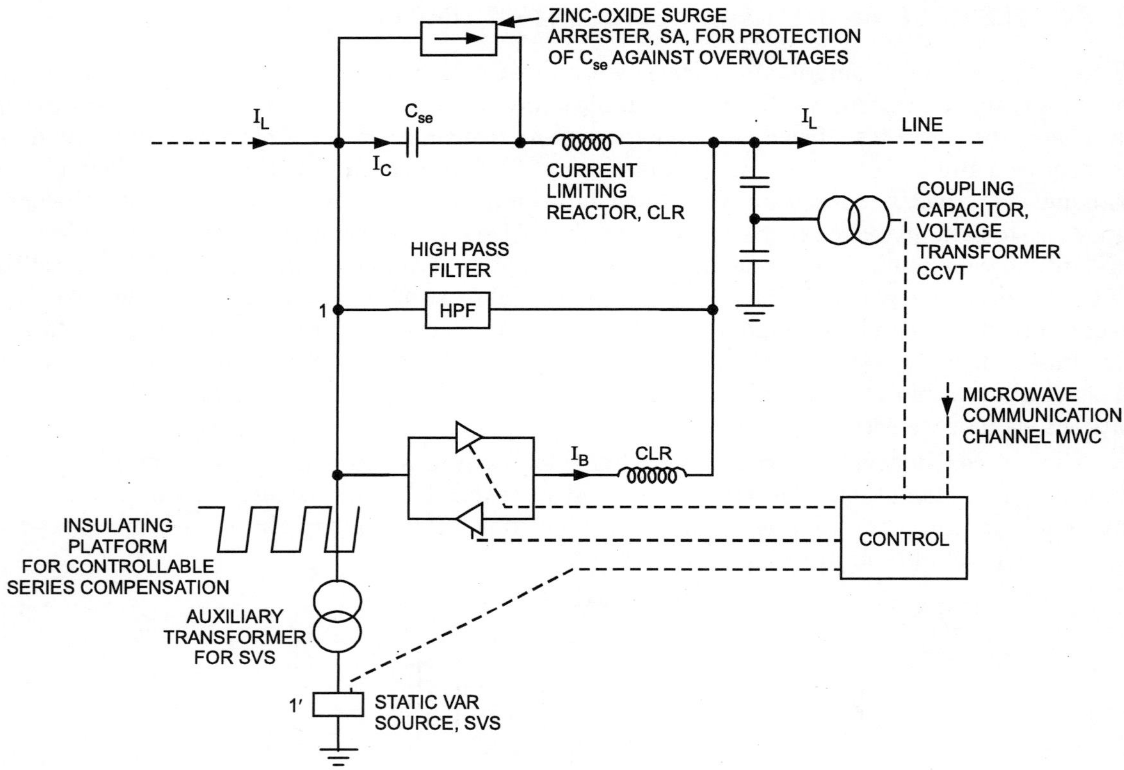

Fig. 13.5. ***Details of controllable Series Compensation in The FACT Installation (Only One Phase)***

In a typical FACTS, controllable series capacitors and controllable SVS combinations are installed at an interval of about 250 km along a long ac transmission line.

Single line schematic diagram of FACTS is shown in Fig. 13.4. Figure 13.5 shows a schematic diagram of the details of controllable series capacitors 1. All the equipment of controllable series capacitors are installed on insulating platform. The control unit senses the phase-to-ground voltage V at the location of controllable series capacitors and adjusts phase angle of thyristor control switch (TCS) and so bypass current I_B is controlled. By varying I_B, I_C is controlled because $I_L = I_C + I_B$. This results in variable series compensation which can be controlled from load control centre through compensation, the power flow P is controlled. By controlling static VAR source (SVS), the voltage at sending end, receiving end and at intermediate substations are controlled.

Series and shunt compensations are controllable and control is fast and over a wide range. Hence FACTS improves the dynamic stability of the ac transmission system.

By simultaneous control of 1, 1, ... and 1′, 1′ ... the power flow P, sending end voltage V_S, receiving end voltage V_R and power angle δ between V_S and V_R are controlled. The oscillations in P, V, δ are damped by controlling 1 and 1′.

EXERCISES

1. Give the advantages, drawbacks and limitations of EHV ac transmission.
2. Answer the following :
   Relative merits and demerits of EHV transmission system.
   [U.P. Technical Univ. Elements of Power System 2009-10]
3. What are the problems involved in transmission of power using EHV ac system ? Discuss in detail. Also state how these problems are being solved.
4. What are disadvatnages of HVAC transmission.
   [U.P. Technical Univ. Elements of Power System 2007-08]
5. Discuss the various problems of EHV power transmission.
   [U.P. Technical Univ. Elements of Power System 2005-06]
6. Describe the design considerations of EHV transmission lines.
   [U.P. Technical Univ. Elements of Power System 2007-08]
7. Discuss the points which are considered while designing a transmission line.
   [U.P. Technical Univ. Elements of Power System 2008-09]
8. Describe the various conductor configurations and choice of number of circuits for EHV transmission lines. **[U.P. Technical Univ. Elements of Power System 2009-10]**
9. Discuss the advantages of using 'bundled conductors' in EHV overhead transmission lines.
   [U.P.S.C. I.E.S. Elec. Engineering II, 1998]
10. What is series compensation ? What for it is provided ? Give its advantages and problems associated with it.
11. Identify the utility of FACTS devices in power network.
    [U.P. Technical Univ. Elements of Power System 2005-06]
12. Explain the principle and usefulness of flexible ac transmission system.
    [U.P. Technical Univ. Elements of Power System 2003-04]
13. Discuss the principle and advantages of using Flexible AC transmission (FACT) system.
    [U.P. Technical Univ. Elements of Power System 2002-03]
14. Explain in brief the role of flexible ac transmission systems.
    [U.P. Technical Univ. Elements of Power System 2006-07]
15. What are the various FACTS controllers ? Explain the operating principle of any one of them. **[U.P. Technical Univ. Elements of Power System 2006-07]**
16. How FACTS controllers make ac transmission flexible ? Describe working principle of TCSC and show how it enhances power transfer capability of lines.
    [U.P. Technical Univ. Elements of Power System 2004-05]
17. Write short notes on the following :
    (*a*) Necessity of EHV ac transmission.
    (*b*) Problems associated with EHV ac transmission.
    (*c*) Limitations of EHV ac transmission.
    (*d*) Special features of EHV ac transmission.
    (*e*) Compensation in EHV transmission system.
    (*f*) Flexible AC transmission system (FACTS).

SHORT ANSWER TYPE QUESTIONS WITH ANSWERS

Q. 1. What are the transmission voltages that are used in India ?

Ans. 33, 66, 110, 132, 220 and 400 kV. 765/800 kV is also in used in India.

Q. 2. What are the advantages of adopting extra high voltage / ultra high voltage for transmission of ac electric power.

Ans. Advantages of adopting of EHV/UHV for transmission of ac electric power are reduced line losses, high transmission efficiency, improved voltage regulation, reduced conductor material requirement, flexibility for future system growth, increase in transmission capacity of the line, increase of SIL and reduction in ROW.

Q. 3. Why does surge impedance loading (SIL) increase with the increase in voltage level ?

Ans. SIL varies as the square of the operating voltage, so SIL increases with the increase in voltage level.

Q. 4. What are the problems associated with an EHV transmission ?

Ans. The problems associated with EHV transmission are corona loss and radio interference, requirements of heavy supporting structures and erection difficulties and insulation requirements.

Q. 5. On what basis the conductor size is determined for transmission lines up to 220 kV ?

Ans. For overhead transmission lines operating at voltages up to 220 kV, the conductor size is determined on the basis of its current carrying capacity.

Q. 6. On what basis the conductor size of EHV line is determined ?

Ans. The conductor size of an EHV line is decided on the basis of corona.

Q. 7. What are the sources of audible noise generation in EHV transmission systems ?

Ans. The sources of noise generation in EHV transmission systems are corona, humming of transformers, cooling systems and mechanical and electrical auxiliaries.

Q. 8. What material is universally employed for overhead transmission lines ?

Ans. ACSR conductors.

Q. 9. What is economical value of span for 400 kV transmission line ?

Ans. 350–400 m.

Q. 10. What for series and shunt compensation provided in EHV lines ?

Ans. Series compensation is provided in EHV lines to artificially reduce the series reactance of line so as to improve stability, voltage regulation and transmission efficiency where as shunt compensation is provided to artificially reduce the line susceptance so as to improve voltage regulation under light load conditions.

❑❑❑

14

CHAPTER

HIGH VOLTAGE DIRECT CURRENT (HVDC) TRANSMISSION

INSIDE THIS CHAPTER

14.1. Introduction **14.2.** Principle of HVDC System Operation **14.3.** Advantages of HVDC Transmission **14.4.** Limitations of HVDC Transmission **14.5.** Comparison Between EHV-AC and HVDC Transmission **14.6.** Types of DC Links **14.7.** Ground Return **14.8.** Earth Electrode and Station Earth **14.9.** HVDC Transmission Systems **14.10.** Converter Station Equipment **14.11.** Transient Overvoltages and Insulation Coordination **14.12.** Corona in HVDC Lines **14.13.** Protection Aspect **14.14.** HVDC Circuit Breaking.

14.1. INTRODUCTION

In the initial stages dc used to be employed for generation, transmission and distribution of electric power. The first 110 V dc Central Electric Station was installed by Edison in New York in 1882. With the introduction of transformers and the 3-ϕ ac system the situation changed in favour of ac, especially where the electric energy used to be harnessed from water power which is usually available far from the load centres. For many reasons, such as facility of transformation of voltages from one value to another, better performance of ac motors and superiority of ac generators in comparison to dc generators, the power was generated, transmitted, distributed and utilized in the form of ac. The supporters of dc, however, did not forget the advantages of dc transmission and extensive research has been carried out in this field and as a result it has staged a sort of comeback to the field of electric power transmission at high and extra high dc voltages.

The earlier dc schemes used mercury-arc convertors. Extensive research has been carried out especially in Sweden for the development of high voltage converters. With the successful development of high power (50 kV, 100 A) thyristors in early 1970's the HDVC transmission has become technically and commercially viable alternative to EHV/UHV AC transmission particularly for distance bulk power transmission, cable transmission and system interconnection. For these applications HVDC transmission systems have a distinct superiority over EHV AC transmission systems and are being increasingly preferred. However, HVDC transmission cannot be substitute of ac transmission because for backbone ac network, generation, transmission and distribution is definitely superior. The choice of transmission systems and operating voltages for a transmission line is made from HV-AC (up to 220 kV), EHV-AC (between 400 kV and 765 kV), UHV-AC (exceeding 765 kV) and HVDC (up to 1,600 kV) on the basis of technical and economical studies for each particular line and associated ac systems.

The first dc link was set up in 1954 between Swedish mainland and the island of Gotland. This was a monopolar, 100 kV, 20 MW, cable system making use of sea return. The England-France cross-channel dc link was commissioned in 1961. This was a bipolar, ± 100 kV, 160 MW, cable system over a distance of about 65 km. Since then more and more HVDC systems have been set up in the world.

HVDC transmission has also been introduced in India. ± 500 kV has been selected as the voltage for HVDC transmission. A ± 500 kV, 1500 MW, 810 km bipolar HDVC line has already been set up between Rihand and Delhi. It is designed to operate in the bipolar, monopolar-ground return, and monopolar metallic return modes. Another HVDC system set up in India is Vindhyachal HVDC back to back system. This back to back link is for exchange of power between Northern and Western regions. Each block of 250 MW is capable of operating independently in either direction and can transfer power in the range of 25-250 MW depending on the conditions of the system. In addition, Chandarpur back to back project (2 × 250 MW), Chandarpur-Padghe bipolar system (± 500 kV, 1,500 MW, 736 km), Jeypore back to back project (2 × 250 MW) and Mau back to back project (2 × 250 MW) are also proposed.

At present the world has over 50 HVDC schemes in operation for a total capacity of more than 50,000 MW and the capacity is increasing by about 2,000 MW every year.

14.2. PRINCIPLE OF HVDC SYSTEM OPERATION

When using direct current to provide an asynchronous link between two ac systems, it is necessary to have two convertor stations one at each end, connected by a dc transmission line. The main equipments in a convertor station are transformers and thyristor valves. Chokes and filters are provided at each end to ensure smooth direct current and suppress harmonics. At the sending end the thyristor valves act as rectifiers to convert ac into dc which is transmitted over the line. At the receiving end the thyristor valves act as inverters to convert dc into ac which is utilized at the receiving end.

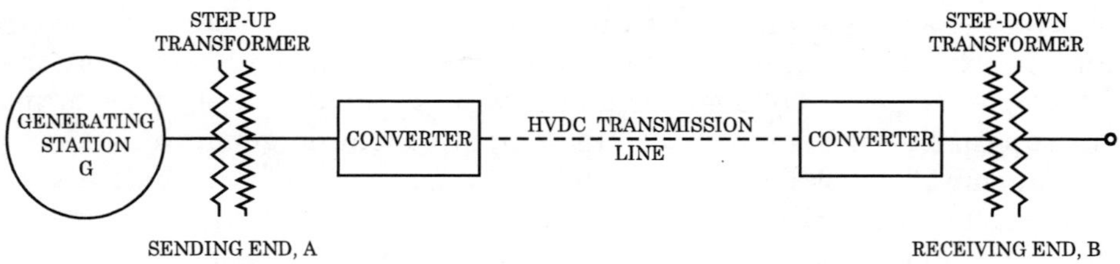

Fig. 14.1. ***Line Diagram of typical HVDC Transmission System***

Single line diagram of a HVDC transmission system is shown in Fig. 14.1, where A and B are the two converter stations. Converter station A is supplied from the generating station G. In converter station at the sending end the voltage is stepped up to appropriate value by step-up transformer and then converted into direct current by the thyristor valves. Thus at the start of transmission line, we have high voltage direct current. This rectified current flows along the transmission line to the receiving-end converting station B, where it is converted into 3-phase ac current by the thyristor valves and then stepped down by the step-down transformer to low voltage for further distribution.

The power despatched from the generating station P_S less the power received at the receiving end P_R *i.e.* $(P_S - P_R)$ represents the power losses due to conversion and transmission. The convertor at the sending end acts as a rectifier and is suitable for power frequency (*i.e.* frequency of generator) on its ac side while the converter at the receiving end acts as an inverter and its frequency is determined by the frequency of the load system. This frequency is independent of the sending-end frequency provided the two ends A and B are not additionally connected by the 3-phase lines.

The dc output voltage magnitude is controlled by varying the firing angle of the thyristor valves in the converter. In rectifier the firing angle is between 0° and 90° while in inverter it is between 90° and 180°. As the dc output voltage is a function of cosine of the firing angle hence the converter voltage becomes negative when firing angle α exceeds 90°. This makes the converter to operate as an inverter. The two converters at sending end and receiving end are identical and whether they have to work as rectifier or inverter is determined by the direction of power flow.

In practical HVDC converter stations three-phase bridge converters are employed at both ends (sending as well as receiving ends). Reversible operation of converters as well as bidirectional power flow in HVDC link is possible simply by the control of firing angle.

14.2.1. Power Control in HVDC System

In dc systems the power transferred from one station to another station is governed only by the magnitudes of terminal dc voltages at the two ends while in ac transmission systems the power transfer is governed by phasor difference (magnitude as well as phase) of the voltages at the two ends. Thus the controllability of HVDC power is fast and stable. The current flows from higher voltage to lower voltage which are set by adjustment of firing/extinction angles of the two converters (rectifier and inverter).

If V_S is the voltage at the sending and V_R is the voltage at the receiving end then the line current is given as

$$I_{dc} = \frac{V_S - V_R}{R} \qquad ...(14.1)$$

where R is the resistance of the entire transmission link

The sending-end voltage is given as

$$V_S = \left[\left(\frac{3\sqrt{2}}{\pi} V_{ac\,S}\right)\cos\alpha - \frac{3X_{CS}}{\pi} I_{dc}\right] \qquad ...(14.2)$$

and the receiving-end voltage is given as

$$V_R = \left[\left(\frac{3\sqrt{2}}{\pi} V_{ac\,R}\right)\cos\beta - \frac{3X_{CR}}{\pi} I_{dc}\right] \qquad ...(14.3)$$

where α is the firing angle of rectifier, β is the extinction angle of inverter, $V_{ac\,S}$ is the ac side line-to-line rms voltage at the sending end, $V_{ac\,R}$ is the ac side line-to-line rms voltage at the receiving end, X_{CS} is the commutation reactance at the sending end and X_{CR} is the commutation reactance at the receiving end.

Thus, the power transferred is given as

$$P = V_S I_{dc} = \frac{V_S - V_R}{R} V_S \text{ watts} \qquad ...(14.4)$$

Tap-changers on the ac side take care of voltage variations on ac side and dc power is controlled by controlling the sending-end and receiving-end voltages V_S and V_R which is possible by control of firing and extinction angles α and β respectively.

14.3. ADVANTAGES OF HVDC TRANSMISSION

DC transmission possesses many technical and economic advantages over ac transmission. Some of these are given below:

1. Cheaper in Cost. Bipolar HVDC transmission lines require two-pole conductors while ac system requires 3 conductors to carry power.

HVDC transmission can utilize earth return and, therefore, does not require a double circuit while EHV-AC transmission always require a double circuit.

The potential stress on the insulation in case of dc system is $\frac{1}{\sqrt{2}}$ times of that in ac system for the same operating voltage. Hence for the same operating voltage less insulation is required.

The phase-to-phase clearances, phase-to-ground clearances and tower size are smaller in case of dc transmission as compared to those for ac transmission. Because of lesser load on the supporting structures tower designs are simpler and cheaper. Requirement of 'right-of-way' for a dc line is about 20-40 per cent lesser than that for an ac line of the same power transmission capability.

EHV-AC transmission needs intermediate substations at an interval of 300 km for compensation but HVDC transmission system does not require any such intermediate substation for compensation.

However, the dc line terminal equipment because of high cost of converters is considerably more expensive than ac line terminal equipment. HVDC transmission becomes economical over ac transmission above the break-even distance when the saving in overall costs of conductors, towers etc. compensates the additional cost of the terminal equipment (rectifiers and inverters). The break-even distance is generally different for different projects. For overhead lines it may be about 600–800 km while for submarine cables and underground cables it may lie in the range of 20 to 50 km and 50 to 100 km respectively. Sometimes it becomes necessary to adopt dc transmission, even for distances smaller than the break-even distances, because of technical advantages.

2. An HVDC line can be built in stages. The dc line can be built as a monopolar line with ground return in the initial stage and may be converted into a bipolar line on a later date when the load requirement increases.

3. No Skin Effect. There is no skin effect in dc, so there is a uniform distribution of current over the section of the conductor. Thus there is full utilization of line conductor in case of dc transmission while it is not so in case of ac transmission.

4. Lower Transmission Losses. HVDC transmission system needs only two conductors and, therefore, the power losses in a dc line are lesser than the losses in ac line of the same power transfer capability.

5. Voltage Regulation. There is no inductance, hence the voltage drop due to inductive reactance does not exist in dc transmission line. Thus voltage regulation is better in case of dc transmission.

6. Line Loading. The permissible loading on an EHV-AC line is limited by transient stability limit and line reactance to almost one-third of thermal rating of conductors. No such limit exists in case of HVDC lines.

7. Surge Impedance Loading. Long EHV-AC lines are loaded to less than 80 percent of natural load. No such condition is applicable to HVDC lines.

8. Greater Reliability. A two-conductor bipolar dc line is more reliable than a 3-wire 3-phase ac line because the dc line may be operated in a mono-polar mode with ground return when the other line develops a fault.

9. Rapid Change of Energy Flow. The control of converter (thyristor) valves permit rapid changes in magnitude and direction of power flow when the two ac systems are interconnected by a dc line. Thus a more flexible coordination of system control at the two ends, a more economical use of cheap power generation in either of the two ac systems, and reduction of system reserve and standby capacity are available. Transient stability is also increased.

10. Independent Control. The ac systems interconnected by a dc line can be controlled independently. They can be completely independent as regards frequency, system control, short-circuit rating, future extension etc.

Independence of frequencies of sending-end and receiving-end networks, makes it possible to generate the power at one frequency (say 50 Hz) and to utilise it at some other frequency (say 60 Hz).

11. Lesser Dielectric Power Loss and Higher Current Carrying Capacity. The cables have lesser dielectric power loss with dc in comparison with ac and, therefore, have higher current carrying capacity.

12. Negligible Sheath Losses. In case of dc, only leakage current flows in the cable sheath where as in case of ac charging, circulating and eddy currents flow through the sheath of cable. Thus in case of dc transmission with cables, the sheath losses in the cables are negligible.

13. Higher Natural Dielectric Strength and Longer Life of Cable Insulation. The natural dielectric strength of cable insulation with dc is substantially greater than that with ac. Phenomenon of dielectric fatigue is also absent, therefore the cable insulation has longer life.

14. Absence of Charging Current and Limitations of Cable Lengths. Because of large charging currents, the use of EHV-AC for underground transmission over long distances is prohibited but because of absence of charging current in dc systems there is no limit on the length of the cable.

In ac systems there is limit on length of cable depending upon rated voltage—about 60 km in case of 145 kV, 40 km in case of 245 kV and 25 km in case of 400 kV.

15. Low Short-Circuit Currents. In ac transmission, addition of parallel lines results in larger short-circuit currents in the system. This is due to reduced equivalent reactance. When an existing ac system is interconnected with another ac system by ac transmission line, the fault level of both the systems increases. Sometimes this necessitates the replacement of the existing circuit breakers by the ones of higher rating. However, when the two ac systems are interconnected by a dc line, the contribution of a dc line to the short-circuit current is only up to the rated current of the dc line.

During fault in dc line, grid control of converter can drastically reduce the fault current.

16. Lesser Corona Loss and Radio Interference. The corona loss is directly proportional to $(f + 25)$, where f is the supply frequency. So corona loss in dc system are lesser than those in ac systems for the same conductor diameter and operating voltage. This is because in dc system f is equal to zero. Corona loss and radio interference are directly related and, therefore, radio interference in case of dc is less as compared to dc. Also corona and radio interference slightly decrease by foul weather conditions (snow, rain or fog) in case of dc where as they increase appreciably in case of ac system.

17. Higher Operating Voltages. Modern high voltage transmission lines are designed on the basis of expected switching surges rather than lightning surges because the former are more severe in comparison to the latter. The level of switching surges due to dc is lower than that due to ac and, therefore, the same size of conductors and string insulators can be employed for higher voltages in case of dc as compared to ac.

18. Reactive Power Compensation. Long distance ac power transmission is feasible only with the use of series and shunt compensation, provided at intervals along the line (Art. 13.4-15). For such lines shunt compensation is required to absorb the line charging VARs during light load conditions and series compensation is provided for stability reasons. HDVC lines do not require any reactive power compensation like EHV-AC lines. It is because of absence of charging currents and unity power factor operation.

19. No Stability Limit. There is no stability limit in dc systems and, therefore, transmission line may be of any length whereas in ac systems the theoretical length of a transmission line which can be operated without loss of stability is only 1.43 times its natural impedance (mostly it is about 500 km). However, with series compensation the length of ac line can be somewhat longer than this value.

20. Use as a Asynchronous Link Between Two AC Systems. Parallel operation of ac with dc which increases the stability limit of the system or interconnection of two large ac systems by a dc transmission line. Here the dc line is a asynchronous link between two rigid (frequency constant) systems where otherwise slight difference in frequency of the two large systems would produce serious problems of power transfer control in the small capacity link.

14.4. LIMITATIONS OF HVDC TRANSMISSION

Although dc transmission has so many advantages over ac transmission, yet it will never completely overtake ac transmission, at the most it will work in conjunction with it. The problems associated with and limitations of dc transmission are given below:

1. Costly Terminal Equipment. The converters required at both ends of the line have proved to be reliable but they are much more expensive than the conventional ac equipments.

The converters have very little overload capacity and absorb considerable reactive power—the reactive power can be as high as 25% of true power.

It is obvious that no reactive power can be transmitted by HVDC transmission system and, therefore, equipment such as static or synchronous capacitors are installed at the receiving end to generate reactive power (MVAR) according to the demand of the load and converting apparatus.

The converters produce lot of harmonics both on dc and ac sides which may cause interference with the audio-frequency communication lines. Filtering and smoothing equipments are provided at the convertor stations in order to remove ripples from the dc output. It is also possible to provide filters on the ac side to absorb the harmonic currents. Thus there is considerable increase in cost of the converter substation.

HVDC converters need complex cooling systems.

HVDC circuit-breaker system comprises several components including the main circuit-breaker capacitors, reactors etc., and the total cost is likely to be several times that of an ac circuit breaker of equivalent rating.

2. More Maintenance of Line Insulators. Maintenance of insulators in HVDC transmission line is more.

3. Circuit breaking in multi-terminal dc systems is difficult and costlier.

4. Voltage Transformation. Voltage transformation is not easier in case of dc and hence it has to be accomplished on the ac side of the system. DC system cannot be employed for distribution, sub-transmission and backbone transmission.

14.5. COMPARISON BETWEEN EHV-AC AND HVDC TRANSMISSION

Comparison between EHV-AC and HVDC for long distance high power transmission is given in tabular form in Table 14.1.

The dc transmission links so far used mostly are based on the factors like long river crossings, frequency conversions and asynchronous ties between large *ac* systems. To give an approximate idea of economic distance for HVDC transmission is that for a distance of 400 km the power of at least 100 MW is required to be transmitted. With the development of converting equipment, the break-even distances will be reduced.

Table 14.1. Comparison of Long Distance Overhead EHV-AC and HDVC Transmission.

S.No.	Characteristics	EHV-AC System	HVDC System	Remarks
1.	**Capital Cost.**			
	Line cost	Higher	Lower	HVDC system becomes economical above 800 MW.
	Substation Cost	Lower	Higher	Choice based on economics.
	Number of circuits	More	One	
	Intermediate Substations	Required	Not required	
2.	Skin Effect	Present	Absent	
3.	Ground Return	Not possible	Possible	Only single bipolar HVDC line adequate in most cases.
4.	Line Losses	High	Low	HVDC substation has power losses.
5.	Power Transfer	Limits imposed by power angle and line inductive reactance.	No limit	Single HVDC link adequate up to 3,000 MW
6.	Voltage Control	Difficult for long lines due to series inductance and shunt capacitance	Easier	—
7.	Compensation Requirements	Series and shunt compensations are necessary	No compensation is required	
8.	Corona and Radio Interference	More	Less	
9.	Reliability	Lesser	More	
10.	Short-circuit Currents	Large	Small	
11.	Control System	Simpler, cheaper and limitations of control	Difficult, costlier but fast and accurate control	For interconnections, HVDC system is advantageous due to superior control
12.	Stability of AC Networks	Very low due to line inductive reactance	Much higher	AC line is loaded up to 50% of its thermal rating while dc line is loaded up to 90% of its thermal rating

14.6. TYPES OF DC LINKS

The dc links may be classified into

(i) monopolar link
(ii) bipolar link and
(iii) homopolar link.

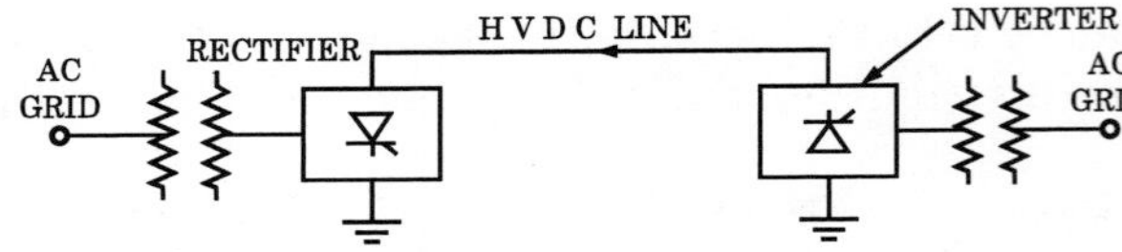

Fig. 14.2. ***Monopolar DC Link***

1. Monopolar DC Link. As the name suggests, monopolar link has only one conductor and return path is provided by permanent earth or sea. The line usually operates with negative polarity with respect to ground so as to reduce corona loss and radio interference. The monopolar *dc* link is shown in Fig. 14.2.

The earth electrodes are designed for continuous rated current operation and for any overload capacity required in the specific case. The sea or ground return is permanent and of continuous rating. The ground return path has a low resistance and, therefore, low power loss in comparison with a metallic line conductor of economical size and equal length provided the ground electrodes are of proper design. Monopolar line is more economical than a bipolar line because the ground return saves the cost of the one metallic conductor and losses in it.

Monopolar HVDC links were used only for low power rating and mainly for cable transmission. In some cases the monopolar lines installed earlier are converted into bipolar systems by adding additional substation pole and transmission pole.

Monopolar HVDC line has only the rating equal to half of corresponding bipolar line rating and is, therefore, not economically competitive with EHV ac schemes for submarine cables longer than 25 km and of power rating of about 250 MW. For such cable transmission high voltage ac scheme is not technically feasible due to large charging currents with ac cables beyond thermal limit. Bipolar cable is not justified for ratings up to about 500 MW. Recent HVDC cable schemes are bipolar.

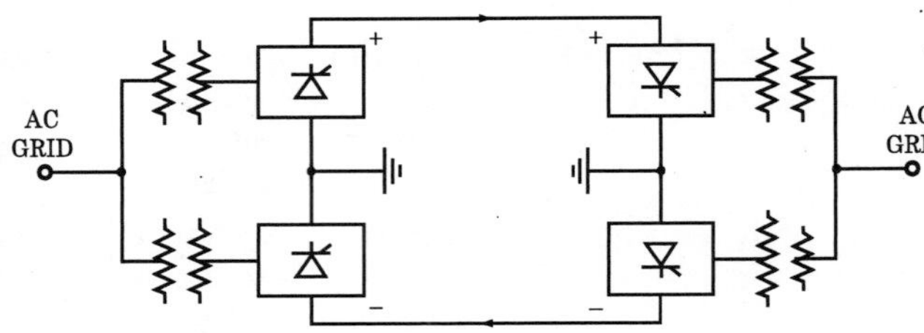

Fig. 14.3. ***Bipolar DC Link***

2. Bipolar DC Link. This is most widely used *dc* link for overhead long distance HVDC transmission systems and also for back-to-back HVDC system. This link has two conductors—one operating with positive polarity and the other with negative polarity with respect to the earthed tower structure. There are two converters of equal voltage rating and connected in series at each end of the *dc* line, as shown in Fig. 14.3. The neutral points, *i.e.* the junction between converters may be grounded at one end or at both the ends. If it is grounded at both ends each pole can operate independently. The rated voltage of a bipolar link is expressed as ± 500 V. Power rating of one pole is about half of bipole power rating. The earth carries only a small out-of-balance current during the normal operation. When the currents in the two conductors are equal, the ground current is zero. During fault or trouble on one of the lines, the other line along with ground return can supply half of the rated load. Thus continuity of supply is maintained. After taking corrective measures, the system is switched over to normal bipolar operation. Thus the reliability of a bipolar line is equal to that of a double circuit 3-phase line although it has only two conductors instead of 6 for 3-phase line. Example of bipolar HVDC link is Ranchi-Delhi single bipolar overhead line of length 810 km, capacity 1,500 MW and operating at ± 500 kV for transmission of bulk power. This line is designed to operate in bipolar mode under normal conditions, both the poles sharing the load equally with negligible current (less than 10 A) in the ground return path. In case of fault on one of the poles, the system automatically switches over to monopolar ground return mode supplying 50% of the rated load. Thereafter the system may be changed over to bipolar metallic return mode when the other conductor is used as return conductor.

3. Homopolar DC Link. A homopolar link has two or more conductors having the same polarity, usually negative, and always operates with ground as the return conductor. In case of a fault on any one of the conductors, the converter equipment can be reconnected so that the healthy conductor can supply power. Such a scheme is very complicated and is preferred to a bipolar link provided continuous ground return does not pose additional problems. Such a scheme may be used for the following :

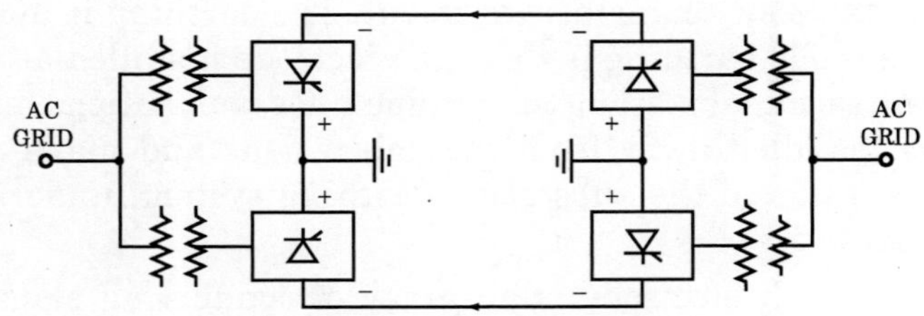

Fig. 14.4. ***Homopolar DC Link***

Two homopolar overhead lines supplying to a common monopolar cable termination.

One overhead transmission tower carrying insulator strings supporting two homolar transmission line conductors.

Thus homolar dc link has limited applications.

14.7. GROUND RETURN

A ground return means the ground or sea water or both as the return conductor. Most *dc* transmission lines use ground return path for reasons of economy and reliability. The monopolar link and homopolar link use the ground return continuously for carrying the return current where as the bipolar link uses ground return for short duration of emergency (when one of the lines is under repair or maintenance). For the same length of transmission the resistance offered by the ground in case of *dc* is much less than that in case of ac transmission. This is because the direct current in the earth, unlike alternating current, does not follow closely the route of the line but spreads over a very large cross-sectional area in both depth and width. The resistance of ground return path in case of *dc* is independent of the length of lines (for long lines) and is essentially the sum of grounding resistances of the earthing electrodes at the two ends of the line.

Besides the advantage of economy, as already mentioned in Art. 14.6(1), there are two more definite advantages of using ground return. The first advantage is that a dc line can be built in two stages—the line can be built as a monopolar line with ground return in the initial stages and may be converted into a bipolar line on a later date when the load requirement increases. Thus a considerable part of the total investment can be deferred until the second stage. The second advantage is the reliability of the system *i.e.* in the event of an outage of one conductor of the bipolar line, it can be operated temporarily at almost half of its rated power, by using the healthy line and the ground, as already mentioned in Art. 14.6(2).

Ground return lines have some problems also. The main problems are designing of ground electrodes for low resistance and low cost of installation and maintenance, location and screening of electrodes so that ground currents cause negligible electrolytic corrosion of buried or immersed metallic structures, prevention of interference of the ground currents with the operation of other services such as ac power transmission, railway signals, ship's compasses etc. and control of surface gradient near the electrodes for the safety of persons, live stock and fish.

14.8. EARTH ELECTRODE AND STATION EARTH

The mid-point of converters, called the *neutral point,* in each station is grounded with a suitable switching arrangement. The earthing is independent of the station earthing. This electrode earthing is through electrode installed at a safe distance (about 5 to 25 km) from the terminal station, major pipe lines, substations and populated areas. The earth electrode is installed away from the substations and major pipe lines so as to avoid the galvanic corrosion of the substation earthing system, underground pipes, buried cables and structures.

Earth electrode is an array of conducting elements placed in the earth or sea which provide a low resistance path between the dc circuit and the earth and which is capable of carrying continuous current for some extended period. The earth electrode at one of the stations acts as an anode* while at the other station it acts as a cathode*. The terminal station is connected to the earth electrode through an insulated cable, called the *earth electrode line.*

Station earth is an array of conducting elements placed in earth at the substation location and which provides connection between the earthed parts of the station equipment and neutral of converter transformers and the earth. Station earth is provided for the protection of equipment from overvoltages and safety of personnel. There is no direct metallic connection between earth electrode and station earth.

Graphite electrodes buried in a pit filled with crushed coke are used. The electrode transfers current to coke which distributes it to earth. This arrangement reduces the loss of material due to corrosion to a negligible value. Iron and graphite cannot be employed as material for earth electrode because iron has high rate of corrosion and graphite when buried in earth causes significant loss of material due to corrosion.

The design aspects of earth electrode are low earth resistance, low current density (not exceeding 1.5 A/m^2) at electrode surface, temperature rise of electrode and surroundings not exceeding 60° C and step voltage on the ground surface above the electrode not exceeding the safe limits.

An earth electrode may be a straight electrode, a ring electrode or a radial star electrode buried horizontally in earth at a depth of about one metre. The radial star arrangement is commonly employed as it uses land area more effectively.

14.9. HVDC TRANSMISSION SYSTEMS

No doubt, 3-phase ac systems have a definite superiority over HVDC systems for generation, transmission, distribution and utilisation of electrical power and are, therefore, universally employed.

However, HVDC transmission is a strong alternative to EHV-AC transmission in some of the applications such as long distance high power transmission by overhead lines, under water transmission,transmission by underground cables, system interconnection by means of overhead lines or underground cables or back-to-back HVDC coupling stations, frequency conversion (50/60 Hz) and hence preferred.

---

* At anode, direct current enters earth from the terminal station while at cathode direct current returns from earth to HVDC circuit.

On the basis of applications, HVDC transmission systems can be classified as follows:

1. Long Distance High Power Transmission. DC overhead lines are cheaper than three-phase transmission lines of the given transmission capacity. With large distances, the saving in cost becomes greater than the additional expenditure on the terminal equipment. HVDC becomes favourable above 800 km, 1,000 MW when the cost of EHV-AC line/substation exceeds that of equivalent HVDC line/substation.

A two-terminal dc link employing overhead lines is used for transmission of large blocks of relatively cheap power from remote sources (say hydroelectric) to load centres situated far away from the generating stations.

2. Underwater Transmission. In case of 3-ϕ transmission, the maximum length of the underwater cable is 25-40 km. This is due to the problem of charging current, and in case of oil-filled cables it is difficult to ensure oil flow over long distances. For longer distances HVDC is the only solution. In the absence of dielectric losses, the dc cables can be of the impregnated type for voltages of at least up to 400 kV, and practically no limit is set to its length.

3. Underground Transmission. Sometimes, the power transmission in congested areas has to be done through underground cables. Wherever the distances involved exceed 40 km, dc transmission is the only feasible solution. DC transmission in such cases offers definite advantages, since the equipment for compensating charging kVA is not required and also there is no reduction in the transmission capacity of cables. Moreover, there is no dielectric heating of the insulation or inductive heating of armouring. Hence the heat dissipation problem in dc cables is far less as compared to the 3-ϕ ac cables. The extra cost of the converters is compensated by saving in cable costs at distances from 50–100 km depending on the conditions in each individual case.

4. DC Link in Parallel With AC Link. For interconnection between two ac systems having their own load frequency control, HVDC links have several advantages over ac links. HVDC links form an asynchronous-tie *i.e.* the two ac systems interconnected by HVDC tie-line need not in synchronism with each other.

HVDC interconnection is superior to EHV-AC interconnection in many respects and is preferred due to its technical superiority. With HVDC interconnection, power flow can be controlled, the frequency disturbances are not transferred, short-circuit levels remain unaffected at both ends, transient stability of an existing ac link can be considerably improved. A dc link has an extremely rapid speed of response (of the order of 50 ms) and, therefore, it can change its transmitted power much more rapidly than, say a turbine generator can adjust its output. Thus the system stability can be significantly improved.

HVDC interconnection can provide a weak tie (of lesser capacity) between strong and a weak ac network. This is difficult with ac interconnection.

HVDC system control can be modified to dampen oscillations in load or power angle δ and thereby improve the stability of an existing ac link.

Most important task of interconnector is transfer of required amount of power in required direction and to assist the interconnected ac network to maintain transient stability. AC interconnectors have severe limitations where as dc interconnectors do not.

5. HVDC Back-To-Back System. Such a system provides an asynchronous interconnection between two adjacent ac systems through converter station and zero (or negligible) length dc line. Such schemes are employed for coupling ac systems of different frequencies (*e.g.* 50/60 Hz) or of different system controls. The back-to-back coupling stations can be located at any suitable location, where two networks meet geographically and exchange of required amount of electrical power is desired.

Vindhyachal HVDC back-to-back link is for exchange of power between Northern and Western regions. Each block of 250 MW is capable of operating independently in either direction and can transfer power in the range of 25 MW to 250 MW depending on the system conditions.

6. Multi-Terminal HVDC Interconnection or Links. This is the new HVDC possibility. Three or more ac networks can be interconnected asynchronously by means of a multi-terminal HVDC link. Power flow from each connected ac network can be controlled suitably. Large power can be transferred.

Almost all dc links set up so far are two-terminal links except *Kingsnorth-London scheme* in which each half of the sending-end power is independently controlled and delivered to two different receiving-end stations. One of the essential requirements in a multi-channel dc scheme is the use of large dc circuit breakers.

14.10. CONVERTER STATION EQUIPMENT

In an HDVC transmission system or HVDC back-to-back coupling station, ac is converted into dc by means of a combination of converter-transformers and converter valves. The conversion of ac into dc is called *rectification.* At the receiving end of the HVDC transmission line, dc is converted back to ac by means of converter valves and converter transformers. The conversion of dc into ac is called the *inversion.*

An HVDC link comprises an ac substation and conversion substation at each end of an HVDC transmission line but in case of back-to-back HVDC link, there is only a conversion substation between two ac substations and there is no dc transmission line.

Thyristor valves are the most important converter station equipment and the other main equipments include converter transformers, dc reactor, harmonic filtering equipment, control equipment and reactive power compensation equipment.

14.10.1. Thyristor Valves

Since, the early 1970's, the thyristor valves are universally employed in all new HVDC schemes while earlier mercury-arc valves were used for conversion/inversion purposes. A thyristor valve has a better operating life and higher consistency of performance than those of a mercury-arc valve. The use of thyristors has resulted in a considerable simplification in the design of conversion substation. The ratings of thyristors have increased remarkably during the last one decade and the current ratings of the thyristors available now are of the magnitudes of line current required for transmission. However, the voltage capability of individual thyristor is quite small in comparison to the line voltage. Thus for obtaining the proper voltage rating it becomes imperative that a large number (typically 5 to 200) of thyristors are connected in series. A thyristor valve is always provided with a number of additional or redundant thyristors in series so that even if some of the thyristors fail, the valve operation remains unaffected. The valves used are of indoor design, air-insulated and air/water cooled. Thyristor assemblies are protected from $\frac{dv}{dt}$ and $\frac{dI}{dt}$ protection. Six-pulse or twelve-pulse converters are used in modern HVDC schemes—12-pulse is more preferred as it contains minimum ripple. There are two 3-phase transformers connecting each 12-pulse converter to the ac bus-bars. One of the transformers is star/star connected and the other is star/delta connected so as to give a phase shift of 30° between the lower and upper 6-pulse bridge of the converter.

The structural design of the converters is very simple. Each thyristor module consists of the thyristors and the converter pole is composed of series connection of several thyristor modules. In addition to the thyristor elements, each module contains damping and over-voltage protection circuitry, voltage dividers and firing control circuits. The modules are mounted in layer form in a box-like structure to form the unit.

Basically, the converters are of 12-pulse operation. However, the use of a single 12-pulse converter per pole as well as multiple valves per pole are available in practice. The former design, though, is very simple in construction and operation but it needs a heavy transformer causing transportation and erection problems. Use of single phase transformers has been attempted in the West to overcome this problem.

Use of four valve units per phase of a 12-pulse converter configuration is common. A typical *Quadruple valve* comprises four valves placed vertically one above the other to form one limb of the converter. Such an arrangement provides the most compact and economical layout of the valves and the valve housing. Three quadruple valves constitute a 12-pulse converter. Valve surge arresters may be directly mounted on the valve structure.

Thyristors produce heat of 30-40 W/cm^2 and, therefore, need effective cooling system. The valves are cooled by air, SF_6 gas, oil or a combination. The cooling system and insulation are independent. The temperature of silicon wafer joint should be held below critical value (90°–125° C) to prevent change in characteristics and damage to thyristors.

The schematic unit of a 12-pulse bridge is shown in Fig. 14.5.

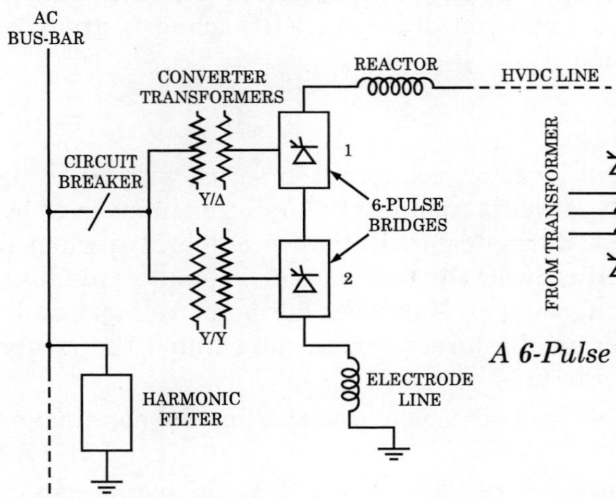

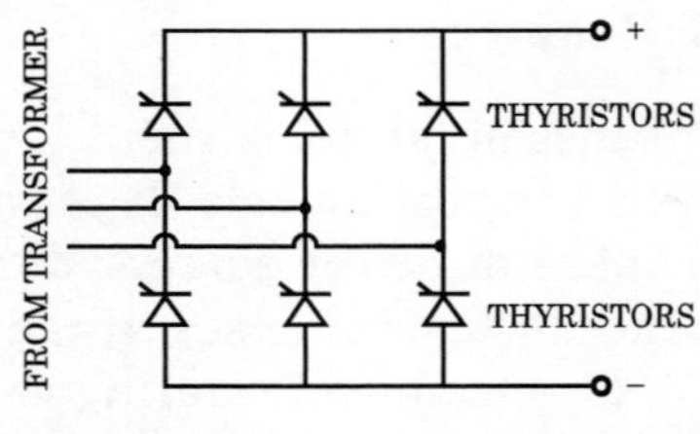

A 6-Pulse Thyristor Converter Bridge

Fig. 14.5. ***Unit 12-Pulse Converter***

In practice two 12-pulse converters per pole (connected in series or in parallel) are used so as to enhance the reliability. In parallel connection, a smoothing reactor is required for each 12-pulse converter and current balancing controls are required to be used. A parallel configuration gives less line loss during the period when full-line currents are not yet established.

The thyristors are required to be triggered in a desired sequence and desired instants. Though conventional firing circuitry is still in use but the most modern method of triggering of thyristors is called the "*optronics*" which employs light guide system. In this system the electrical pulses are converted into light signal pulses by fibre-optic techniques. The light pulses are transmitted through glass-fibre optical cables up to the individual thyristor. Each thyristor is provided with a small separate thyristor which is triggered by the light pulse. Thereby the main thyristor is triggered.

For the protection of thyristors the protections provided are *dv/dt* protection, over-temperature protection, forward voltage firing, forward recovery protection and current sharing protection between two parallel thyristors.

14.10.2. Converter Transformers

Converter transformers are connected between converter valves and the ac bus-bars and their main function is to transform the ac voltage to a suitable value for feeding the converter. The other functions served by the converter transformers is to supply the reactive power to the converter through tap changing; control of fault level by suitable reactance offered by converter transformers and help in harmonic suppression. Twelve-pulse operation is feasible due to transformer connections—by suitable star-star and star-delta connections the required phase shift of 30° for 12-pulse operation is achieved.

Converter transformers are either single phase units, or three-phase units with either two winding type or three winding type. These are specially designed so as to withstand direct voltage stresses. Moreover, in a converter transformer the currents flowing through the winding have high harmonic contents. Thus the eddy current losses are high. Noise due to magnetostriction is higher and may require special design tank to suppress the same. On load tap changing is normally used to reduce the demand of reactive power in steady-state operation. The number of tap-changer operations is higher in HVDC systems and, therefore, proper quality control of components of tap-changers is required. Converter transformers employed in large HVDC schemes are of high MVA ratings (about 300 MVA or even more).

14.10.3. DC Reactor

In HVDC transmission systems the dc reactor, connected in series with each pole of the converter plays an important role. It prevents commutation failures in inverter by limiting the rate of increase of direct current during commutation in one bridge when the direct voltage of another bridge collapses. It reduces the incidence of commutation failures in the inverter during dips in the alternating voltage. It reduces harmonic voltages and currents in the dc line, smoothens the ripple in the direct current and limits the current in the rectifier following a short-circuit in the line.

The dc reactor is usually of air-core and oil-cooled type and has the non-linear magnetic characteristics.

The value of inductance of the reactor is very important. With the increase in inductance the current waveform on the dc side improves but the control response slows down and the resonance frequency falls making the stabilization of current control more difficult. The limitation of dc line short-circuit currents is normally the decisive factor and leads to smoothing reactances of between 0.4 H and 1.0 H.

14.10.4. Harmonic Filtering Equipment

The 3-phase bridge converter employed in HVDC transmission should convert pure ac sinusoidal wave form to pure dc form but in practice the operation of converter generates harmonic currents and harmonic voltages on ac side as well as dc side. These harmonics do not interfere with converter operation but they flow through ac and dc lines and thereby produce several harmful effects such as overheating of capacitors and generators, overvoltages at points in the networks, interference with protective gear, interference with nearby communication systems, radio interference and television interference. These disturbances are not confined to the vicinity of the converter station but spread over the ac network and dc line and surrounding residential areas.

An n-pulse converter generates harmonics predominantly of the order of $nx \pm 1$ on ac side and nx on dc side where x is an integer and n is usually 6 or 12. Thus in case of a 12-pulse valve group operation, the orders of main harmonics are 11th, 13th, 23rd and 25th on the ac side and 12th and 24th on the dc side. The amplitudes of harmonics decrease with increasing order. It is necessary to provide filters on both the ac and dc sides.

Harmonic filters provided on the ac side serve the following purposes :

1. Harmonic voltages and currents in the ac power network are reduced to acceptable levels.

2. All or a part of the reactive power required by the converter is provided, the additional reactive power being supplied by the shunt capacitor banks or by the ac power system.

The ac filters employed in HVDC systems are tuned filter which is sharply tuned to a harmonic frequency, damped filter which, if shunt connected, offers a low impedance over a broad band of frequencies (say 11th and harmonics of higher orders). The adoption of 12-pulse valve group operation has resulted in reduction of harmonic current generation and considerable simplification in ac filters. In a 12-pulse valve group operation only two damped filters are employed while 6-pulse valve group operation requires four tuned filters and one damped filter. Damped filters are simpler to design and have a lower risk of the resonant overvoltages and currents in comparison to tuned filters.

In bipolar operation, under ideal conditions, the induced voltages would be negligible and filters are not required. However, in practice, assuming certain unbalance (in transformer reactances etc.), the induced voltage level would increase. So use of high pass dc filter tuned to 12th harmonic is usually made for a 12-pulse converter circuit.

14.10.5. Control Equipment

The control of firing angle is very important in HVDC systems. This is accomplished by optic fibre based hardware circuitry.

An important difference between the operations of a dc system and an ac system is that the power transmitted over a dc link is always controlled. In most practical HVDC systems the valves of the two converter stations are controlled in such a way that the rectifier end controls the current while the inverter end controls the voltage. These controls allow the link to maintain constant power.

14.10.6. Reactive Power Compensation Equipment

HVDC converter stations need reactive power, particularly because of converter control and commutation process. The reactive power requirement of a rectifier varies as the sin of the firing angle (*i.e.* $\sin \alpha$) whereas the power requirement of the inverter varies as the sin of extinction angle (*i.e.* $\sin \beta$). Moreover, converter transformers consume reactive power. The reactive power requirement is usually in the range of 50 to 60 per cent of the transmitted power. As no reactive power is transmitted over the dc line and maintenance of reactive power balance at the two ends is essential to keep ac voltage within specified limits, installation of reactive power compensation equipment in the HVDC terminal stations becomes imperative. This equipment may consist of ac filters, static shunt capacitors, synchronous condensers, static VAR compensators, etc. A combination of these equipments is also used.

14.11. TRANSIENT OVERVOLTAGES AND INSULATION COORDINATION

Convertor overvoltages are due to transient in converter itself, disturbance on dc side, disturbance on ac side, and transient due to external agents.

Transient overvoltages may occur in the converter itself. This occurs due to convertor's own problems like commutation failure or repetitive misfire. Sudden blocking of inverter firing may also cause overvoltage across the converter.

Disturbance on dc side that causes overvoltages may be due to short-circuit of one pole in a bipolar link and transient overvoltage induced in the healthy pole. Re-energization of thyristor bridges can cause large transient overvoltages.

Severe overvoltages may be caused, on the ac side, due to resonance between the filter and the system for a low order harmonic. The other factors that may cause overvoltages on the ac side, are transformer switching, load rejection, ac system faults etc. Transformer saturation due to harmonic current may also result in overvoltage.

Lightning and switching surges are external agents that may cause transient overvoltages in the HVDC systems. Also the switching surges and lightning surges in the ac system can be transferred to the dc side through the converter transformers.

Zinc oxide surge arresters (ZnO arresters)* are connected in dc yard, valve hall, ac yard filter circuits, etc. HVDC system require a large number of arresters. These are coordinated with surge arresters in ac yard.

The rated voltage switching impulse withstand characteristics, lightning impulse withstand characteristics, test voltages, test procedures, connection surge arresters, etc. form a part of insulation coordination studies.

14.12. CORONA IN HVDC LINES

In dc corona, the charges released from one conductor are carried to the ground or the other conductor because of the opposite polarity. Thus the movement of charged particles is characterised by electric field distribution rather than the surface gradient. In case of ac voltage, the peak value of voltage wave is $\sqrt{2}$ times the rms value but in case of dc there is no such factor and, therefore, in HVDC lines the corona loss is smaller. In case of HVDC lines, corona over insulators and metal surface are negligible. Corona loss is HVDC lines during bad weather conditions does not increase abruptly as in case of ac line. Corona loss in case of dc lines does not increase with the increase in operating voltage as rapidly as in case of ac lines.

In case of ac lines, the ions going away from the conductor surface due to like polarity of charging during a half cycle get attracted towards the conductor during the next half cycle of ac wave. But in case of dc lines, the ions with the same polarity as that of the conductor get a time to go away from the conductor. Thus the corona phenomenon acts differently with ac and dc lines.

However, the corona loss in HVDC system is an important factor and cannot be neglected in the design of the line. The corona behaviour of a monopolar HVDC line is different from that of a bipolar line. This is due to the difference in release of the charge from the vicinity of the conductor surface. Corona loss in the monopolar HVDC line is minimum.

---

* Zinc oxide surge aresters have superior characteristics and high discharge capability.

The critical line voltages V_C for monopolar and bipolar HVDC line are given by Eqs. (14.5) and (14.6) respectively.

For monopolar HVDC line, $V_C = E_C \, r \, m \, \log_e \frac{2h}{r}$ kV (dc) ...(14.5)

and for bipolar HVDC line, $V_C = 2 \, E_C \, r \, m \, \log_e \frac{D}{r} \frac{1}{\sqrt{1+\left(\frac{D}{2h}\right)^2}}$ kV ...(14.6)

where E_C is critical stress in kV/cm (14kV/cm in case of dc), m is the surface irregularity factor (0.8–0.92), h is the height of the conductor above earth in cm, r is conductor radius in cm and D is the pole to pole spacing in cm.

The corona loss in a bipolar HVDC line is given as

$$P_{dc} = P_0 + K\,(E - E_C)^2 \text{ kW/km} \quad ...(14.7)$$

where P_0 is corona loss at critical gradient (0.3–0.5 kW/km), E is actual surface stress in kV/cm and K is a constant.

14.13. PROTECTION ASPECT

The malfunctioning of a converter may occur due to the following reasons.

1. Misfire and back fire in converter valves.
2. Commutation failure, the disturbance that usually occur in inversion operation.
3. Short-circuit in converter station.

Many of the valve faults are of temporary nature and can be eliminated by interrupting conduction temporarily. The faults can be eliminated by the control action used in controlling firing angle.

The faults (short-circuit or earth faults) occurring on ac side of the HVDC system cause dip in the voltage at the terminals of the converter and are required to be cleared in the shortest possible time duration. It is necessary for keeping dc power transfer normal. Inverse relays and differential protections are employed for clearing the faults that are generated in the filter banks.

14.14. HVDC CIRCUIT BREAKING

The process of arc extinction in ac circuit breakers is easier because of natural zero value of alternating currents. But direct current being a steady unidirectional current does not have a natural zero and, therefore, it is difficult to interrupt large direct currents at high voltages. The lack of dc circuit breakers has limited the network of dc lines.

Almost all dc transmission projects till this date are two terminal projects and it is not difficult to interrupt the fault currents. The transient faults can be cleared using valve control methodology while for clearing permanent faults combination of valve control, fault locators and isolating switches be employed.

A vacuum interrupter is employed for breaking direct currents in HVDC systems. An auxiliary device is attached to the vacuum interrupter such that the arc current is opposed by the auxiliary device current (oscillatory discharge of capacitor). However, in practice, the fault current is shaped before being interrupted. Commutation of load current from a saturable reactor to an R-C circuit is made such that after commutation the current is opposed by the high inductance of the reactor. Researches on HVDC circuit breakers are still in progress and use of these circuit breakers is limited to multi-terminal dc systems only.

EXERCISES

1. What is the principle of HVDC system operation ? Discuss the technical and economic advantages of dc systems over ac systems.
2. Discuss the control philosophy of HVDC transmission system.
[U.P. Technical Univ. Elements of Power System 2005-06]
3. Briefly discuss the advantages and the operational problems of HVDC transmission.
4. Enumerate the advantages of using HVDC for transmission of large amounts of power over long distances. Also point out the difficulties and limitations, if any, of such dc systems of transmission.
5. What are the problems in EHVAC transmission ? How are they overcome in HVDC transmission ? **[U.P. Technical Univ. Elements of Power System 2002-03]**
6. Explain the main problems with EHVAC system. How do these limitations avoided in HVDC transmission. **[U.P. Technical Univ. Elements of Power System 2004-05]**
7. Discuss the merits and demerits of HVDC transmission.
[U.P. Technical Univ. Elements of Power System 2005-06]
8. Write merits and demerits of high voltage dc transmission systems.
[U.P. Technical Univ. Elements of Power Sysetm 2006-07]
9. Discuss the advantages and limitations of HVDC systems over High Voltage AC systems.
[U.P. Technical Univ. Elements of Power System 2003-04]
10. Discuss merits and demerits of HVDC transmission over EHVAC transmission.
[U.P. Technical Univ. Elements of Power System 2007-08]
11. In what way HVDC transmission system is superior to HVAC transmission network.
[U.P. Technical Univ. Elements of Power System 2005-06]
12. Discuss types of HVDC links.
[U.P. Technical Univ. Elements of Power System 2004-05; 2007-08]
13. Classify the HVDC links and show them with the help of neat sketch.
[U.P. Technical Univ. Elements of Power System 2005-06]
14. Discuss layouts, relative merits and demerits of different types of HVDC links.
[U.P. Technical Univ. Elements of Power System 2008-09]
15. Give a comparisons of monopolar and bipolar HVDC links.
[U.P. Technical Univ. Elements of Power System 2006-07]
16. What is ground return and why is it employed ? Give the problems associated with the use of ground as the return conductor.
17. Give the applications of HVDC transmission.
18. Write short notes on the following :
    (*i*) HVDC power transmission.
    (*ii*) Special features of HVDC transmission.
    (*iii*) Techno-economic aspects of HVDC transmission.
    (*iv*) EHV DC transmission versus EHV AC transmission.
    (*v*) Terminal components of HVDC transmission system.

SHORT ANSWER TYPE QUESTIONS WITH ANSWERS

Q. 1. Why is voltage regulation better in case of dc transmission ?

Ans. Because of absence of inductance in dc systems.

Q. 2. Why there is no limit on the length of the cable for dc system while it is in case of ac system ?

Ans. Because of absence of charging currents in dc system.

Q. 3. Why HVDC line do not require any reactive power compensation ?

Ans. Because of absence of charging currents and unity power factor operation.

Q. 4. Why dc systems invariably use ground or sea return ?

Ans. Because of economy and reliability.

Q. 5. Why iron and graphite cannot be employed as material for earth electrode ?

Ans. Because iron has high rate of corrosion and graphite when buried in earth causes significant loss of material due to corrosion.

Q. 6. Why dc transmission is economical and preferable over ac transmission for large distances only ?

Ans. Because with larger distances, the saving in cost of dc overhead lines becomes greater than the additional expenditure on terminal equipment.

Q. 7. What is the most modern method of triggering of thyristors ?

Ans. Optronics employing light guide system.

Q. 8. What is normal value of inductance of the dc reactor ?

Ans. The value of inductance of the dc reactor should be such that a resonance of the dc circuit does not occur at power frequency and is in the range of 0.4–1.0 H.

Q. 9. What are the equipments that supply reactive power in HVDC converter stations ?

Ans. AC filters, static shunt capacitors, synchronous condensers, static VAR compensators etc.

Q. 10. Can corona loss be ignored in the design of HVDC transmisson lines ?

Ans. No.

❑❑❑

UNIVERSITY EXAMINATION PAPERS

UNIVERSITY EXAMINATION PAPERS

B.TECH.

FIFTH SEMESTER EXAMINATION, 2004-2005
ELEMENTS OF POWER SYSTEM

Time : 3 Hours *Maximum Marks: 100*

Note : *Attempt ALL questions.*

Q. 1. Attempt *any four* parts of the following : *(5 × 4 = 20)*

(a) A synchronous generator G_1 is feeding power to bus 1 through transformer T_1. Bus 2 receives power through a tie-line L_1 connected between buses 1 and 2 and feeds a set of four motors M_1 to M_4 through transformer T_2. A local load $S_1 = P_1 + jQ_1$ is also fed by synchronous generator at bus 1. Draw single line diagram of the system.

(b) Compare the conductor cost of bipolar dc system and three phase, three wire system.

(c) What are the limitations of Kelvin's law ?

(d) Describe ACSR conductors used in transmission lines.

(e) Show how 'skin effect' increases effective resistance of the conductor.

(f) What is difference between stranded conductors and bundled conductors ?

Q. 2. Attempt *any four* parts of the following : *(5 × 4 = 20)*

(a) Calculate inductance of a 500 m long, 1-ϕ, 2 wire transmission line. Each wire has identical area of cross-section of 10 cm^2 and conductors are separated by a distance of 5 m.

(b) What are the advantages of transposition of conductors ?

(c) Show that presence of earth increases capacitance of transmission lines.

(d) A short three-phase transmission line of parameters R = 0.4 Ω and X = 0.4 Ω is delivering 2,000 kVA to a load at a pf of 0.8 lagging at the receiving end of the line. If the load voltage is 3,000 V, determine voltage regulation and efficiency of the line.

(e) A long transmission line is terminated by an impedance equal to its characteristic impedance. Show that,

$$\frac{V_R}{V_S} = e^{-yl}$$

Where, V_R = Receiving end voltage
V_S = Sending end voltage
Y = Propagation constant
l = Length of line

(*f*) Describe 'Ferranti effect' in long transmission lines. How can it be compensated ?

Q. 3. Attempt *any two* parts of the following : (*10 × 2 = 20*)

(*a*) (*i*) Explain the role of bundled conductors in reducing corona loss.

(*ii*) A certain 3-ϕ equilaterally spaced transmission line has a total corona loss of 53 kW at 106 kV and a loss of 98 kW at 110.9 kV. What is the disruptive critical voltage between lines ? What is the corona loss at 113 kV ?

(*b*) Describe different types of insulators used in transmission lines with their applications.

(*c*) What do you mean by 'string efficiency' ? Describe different methods of improving string efficiency.

Q. 4. Attempt *any two* parts of the following : (*10 × 2 = 20*)

(*a*) A transmission line conductor is having a diameter of 20 mm and weighs 1.0 kg/m. The span is 280 m. The wind pressure is 40 kg/m^2 of projected area with ice coating of 10 mm. The ultimate strength of conductor is 10,000 kg. Calculate the maximum sag if the factor of safety is 2 and ice weighs 910 kg/m^3.

(*b*) Develop the expression of insulation resistance of a cable. Calculate the insulation resistance for a 5 km length of a 1-core cable. Specific resistance of insulation (impregnated paper) is 5 × 10^{14} ohm-cm, insulation thickness is 1 cm and radius of conductor is 2.5 cm.

(*c*) What do you mean by grading of cables ? Discuss different methods of cable grading.

Q. 5. Attempt *any two* parts of the following : (*10 × 2 = 20*)

(*a*) (*i*) What is the necessity of neutral grounding ?

(*ii*) Write a note on 'earthing transformer'.

(*b*) Describe different types of HVDC links.

(*c*) How FACTS controllers make AC transmission flexible ? Describe working principle of TCSC and show how it enhances power transfer capability of lines.

B.TECH.

FIFTH SEMESTER EXAMINATION, 2005-2006

ELEMENTS OF POWER SYSTEM

Time : 3 Hours *Maximum Marks: 100*

Note : (*i*) *Attempt ALL questions.*

(*ii*) *Be precise in your answer.*

Q. 1. Attempt *any four* parts of the following : (*5 × 4 = 20*)

(*a*) Draw the schematic diagram of power system starting from boiler to the Grid structure.

(*b*) Identify the role of circuit breaker in protection of power network.

(*c*) What criterion is adopted for selection of transmission line operating voltage in case of long transmission lines ?

(*d*) In what way HVDC transmission system is superior to HVAC transmission network?

(*e*) Mention the contribution of bundle conductors in performance of transmission line.

(*f*) Explain the concept of bus bar in power network alongwith suitable diagram.

Q. 2. Attempt *any four* parts of the following : (*5 × 4 = 20*)

(*a*) Identify the criterion for classification of transmission line. Which transmission line representation is close to actual system dynamics ?

(*b*) Derive an expression for inductance of a single phase transmission line.

(*c*) Explain the concept of regulation in long transmission line. If regulation is very high then what can be seen adversely in power network ?

(*d*) A 50 Hz, 3-phase transmission line of length 30 km has a total series impedance of $(40 + j125)$ ohms and shunt admittance of 10^{-3} mho. The load is 50 MW at 220 kV with 0.8 lagging power factor. Find the sending end voltage. (Use Nominal-π representation).

(*e*) Derive the A, B, C, D constants for the transmission line represented by nominal-T section.

(*f*) A 3-phase, 50 Hz transmission line has flat horizontal configuration with 3.5 m between adjacent conductors. The conductors are hard drawn seven strand copper wire (outside conductor dia = 1.05 cm). The voltage of the line is 110 kV. Find the capacitance to neutral and the charging current per km.

Q. 3. Attempt *any two* parts of the following : (*10 × 2 = 20*)

(*a*) Explain the situation under which corona phenomenon starts in high voltage power transmission lines. Identify the factors that affect the corona alongwith reasoning.

(*b*) Define critical disruptive voltage. A 3-phase 220 kV, 50 Hz transmission line consists of 1.2 cm radius conductors spaced 2 m at the corners of an equilateral triangle. Calculate the critical disruptive voltage between the lines. Irregularly factor = 0.96; Temperature = 20°C. Barometric pressure = 72.2 cm of mercury. Dielectric strength of air = 21.1 kV (rms)/cm.

(*c*) Define string efficiency.

A string of 6 insulator units has a self-capacitance equal to 10 times the pin-to-earth capacitance. Find (*i*) the voltage distribution across various units as a percentage of total voltage across the string, (*ii*) the string efficiency.

Q. 4. Attempt *any two* parts of the following : (*10 × 2 = 20*)

(*a*) Mention criterion of the cable classification alongwith justification. Draw the Extra High Tension Cable schematic diagram and explain the various terminology.

(*b*) A 33 kV, 3-phase, 50 Hz underground line, 3.4 km long, uses three single core cables. Each cable has a core diameter of 2.5 cm and the radial thickness of insulation is 0.6 cm. The relative permittivity of the dielectric is 3.1. Find (*i*) maximum stress and (*ii*) total charging kVAR.

(*c*) Derive the expression of vertical sag in situations where the supports are at different levels. Consider the effect of wind and ice also.

Q. 5. Attempt *any two* parts of the following : *(10 × 2 = 20)*

(*a*) Identify the criterion of grounding at 400 kV substation. If the total safety is to be ensured what will be the layout of grounding ? Explain alongwith diagram.

(*b*) (*i*) Discuss the control philosophy of HVDC transmission system.

(*ii*) Discuss the merits and demerits of HVDC transmission.

(*c*) (*i*) Identify the utility of FACTS devices in power network.

(*ii*) Classify the HVDC links and show them with the help of neat sketch.

B.TECH.

FIFTH SEMESTER EXAMINATION, 2006-2007

ELEMENTS OF POWER SYSTEM

Time : 3 Hours *Maximum Marks: 100*

Note : (*i*) *Attempt ALL questions.*

(*ii*) *All questions carry equal marks.*

(*iii*) *In case of numerical problems assume data wherever not provided.*

(*iv*) *Be precise in your answer.*

Q. 1. Attempt *any four* parts of the following : *(5 × 4 = 20)*

(*a*) Describe in brief the different power system components.

(*b*) Compare the different supply systems with their choice of transmission voltage.

(*c*) Differentiate between skin effect and proximity effect.

(*d*) What is the percentage saving in feeder copper if the line voltage in a 2-wire dc system be raised from 220 volts to 400 volts for the same power transmitted over same distance and having the same power loss ?

(*e*) State Kelvin's law for obtaining the size of the conductor for transmission. Discuss the limitations.

(*f*) Explain different types of conductors used for transmission lines. Give their properties.

Q. 2. Attempt *any four* parts of the following : *(5 × 4 = 20)*

(*a*) Calculate the loop inductance per km of a single phase line comprising of 2 parallel conductors 1 metre apart and 1 cm in diameter, when the material of conductor is :

(*i*) Copper and

(*ii*) Steel of relative permeability 50. Prove the formula used.

(*b*) What is a bundled conductor and why is it used ?

(*c*) Calculate the capacitance of a 100 km long, 3 phase, 50 Hz overhead transmission line consisting of 3 conductors each of diameter 2 cm and spaced 2.5 m at the corners of equilateral triangle.

(*d*) A 3-phase line, 3 km long delivers, 3000 kW at a power factor of 0.8 lag to load. If the voltage at supply end is 11 kV, determine the voltage at the load end and the efficiency of transmission. The resistance and reactance per km of each conductor are 0.4 Ω and 0.8 Ω and respectively.

(*e*) Explain surge impedance loading with respect to an overhead transmission line. Show that $Z_c = \sqrt{Z_{oc} Z_{sc}}$.

(*f*) A 3 phase overhead line has resistance and reactance per phase of 5 Ω and 25 Ω. The load at the receiving end is 15 MW, 33 kV, 0.8 pf lagging. Estimate the capacity of the compensation equipment needed to supply this load with a sending end voltage of 33 kV.

Q. 3. Attempt *any two* parts of the following : (*10* × *2* = *20*)

(*a*) Explain the methods of equalizing the potential across the string insulator. A string of eight suspension insulators is to be fitted with a grading ring. If the pin to earth capacitance are equal to C, find the values of line to pin capacitances that would give a uniform voltage distribution over the string.

(*b*) Describe pin type, suspension type and strain type insulators.

(*c*) Explain corona loss. How is disruptive critical voltage estimated ? What are the factors that effect the corona loss ? Give advantages and disadvantages of it.

Q. 4. Attempt *any two* parts of the following : (*10* × *2* = *20*)

(*a*) Show that the most economical size of conductor in a single core cable is obtained when radius of cable sheath (R) equals *e*. *r* where *e* is the base of natural log and *r* is the radius of conductor. Explain dielectric loss and heating of a cable.

(*b*) Calculate the kVA taken by a 10 km long, 3 phase 3 core cable, if the capacitance measured between any two cores is 0.3 μF/km when it is connected to 10 kV, 50 Hz bus-bar.

(*c*) An overhead line has a span of 200 metres, the line conductor weighs 0.7 kg per metre. Calculate the maximum sag if the maximum allowable tension in the line is 1,400 kg. Prove formula used.

Q. 5. Attempt *any four* parts of the following : (*5* × *4* = *20*)

(*a*) Explain in brief the role of flexible ac transmission systems.

(*b*) What are the various FACTS controllers ? Explain the operating principle of any one of them.

(*c*) Write merits and demerits of high voltage dc transmission systems.

(*d*) What is an earthing transformer ? Explain the same in brief.

(*e*) Explain the role of earthing transformer in neutral grounding.

(*f*) What are the various methods of neutral grounding ?

B.TECH.

FIFTH SEMESTER EXAMINATION, 2007-2008

ELEMENTS OF POWER SYSTEM

Time : 3 Hours *Maximum Marks: 100*

Note : *Attempt ALL questions.*

Q. 1. Attempt *any two* parts of the following : *(2 × 10 = 20)*

(*a*) Show that at higher voltage and at higher power factor, the efficiency of transmision line will be increased. What are the limiting factors of high voltage transmission lines ?

(*b*) Describe the main limitations of Kelvin's law. The cost of a 3-phase overhead transmission line having cross-sectional area A cm^2 is Rs. (500 + 2,000 A) per km. Calculate the most economical current density for the conductor if the rate of interest and depreciation is 12% per annum. The cost of energy wasted is Rs 0.05/kWh.

The resistance of conductor is $\frac{0.17}{A}$ ohm/km. Take load factor for losses = 12%.

(*c*) Differentiate the followings :

(1) Skin effect and proximity effect.

(2) Circuit breaker and isolator.

(3) Composite and stranded conductors.

Q. 2. Attempt *any two* parts of the following : *(2 × 10 = 20)*

(*a*) A 50 Hz, 3-phase line (transposed) of one ACSR moose conductor (over all dia. = 31.8 mm) per phase has flat horizontal spacing of 10 m between adjacent conductors. Compare the inductive and capacitive reactances in ohm per kilometre per phase of this line with that of a line (transposed) using a three-conductor bundle of ACSR hynx conductors (each having overall dia. = 19.6 mm) having 10 m spacing measured from the centre of the bundles. The bundle conductors in each phase are arranged in an equilateral triangle formation with spacing between conductors in the bundle as 40 cm.

(*b*) The ABCD constants of a lossless three phase, 400 kV transmission line are,

$$A = D = 0.90 + j0$$

$$B = j190$$

$$C = j0.001$$

Obtain the sending end quantities and the voltage regulation when line delivers a load of 750 MVA at 0.8 lagging power factor at 400 kV.

To improve the line performance, series capacitors, each having reactance $X_c = 50\ \Omega$, are installed at both ends in each phase of the transmission line. Determine the compensated ABCD constants.

(c) Starting with flux linkage concept, find the inductance of 3-phase double circuit line with unsymmetrical spacing.

Q. 3. Attempt *any two* parts of the following : (2 × 10 = 20)

(a) What is corona ? Show that the maximum critical disruptive voltage occurs when the radius of conductor is *d/e* where d is the distance between conductors.

(b) A string insulator has 4 units and each unit is rated for 12 kV. Find the maximum line voltage on which it can be operated safely. The mutual capacitance of unit is 10 times the capacitance between the pin to earth. Find the string efficiency.

(c) A single phase 50 Hz overhead power line (conductors *a* and *b*) is symmetrically supported on a horizontal cross-arm. A telephone line (conductor *c* and *d*) is also symmetrically supported on a horizontal cross-arm directly below the power line. Show that the mutual inductance per unit length between circuits *a-b* and *c-d* is given by

$$4 \times 10^{-7} \ln \sqrt{\frac{D_{ad}D_{bc}}{D_{ac}D_{bd}}} \text{ H/m}$$

where for examples, D_{ad} denotes the distance in metres between conductors *a* and *d*.

Q. 4. Attempt *any two* parts of the following : (2 × 10 = 20)

(a) Derive the expression for the lowest point of the catenary curve if the supports are at different levels. Given that weight of conductor = 0.35 kg/m, Maximum allowable strength = 800 kg, span length = 160 m and safety factor = 2, find the minimum point of catenary if supports are at 70 m and 65 m.

(b) Prove that the ratio of gradient with and without intersheath will be (2/1 + α), when there is only one sheath, where $\alpha = \frac{r_1}{r} = \frac{R}{r_1}$. The core radius is r and outer sheath radius is R and intersheath is at r_1 distance from centre.

(c) What is stringing chart ? How is it plotted for different temperatures ?

Q. 5. Attempt *any four* parts of the following : (4 × 5 = 20)

(a) What are the various advantages of neutral grounding ?

(b) What is use of earthing transformer ?

(c) Describe the design considerations of EHV transmission lines.

(d) What are advantages of bundled conductors ?

(e) What are disadvantages of HVAC transmission ?

(f) Discuss various types of HVDC links.

B.TECH.

FIFTH SEMESTER EXAMINATION, 2008-2009

ELEMENTS OF POWER SYSTEM

Time : 3 Hours *Maximum Marks: 100*

Note : *Attempt ALL questions.*

Q. 1. Attempt *any four* parts of the following : *(5 × 4 = 20)*

(*a*) Draw single line-diagram of a three-bus system having generator G_1 connected to bus-1 through transformer T_1, generator G_2 connected to bus-2 through transformer T_2, four synchronous motors M_1 to M_4 connected to bus-3 through transformer T_3, transmission lines TL_1, TL_2 and TL_3 connected between buses 1-2, 2-3 and 1-3 respectively.

(*b*) What is the difference between isolator and circuit breaker ?

(*c*) Compare the conductor requirement for 1-ϕ ac system with 3-ϕ ac system. Assume same maximum potential differnece between conductor and earth.

(*d*) The cost of a 3-ϕ overhead line having cross-section area '*a*' cm^2 is Rs. (500 + 2,000*a*) per kilometre. Calculate the most economical current density for the conductor if the rate of interest and depreciation is 12% per annum and the cost of energy wasted is 5 paise per kWh. The resistivity of conductor may be taken as 1.7 μΩ-cm. Take load factor of 12%.

(*e*) What advantages you get by using :

(*i*) Stranded conductors.

(*ii*) Bundled conductors.

(*f*) Explain with justification whether the following statement is TRUE or FALSE "Skin effect increases effective resistance and reactance of conductor."

Q. 2. Attempt *any two* parts of the following : *(10 × 2 = 20)*

(*a*) Determine capacitance of the arrangement shown in the figure when (i) the effect of earth is neglected and (*ii*) the effect of earth is considered. The heights of conductors from ground are 10 metres and radius of each of the conductor is 2 cm.

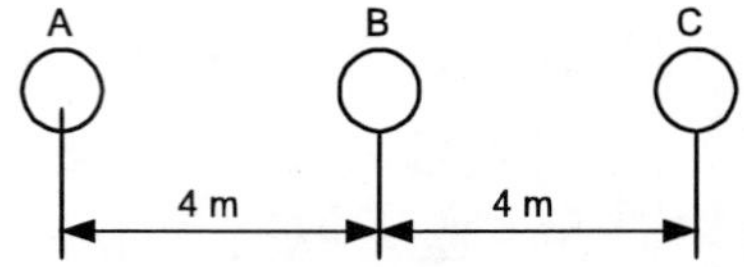

(*b*) The A, B, C, D constants of a three-phase transmission line are A = D = 0.936 + *j*0.016, B = 33.5 + *j*138 Ω and C = (–0.9280 + *j*901.223) × 10^{-6} mho.

The load at the receiving end is 40 MW at 200 kV with pf of 0.86 lagging. Find the magnitude of the sending end voltage, current, power, line frequency and the voltage regulation. Assume that the magnitude of the sending end voltage remains constant.

(*c*) Obtain equivalent-π model of a long transmission line.

Q. 3. Attempt *any two* parts of the following : (*10 × 2 = 20*)

(*a*) A conductor with 2.5 cm dia is passed centrally through a porcelain bushing having $\varepsilon_r = 4$ and internal and external diameters of 3 cm and 9 cm respectively. The voltage between the conductor and an earthed clamp surrounding the porcelain is 20 kV rms. Determine whether corona will be present in the air-space round the conductor.

(*b*) Discuss advantages and disadvantages of corona. How can corona loss be minimized ?

(*c*) Determine the maximum voltage that the string of suspension insulators in the figure can withstand if the maximum voltage per unit is 17.5 kV.

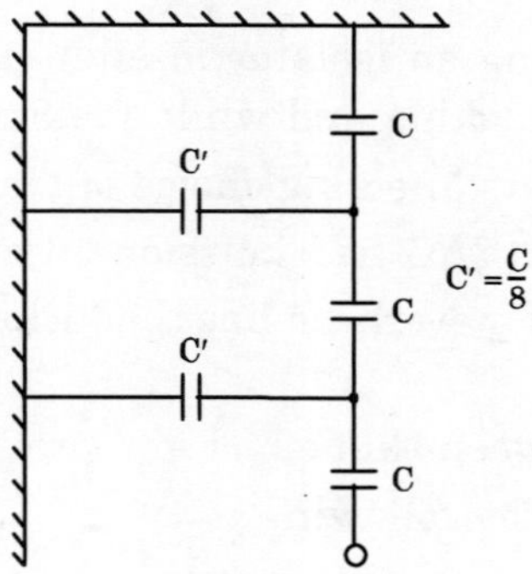

Q. 4. Attempt *any two* parts of the following : (*10 × 2 = 20*)

(*a*) Derive expressions for sag and tension in a power conductor strung between two supports at equal heights taking into account the wind and ice loading also.

(*b*) Explain the necessity of a stringing chart for a transmission line and show how such a chart can be constructed.

(*c*) What is the purpose of using intersheaths in a cable ? Show that in a cable using two intersheaths the maximum stress in the (homogeneous) dielectric is reduced to $\dfrac{3}{(1+\alpha+\alpha^2)}$ of maximum stress without intersheaths where $\alpha = \dfrac{d_1}{d} = \dfrac{d_2}{d_1} = \dfrac{D}{d_2}$

(D and d are the internal sheath diameter and core diameter and d_1 and d_2 are the internal diameters of intersheaths 1 and 2 respectively

Q. 5. Attempt *any two* parts of the following : (*10 × 2 = 20*)

(*a*) Discuss following methods of neutral grounding :

(*i*) Peterson coil grounding.

(*ii*) Resonant grounding.

(*iii*) Solid grounding.

(*b*) Discuss the points which are considered while designing a transmission line.

(*c*) Discuss layouts, relative merits and demerits of different types of HVDC links.

B.TECH.

FIFTH SEMESTER EXAMINATION, 2009-2010

ELEMENTS OF POWER SYSTEM

Time : 3 Hours *Maximum Marks: 100*

Note : *Attempt ALL questions.*

Q. 1. Attempt *any four* parts of the following : *(5 × 4 = 20)*

(*a*) What do you understand by single line diagram of a power system ? What is the need of this diagram ?

(*b*) Justify the need of using an isolator in addition to a circuit breaker ? How are they operated while breaking and while making a circuit ?

(*c*) Discuss about the factors affecting choice of transmission voltage level.

(*d*) Compare HVDC with HVAC transmission on at least 5 major grounds.

(*e*) Which material is used for overhead line conductor ? Discuss the reasons for selecting that material.

(*f*) Explain difference between skin effect and proximity effect.

Q. 2. Attempt *any two* parts of the following : *(10 × 2 = 20)*

(*a*) Starting from first principles, derive the expression for inductance of a 3 phase unsymmetrical spaced transposed transmission line.

(*b*) Explain briefly about the following :

(*i*) GMD.

(*ii*) Bundled conductors.

(*iii*) Double circuit lines.

(*iv*) Transposition.

(*v*) SIL.

(*c*) Derive the expressions for sending end voltage and current in terms of receiving end voltage and current for long transmission lines. What do you understand by the "Characteristic Impedance" and "Propagation Constant" in it ?

Q. 3. Attempt *any two* parts of the following : *(10 × 2 = 20)*

(*a*) Write and explain the expression for power loss due to corona. What factors affect the corona loss ?

(*b*) In a string insulator having 5 units, the self capacitance of insulator units is 6 times to that between a unit and earth. Obtain the potential distribution across each unit in the string as percentage of potential of conductor with respect to earth. Also find the string efficiency.

(*c*) Describe about the following in brief :

(*i*) String insulators.

(*ii*) Pin insulators.

(*iii*) Strain insulators.

(*iv*) Shackle insulators.

(*v*) Methods of reducing corona.

Q. 4. Attempt *any four* parts of the following : (*5* × *4* = *20*)

(*a*) What is the use of a sag template ? Describe it in detail.

(*b*) Derive the expression for sag when the conductor is supported by supports at equal height.

(*c*) Why is capacitance grading required in cables ? How is it done ?

(*d*) Discuss in brief about the types of cables.

(*e*) Why do the vibrations get generated in conductors ? How are they damped ?

(*f*) Derive the formula for insulation resistance of a cable. Calculate insulation resistance of 5 km length of single core cable whose insulation resistance is 5×10^{14} ohm-cm, insulation thickness is 1 cm and radius of conductor is 1.25 cm.

Q. 5. Attempt *any two* parts of the following : (*10* × *2* = *20*)

(*a*) What is the need of grounding the neutral ? Describe briefly the various grounding techniques.

(*b*) Describe the various conductor configurations and choice of number of circuits for EHV transmission lines.

(*c*) Answer the following :

(*i*) Resonant grounding.

(*ii*) Relative merits and demerits of EHV transmission system.

❑❑❑

B.TECH.

(SEMESTER V) ODD SEMESTER THEORY EXAMINATION, 2010-2011

ELEMENTS OF POWER SYSTEM

Time : 3 Hours *Maximum Marks: 100*

Note : *Attempt ALL questions.*

Q. 1. Attempt *any two* parts : *(10 × 2 = 20)*

(*a*) What are the factors which are responsible for the presence of skin effect in A.C. Transmission ? Suggest some measures to reduce it.

(*b*) Bring out the relative advantages and disadvantages of overhead and underground systems.

(*c*) Derive expressions for calculating the economic voltage and economic conductor cross section of a line. Comment on the results.

ANSWERS

1. (*a*) Refer to Art. 3.5.

(*b*) Refer to Art. 2.1.2.

(*c*) Refer to Arts. 2.7 and 3.7 and 3.7.1.

Q. 2. Attempt *any two* parts : *(10 × 2 = 20)*

(*a*) Describe the methods of images for the calculation of capacitance of a line. Derive the expression for the inductance per meter length of a 3-phase composite conductor line.

(*b*) A 3-phase 132 kV, 100 km, 50 Hz, single circuit line has horizontal spacing with 3.5 m between adjacent conductors. The conductors diameter is 1.2 cm. Find the line capacitance per phase and the charging current per phase

(*i*) without effect of earth.

(*ii*) with effect of earth.

(*c*) A 3-phase, 50 Hz line has a resistance of 8.0 ohm and a reactance of 11 ohm. It supplies a 0.8 lagging pf load at a voltage of 11 kV. Find the load at which the source-end voltage equals the load-end voltage when a 25 μF capacitor is connected across the load in each phase.

ANSWERS/SOLUTION

2. (*a*) Refer to Arts. 4.11.1 and 4.2.3.

(*b*)

Spacing, $d_1 = 3.5$ m

Spacing, $d_2 = 3.5$ m

Spacing, $d_3 = 7$ m

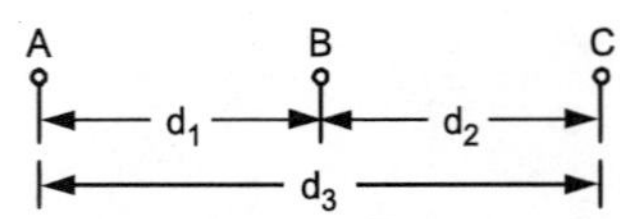

Fig. 1

Conductor radius, $r = \frac{1.2}{2} = 0.6 \text{ cm} = 0.006 \text{ m}$

(*i*) Capacitance per phase (neglecting earth effect)

$$C_N = \frac{2\pi\varepsilon_0}{\log_e \frac{\sqrt[3]{d_1 d_2 d_3}}{r}} \times l = \frac{2\pi \times 8.854 \times 10^{-12} \times 100 \times 1{,}000}{\log_e \frac{\sqrt[3]{3.5 \times 3.5 \times 7}}{0.006}} = 0.843\ \mu\text{F} \quad \textbf{Ans.}$$

Charging current per phase, $I_C = 2\pi f\, C\, V_{ph} = 2\pi \times 50 \times 0.843 \times 10^{-6} \times \frac{132{,}000}{\sqrt{3}}$

$$= 20.2\ \text{A} \quad \textbf{Ans.}$$

(*ii*) Assume height of conductors from ground to be 10 m

From Fig. 2

$$h_1 = h_2 = h_3 = 20\ \text{m}$$

$$h_{12} = h_{23} = \sqrt{20^2 + 3.5^2} = 20.3\ \text{m}$$

$$h_{31} = \sqrt{20^2 + 7^2} = 21.19\ \text{m}$$

$$r = 0.006\ \text{m}$$

$$D_{eq} = \sqrt[3]{d_1 d_2 d_3} = \sqrt[3]{3.5 \times 3.5 \times 7} = 4.41\ \text{m}$$

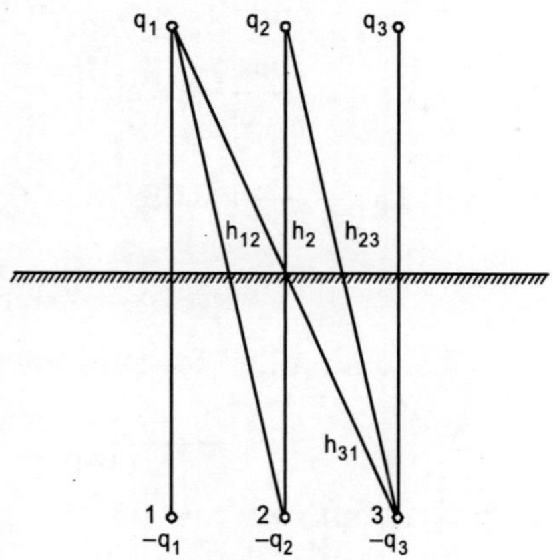

Fig. 2

Capacitance per phase with effect of earth,

$$C_N = \frac{2\pi\varepsilon_0}{\log_e \frac{D_{eq}}{r} - \log_e \sqrt[3]{\frac{h_{12} h_{23} h_{31}}{h_1\, h_2\, h_3}}} \times l$$

$$= \frac{2\pi \times 8.854 \times 10^{-12} \times 100 \times 1{,}000}{\log_e \frac{4.41}{0.006} - \log_e \sqrt[3]{\frac{20.3 \times 20.3 \times 21.19}{20 \times 20 \times 20}}} = 0.846\ \mu\text{F} \quad \textbf{Ans.}$$

Charging current per phase,

$$I_C = 2\pi f\, CV_{ph} = 2\pi \times 50 \times 0.846 \times 10^{-6} \times \frac{1{,}32{,}000}{\sqrt{3}} = 20.255\ \text{A} \quad \textbf{Ans.}$$

(*c*) Line parameters are :

Resistance, $R = 8\ \Omega$

Reactance, $X = 11\ \Omega$

Current drawn by 25 μF capacitor connected across the load.

$$I_C = 2\pi f\, CV_{RP} = 2\pi \times 50 \times 25 \times 10^{-6} \times \frac{11{,}000}{\sqrt{3}} = 49.88\ \text{A leading } V_{RP} \text{ by } 90°$$

Load power factor $\cos \phi_R = 0.8$ lagging

$$\phi_R = \text{Cos}^{-1}\ 0.8 = -36.87°$$

Let the load and line currents be I_R and I_L respectively

So $\qquad \mathbf{I_R} = I_R \angle -36.87°\ \text{A}$

Voltage drop given as $I_L R \cos\theta + I_L X \sin\theta$ must be equal to zero because sending-end and receiving-end voltages are equal.

So $I_L R \cos\theta + I_L X \sin\theta = 0$

$$\text{or } \theta = \text{Tan}^{-1}\frac{R}{X} = \text{Tan}^{-1}\frac{8}{11} = 36° \text{ (leading)}$$

So line current, $I_L = I_L \angle 36°$ A

From phasor diagram shown in Fig. 3

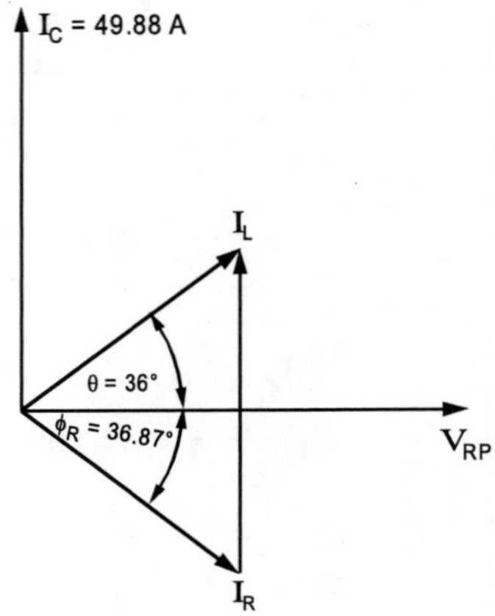

$$I_L \cos\theta = I_R \cos\phi_R$$

$$\text{or } I_L = I_R \frac{\cos\phi_R}{\cos\theta}$$

$$\text{and } I_L \sin\theta = I_C - I_R \sin\phi$$

$$\text{or } I_R \frac{\cos\phi_R}{\cos\theta} \times \sin\theta = 49.88 - I_R \times 0.6$$

$$\text{or } I_R\left(\frac{0.8}{0.809} \times 0.58778\right) + 0.6 I_R = 49.88$$

$$\text{or } I_R = \frac{49.88}{1.1812} = 42.227 \text{ A}$$

Load supplied for zero volt drop in line

$$= \frac{\sqrt{3}\, V_{RL}\, I_R \cos\phi_R}{1{,}000} = \frac{\sqrt{3} \times 11{,}000 \times 42.227 \times 0.8}{1{,}000} = 643.6 \text{ kW} \quad \textbf{Ans.}$$

Q. 3. Attempt *any two* parts : *(10 × 2 = 20)*

(*a*) Draw a phasor diagram of a nominal π transmission line finding out its A, B, C, D constants.

(*b*) Define the following :

(*i*) Critical disruptive voltage.

(*ii*) Flashover and puncture voltage

(*iii*) Visual critical disruptive voltage.

(*c*) Compare all the types of insulators. A string insulator has 4 units and each unit of the string is having capacitance "C" the pin to earth capacitance is C/10 find the voltages across each unit of the string and string efficiency.

ANSWERS/SOLUTION

3. (*a*) Please refer to Arts. 5.5.3 and 5.9 (*iv*).

(*b*) Refer to Arts. 7.5, 7.6 and 7.7.

(*c*) Refer to Art. 9.4 and Example 9.6.

Q. 4. Attempt *any two* parts : *(10 × 2 = 20)*

(*a*) An overhead stranded galvanized steel conductor has a 183 m span. The conductor has 37 strands each of 0.259 cm diameter. The weight of conductor is 7.15 N/m and the breaking strength is 67,700 N. The factor of safety should be 2.5. Calculate the sag under ice and wind condition if the radial thickness of ice is 0.96 cm and the wind load is 382 N/m^2 of projected area (coated with ice). The weight of ice is 8,920 N/m^3.

(*b*) Explain the concept of grading in the underground cable. Derive the expression for dielectric loss and dielectric power factor.

(*c*) A 2-km long 3-crore, 3-phase cable has capacitance 0.5 μF/km between two conductors bunched with sheath and the third conductor the capacitance between the conductors is also measured when bunched together and the sheath and found to be 0.75 μF/km. Determine :

(*i*) Capacitance between phases.

(*ii*) Capacitance between conductor and sheath.

(*iii*) Effective per phase capacitance.

(*iv*) Capacitance between two conductors connecting a third conductor to the sheath.

(*v*) Charging current if the supply voltage is 11 kV, 50 Hz.

ANSWERS/SOLUTION

4. (*a*) Stranded conductors usually have a central wire around which there are successive layers of 6,18,36 wires for n layers, the total number of individual wires is $3n(n + 1) + 1$. If the diameter of each strand is d then the diameter of the stranded conductor will be $(2n + 1)d$.

$$\text{Thus } 3n(n+1)+1 = 37$$
$$\text{or } 3n^2 + 3n - 36 = 0$$
$$\text{or } n^2 + n - 12 = 0$$
$$\text{or } (n+4)(n-3) = 0$$
$$\text{or } n = -4 \text{ or } 3$$
$$\text{Rejecting negative value, } n = 3$$

The number of strands in different layers are 1, 6, 12, 18, Total 37

$$\text{Overall diameter of conductor, D} = (2n+1)d = (2 \times 3 + 1)d$$
$$= 7 \times d = 7 \times 0.259 = 1.813 \text{ cm}$$

$$\text{Ultimate strength} = 67{,}700 \text{ N}$$

$$\text{Working strength, T} = \frac{67{,}700}{2.5} = 27{,}080 \text{ N}$$

$$\text{Weight of conductor per metre, } N_C = 7.15 \text{ N/m}$$

Projected area of conductors per metre length to wind pressure

$$= d \times 1 \times 100 = 1.813 \times 100$$
$$= 181.3 \text{ cm}^2 = 181.3 \times 10^{-4} \text{ m}^2$$

$$\text{Wind load per metre, } w_w = \text{Wind pressure} \times \text{projected area}$$
$$= 382 \times 181.3 \times 10^{-4} = 6.92566 \text{ N}$$

$$\text{Weight of ice coating/m, } w_i = 8{,}920 \times \pi r(\text{D} + r)$$
$$= 8{,}920 \times \pi \times 0.96 \times 10^{-2} \times (1.813 + 0.96) \times 10^{-2}$$
$$= 7.46 \text{ N}$$

Resultant weight per metre length of conductor

$$w_r = \sqrt{w_c + w_i)^2 + w_w^2}$$

$$= \sqrt{(7.15 + 7.46)^2 + (6.92566)^2} = 16.1684 \text{ N}$$

$$\text{Total sag, } S = \frac{w_r L^2}{8T} = \frac{16.1684 \times 183^2}{8 \times 27{,}080} = 2.5 \text{ m}$$

$$\text{Vertical sag} = S \times \frac{w_c + w_i}{w_r} = 2.5 \times \frac{7.15 + 7.46}{16.1684} = 2.259 \text{ m} \quad \textbf{Ans.}$$

(*b*) Refer to Arts. 11.11 and 11.13.

(*c*) Capacitance between all three conductors joined and sheath [Fig. 11.20 (*a*)] gives $3C_S$

$$\text{So, } 3C_S = 0.75 \ \mu\text{F/km} \times 2 \text{ km} = 1.50 \ \mu\text{F}$$

(*ii*) or Capacitance between conductor and sheath, $C_S = \frac{1.50}{3} = 0.5 \ \mu\text{F}$ **Ans.**

Capacitance between two conductors bunched with sheath and the third conductor [Fig. 11.20(*c*)],

$$2C_C + C_S = 0.5 \times 2 = 1 \ \mu\text{F}$$

(*i*) or Capacitance between phases, $C_C = \frac{1 - C_S}{2} = \frac{1 - 0.5}{2} = 0.25 \ \mu\text{F}$ **Ans.**

(*iii*) Effective per phase capacitance, $C_N = 3C_C + C_S = 3 \times 0.25 + 0.5 = 1.25 \ \mu\text{F}$ **Ans.**

(*iv*) Capacitance between two conductors connecting a third conductor sheath,

$$= \frac{1}{2} C_N = \frac{1.25}{2} = 0.625 \ \mu\text{F} \quad \textbf{Ans.}$$

(*v*) Charging current per phase, $I_C = 2\pi f V_P C_N$

$$= 2\pi \times 50 \times \frac{11{,}000}{\sqrt{3}} \times 1.25 \times 10^{-6} = 2.494 \text{ A} \quad \textbf{Ans.}$$

Q. 5. Attempt *any two* parts : (*10 × 2 = 20*)

(*a*) Write short notes on :

(*i*) audible noise and radio interference

(*ii*) effect of wind and ice loading on the mechanical design of a line.

(*b*) Compare EHV ac and HVDC system in detail.

(*c*) Classify various methods of neutral grounding. Explain any one of them.

ANSWERS

3. (*a*) (*i*) Refer to Chapter 8.

(*ii*) Refer to Art. 10.4.2 page 246.

(*b*) Refer to Art. 14.3.

(*c*) Refer to Arts. 12.5, 12.7, 12.8, 12.9 or 12.10.

❑❑❑

B.TECH.

(SEMESTER V) THEORY EXAMINATION, 2011-2012

ELEMENTS OF POWER SYSTEM

Time : 3 Hours *Maximum Marks: 100*

Note : *Attempt ALL questions.*

Q. 1. Attempt any ***two*** of the following : *(10 × 2 = 20)*

(*a*) What are the principal components of an electric supply system ? Draw and explain a typical power supply network.

(*b*) Name and explain briefly the four parameters on which the performance of a transmission line depends. Also discuss the advantages and limitations of high transmission voltage.

(*c*) Distinguish between ac and dc resistances of a conductor. Why the two differ ? Derive an expression for internal flux linkage due to a single current carrying conductor.

ANSWERS

1. (*a*) Refer to Arts. 1.7 and 1.3.

(*b*) Refer to Arts. 5.1 and 2.7.

(*c*) Refer to Arts. 2.5 and 4.11.

Q. 2. Attempt any ***two*** of the following : *(10 × 2 = 20)*

(*a*) Give the concept of 'Self GMD'. Derive an expression for inductance of an unsymmetrical, transposed line using solid round conductors.

(*b*) A single phase transmission line has two parallel conductors, each of 1.2 cm diameter and 2.5 meters apart. Calculate the loop inductance per km length of the line if the material of the conductor is steel with relative permeability of 200.

(*c*) A long transmission line is open circuited at the receiving end. Will there be any current in the line at the sending end ? Explain your answer.

ANSWERS/SOLUTION

2. (*a*) Refer to Arts. 4.2.3 and 4.3.

(*b*) Loop inductance per km of line with steel conductor

$$= \left(\mu + 4 \log_e \frac{d}{r}\right) \times 10^{-7} \times 1{,}000 \text{ H} = \left(200 + 4 \log_e \frac{250}{0.6}\right) \times 10^{-4} \text{ H} = 22.4 \text{ mH} \textbf{ Ans.}$$

(*c*) Refer to Art. 5.5.

Q. 3. Attempt any ***two*** of the following : *(10 × 2 = 20)*

(*a*) What should be the desirable characteristics of insulating materials used in cables? Discuss the classification of cables according to voltage levels.

(*b*) Derive an expression for sag and tension in a power conductor strung between two supports at equal heights taking into account the wind and ice loadings also.

(*c*) Discuss the following methods of neutral grounding :

(*i*) Peterson coil grounding (*ii*) Resistance grounding.

ANSWERS

3. (*a*) Refer to Arts. 11.5 and 11.6

(*b*) Refer to Art. 10.4.2.

(*c*) (*i*) Refer to Art. 12.10.

(*ii*) Refer to Art. 12.8.

Q. 4. Attempt any ***two*** of the following : *(10 × 2 = 20)*

(*a*) Define 'string efficiency'. Explain different methods of improving string efficiency.

(*b*) A string of 5 insulators is connected across a 100 kV line. If the capacitance of each disc to earth is 0.1 of the capacitance of the insulator, calculate :

(*i*) the distribution of voltage on the insulator discs and

(*ii*) the string efficiency.

(*c*) A certain 3-phase equilateral transmission line has a total corona loss of 53 kW at 106 kV and a loss of 98 kW at 110.9 kV. What is the disruptive critical voltage ? What will be the corona loss at 113 kV ?

ANSWERS/SOLUTION

4. (*a*) Refer to Answer of Q. 5 at page 232 and Art. 9.7.

(*b*) Voltage between line conductor and earth, V = 100 kV

Number of units in the insulator string, n = 5

Shunt capacitance to mutual capacitance ratio, k = 0.1

(*i*) Voltage across the top unit,

$$V_1 = \frac{V}{5+20k+21k^2+8k^3+k^4} \quad \text{Refer to Eq. } (v) \text{ of Example 9.1}$$

$$= \frac{100}{5+20\times0.1+21\times(0.1)^2+8\times(0.1)^3+0.1^4} = \frac{100}{7.2181} = 13.854 \text{ kV} \quad \textbf{Ans.}$$

Voltage across second unit from top, $V_2 = 1.1\ V_1 = 1.1 \times 13.854 = 15.2395$ kV **Ans.**

Voltage across 3rd unit from top, $V_3 = 1.31\ V_1 = 1.31 \times 13.854 = 18.1488$ kV **Ans.**

Voltage across 4th unit from top, $V_4 = 1.651\ V_1 = 1.651 \times 13.854 = 22.873$ kV **Ans.**

Voltage across 5th unit from top, $V_5 = 2.1571\ V_1 = 2.1571 \times 13.854 = 29.885$ kV **Ans.**

(*ii*) String efficiency $= \dfrac{V}{nV_n} \times 100 = \dfrac{100}{5 \times 29.885} \times 100 = 66.9232\%$

(*c*) Refer to Solution of Example 7.7.

Q. 5. Attempt any ***two*** of the following : *(10 × 2 = 20)*

(*a*) Give reasons for the following :

(*i*) ACSR conductors are preferred for transmission and distribution lines.

(*ii*) Suspension insulators are preferred for high voltage power transmission.

(*b*) Evaluate the generalized circuit constants for :

(*i*) short transmission line

(*ii*) medium transmission line—nominal T method.

(*c*) Write short notes on the following :

(*i*) Sag template. (*ii*) Proximity effect.

(*iii*) Grading of cables. (*iv*) Power system elements.

ANSWERS

5. (*a*) (*i*) Refer to Art. 3.3(3).

(*ii*) Refer to Art. 9.4(2).

(*b*) (*i*) Refer to Art. 5.9 (*i*).

(*ii*) Refer to Art. 5.9 (*iii*).

(*c*) (*i*) Refer to Art. 10.6.

(*ii*) Refer to Art. 3.6.

(*iii*) Refer to Art. 11.11.

(*iv*) Refer to Art. 3.1.

❑❑❑

B.TECH.

(SEMESTER V) ODD SEMESTER THEORY EXAMINATION, 2012-2013

ELEMENTS OF POWER SYSTEM

Time : 3 Hours *Maximum Marks: 100*

Note : *Attempt **all** questions.*

Q. 1. Attempt any ***four*** parts of the following : **(5 × 4 = 20)**

(*a*) Draw and explain the layout of a power system network from generation to distribution showing all the voltage levels at various intermediate stages.

(*b*) Explain in short the following components of power system :

(*i*) Substation

(*ii*) Feeder

(*iii*) Distributor

(*iv*) Service mains

(*v*) Tie line.

(*c*) State and prove Kelvin's law for size of conductor for transmission. Discuss its limitations.

(*d*) In reference to the ac conductors explain the difference between skin effect and proximity effect. What remedy can be suggested to reduce both these effects ?

(*e*) Explain in brief various types of conductors with one advantage of each type.

(*f*) Explain how bundled conductors are superior than solid or stranded conductors.

ANSWERS/SOLUTION

1. (*a*) Refer to Art. 1.3.

(*b*) Refer to Art. 1.2.

(*c*) Refer to Arts. 3.7 and 3.7.1.

(*d*) Refer to Arts. 3.5 and 3.6.

(*e*) Refer to Art. 3.3.

(*f*) Refer to Art. 4.5.

Q. 2. Attempt any ***two*** parts of the following : **(10 × 2 = 20)**

(*a*) Deduce an expression for line to neutral capacitance for a 3-ϕ overhead transmission line when the conductors are :

(*i*) Symmetrically placed

(*ii*) Unsymmetrically placed but transposed.

(*b*) A 50 Hz, 3-ϕ, 275 kV, 400 km transmission line has the following parameters (per phase) :

Resistance = 0.035 Ω/km, L = 1 mH/km, C = 0.01 μF/km.

If the line is supplied at 275 kV, determine the MVA rating of a shunt reactor having negligible losses that would be required to maintain 275 kV at the receiving end, when the line is delivering no load. Use nominal π method.

(*c*) A 3-ϕ, 132 kV, 100 km, 50 Hz single circuit line has horizontal spacing with 3.5 m between adjacent conductors. The conductors diameter is 1.2 cm. Find :

(*i*) Inductance of the conductor

(*ii*) Capacitance per phase and charging current with effect of earth.

ANSWERS/SOLUTION

2. (*a*) Refer to Arts. 4.8, 4.8.1, 4.8.2 and 4.8.3.

(*c*) (*i*) Radius of each conductor, $r = \dfrac{1.2}{2} = 0.6$ cm

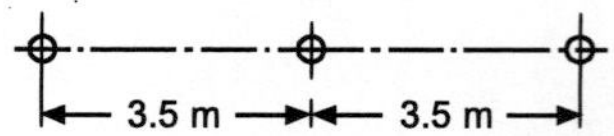

Fig. 1

$$d = 3.5 \text{ m}$$

$$\text{GMR, } r' = 0.7788 \times 0.6 = 0.46728 \text{ cm}$$

From Eq. (4.20), (4.21) and (4.22) respectively

$$L_A = 0.2\left[\log_e \frac{d}{r'} + \frac{1}{2}\log_e 2 - j0.866 \log_e 2\right] \text{ mH/km}$$

$$= 0.2\left[\log_e \frac{3.5}{0.0046728} + \frac{1}{2}\log_e 2 - j0.866 \log_e 2\right] \text{ mH/km}$$

$$= 0.2[6.619 + 0.346 - j0.866 \times 0.6931] \text{ mH/km}$$

$$= (1.393 - j0.12) \text{ mH/km}$$

$$= (1.393 - j0.12) \times 100 \text{ mH} = (139.3 - j12) \text{ mH} \quad \textbf{Ans.}$$

$$L_B = 0.2 \log_e \frac{d}{r'} \times \text{length of line in km}$$

$$= 0.2 \log_e \frac{3.5}{0.0046728} \times 100 \text{ mH} = 132.375 \text{ mH} \quad \textbf{Ans.}$$

$$L_C = 0.2\left[\log \frac{3.5}{0.0046728} + \frac{1}{2}\log_e 2 + j0.866 \log_e 2\right] \times 100 \text{ mH}$$

$$= (139.3 + j12) \text{ mH}$$

(*ii*) The height of the conductors from ground is not given. Assume it to be 10 m From Fig. 4.40 page 115, we have

$$h_1 = h_2 = h_3 = 20 \text{ m}$$

$$h_{12} = h_{23} = \sqrt{20^2 + 3.5^2} = 20.3 \text{ m}$$

$$h_{31} = \sqrt{20^2 + 7^2} = 21.19 \text{ m}$$

$$r = 0.6 \text{ cm}$$

$$D_{eq} = \sqrt[3]{d_1 d_2 d_3} = \sqrt[3]{3.5 \times 3.5 \times 7} = 4.41 \text{ m}$$

Capacitance per phase

$$C_N = \frac{2\pi\varepsilon_0}{\log_e \dfrac{D_{eq}}{r} - \log_e \sqrt[3]{\dfrac{h_{12}h_{23}h_{31}}{h_1 h_2 h_3}}} \times 100 \times 1{,}000 \text{ F}$$

$$= \frac{2\pi \times 8.854 \times 10^{-12} \times 10^5}{\log_e \dfrac{4.41}{0.006} - \log_e \sqrt[3]{\dfrac{20.3 \times 20.3 \times 21.54}{20 \times 20 \times 20}}}$$

$$= \frac{55.6313 \times 10^{-7}}{6.6 - 0.03465} = 0.847 \ \mu\text{F} \quad \textbf{Ans.}$$

Charging current per phase, $I_C = 2\pi f C_N V_{ph}$

$$= 2\pi \times 50 \times 0.847 \times 10^{-6} \times \frac{132 \times 1{,}000}{\sqrt{3}}$$

$$= 20.27 \text{ A} \quad \textbf{Ans.}$$

Q. 3. Attempt any ***two*** parts of the following : **(10 × 2 = 20)**

(*a*) A particular 3-ϕ transmission line has total corona loss of 57 kW at 110 kV and corona loss of 99 kW at 114.8 kV. Calculate the critical disruptive voltage per phase and corona loss at 120 kV.

(*b*) Write short notes on:

(*i*) Critical Disruptive Voltage

(*ii*) Audible Visual Corona

(*iii*) Necessity of Insulators.

(*c*) (*i*) An insulator string consists of three units, each having a safe working voltage of 20 kV, the ratio of self to shunt capacitance is 6 : 1. Find the maximum safe working voltage of the string and the string efficiency.

(*ii*) A string insulator has 4 units and each unit of string is having capacitance 'C' the pin to earth is C/10, find voltages across each unit of string.

ANSWERS/SOLUTION

3. (*a*) $P_C \propto (V_{ph} - V_{do})^2$ Refer to Eq. (7.8)

$$\therefore \quad 57 \propto \left(\frac{110}{\sqrt{3}} - V_{do}\right)^2 \propto (63.5085 - V_{do})^2 \quad \text{...}(i)$$

$$\text{and} \quad 99 \propto \left(\frac{114.8}{\sqrt{3}} - V_{do}\right) \propto (66.2798 - V_{do})^2 \quad \text{...}(ii)$$

Dividing expression (*ii*) by expression (*i*), we have

$$\frac{99}{57} = \frac{(66.2798 - V_{do})^2}{(63.5085 - V_{do})^2}$$

or $V_{do} = 54.79$ kV

Total corona loss at 120 kV

$$P_C \propto \left(\frac{120}{\sqrt{3}} - V_{do}\right)^2 \propto (69.282 - V_{do})^2 \quad \text{...}(iii)$$

Dividing expression (*iii*) by expression (*i*), we have

$$P_C = \left(\frac{69.282 - 54.79}{63.5085 - 54.79}\right)^2 \times 57 = 157.5 \text{ kW} \quad \textbf{Ans.}$$

Disruptive critical voltage between lines = $\sqrt{3} \times V_{do} = \sqrt{3} \times 54.79 = 94.9$ kV **Ans.**

(*b*) (*i*) Refer to Art. 7.6.

(*ii*) Refer to Art. 7.1.

(*iii*) Refer to Art. 9.1.

(c) (i) Number of units, $n = 3$

Ratio of shunt capacitance to mutual capacitance, $K = \frac{1}{6}$

Voltage across the bottom most unit, V_3 = Safe working voltage of the unit = 20 kV

So, Voltage across top most unit, $V_1 = \frac{V_3}{1+3K+K^2} = \frac{20}{1+3\times\frac{1}{6}+\left(\frac{1}{6}\right)^2} = 13.09$ kV

Voltage across middle unit, $V_2 = V_1(1 + K) = 13.09\left(1+\frac{1}{6}\right) = 15.27$ kV

Maximum safe working voltage of the string,

$$V = V_1 + V_2 + V_3 = 13.09 + 15.27 + 20 = 48.36 \text{ kV}$$

$$\text{String efficiency, } \eta = \frac{V}{nV_n}\times 100 = \frac{48.36}{3\times 20}\times 100 = 80.6\% \quad \textbf{Ans.}$$

(ii) Let the voltage between conductor and earth be V

Ratio of shunt to mutual capacitance,

$$K = \frac{C_1}{C} = \frac{1}{10} = 0.1$$

Voltage across top unit, $V_1 = \frac{V}{4+10K+6K^2+K^3}$ ...Refer to Eq. (9.6)

$$= \frac{V}{4+10\times 0.1+6\times(0.1)^2+(0.1)^3} = \frac{V}{5.061} = 0.1976\text{ V} \quad \textbf{Ans.}$$

Voltage across the second unit from top,

$$V_2 = (1 + K)V_1 = 1.1 \times 0.1976 = 0.2174 \text{ V} \quad \textbf{Ans.}$$

Voltage across the third unit from top,

$$V_3 = (1 + 3K + K^2)V_1$$
$$= (1 + 3 \times 0.1 + (0.1)^2) \times 0.1976 = 0.2588 \text{ V} \quad \textbf{Ans.}$$

Voltage across the bottom unit,

$$V_4 = (1 + 6K + 5K^2 + K^3)V_1$$
$$= (1 + 6 \times 0.1 + 5 \times (0.1)^2 + (0.1)^3) \times 0.1976$$
$$= 0.3262 \text{ V} \quad \textbf{Ans.}$$

Q. 4. Attempt any ***two*** parts of the following : **(10 × 2 = 20)**

(a) (i) Derive the expressions of sag and tension when supports are at unequal level.

(ii) A transmission line has a span of 275 m between level supports the conductor has an effective diameter of 1.96 cm and weight of 0.865 kg/m. Its ultimate strength is 8060 kg. If the conductor has ice coating of radial thickness 1.27 cm and is subjected to a wind pressure of 3.9 gm/cm^2 of projected area, calculate sag for a safety factor of 2, weight of 1cc of it is 0.91 gm.

(b) Define dielectric stress in a single core cable. A cable is graded with three dielectrics of permittivities 4, 3 and 2. The maximum permissible potential gradient for all dielectrics is same and equals to 30 kV/cm. The core diameter is 1.5 cm and sheath diameter is 5.5 cm. Find the working voltage of cable.

(*c*) Write in brief on any **two** :

(*i*) Vibration dampers, stringing chart

(*ii*) Methods of grading of cable

(*iii*) Flashover and puncture voltage in insulator.

ANSWERS/SOLUTION

4. (*a*) (*i*) Refer to Art. 10.4.2.

(*ii*) Working tension, $T = \dfrac{\text{Ultimate strength}}{\text{Factor of safety}} = \dfrac{8{,}060}{2} = 4{,}030 \text{ kg}$

Weight of ice coating per metre length,

$$w_i = \text{Density of ice} \times \pi r(D + r)$$

$$= \frac{0.91 \times (100)^3}{1{,}000} \times \pi \times 1.27 \times 10^{-2} \times (1.96 \times 10^{-2} + 1.27 \times 10^{-2})$$

$$= 1.1727 \text{ kg}$$

Wind force per metre length of conductor,

$$w_w = p \times (D + 2r)$$

$$= \frac{3.9 \times (100)^2}{1{,}000} \times (1.96 \times 10^{-2} + 2 \times 1.27 \times 10^{-2}) = 1.755 \text{ kg}$$

Resultant force per metre length,

$$w_r = \sqrt{(w_c + w_i) + w_w^2} = \sqrt{(0.865 + 1.1727)^2 + (1.755)^2} = 2.689 \text{ kg}$$

$$\text{Maximum sag, } S = \frac{w_r L^2}{8T} = \frac{2.689 \times (275)^2}{8 \times 4{,}030} = 6.31 \text{ m} \quad \textbf{Ans.}$$

(*b*) Diameter of conductor, $d = 1.5$ cm

Diameter over the insulation, D = 5.5 cm

Maximum potential gradient, $g_{max} = 30$ kV/cm

Let the diameters over the insulations of relative permittivities 4 and 3 be d_1 and d_2 respectively.

As the maximum stress in all of the three dielectrics is the same,

$$\varepsilon_{r_1} d = \varepsilon_{r_2} d_1 = \varepsilon_{r_3} d_2$$

or
$$d_1 = \frac{\varepsilon_{r_1}}{\varepsilon_{r_2}} d = \frac{4}{3} \times 1.5 = 2 \text{ cm}$$

and
$$d_2 = \frac{\varepsilon_{r_1}}{\varepsilon_{r_3}} d = \frac{4}{2} \times 1.5 = 3 \text{ cm}$$

Permissible peak voltage for the cable

$$= \frac{g_{max}}{2}\left[d \log_e \frac{d_1}{d} + d_1 \log_e \frac{d_2}{d_1} + d_2 \log_e \frac{D}{d_2}\right]$$

$$= \frac{30}{2}\left[1.5 \log_e \frac{2}{1.5} + 2 \log_e \frac{3}{2} + 3 \log_e \frac{5.5}{3}\right]$$

$$= 15(0.4315 + 0.8109 + 1.8184) = 45.912 \text{ kV}$$

∴ Working voltage (rms) for the cable = $\frac{45.912}{\sqrt{2}}$ = 32.46 kV **Ans.**

(*c*) (*i*) Refer to Arts. 10.8 and 10.5.

(*ii*) Refer to Art. 11.11.

(*iii*) Refer to Art. 9.5.

Q. 5. Attempt any ***two*** parts of the following : **(10 × 2 = 20)**

(*a*) What is grounding ? Why is earthing necessary ? Give various advantages of grounding.

(*b*) (*i*) Differentiate between various types of grounding.

(*ii*) Explain the problems associated with EHVAC transmission in detail.

(*c*) (*i*) What is the principle of operation of an HVDC transmission line ?

(*ii*) Define the following :

(*a*) Surge impedance loading

(*b*) Types of dc link.

ANSWERS/SOLUTION

5. (*a*) Refer to Art. 12.1.

(*b*) (*i*) Refer to Arts. 12.6, 12.7, 12.8, 12.9 and 12.10.

(*ii*) Refer to Art. 13.3.

(*c*) (*i*) Refer to Art. 14.2.

(*ii*) (*a*) Refer to Art. 6.5.

(*b*) Refer to Art. 14.6.

B.TECH.

(SEMESTER V) ODD SEMESTER THEORY EXAMINATION, 2013-2014

ELEMENTS OF POWER SYSTEM

Time : 3 Hours *Maximum Marks: 100*

Note : (*1*) *Attempt **ALL** questions.*
(*2*) *All questions carry equal marks.*

Q. 1. Attempt any ***four*** parts of the following : **(5 × 4 = 20)**

(*a*) Explain the advantages of HVDC transmission over EHVAC transmission.

(*b*) A 50 km long transmission line supplies a load of 5 MVA at 0.8 power factor lagging at 33 kV. The efficiency of transmission is 90 per cent. Calculate the volume of conductor aluminium required for the line when single phase, 2-wire system is used.

(*c*) Explain different types of conductors in power system.

(*d*) Explain Kelvin's Economy Law and derive the condition for most economical cross-sectional area of the conductor.

(*e*) Explain skin effect and proximity effect.

(*f*) Draw and explain single line diagram of power system.

ANSWERS/SOLUTION

1. (*a*) Refer to Arts. 14.3 and 14.5.

(*b*)

$$\text{Line current, } I_L = \frac{\text{P in MVA} \times 10^6}{V_L \text{ in kV} \times 10^3} = \frac{5 \times 10^6}{33 \times 10^3} = 151.52 \text{ A}$$

$$\text{Power supplied, P} = \text{Load in MVA} \times \cos\phi = 5 \times 0.8 = 4 \text{ MW}$$

$$\text{Power losses in line, W} = P(1 - \eta) = 4(1 - 0.9) = 0.4 \text{ MW or } 400 \text{ kW}$$

$$\text{Line losses per conductor} = \frac{W}{2} = \frac{400}{2} = 200 \text{ kW}$$

$$\text{Resistance per conuctor, R} = \frac{\text{Line loss per conductor}}{I^2} = \frac{200 \times 1{,}000}{151.52^2} = 8.71\ \Omega$$

Resistance of transmission line per conductor $R = \rho\dfrac{l}{a}$

where $\rho = 2.6 \times 10^{-8}$ Ω-m, $l = 50$ km $= 50{,}000$ m and $R = 8.71$ Ω

or Area of cross section of conductor,

$$a = \rho\frac{l}{R} = \frac{2.6 \times 10^{-8} \times 50{,}000}{8.71} = 1.4925 \times 10^{-4} \text{ m}^2$$

Volume of aluminium conductor required for a 2-wire line of 50 km length

$$= 2al$$

$$= 2 \times 1.4925 \times 10^{-4} \times 50 \times 1{,}000 = 14.925 \text{ m}^3 \quad \textbf{Ans.}$$

(*c*) Refer to Art. 3.3.

(*d*) Refer to Art. 3.7.

(*e*) Refer to Arts. 3.5 and 3.6.

(*f*) Refer to Art. 1.3.

Q. 2. Attempt any ***two*** parts of the following : **(10 × 2 = 20)**

(*a*) Deduce an expression for the total inductance of a single phase line.

(*b*) A three phase 50 Hz transmission line consists of three equal conductors of radii π, placed in a horizontal plane, with a spacing of 6 m between the middle and each outer conductors. Determine the inductive reactance per phase per km of the transposed line if the radius of each conductor is 12.5 mm.

(*c*) What is the effect of earth on line capacitance ? Explain the method of images to calculate the capacitance of two wire single phase line.

ANSWERS/SOLUTION

2. (*a*) Refer to Art. 4.2.1.

(*b*) Radius of each conductor,

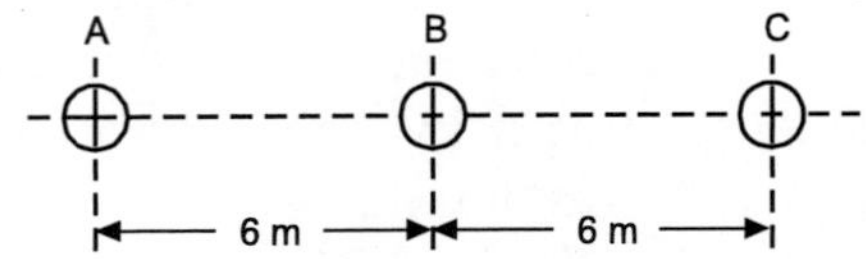

Fig. 1

$$r = 12.5 \text{ mm} = 1.25 \text{ cm}$$

$$\text{GMR, } r' = 0.7788 \times 1.25 = 0.9735 \text{ cm}$$

Average inductance per phase per km may be given as

$$L_{av} = \frac{L_A + L_B + L_C}{3}$$

Substituting the values of L_A, L_B and L_C from Eqs. (4.20), (4.21) and (4.22) respectively, we have

$$L_{av} = \frac{1}{3} \times 2 \times 10^{-7} \left[\log_e \frac{d}{r'} + \frac{1}{2} \log_e 2 - j0.866 \log_e 2 + \log_e \frac{d}{r} + \log_e \frac{d}{r} + \frac{1}{2} \log_e 2 + j0.866 \log_e 2 \right]$$

$$= 2 \times 10^{-7} \left[\log_e \frac{d}{r'} + \frac{1}{3} \log_e 2 \right] \text{ H/m}$$

$$= 2 \times 10^{-4} \left[\log_e \frac{d}{r'} + \frac{1}{3} \log_e 2 \right] \text{ H/km}$$

$$= 2 \times 10^{-4} \left[\log_e \frac{600}{0.9735} + \frac{1}{3} \log_e 2 \right] \text{ H/km} = 2 \times 10^{-4}(6.4237 + 0.2310)$$

$$= 1.331 \text{ mH/km}$$

Inductive reactance per phase per km

$$= 2\pi f L = 2\pi \times 50 \times 1.331 \times 10^{-3} = 0.418\ \Omega \quad \textbf{Ans.}$$

(*c*) Refer to Arts. 4.11, 4.11.1.

Q. 3. Attempt any ***two*** parts of the following : **(10 × 2 = 20)**

(*a*) Derive A, B, C and D parameters for Nominal π model of a medium line and draw its phasor diagram.

(*b*) Explain the phenomenon of corona and factors affecting corona.

(*c*) A string of suspension insulators consists of four units. The capacitance between each link pin and earth is one tenth of the self capacitance of a unit. The voltage between the line conductor and earth is 100 kV. Find :

(*i*) the voltage distribution across each unit and

(*ii*) string efficiency.

ANSWERS/SOLUTION

3. (*a*) Refer to Art. 5.9(*iv*).

(*b*) Refer to Arts. 7.1 and 7.2.

(*c*) (*i*) Ratio of shunt to mutual capacitance,

$$K = \frac{C_1}{C} = \frac{1}{10} = 0.1$$

Voltage between conductor and earth, V = 100 kV

Voltage across top unit, $V_1 = \dfrac{V}{4 + 10K + 6K^2 + K^3}$ ...Refer to Eq. (9.6)

$$= \frac{100}{4 + 10 \times 0.1 + 6 \times (0.1)^2 + (0.1)^3} = 19.759 \text{ kV} \quad \textbf{Ans.}$$

Voltage across the second unit from the top,

$$V_2 = (1 + K)V_1 = (1 + 0.1) \times 19.759 = 21.735 \text{ kV} \quad \textbf{Ans.}$$

Voltage across the third unit from the top

$$V_3 = (1 + 3K + K^2)V_1$$
$$= (1 + 3 \times 0.1 + 0.1^2) \times 19.759 = 25.884 \text{ kV} \quad \textbf{Ans.}$$

Voltage across the fourth unit from the top (*i.e.*, bottom unit),

$$V_4 = (1 + 6K + 5K^2 + K^3)\, V_1$$
$$= [1 + 6 \times 0.1 + 5 \times (0.1)^2 + (0.1)^3] \times 19.759$$
$$= 32.622 \text{ kV} \quad \textbf{Ans.}$$

(*ii*) String efficiency, $\eta = \dfrac{V}{4V_4} \times 100 = \dfrac{100}{4 \times 32.622} \times 100 = 76.635\%$ **Ans.**

Q. 4. Attempt any ***two*** parts of the following : **(10 × 2 = 20)**

(*a*) Explain catenary method for the calculation of sag and tension in transmission line.

(*b*) A transmission line conductor at a river crossing is supported from two towers at heights of 30 m and 90 m above water level. The horizontal distance between the towers is 270 m. If the tension in the conductor is 1800 kgf and the conductor weighs 1.0 kgf per meter find the clearance between the conductor and the water at a point midway between the towers. Assume parabolic configuration.

(*c*) Explain capacitance grading of the cables used in power system.

ANSWERS/SOLUTION

4. (*a*) Refer to Art. 10.4.2.

(*b*) Difference in level betwen the two supports, $h = 90 - 30 = 60$ m

Distance of lowest point of conductor, O from the support at low level,

$$x = \frac{L}{2} - \frac{Th}{wL} = \frac{270}{2} - \frac{1800 \times 60}{1.0 \times 270} = 135 - 400 = -265 \text{ m}$$

The negative value of x shows that support A is on the same side of O as support B (Fig. 2)

Distance of mid point P from O

$$= \frac{L}{2} - x = \frac{270}{2} - (-265) = 400 \text{ m}$$

Distance of point B from O

$= L - x = 270 + 265 = 535$ m

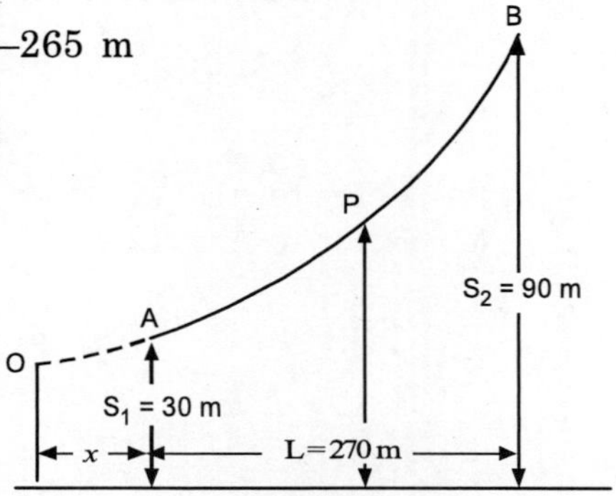

Fig. 2

$$\text{Height of midpoint P above O, } S_{mid} = \frac{w\left(\frac{L}{2} - x\right)^2}{2T} = \frac{1 \times 400^2}{2 \times 1800} = 44.44 \text{ m}$$

$$\text{Height of point B from O, } S_B = \frac{w(L - x)^2}{2T} = \frac{1 \times 535^2}{2 \times 1,800} = 79.507 \text{ m}$$

Hence, midpoint is (79.507 – 44.44) *i.e.*, 35.067 m below point B or (90 – 35.067) *i.e.*, 54.933 m above the water level **Ans.**

(*c*) Refer to Art. 11.11(*i*).

Q. 5. Attempt any ***two*** parts of the following : **(10 × 2 = 20)**

(*a*) Derive the condition for most economical size of cable using voltage gradient method. The test results for 1 km of a three-phase metal sheathed belted cable gave a measured capacitance of 0.7 μF between one conductor and the other two conductors bunched together with the earth sheath and 1.2 μF measured between the three bunched conductors and the sheath. Find :

(*i*) the capacitance between any pair of conductors, the sheath being isolated and

(*ii*) the charging current when the cable is connected to 11 kV, 50 Hz supply.

(*b*) Explain advantage of Neutral grounding and Peterson Coil grounding.

(*c*) Explain the phenomena of Neutral grounding by grounding transformer.

ANSWERS/SOLUTION

5. (*a*) Refer to Art. 11.10.

Capacitance between all the conductors joined and sheath [Fig. 11.20(*a*)] gives $3C_S$

So, $3C_S = 1.2\ \mu F$

$$\text{or } C_S = \frac{1.2}{3} = 0.4\ \mu F$$

Capacitance between two conductors joined to sheath and third conductor [Fig. 11.20(*c*)] gives $2C_C + C_S$

So, $2C_C + C_S = 0.7\ \mu F$

$$\text{or } C_C = \frac{0.7 - C_S}{2} = \frac{0.7 - 0.4}{2} = 0.15\ \mu F$$

(*i*) Capacitance between any pair of conductors, the sheath being isolated

$$= \frac{1}{2}C_N = \frac{3C_C + C_S}{2} = \frac{3 \times 0.15 + 0.4}{2} = 0.425\ \mu F \textbf{ Ans.}$$

(*ii*) Charging current per phase,

$$I_C = 2\pi f V_P C_N = 2\pi \times 50 \times \frac{11,000}{\sqrt{3}} \times (3 \times 0.15 + 0.4) \times 10^{-6}$$

$$= 1.696 \text{ A} \textbf{ Ans.}$$

(*b*) Refer to Art. 12.10.

(*c*) Refer to Art. 12.12.

❑❑❑

Win Prizes

ATTENTION READERS

Win Prizes

We request for your frank assessment regarding various aspects of the book as given below :

ELEMENTS OF POWER SYSTEM

By : J.B. GUPTA

Please Fill up the Sheet in Neat and Clean Writing

(i) What topics of your syllabus, important for your examination point of view, are not covered in this book ?

....

....

....

....

(ii) What are the chapters wherein the treatment of the subject matter is not systematic or organised or updated ?

....

....

....

....

(iii) Have you come across misprints/mistakes/factual inaccuracies in the book ? Please specify the chapters and the page numbers.

....

....

....

....

(iv) What other books on the same subject have you found/heard better than the present book ? Please specify in terms of price and quality.

....

....

....

....

(v) Further suggestions and comments for the improvement of this book.

....

....

....

....

(PS : If need be please attach a separate sheet to write your views)

OTHER DETAILS

(i) Who recommended you this book ? (Please tick (✓) in the box near the option relevant to you.

☐ Teacher ☐ Friends ☐ Bookseller

....

....

....

....

(ii) Name of the recommending teacher, his designation and address:

....

....

....

....

(iii) Name and address of the bookseller you purchased the book from :

....

....

....

....

(iv) Name and address of your institution (Please mention the University or Board as the case may be)

....

....

....

....

(v) Your name and complete address :

....

....

....

....

(vi) Course you are enrolled in :

....

....

....

....

The best assessment will be awarded each month : The award will be in the form of our publications as decided by the Editorial Board.

Please mail the filled up coupon at the vearliest to :

S.K. KATARIA & SONS®

Publisher of Engineering Books

4885/109, Prakash Mahal, Dr. Subhash Bhargav Lane,

Opposite Delhi Medical Association, Daryaganj, New Delhi–110002 (INDIA)

Phone: +91-11-43551243; *Mobile:* +91-9871775858, 7678425209

e-mail: katariabook@gmail.com, katariabooks@yahoo.com

Website: www.skkatariaandsons.com